Pile Driving by Pile Buck

Don C. Warrington, P.E.
Editor

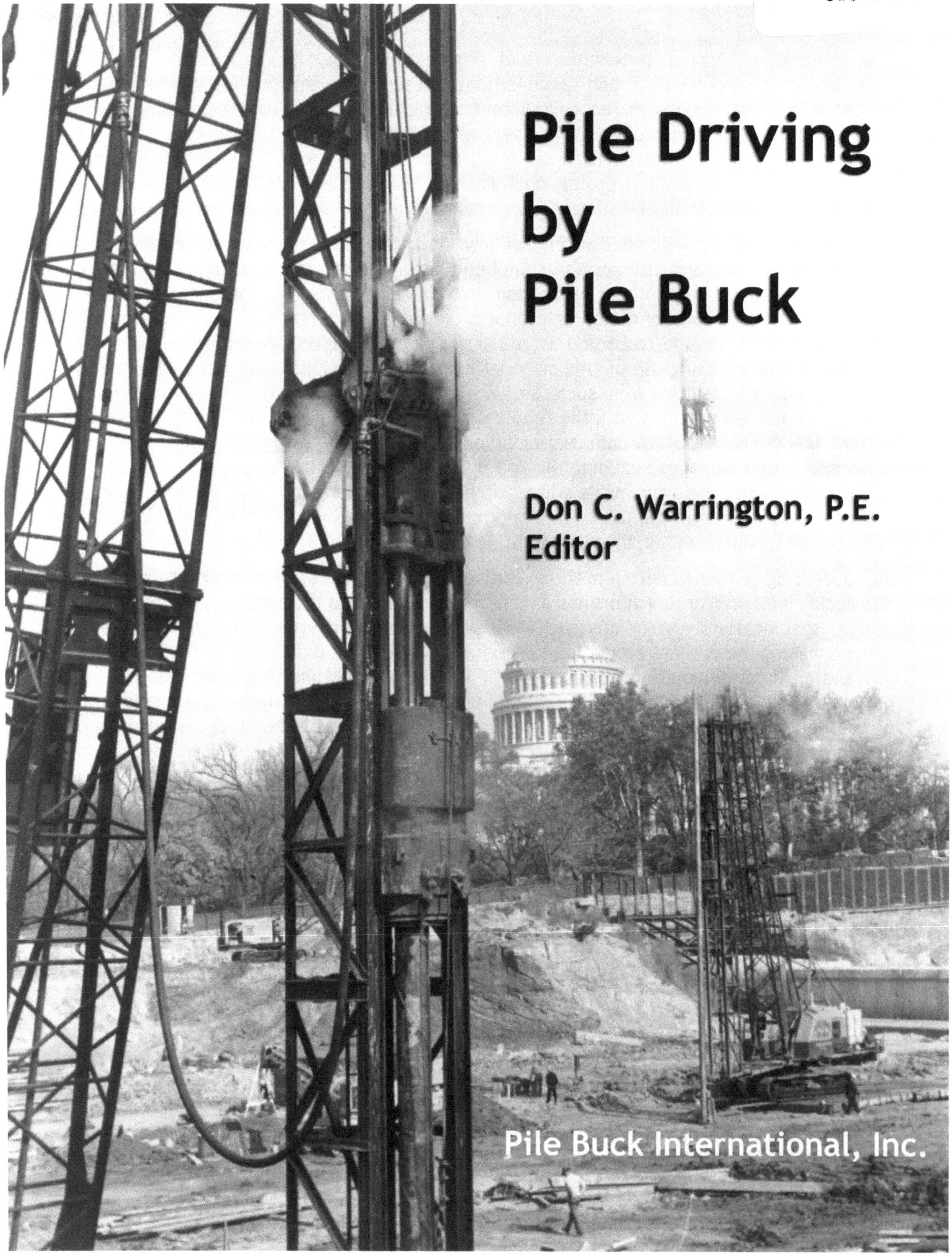

Pile Buck International, Inc.

Notice

The information, including technical and engineering data, figures, tables, designs, drawings, details, procedures and specifications, presented in this publication have been prepared in accordance with recognized contracting and/or engineering principles, and are for general information only. While every effort has been made to insure its accuracy, this information should not be used or relied upon for any specific application without independent, competent and professional examination and verification of its accuracy, suitability and applicability by a licensed professional.

This book is provided without warranty of any kind. Pile Buck®, Inc. and/or its editors hereby disclaim any and all express or implied warranties of merchantability, fitness for any general or particular purpose or freedom from infringement of any patent, trademark, copyright in regard to information or products contained or referred to herein. Nothing herein contained shall be construed as granting a license, express or implied, under any patents. Anyone making use of this material does do at his or her own risk and assumes any or all liability resulting from such use. The entire risk as to quality or usability of the material contained within is with the reader. In no event will Pile Buck®ii, Inc., and/or its editors, be held liable for any damages including lost profits, lost savings or other incidental or consequential damages including the use or inability to use the information contained herein. Pile Buck®, Inc., and/or its editors do not insure anyone utilizing this book against any liability arising from the use of this information and hereby will not be held liable for "consequential damages" of any kind resulting from such use.

All advertising contained herein in the exclusive representation of those registered herein. Pile Buck®, Inc., and/or its editors make no representation as to the accuracy, performance, design, specifications and/or any such "claims" made by advertisers contained herein. Anyone making use of these products and services does so at his or her own risk and assumes any and all liability resulting from such use. In no event will Pile Buck, Inc., and/or its editors be held liable for any damages including lost profits, lost savings or other incidental or consequential damages arising from the use or inability to use the products advertised herein. Pile Buck, Inc., and/or its editors do not insure anyone from liability arising from the use of these products and hereby will not be liable for "consequential damages" of any kind resulting from such use.

Additional copies of this book can be ordered from Pile Buck at http://www.pilebuckinternational.com.

Forward

Pile Driving by Pile Buck is designed to be the comprehensive guide to driven piles and the equipment used to drive and test them. Driven piles are the oldest type of deep foundation in use; they not only predate the Industrial Revolution, they predate literate civilization. Many structures have stood for hundreds and thousands of years on driven piles, largely hidden from view and the attention of those who use the structures.

In spite of the advances of other types of deep foundations, driven piles remain a viable foundation option because:

a) the quality of the pile is known before installation, not after; and

b) b) the load bearing capability of a pile can be estimated from the driving history, using the hammer as a measuring instrument.

In addition, driven piles are used in soils where the use of in situ installation of any kind is difficult if not impossible due to the nature of the soil.

This book is divided into six chapters:

1. *Introduction*: This includes a brief history of pile driving and installation equipment, and a discussion on the proper use of driven piles to foundation applications.
2. *Materials:* An overview of the different types of piles, their advantages and disadvantages, along with information on pile accessories.
3. *Installation Equipment:* An overview of the different types of pile hammers, both impact and vibratory, with their distinctive characteristics.
4. *Pile Design:* A description of methods of estimating the capacity of the pile before driving, both static and dynamic. These include step-by-step methods of computing the load bearing capability of piles. This section is intended to focus on issues that pertain to piles alone; information on soil mechanics and other matters can be found in other Pile Buck publications.
5. *Means and Methods:* A detailed description of the installation of piles, including all aspects of job organization, safety, mobilization, actual installation, and job completion.
6. *Construction Monitoring:* A description of load testing techniques for piles, both static and dynamic.

Although some theoretical issues are discussed in this book, this book focuses on issues that owners, engineers, inspectors, contractors, equipment manufacturers and material suppliers deal with on a daily basis in the successful installation of driven pile foundations.

This book is a compilation of material from a variety of sources assembled to be more seamless and cohesive than by simply replicating the sources and placing them into one or more volumes. The sources of this book are listed at the end. Acknowledgements for many of the drawings and photographs are listed with each chapter.

We would also like to thank the following for their review of the manuscript, which resulted in many helpful suggestions and a better book:

- Jerry DiMaggio, FHWA
- Mohamad Hussein, GRL Engineers (we also thank GRL for their permission to use GRLWAVE in our text)
- Mark Lee, Dawson Construction Plan
- Glenn Lockie, Sure-Lock

Finally, this Editor would like to thank Mr. Chris Smoot of Pile Buck both for selecting him for this project and for the vision in making this long overdue reference book a reality.

Don C. Warrington, P.E.

Table of Contents

Chapter 1. Introduction

1.1. History

Piles are braced, structural columns that are driven, pushed or otherwise installed into the ground. Pre-literate man found that pile foundations were very useful in that they allowed construction of a shelter high above the water or the land and out of reach of marauding animals and warring neighbors. Driven piles are the oldest type of deep foundation in existence. They enable the placement of structures in areas that are otherwise unsuitable because of the subsurface condition.

Pile driving is the process of installing a pile into the ground without previous excavation. Historically, the oldest method of driving a pile, and the method most often used today, is by a hammer. No doubt, the earliest bearing piles were driven by hand using a wooden mallet of some sort.

1.1.1. Pile Driving in the Roman World

Although the Romans were neither the most innovative builders nor the ones with the most impressive single structures, (the Egyptians with the pyramids claim that honor) they were the most accomplished planners of infrastructure in the ancient world. Because of the varied soils they encountered around the Mediterranean Sea, they also drove many piles to support the military and civil works which they undertook.

The Latin word for pile driver is fistuca or festuca, which is defined as follows:

> FISTU'CA, an instrument used for ramming down pavements and threshing floors, and the foundations of buildings (Cato, R.R. 18, 28; Plin. H.N. xxxvi.25. s61; Vitruv. iii.3 s4 §1, x.3 s2 §3); and also for driving piles (Caes. B.G. iv.17). When used for the former purpose, that of making earth solid, it was no doubt a mere log of wood (shod perhaps with iron), with handles to lift it up; just like a paviour's rammer. But in the case cited from Caesar, where it was used for driving the piles of his bridge over the Rhine, it is almost evident that it must have been a machine, something like our pile-driving engine or monkey), by which a heavy log of wood, shod with iron, was lifted up to a considerable height and then let fall on the head of the pile.[1]

One of the oldest bridges in Rome was the "Pons Sublicius" or "bridge of piles." In 509 B.C. the Romans expelled their last king and established a republic. The kings took refuge at the court of Lars Porsena, the king of neighboring Clusium; three years later, they attempted to retake the city. They had not prepared to deal with the bridge and its defender, Horatius Cocles:

> Once more Horatius stood alone; with defiance in his eyes he confronted the Etruscan chivalry, challenging one after another to single combat, and mocking them all as tyrants' slaves who, careless of their own liberty, were coming to destroy the liberty of others. For a while they hung back, until shame at the unequal battle drove them to action, and with a fierce cry they hurled their spears at the solitary figure which barred their way. Horatius caught the missles on his shield and, resolute as ever, straddled the bridge and held his ground. The Etruscans moved forward, and would have thrust him aside by the sheer weight of numbers, but their advance was suddenly checked by the crash of the falling bridge and the simultaneous shout of triumph from the Roman soldiers who had done their work in time. The Etruscans could only stare in bewilderment as Horatius, with a prayer to Father Tiber to bless him and his sword, plunged fully armed in the water and swam, through the missiles which fell thick about him, safety to the other side where his friends were waiting to receive him.[2]

At the end of the Roman Republic, we have what is probably the most spectacular documented application of piles in the ancient world. This was done by Julius Caesar's army in crossing the Rhine River during his extended campaign in Gaul. He described the job as follows:

> ...(Caesar) caused pairs of balks a foot and a half thick, sharpened a little way from the base and measured to suit the depth of the river, to be coupled together at an interval of two feet. These he lowered into the river by means of rafts, and set fast, and drove home by rammers; not, like piles, straight up and down, but leaning forward at a uniform slope, so that they inclined in the direction of the stream. Opposite to these, again, were planted two balks coupled in the same fashion, at a distance of forty feet from base to base of each pair, slanted against the force and onrush of the stream. These pairs of balks had two-foot transoms let into them atop, filling the interval at which they were coupled, and were kept apart by a pair of braces on the outer side at each end. So, as they were held apart and contrariwise clamped together, the stability of the structure was so great and its character such that, the greater the force and thrust of the water, the tighter were the balks held in lock. These trestles were interconnected by timber laid over at right angles, and floored with long poles and wattlework. And further, piles were driven in aslant on the side facing downstream, thrust out below like a buttress and close joined with the whole structure, so as to take the force of the stream; and others likewise at a little distance above the bridge, so that if trunks of trees, or vessels, were launched by the natives to break down the structure, these fenders might lessen the force of such shocks, and prevent them from damaging the bridge.[3]

1.1.2. Other Pile Driving before the Nineteenth Century

Piles in the Roman world were exclusively wood piles; this was to remain the case for all piling until the end of the nineteenth century. These wood piles were driven by drop hammers on small wooden rigs. A variety of these rigs is shown in Figure 1-1. A more elaborate rig of this type from the eighteenth century is shown in Figure 1-2.

Figure 1-1 Ancient and Medieval Pile Driving Rigs

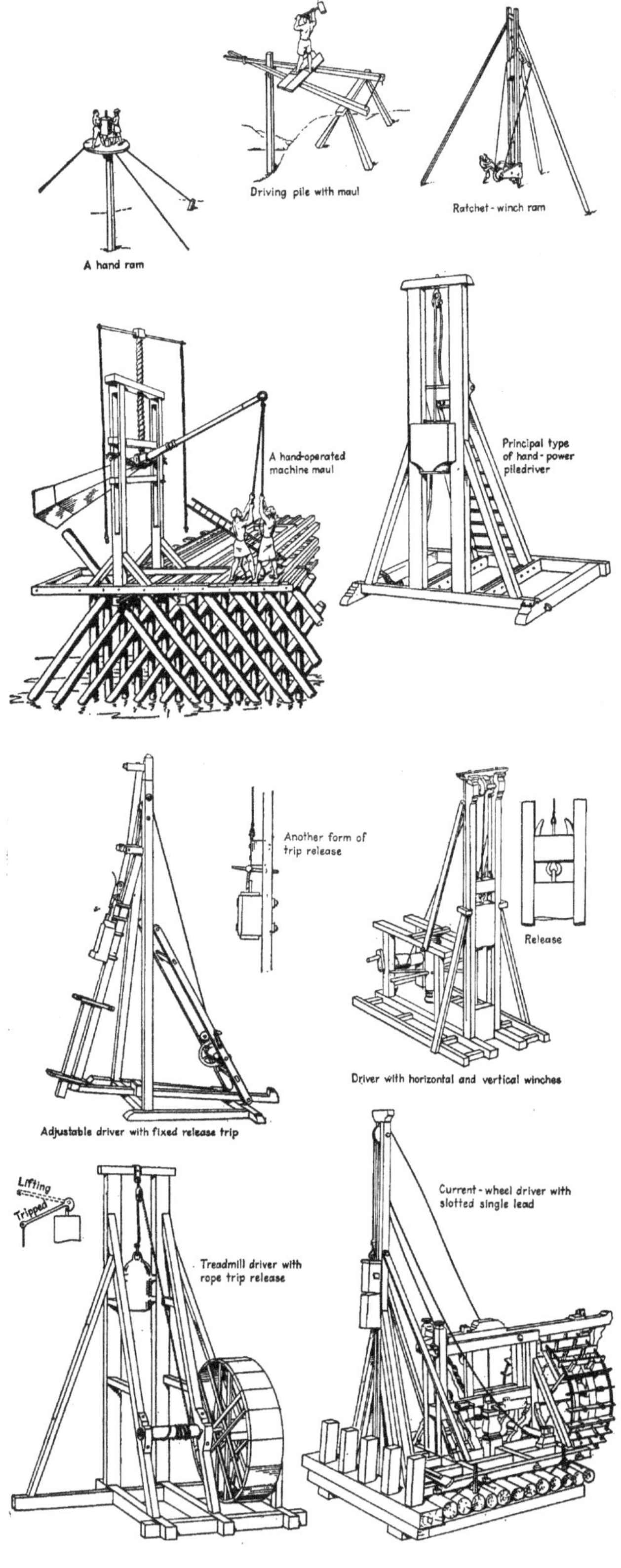

Figure 1-2 English Pile Driving Rig from the Eighteenth Century

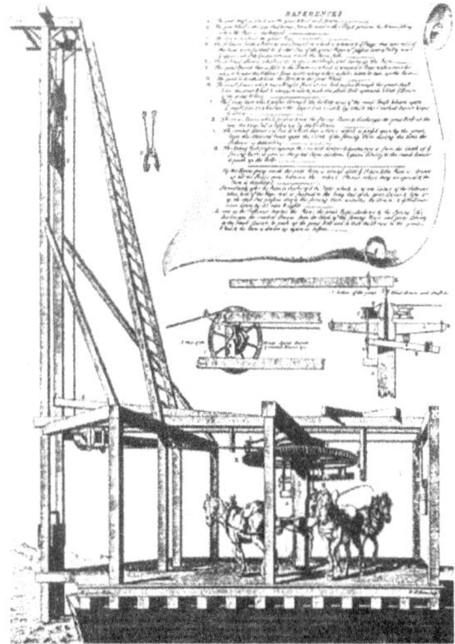

For thousands of years the Chinese and other oriental builders used a stone block as a hammer. It was lifted by ropes and stretched taut by human beings, who were arranged in a star pattern about the pile head. The rhythmic pulling and stretching of the ropes flipped the stone block up and guided the downward blow upon the pile head.

The City of Venice was built in the marsh delta of the Po River because the early Italians wanted to live in safety from the warring Huns of Central Europe. The buildings of Venice are supported on timber piles, driven centuries ago through the soft mud onto a layer of boulders below. When the bell tower of St. Mark's, built in 900 A.D., fell in 1902, the timber piles in the foundation were found to be in such a good state of preservation that they were used to support the reconstructed tower.

1.1.3. Advances in the Nineteenth Century

The nineteenth century saw real progress on several fronts for driven piles. To begin with, steam replaced muscle power for the turning of the winches. Other developments at that time include the steam hammer, concrete piles and the first dynamic pile driving formula.

1.1.3.1. Naysmith's Steam Hammer

Drop hammers had served civilization well for centuries; however, they are slow and require some skill from the operator in use. The development of the steam engine in the eighteenth century in both Britain and Russia showed the potential of steam power. The Scottish inventor James Naysmith developed the first steam hammer. In 1839 he sketched out the first design of the steam hammer; he described it as follows:

> My Steam Hammer as thus first sketched, consisted of, first, a massive anvil on which to rest the work; second, a block of iron constituting the hammer or blow-giving portion; and, third, an inverted steam cylinder to whose piston-rod the hammer-block was attached. All that was then required to produce a most effective hammer was simply to admit steam of sufficient pressure into the cylinder, so as to act on the under-side of the piston, and thus to raise the hammer-block attached to the end of the piston rod. By a very simple arrangement of a slide valve, under the control of all attendant, the steam was allowed to escape and thus permit the massive block of iron rapidly to descend by its own gravity upon the work then upon the anvil.
>
> Thus, by the more or less rapid manner in which the attendant allowed the steam to enter or escape from the cylinder, any required number or any intensity of blows could be delivered. Their succession might be modified in an instant. The hammer might be arrested and suspended according to the requirements of the work. The workman might thus, as it were, think in blows. He might deal them out on to the ponderous glowing mass, and mould or knead it into the desired form as if it were a lump of clay; or pat it with gentle taps according to his will, or at the desire of the forgeman.[4]

This hammer was originally intended as a forge hammer for steel production. In 1845, however, Naysmith designed and built a steam hammer for driving piles at the Royal dockyards at Devonport, England.

He describes this first job as follows:

> Some preliminary pile-driving had been done in the usual way, in order to make a stage or elevated way for my pile-driver to travel along the space where the permanent piles were to be driven. I arranged my machines so that they might travel by their own locomotive powers along the whole length of the coffer dam, and also that they should hoist up the great logs of Baltic timber which formed the Piles into their proper places before being driven. The entire apparatus of the machine was erected on a strong timber platform, and was placed on wheels, so that it might move along the rails laid down upon the timber way. The same boiler that supplied the steam hammer part of the apparatus served to work the small steam-engine fixed to the platform for its locomotion, and also to perform the duty of rearing the next pile which had to be driven. The steam was conveyed to the hammer cylinder by the jointed pipe seen in the annexed engraving. The pipe accommodated itself to any elevation or descent of the hammer. The whole weight of the cylinder, hammer-block, and guide box, supported by the shoulders of the pile, amounting to seven tons in all, rested upon the shoulders of the pile as a "persuader;" and the eighty blows per minute of the four-ton hammer came down with tremendous energy upon the top of the pile head. No soil, that piles could penetrate, could resist such effective agencies... There was a great deal of curiosity in the dockyard as to the action of the new machine. The pile-driving machine-men gave me a good-natured challenge to vie with them in driving down a pile. They adopted the old method, while I adopted the new one. The resident managers sought out two great pile logs of equal size and length--70 feet long and 18 inches square. At a given signal we started together. I let in the steam, and the hammer at once began to work. The four-ton block showered down blows at the rate of eighty a minute; and in the course of four and a half minutes my pile was driven down to the required depth. The men working at the ordinary machine had only begun to drive. It took them upwards of twelve hours to complete the driving of their pile![5]

In the U.S., steam hammers came into use after 1875. After manufacturing other designs, Vulcan Iron Works developed in 1887 the first "#1" hammer designed by the Warrington brothers, George, James N. and William H. This hammer (and its derivatives, the Raymond and Conmaco hammers) became the most popular steam hammers in the U.S. The McKiernan Terry Corporation (MKT) also developed a successful line of steam hammers, along with the Union and Industrial Brownhoist hammers. In Europe steam hammers were developed by such manufacturers as BSP (U.K), Menck + Hambrock (Germany), and Nilens (Belgium).

The original steam hammers were single acting, i.e., they relied solely on the drop of the ram through the gravity field for the kinetic energy used to drive the pile. The twentieth century saw the development of hammers with downward assist of some kind, where the steam (and later the compressed air) was used to accelerate the ram downward more than gravity. These fell into two types. The first were the compound hammers, which used the air or steam expansively on the downstroke. They included first the Vulcan "California" series hammers and later the MKT "C" series hammers. The second were the double or differential acting hammers, which simply used the air or steam at full pressure to help accelerate the ram downward. Vulcan developed the differential acting hammer in the 1930's in both closed and open type; Raymond developed its own differential acting hammers, first steam and then hydraulic. MKT's successful double-acting "B" series is still commonly used to drive sheet piling. European manufacturers also developed closed-type double acting hammers. Figure 1-3 shows such a hammer, in this case a Nilens T-3.

Figure 1-3 Nilens Double-Acting Steam Hammer

1.1.3.2. Raymond's Concrete Piles

Under the proper conditions, timber piles are a very durable pile type, but timber piles are subject to varying degrees of deterioration. Moreover, timber piles are limited in length and size by the trees they come from.

The French engineer François Hennebique was one of Europe's premier bridge builders. In 1897, he introduced the use of reinforced concrete piles. The U.S. was not far behind and in 1901, A.A. Raymond first used concrete piles in a building foundation in Chicago. Raymond went on to found the Raymond Concrete Pile Company, which for most of the twentieth century was the greatest pile driving organization in the world, even extending its activities to the construction of offshore oil platforms with steel pipe piles.

Timber piles were usually driven to less than 50 kips (222 kN) allowable capacity, but the new concrete piles were designed for 60 kips (267 kN) and higher. This meant that fewer piles and smaller footings could be utilized for the same imposed loads. Technological advances in the cement and concrete industries made concrete piles cost competitive and, because of this, their use became prevalent.

1.1.4. Steel Piles

The beginning of the twentieth century also saw the start of the use of steel piling as well, both H-piles and pipe. Both of these steel shapes existed for structural use and were adapted to piles. In the case of pipe, though, the evolution of tubular steel piling went in two distinct directions:

- Without concrete fill, as open ended or closed ended pipe piles: These are typically used in applications where lateral or tensile loads predominate, such as offshore oil platforms and seismic or scour resisting members.
- With concrete fill: This includes a wide variety of piles, such as caissons, bulb piles, Monotube piles and (when the steel is thin enough) shell piles driven with mandrels, such as the Raymond Step-Taper® piles.

H-piles were an outgrowth of I-beams driven to meet a serious problem: scour undermining of bridge piers and abutments in compact sand and gravel. The piles could withstand both hard driving and were able to be driven deep enough to adequately resist scour. Bethlehem Steel rolled the first H-pile in 1908 and this replaced the I-beams used before that time.[6]

1.1.5. Advances in Rigs

No history of piling would be complete without some mention of the advances in pile driving rigs.

The most popular type of pile driving rig before the advent of crane-mounted rigs was the skid rig, as shown in Figure 1-4. Skid rigs dominated the industry in an era when the motto was frequently "when in doubt, drive piles."

Figure 1-4 Skid Rig with Vulcan Hammer driving Monotube Piles

An advance with the skid rig is the rotating rig. Raymond built its first rotating rig in 1913 and continued to build them until 1929 and use them into the 1950's. They consisted of a bedsill with a 12' (3.66 m) rail circle and a turntable with a 13'-8" (4.17 m) fixed radius from kingpin to the centerline of the turntable. The leaders were mounted on the front of the turntable, were of rugged construction and were 56' (17 m) in height. A larger and heavier version of this was the Long Swing Driver; this is shown in Figure 1-5.

Figure 1-5 Raymond Long Swing Type Driver

The advent of large mobile cranes, with their greater maneuverability, led to a new era in pile driving rigs. A typical mobile rig with fixed leaders is shown in Figure 1-6.

Figure 1-6 Pile Driving Rig using Mobile Crane

1.1.6. Pile Dynamics

Pile driving is an "interdisciplinary" process in that it is a combination of geotechnical engineering (the interaction with the soils), engineering mechanics (the dynamics of moving bodies) and structural engineering (the stresses during driving and after installation.) Nowhere is that better illustrated than in the subject of pile dynamics.

1.1.6.1. Dynamic Formulae

The first attempt to model the dynamics of pile driving and to make this modeling useful to practitioners were the dynamic formulae. These formulae use Newtonian impact mechanics to model the motion of the pile; these results could generally be expressed in a simple formula, which could be readily applied to the work at hand.

The most widely used dynamic formula is the Engineering News formula, although many others (and their local variations) have made their way into codes and other standards of practice (Hiley, Gates, Danish, etc.) Many foundations have been installed using these formulae as a basis of pile control and acceptance.

The dynamic formulae have been pilloried extensively since the wave equation became practical to use. Their three main weaknesses are

- Modeling of the pile as one rigid mass;
- Inadequate modeling of the driving system (hammer, cushion, helmet);
- Inadequate modeling of the soil as it interacts with the pile.

These weaknesses and others were not as apparent when timber piles installed using drop hammers was the norm in pile driving. With the introduction of concrete and steel piles these deficiencies became critical, especially when concrete piles began to exhibit tension cracking, a phenomenon the dynamic formulae could neither anticipate nor quantify.

1.1.6.2. Application of Stress-Wave Theory to Piles

It seems difficult to believe that such a simple looking operation as pile driving can involve such a complicated phenomenon as stress-wave propagation. The history of the theory's application, however, is important in the continued viability of driven piles as a deep foundation. The theory is discussed later in this book; the following is an overview of the development of this theory with application to piles.

1.1.6.2.1. Isaacs' Work

The use of the wave equation (or stress-wave theory) to model impact pile driving began in Australia with the work of David Victor Isaacs[7]. The dynamic formulae had been developed primarily with timber piles in mind; with the growing usage of concrete piles, it became apparent that, because of the length and properties of timber piles, the dynamic formulae (with their assumption that the pile is a rigid mass) would not be sufficient for concrete piles. Isaacs began by reviewing the dynamic formulae. Part of his review included a discussion of the factor of safety, where he makes a statement that is still relevant:

> It should be remembered, however, that these are not true factors of safety, but include a "factor of ignorance." The author suggests that when the ultimate resistance of any pile has been determined, in fixing the factor of safety...the most unfavourable conditions possible in the supporting strata should be judged (the range of conditions possible being narrowed with better knowledge of the subsurface conditions and of the possibility of disturbance from extraneous sources) and a proportion of the factor of safety -- a "factor of ignorance" -- then allowed in respect to these possible conditions, the manner of determining the ultimate load, and the type of loading to be borne. The remaining proportion of the factor of safety -- or true margin of safety -- should be approximately constant for all classes of loading and foundation conditions involving the same value of loss in case of failure; and the overall factor of safety...will then be equal to the product of the true factor of safety with the "factor of ignorance."

He also describes an experiment where rods are impacted against each other in a pendulum arrangement. As the rods were lengthened, the behavior of the rods deviated more and more from Newtonian impact theory.

He then developed an integration technique that is best described as a semi-graphical one. He developed a mathematical model based on the successive transmission and reflection of waves (similar in principle to the method of images.) A sample solution is given Figure 1-7, in this case showing multiple impacts.

Figure 1-7 Graphical Solution of the Wave Equation

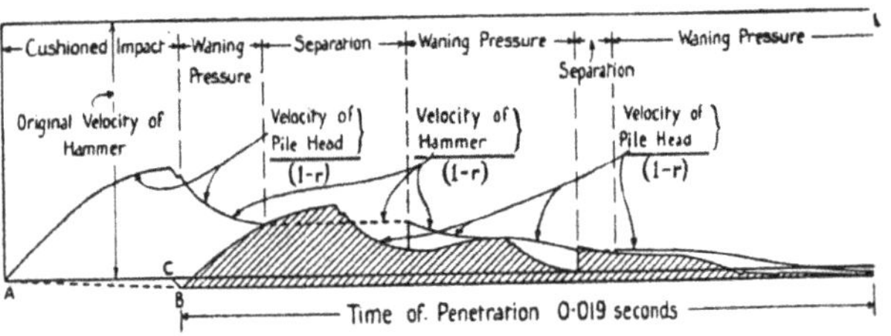

He then constructed a machine to graphically integrate the solution, a diagram of which is shown in Figure 1-8.

Figure 1-8 Graphical Machine for Wave Equation Solution

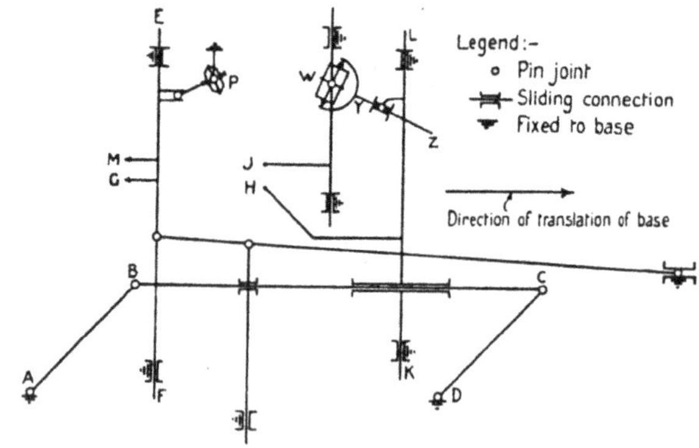

He was then able to solve for the stresses and displacements of the pile during driving. Isaacs developed a set of formulae and charts to make his results accessible for the analysis of piles.

In the course of the investigation, Isaacs dealt with a number of questions that would become central to stress wave analysis of piles, including tension stresses in concrete piles, the effect of ram weight (he concluded that a heavier ram reduced tension stresses,) and the effect of hammer cushion stiffness and drive cap weight.

Isaacs' work also revealed the computational complexity of stress wave analysis, a complexity that insured the dominance of dynamic formulae in pile analysis (with all of their serious limitations) for another half century.

1.1.6.2.2. The Work of Glanville

After Isaacs' work, the centre of work on the subject shifted to the United Kingdom, with the extensive study commissioned by the British Building Research Board under the direction of (later Sir) W.H. Glanville[8] and his colleagues. This study was one of the first comprehensive studies on stress waves in piles. It was commissioned in part to investigate problems encountered in the breakage of concrete piles during driving, at both the head and the toe. The wave equation using d'Alembert's solution were used to develop equations to estimate the stress in the pile during driving, using the method of images. Because of the complexity of the equations, the results were reduced to a series of charts where a quantity of dimensionless stress was plotted against the ratio of hammer weight to pile weight. The charts could then be used to estimate pile stresses and resistance. The charts were applicable to concrete piles only, and this was a serious limitation to such solutions.

In addition to developing a solution to the wave equation, the authors continued Isaacs' work in addressing technical issues and experimental techniques that have enduring interest in pile dynamics. These include the instrumentation and data collection on stresses and forces in piles, including remote data gathering through "portable" equipment in a trailer, further research on the effect of the hammer cushion on the generation and effect of the pile stress wave (these were included in the analytical work,) drop tower testing on cushion material to determine the cushion stiffness, and further work on the relationship of ram weight to pile weight and cross section. An example of the data collected is illustrated in Figure 1-9, which shows actual photographs of the oscilloscope readings recording stresses at various points of the pile during driving. Note that the stress wave impulse at the middle is delayed from the head, as is the "foot" from the middle; this is as predicted by stress wave theory.

Figure 1-9 Recorded stresses in the pile

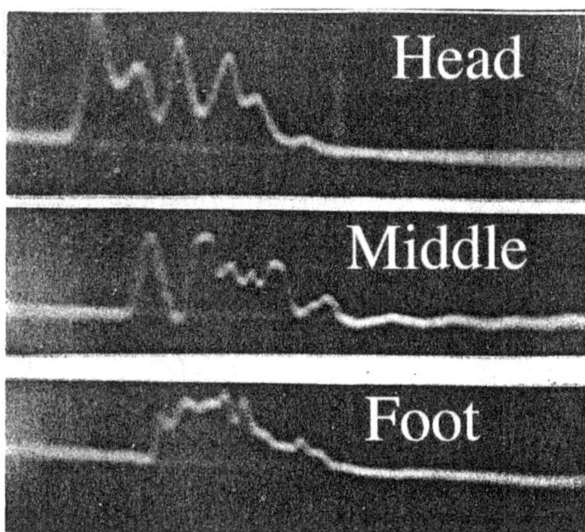

The Second World War interrupted further research; moreover, the difficulties that this study encountered in developing a readily usable prediction technique ended the possibility of using a closed form solution for this application. Research has continued to the present on closed form solutions[9].

1.1.6.2.3. Enter the Numerical Methods

The difficulties in applying the closed form solutions, led to the use of numerical methods to model stress waves in piles and pile behavior itself. Today we say "obvious" but in the 1940's and 1950's, this was not the case, primarily because the practical application of numerical methods required the use of the recently developed digital computer. Moreover geotechnical engineering in general was slower than other specialties to adopt advanced analytical techniques, primarily because the ground itself presented complexities that are in reality only now beginning to be modeled on a reasonable basis. Experience and judgment, essential elements in any engineering practice, were and are especially important in geotechnical applications.

This background clearly illustrates how remarkable the accomplishment of E.A.L. Smith[10] really is. Smith was the Chief Mechanical Engineer for the Raymond Concrete Pile Company. By the time his historic "wave equation" paper had been published, he had already retired. The basic elements of Smith's technique are as follows:

- Division of the pile into a series of springs and masses. The hammer was modeled as a mass with a hammer cushion spring (or capblock, to use the Raymond terminology.) The driving accessory (or follower) was also modeled as a mass, with provision for another cushion for the pile if needed.
- Integration of the model using a first order finite difference technique.
- Modeling of the hammer and pile cushions using a "static" hysteresis technique, defined by an elastic spring constant during compression and modified by a "coefficient of restitution" during rebound.
- Modeling of the soil as a combination of displacement dependent springs and velocity dependent dampers. These were applied both along the pile shaft and at the pile toe as the actual soil resistance distribution required.
- Modeling of the non-linearities of the soil. The soil was given a "yield limit" (a quake); after this deflection the non-dynamic resistance was constant. The model could simulate yielding in some areas of the pile without requiring it in others.

Smith made many of the assumptions for his model (especially for the soils) with a minimum of theoretical basis; nevertheless, the soil model he proposed is still standard in many wave equation programs today. This is a tribute to the durability and basic soundness of his technique.

In spite of the advanced state of Smith's model, it was several years before it was disseminated to practitioners. The first version of Smith's model that was made available to a wider audience was the TTI (Texas Transportation Institute) program, developed at Texas A&M University by L.L. Lowery and T.J. Hirsch[11]. In addition to making Smith's model available, many of the assumptions behind the Smith model were examined to improve the quality of the input and thus confidence in the output. The TTI program is still available and is used especially with offshore oil and gas platform piling.

When Smith developed his wave equation method, he only included hammers whose rams could be modeled as rigid, impacting a hammer cushion. This is principally because Raymond used steam hammers at the time. Both the Smith had the TTI programs had only a rudimentary modeling of the combustion complexities of diesel hammers, which became popular in the U.S. in the 1960's. This deficiency was overcome with the WEAP program, developed by George Goble and Frank Rausche[12]. In addition to advancing the modeling of the diesel hammers, WEAP and its successors gave the wave equation for piling the one thing it needed more than anything else: an organization (GRL) prepared to actualize it through support and dissemination. An example of this technical support was WEAP86, WEAP87 and GRLWEAP, which include a hammer database and PC capability. The need for technical support became more critical with the development of pile monitoring techniques in the field.

There are other wave equation computer codes in use today, such as TNOWAVE, by TNO in the Netherlands. In addition to the finite difference technique, there have been finite-element methods used, either one-dimensional or with the surrounding soil modeled[13].

1.1.6.3 Field Monitoring Techniques

Geotechnical engineering principles are inherently influenced by the high degree of uncertainty created by the use of soil and rock as an engineering material. The wide variability of soils and the difficulty inherent in modeling their behavior makes some kind of in situ verification essential.

With driven piles, dynamic formulae gave the industry just such verification. By relating the blow count of the hammer per foot (or inch, meter, etc.) of pile penetration to the resistance of the pile, the capacity of piles could be estimated. This unusual situation made the pile hammer both an installation tool and a measuring device.

Although the wave equation enabled prediction of performance, relating that prediction to the actual capacity or resistance of the pile involves more than just comparing the wave equation to the static load test. From the beginning, the need for pile installation instrumentation was felt necessary to learn more about the pile's behavior both during driving and when loaded by the structure.

The first comprehensive (and successful) attempt at instrumentation was Glanville's study. Subsequent efforts in this direction took place in Sweden, at the Gubbero site in 1960[14]. Both of these efforts relied on the photographed output of an oscilloscope.

The major step in using stress wave theory to analyze piles during driving and to estimate their static capacity was the development of the Case Method[15]. This method compared the pile force and velocity at a given time with a time 2L/c after that. The static and dynamic soil resistance components were then separated. This method was very simple and could be readily applied in the field, through the measurement of force and acceleration near the pile head using both strain gages and accelerometers.

A more advanced analytical computer model of the measured data became available with CAPWAP[16] (Case Pile Wave Analysis Program). This technique uses similar instrumentation to the Case Method, but allows further refinement of the soil response through in iterative technique. A profile of the soil resistance distribution is thus obtained along with end bearing resistance.

Other organizations (such as TNO) have developed methods of analyzing the return signals of impact. The result in all cases is once again the use of the hammer, this time in conjunction with stress wave theory and modern measuring techniques, as a measuring tool to estimate the pile's capacity as it is being driven.

One further application of stress wave theory in the field is integrity testing. It is used to find the presence of cracks (driven piles) or voids (drilled shafts and auger-cast piles.) There are two variations to this technique:

- Low strain integrity testing, where a small hammer transmits a stress wave along the pile length and the returning echo is analyzed, much like sonar; and
- High strain integrity testing, which is also used to dynamically measure the pile capacity.

1.1.7. Diesel Hammers

Diesel hammers were first developed in Germany in the 1920's. Their two key advantages were that a) they were able to operate without an external power source and b) they tended to be lighter for a comparable striking energy. They were introduced into the U.S. after World War II, first by the Syntron/Link Belt series of closed ended diesel hammers and then the Delmag diesel hammers from Germany. Today they are manufactured by several manufacturers in various parts of the world.

Most diesel hammers manufactured today—and all of these depicted in Chapter 3—are of the tubular, air cooled type. Two variations from these are a) rod type diesel hammers and b) water cooled diesel hammers. Rod type diesel hammer have a ram that runs on columns in a similar manner as air/steam hammers. The combustion chamber is hidden as the air is compressed and the fuel injected and exposed as the ram is thrown upward by the results of the combustion. Rod type diesel hammers are generally restricted today to very small diesel hammers. An example of a rod type diesel hammer is shown in Figure 1-10.

Figure 1-10 Rod Type Diesel Hammer

Water cooled diesel hammers have a water tank or jacket surrounding the combustion chamber. These afford superior cooling capabilities, but their inconvenience to the user has limited their popularity. An example of a water cooled diesel hammer is shown in Figure 1-11.

Figure 1-11 Water Cooled Diesel Hammer

1.1.8. Vibratory Hammers

The first vibratory pile driver used was in the former Soviet Union, a model BT-5 developed and first used under the direction of D.D. Barkan. This hammer had a dynamic force of 214 kN and the eccentrics rotated at 41.67 Hz, powered with a 28 kW electric motor. Used in the construction of the Gorki (now called once again Nizhni-Novgorod) hydroelectric development. This hammer drove 3700 sheet piles 9-12 m long in 2-3 minutes each.

The Soviets then developed a large variety of vibratory pile drivers and soil drilling equipment in the 1950's and afterwards. Although space does not permit a complete catalogue of these machines, two are mentioned here as they are of special interest.

- VPM-170. This hammer had a dynamic force of 1700 kN and the eccentrics rotated at 550 RPM; it was the largest vibratory pile driver the Soviets produced. This was designed to drive pipe piles 1600 mm in diameter into any type of soil except for rocky soils. It could operate at two different frequencies. This hammer is shown in Chapter 3.
- VU-1.6. It had a dynamic force of 958 kN and the eccentrics rotated at 495 RPM. Like the VPM-170, it was designed to drive pipe piles 1600 mm in diameter, in its case to depths up to 30 m; however, it also could remove the plug from the pile during driving. The machine had a large centre hole 1400 mm in diameter, so that the soil could be removed without having to stop and remove the pile driver. The eccentrics were arranged in a square pattern and synchronized by special gearing on the ends of the shafts. This machine is shown in Figure 1-12.[17]

Figure 1-12 Soviet VU-1.6 Vibratory Hammer

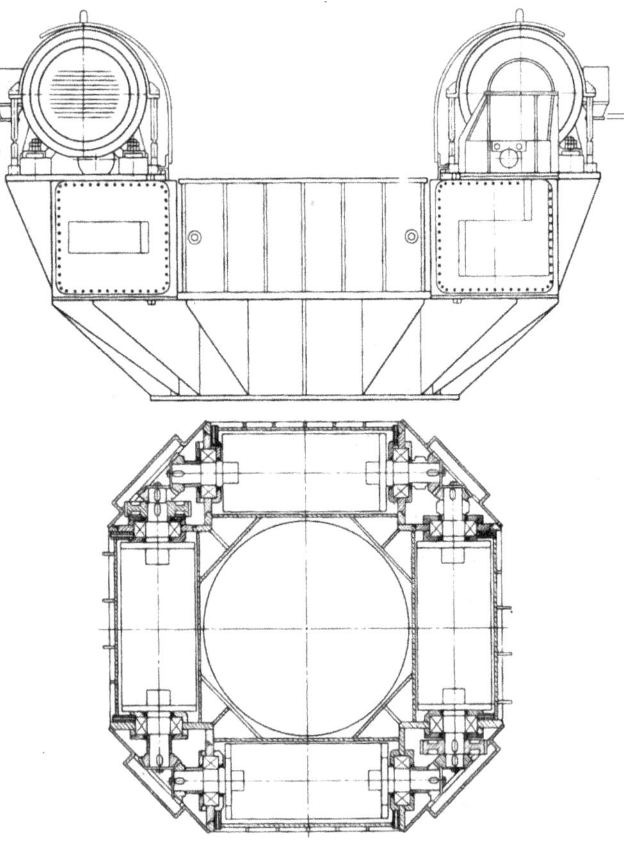

The Soviets first licensed their technology to the Japanese, who developed an extensive array of vibratory hammers. Many of the Japanese hammers were not much different from the Soviet ones. One exception was the Uraga hammer; this hammer was unique in that an electric motor was placed inside of each eccentric, thus making it a very "directly driven" machine. This technology has since spread worldwide, with such concerns as PTC and Tramac in France, Müeller, Tünkers and MGF in Germany, Tomen in Japan and ICE Europe in the Netherlands being well established in the field.

The first American made hydraulic vibratory was the MKT V-10, which they introduced in 1969, although both Vulcan and Foster had introduced Japanese and French vibratory pile drivers respectively in the early 1960's. A diagram of this machine is shown in Figure 1-13. This pioneering machine differs from most current vibratory pile driving equipment in several respects. In common with the practice of the time, the V-10 suspension used steel coil springs to provide dampening for the crane boom and hook, whereas now most machines use rubber springs. The V-10's eccentrics were long (a practice borrowed from vibratory screens and separators) and mounted crossways on the machine. A motor was coupled to one of the eccentrics gears transmitted the power to the rest. Most machines today mount the eccentrics from front to back of the case, and drive them either directly or through a speed changing pinion gear.

Figure 1-13 MKT V-10 Vibratory Hammer

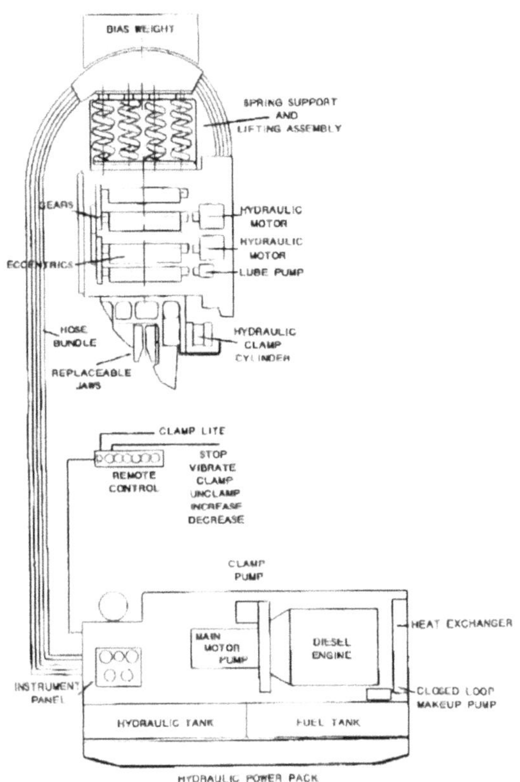

From this beginning, the unique practices in the U.S. have lead to the evolution of a distinctive style of vibratory hammer in the U.S., and this has been followed elsewhere. Today it is embodied to a greater or lesser degree in the APE, ICE (America and Europe), MKT, and HPSI units that are on the market. In addition to the changes mentioned above, these characteristics include slim throat hammers for sheet pile driving, hydraulic drive, and high power motors, pumps, and engines.

Classifying vibratory pile driving equipment can be a complex business, but the most important division can be made based on frequency, with the resultant relationships between dynamic force and eccentric moment.

1.1.9. Impact-Vibration Hammers

The term "impact-vibration hammer" refers to a type of vibratory pile driver that imparts both vibrations and impacts to the pile during operation. Based on theoretical work done during the Second World War, S.A. Tsaplin built the first experimental impact-vibration hammer in the Soviet Union in 1949. The specifications of the hammer and test set-up are shown in Table 1-1. In field tests his impact-vibration hammer was welded to the top of metallic tube 110 mm in diameter, 8 mm wall thickness, 2.6 m long and with a mass of 200 kg. The hammer then drove the tube into a variety of sandy, sandy loam, and clay soils. A comparison was made of the effect of driving by the impact-vibration mode versus the vibration mode, the latter of which was achieved by the complete blocking of the springs. The tests made it possible to establish that the efficiency of the impact-vibration driving is substantially higher with regard to both the maximum driving depth and the pile installation speed. It also demonstrated that the efficiency of the driving increases with increasing amplitude of the vibration exciter vibrations.

Table 1-1 Hammer and Test Setup for Tsaplin Impact-Vibration Hammer

Parameter	Value
Nominal rotational speed, Hz	48
Permissible frequency range w/generator, Hz	40-200
Power, kW	1.6
Exciter mass, kg	75
Overall mass, kg	95
Test Stand	
Material	Reinforced Concrete
Mounting	Rigidly Attached to Hammer
Mass, kg.	500
Generator	
Power Output, kW	50

The first practical application of the impact-vibration hammer took place in the construction of the Stalingrad (now Volgograd) power plant, where in the construction of the anti-filtration wall under the dam "Larssen-5" piles were driven into sandstone of medium firmness to a depth of 13 m. On this and other jobsites, the impact-vibration hammers were able to outperform conventional vibratory hammers, air/steam and diesel hammers. An example of an impact-vibration hammer is shown in Figure 1-14.

Figure 1-14 Impact-Vibration Hammer

These successes led to further use of impact-vibration hammers, not only in Russia but also in other countries, especially those of the EU where manufacturers such as Menck and PTC have taken up the production of these units. There are no impact-vibration hammers manufactured in the U.S. at present.

1.2. Overview Of Pile Foundation Design And Construction

1.2.1. Design Of Pile Foundations

As is the case with most geotechnical design, the design of pile foundations lacks the neat precision of structural design. The interactions among the piles and the surrounding soil are complex. Insertion of piles generally alters the character of the soil and intense strains are set up locally near the piles. The nonhomogenity of soils, along with the effects of the pile group and pile shape, add further difficulties to the understanding of soil-pile interaction.

Broad generalizations about pile behavior are unrealistic. An understanding of the significance of several factors involved is required to be successful in the design of pile foundations. Because of the inherent complexities of pile behavior, it is necessary to use practical semi-empirical methods of design, and to focus attention on significant factors rather than minor or peripheral details. The foundation engineer must have a thorough understanding of foundation loads, subsurface conditions including soil/rock properties and behavior, the significance of special design events, foundation performance criteria, and current practices in foundation design and construction in the area where the work is to be done to arrive at the optimum foundation solution. Pile design is discussed in detail in Chapter 4.

The input of an experienced geotechnical engineer from the planning stage through project design and construction is essential to produce a successful driven pile foundation. The geotechnical engineer who specializes in foundation design is the most knowledgeable person for selecting the pile type, estimating pile length, and choosing the most appropriate method to determine ultimate pile capacity. Therefore, the geotechnical engineer should be involved throughout the design and construction process. In some project phases, i.e. preliminary explorations, preliminary design, and final design, the geotechnical engineer will have significant involvement. In other project phases, such as construction, and post construction review, the geotechnical engineer's involvement may be more of a technical services role. The geotechnical engineer's involvement provides the needed continuity of design personnel in dealing with design issues through the construction stage.

1.2.2. Construction Of Pile Foundations

Construction of a successful driver) pile foundation that meets the design objectives depends on relating the requirements of the static analysis methods presented on the plans to the dynamic methods of field installation and construction control. The tools for obtaining such a foundation must be explicitly incorporated into the plans and specifications as well as included in the contract administration of the project.

It is important that a pile foundation be installed to meet the design requirements for compressive, lateral and uplift capacity. This may dictate driving piles for a required ultimate capacity or to a predetermined length established by the designer. It is equally important to avoid pile damage or foundation cost overruns by excessive driving. These objectives can all be satisfactorily achieved by use of wave equation analysis, dynamic monitoring of pile driving, and static load testing. Commonly used dynamic formulas, such as Engineering News formula, have proven unreliable as pile capacities increased and contractors routinely used more sophisticated pile installation equipment.

Knowledgeable construction supervision and inspection are the keys to proper installation of piles. State-of-the-art designs and detailed plans and specifications must be coupled with good construction supervision to achieve desired results. The actual construction of pile foundations is detailed in Chapter 5.

Post construction review of pile driving results versus predictions regarding pile driving resistances, pile length, field problems, and load test capacities is essential. These reviews add to the experience of all engineers involved on the project and will enhance their skills.

1.2.3. Driven Pile Design-Construction Process

The driven pile design and construction process has aspects that are unique in all of structural design. Because the driving characteristics are related to pile capacity for most soils, they can be used to improve the accuracy of the pile capacity estimate. In general, the various methods of determining pile capacity from dynamic data such as driving resistance with wave equation analysis and dynamic measurements are considerably more accurate than the static analysis methods based on subsurface exploration information. Furthermore, pile driveability is a very important aspect of the process and must be considered during the design phase. If the design is completed, a contractor is selected, and then the piles cannot be driven, large costs can be generated. It is absolutely necessary that the design and construction phases be linked in a way that does not exist elsewhere in construction.

The driven pile design-construction process is outlined in the flow chart of Figure 1-15. This flow chart will be discussed block by block using the numbers in the blocks as a reference and it will serve to guide the designer through all of the tasks that must be completed.

Figure 1-15 Driven Pile Design and Construction Process

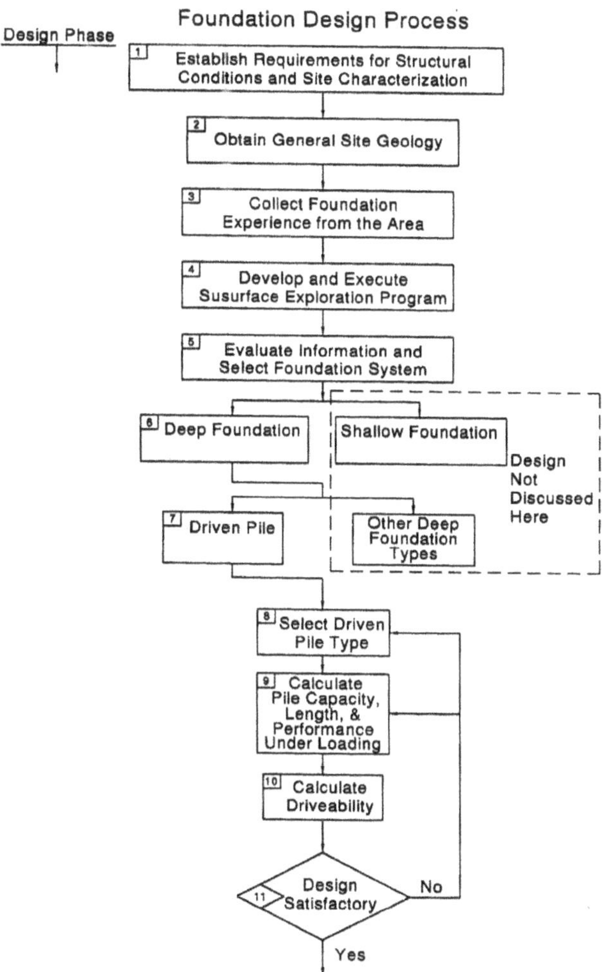

Figure 1-15 Driven Pile Design and Construction Process (continued)

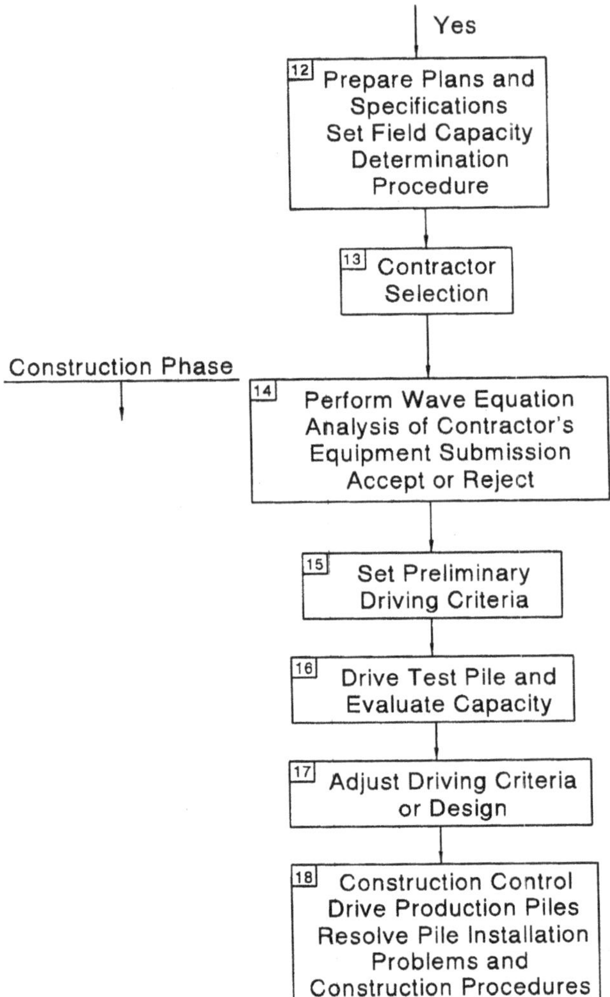

Block 1: Establish Requirements for Structural Conditions and Site Characterization

The first step in the entire process is to determine the general structure requirements.

A. Is the project a new bridge, a replacement bridge, a bridge renovation, a retaining wall, a noise wall, or sign or light standard?

B. Will the project be constructed in phases or all at one time?

C. What are the general structure layout and approach grades?

D. What are site characteristics at the soil surface?

E. Is the structure subjected to any special design events such as seismic, scour, debris, vessel impact, etc.? If there are special design events, the design requirements for the event should be reviewed at this stage so that these considerations can be factored into the site investigation.

F. Are there possible modifications in the structure that may be desirable for the site under consideration?

G. What are the approximate foundation loads? Are there deformation or deflection limitations beyond the usual requirements?

Block 2: Obtain General Site Geology

A great deal can be learned about the foundation requirements with even a very general understanding of the site geology. For small structures, this may involve only a very superficial investigation such as a visit to the site. The foundation design for very large structures may require extensive geologic studies.

Block 3: Collect Foundation Experience from the Area

Frequently there is information available on foundations that have been constructed in the area. This information can be of assistance in avoiding problems. Both subsurface exploration information and foundation construction experience should be sought prior to selecting the foundation type.

Block 4: Develop and Execute Subsurface Exploration Program

Based on the information obtained in Blocks 1-3 it is possible to make decisions regarding the necessary information that must be obtained at the site. The program must meet the needs of the design problem that is to be solved at a cost consistent with the size of the structure. The subsurface exploration program as well as the appropriate laboratory testing must be selected. The results of the subsurface exploration program and the laboratory testing are used to prepare a subsurface profile and identify critical cross sections. These tasks are covered in greater detail in Chapters 4, 5, and 6.

Block 5: Evaluate Information and Select Foundation System

A. The information collected in Blocks 1-4 must be evaluated and a foundation system selected. The first question to be decided is whether a shallow or a deep foundation is required. This question will be answered based primarily on the strength and compressibility of the site soils, the proposed loading conditions, and the project performance criteria. If settlement is not a problem for the structure, then a shallow foundation will probably be the most economical solution. Ground improvement techniques in conjunction with shallow foundations should be evaluated. Shallow and deep foundation interaction with approach embankments must also be considered. If the performance of a shallow foundation exceeds the structure performance criteria, a deep foundation must be used. The design of shallow foundations and ground improvement techniques are not covered in this manual. The problem of selecting the proper foundation system is discussed in detail in Chapter 7.

B. Refined foundation loading information and performance criteria should be established at this time. In Block 1, this issue was considered. Probably the result of that effort has matured in the intervening time (which might be quite long for some projects) and better-defined foundation loads and performance criteria should now be available. The geotechnical engineer must obtain a completely defined and unambiguous set of foundation loads and performance requirements in order to proceed through the foundation design.

Block 6: Deep Foundation

The decision among deep foundation types is now divided between driven piles and drilled shafts. What is really intended is the difference between driven piles and all other deep foundation systems. These other deep foundation systems have been called a drilled shaft but would also include auger cast piles, micropiles and other drilled-in deep foundation systems. The questions that must be answered in deciding between driven piles and other deep foundation systems will center around the relative costs of available, possible systems. In addition, constructability must be considered. This book is concerned with driven piles so the other types of deep foundations will not be discussed here.

Block 7: Driven Pile

Block 8: Select Driven Pile Type

A. At this point on the flow chart, the primary concern is for the design of a driven pile foundation. The pile type must be selected consistent with the applied load per pile. Consider this problem. The general magnitude of the column or pier loads is known from the information obtained in Blocks 1 and 5. However, a large number of combinations of pile capacities and pile types can satisfy the design requirements Should twenty, 1000 kN capacity piles be used to carry a 20,000 kN load, or would it be better to use ten, 2000 kN capacity piles? This decision should consider both the structural capacity of a pile and the realistic geotechnical capacities of the pile type for the soil conditions at the site, the cost of the available alternative piles, and the capability of available construction contractors to drive the selected pile. Of course, there are many geotechnical factors that must also be considered.

B. At this stage the loads must be firmly established. In Block 1, approximate loads were determined. At that time the other aspects of the total structural design were probably not sufficiently advanced to establish the final design loads. By the time that Block 5 has been reached the structural engineer should have finalized the various loads. One of the most common inadequacies that is discovered when foundation problems arise is that the design loads were never really accurately defined.

C. In the former use of the dynamic formula, the pile load specified was a design or working load since a factor of safety was contained in the formula. Modern methods of pile capacity determination always use ultimate loads with a factor of safety selected and applied. This should also be ma de clear in the job specifications so that the contractor has no question regarding the driving requirements.

D. If there are special design events to be considered, they must be included in the determination of the loads. Primarily the structural engineer will evaluate vessel impact and the results of that analysis will give pile design loads for this case. There may be stiffness considerations in dealing with vessel impact since the design requirement is basically a requirement that some vessel impact energy be absorbed.

E. Scour presents a different requirement. The loads due to the forces from the stream must be determined and this determination should be included in the structural engineer's load determination process. The depth of scour must also be determined. In the design process, it must be assured that after scour the pile will still have adequate capacity.

F. In many locations in the country, seismic loads will be an important contributor to some of the critical pile load conditions. Since the 1971 San Fernando Earthquake, much more emphasis has been placed on seismic design considerations in the design of highway bridges. Usually the structural engineer will determine the seismic requirements. Frequently the behavior of the selected pile design will affect the structural response and hence the pile design loads. In this case, there will be another loop in the design process that includes the structural engineer.

G. Pile selection is covered in more detail in Chapter 2.

Block 9: Calculate Pile Length, Capacity, and Performance

For the selected pile type, perform static analyses to determine the length necessary to provide the required compression, uplift and lateral load capacity and to meet performance criteria. The calculation of the geotechnical and structural pile capacity and performance under load is discussed in detail in Chapter 4. It may be necessary to change pile type or number of piles at this stage.

Block 10: Calculate Driveability

At this point, the proposed pile type and length have been chosen to meet the foundation loading and performance requirements. However, the design is not complete until it can be verified that the chosen pile can be driven to the required capacity and penetration depth at a reasonable driving resistance without excessive driving stresses. This analysis is performed using the wave equation program. All of the necessary information is available except the hammer selection. Since the hammer to be used on the job will only be known after the contractor is selected, possible hammers must be tried to make sure that the pile is drivable to the capacity and depth required.

Block 11: Design Satisfactory

At this point in the process, all aspects of the design should be reviewed and if changes are indicated, the flow chart is re-entered at some earlier point and a new design is developed.

Block 12: Prepare Plans and Specifications, Set Field Capacity Determination Procedure

When the design has been finalized, plans and specifications can be prepared and the procedures that will be used to verify pile capacity can be defined. It is important that all of the quality control procedures are clearly defined for the bidders to avoid claims, after construction is underway.

Block 13: Contractor Selection

After the bidding process is complete, a successful contractor is selected.

Block 14: Perform Wave Equation Analysis of Contractor's Equipment Submission

At this point the engineering effort shifts to the field. The contractor will submit a description of the pile driving equipment that he intends to use on the job for the engineer's evaluation. Wave equation analysis is performed to determine the driving resistances that must be achieved in the field to meet the required capacity and pile penetration depth. Driving stresses are determined and evaluated. If all conditions are satisfactory, the equipment is approved for driving. Some design specifications make this information advisory to the contractor rather than mandatory.

Block 15: Set Preliminary Driving Criteria

Based on the results of the wave equation analysis of Block 14 and any other requirements in the design, the preliminary driving criteria can be set.

Block 16: Drive Test Pile and Evaluate Capacity

The test pile(s) are driven to the preliminary criteria developed in Block 15. Driving requirements may be defined by penetration, driving resistance, dynamic monitoring results or a combination of these conditions. The capacity can be evaluated by driving resistance from wave equation analysis, the results of dynamic monitoring, static load test, or a combination of these. Dynamic monitoring and static load testing is described in Chapter 6.

Block 17: Adjust Driving Criteria or Design

A. At this stage the final conditions can be set or, if test results from Block 16 indicate the capacity is inadequate, the driving criteria may have to be changed. In a few cases, it may be necessary to make changes in the design as far back as Block 8. If major changes are required, it will be necessary to repeat Blocks 14, 15, and 16.

B. In some cases, it is desirable to perform preliminary field testing before final design. When the job is very large and the soil conditions are difficult, it may be possible to achieve substantial cost savings by having results from a design stage test pile program, including actual driving records at the site, as part of the bid package.

Block 18: Construction Control

After the driving criteria are set, the production pile driving begins. Quality control and assurance procedures have been established and are applied. Problems may arise and must be, handled as they occur in a timely fashion.

1.2.4. Communication

Good communication between all parties involved in the design and construction of a pile foundation is essential to reach a successful completion of the project. In the design stage, communication and interaction is needed between the structural, geotechnical, geologic, hydraulic, and construction disciplines, as well as with consultants, drill crews and laboratory personnel. In the construction stage, structural, geotechnical and construction disciplines need to communicate for a timely resolution of construction issues as they arise. Figure 1-16 and Figure 1-17 highlight some of the key issues to be communicated in the design and construction stages.

Photo and Figure Credits

- Don C. Warrington, P.E.: Figure 1-4, Figure 1-6, Figure 1-10, Figure 1-11, Figure 1-14.

Figure 1-16 Design Stage Communication

DESIGN STAGE COMMUNICATION						
Subject	Structural	Geotechnical	Hydraulic	Construction	Field Crews	Laboratory
Preliminary Structure Loads and Performance Criteria.	X	X	X			
Determination of Scour Potential.	X	X	X			
Determination of Special Design Event Requirements.	X	X	X			
Review of Past Construction Problems in Project Area.	X	X	X	X		
Implementation of Subsurface Exploration and Testing Programs.	X	X	X		X	X
Determination of Pile Type, Length and Capacity.	X	X				
Effect of Approach Fills on Design.	X	X				
Prepare Plans and Specifications.	X	X	X	X		

Figure 1-17 Construction Stage Communication

CONSTRUCTION STAGE COMMUNICATION			
Subject	Structural	Geotechnical	Construction
Establish Appropriate Methods of Construction Control and Quality Assurance.	X	X	X
Perform Wave Equation Analysis of Contractors Driving System to Establish Driving Criteria.	X	X	X
Perform Static Load Test(s) and/or Dynamic Monitoring and Adjust Driving Criteria.	X	X	X
Resolve Pile Installation Problems / Construction Issues.	X	X	X

Footnotes

[1] William Smith, D.C.L., LL.D.: *A Dictionary of Greek and Roman Antiquities*, John Murray, London, 1875. Article by Philip Smith, B.A., of the University of London on p. 538.

[2] Livy, *The Early History of Rome*, 2.10. Translated by Aubrey de Selincourt. London: Penguin Books, 1960.

[3] Julius Caesar, *De Bello Gallico* (Concerning the War in Gaul), IV, 17, as cited by L. Sprague de Camp, *The Ancient Engineers*. New York: Barnes and Noble, 1993.

[4] James Naysmith, *James Nasmyth: Engineer, An Autobiography*, Chapter 13.

[5] Ibid., Chapter 15.

[6] Bethlehem Steel H-Piles, Bethlehem Steel Company, June 1979

[7] Most of this section - including the drawings and quotation - comes from Isaacs, D.V. "Reinforced Concrete Pile Formulae." Journal of the Institution of Engineers Australia, Vol. 3, No. 9, September 1931, pp. 305-323.

[8] This section - again including the drawing - is largely drawn from Glanville, W.H., Grime, G., Fox, E.N, And Davies, W.W, "An Investigation of the Stresses in Reinforced Concrete Piles During Driving." Department Sci. Ind. Research, British Building Research Board Technical Paper No. 20, 1938

[9] The most recent effort in this regard is that of Warrington, D.C., *Closed Form Solution of the Wave Equation for Piles*. Master's Thesis, University of Tennessee at Chattanooga, 1997. Downloadable from http://www.vulcanhammer.net/wave. This study combines a semi-infinite pile solution and Fourier series to overcome some of the difficulties encountered in the past.

[10] Smith, E.A.L. "Pile-Driving Analysis by the Wave Equation." *Journal of the Engineering Mechanics Division, Proceedings of the American Society of Civil Engineers*. Vol. 86, No. EM 4, August 1960.

[11] Lowery, L.L, Hirsch, T.J., Edwards, T.C., Coyle, H.M. And Samson, C.H. Pile Driving Analysis -- State of the Art. Research Report 33-13. College Station: Texas Transportation Institute, 1969, also Hirsch, T.J., Lowery, L.L., And Carr, L. (). *Pile Driving Analysis -- Wave Equation User's Manual, TTI Program*. U.S. Department of Transportation, Federal Highway Administration, Washington, DC. Report FHWA-IP-76-14 (1976, 4 Vols.)

[12] Goble, G.G., And Rausche, F. *Wave Equation Analysis of Pile Driving, WEAP Program*. U.S. Department of Transportation, Federal Highway Administration, Washington, DC. Report FHWA-IP-76-13 (1976, 4 Vols.) and Goble, G.G., And Rausche, F. () *Wave Equation Analysis of Pile Driving, WEAP86 Program*. U.S. Department of Transportation, Federal Highway Administration, Washington, DC. Report DTFH61-84-C-00100 (1986, 4 Vols.)

[13] An example of this, compared with other methods, is shown in Warrington, D.C. (1997) *Closed Form Solution of the Wave Equation for Piles*.

[14] Fellenius, B.H. "Reflections on Pile Dynamics." *Proceedings of the Fifth International Conference on the Application of Stress-Wave Theory to Piles*, 11-13 September 1996, Orlando, Florida.

[15] Goble, G.G., Rausche, F., And Likins, G.E., Jr. "The Analysis of Pile Driving -- A State of the Art" *Proceedings of the International Seminar of the Application of Stress-Wave Theory on Piles*, Stockholm, Sweden, 4-5 June 1980.

[16] Rausche, F., Goble, G.G., And Likins, G.E., Jr. "Dynamic Determination of Pile Capacity." *Journal of Geotechnical Engineering*, Vol. 111, No. 3, March 1985. New York: American Society of Civil Engineers.

[17] Information on both of these vibratory hammers is taken from Erofeev, L.V., Smorodinov, M.I., Fedorov, B.S., Vyazovikii, V.N., And Villumsen, V.V. *Machines and Equipment for the Installation of Shallow and Deep Foundations*. (In Russian) Second Edition, Mashinostrenie, Moscow, 1985, pp. 95-111.

Chapter 2. Pile Types and Guidelines for Selection

Piles can be categorized in two main types: load bearing piles and sheet piles.

There are numerous types of load bearing piles. Figure 2-1 shows a pile classification system based on type of material, configuration, installation technique and equipment used for installation. Load bearing piles can also be classified based on their method of load transfer from the pile to the soil mass. Load transfer can be by friction, toe bearing or a combination.

2.1. Bearing Piles

2.1.1. Steel Piles

The general category of steel piles includes H-piles and pipe piles. Since steel piles are a manufactured product, their properties are controlled and well known prior to installation. Among all piling materials steel piles are allotted the highest allowable unit working stresses, but not necessarily the highest in proportion to the ultimate strength of the material. Steel piles are generally considered to be high capacity piles but have been historically used for a wide range of loadings.

2.1.1.1. Steel H-Piles

Steel H-piles are a specially designed sub-group of wide flange shapes with equal thickness in the web and flanges. The depth of the section is approximately equal to the width. H-piles are hot rolled from ingots on the same type mill used to manufacture wide flange structural shapes. A table of the various sizes commonly available is shown in Table 2-1. Other wide flange shapes have been used for foundation piles in special situations.

H-piles are very versatile pile type. They can be used for both friction and end bearing applications. They are manufactured as a finished product, which can be driven with standard equipment. Advantages and disadvantages of H-Piles are shown in Table 2-2.

H-piles can be considered for a design load between 80 kips (356 kN) and 500 kips (2224 kN). They function most efficiently for end-bearing or partial end-bearing applications. They are a standard in many states for highway bridge piers and abutments where the job sites are remote, compacted fill approach embankments must be penetrated, battered piles are often required and pile loadings are in the medium to high medium range (80 kips (356 kN) to 180 kips (800 kN)).

H-piles are also commonly applied for high unit dead and live loads associated with multi-storey buildings. The magnitude of these loads generally requires high capacity piles driven to end-bearing conditions to limit settlements and for space and economic reasons. H-piles have carried design loads of over 400 kips (1779 kN) in a number of such applications.

H-piles are good piles in tension - the constant cross-section together with the entrapment of soil between the flanges provides excellent resistance to pullout when that is a factor. H-piles are used as battered tension piles to anchor sheet pile bulkheads. Uplift due to hydrostatic or wind conditions is an important consideration in many foundation designs, and steel H-piles have the ability to work both as compression and tension piles. Their low-displacement characteristic would favor H-piles over displacement piles where ground heave might be a problem.

The ability to resist hard driving enable H-piles to penetrate soils where other piles may not.

2.1.1.1.1. Grades of Steel H-Piles

Steel H-piles are produced to either ASTM Specification A-36 or A-572. The properties for these specifications (and their grades) are shown in Table 2-3. In addition, they can be produced to ASTM A-690 when specified. This formulation is a 50 ksi (345 MPa) steel with improved corrosion performance in the salt-water splash zone. This grade is occasionally specified as "weathering" type steel for appearance-sake above ground.

Experience indicates that corrosion is not a practical problem for steel piles driven in natural soil, due primarily to the absence of oxygen in the soil. However, in fill ground at or above the water table, moderate corrosion may occur and protection may be needed. Commonly used protection methods include coatings applied before driving. Coal-tar epoxies, fusion bonded epoxies, metallized aluminum and Phenolic mastics are used to provide protection. Encasement by cast-in-place concrete or precast concrete jackets or cathodic protection can also provide needed protection for piles extending above ground. Sometimes heavier sections are used to meet long-term service requirements.

Splices are commonly made by full penetration butt welds. The splice should be as strong as the pile. Proprietary splicers are also used for quick splicing H-piles.

A steel load transfer cap is not required if the head of the pile is adequately embedded in a concrete cap.

H-piles may require toe reinforcement for driving through dense soil or soil containing boulders or rock. Toe reinforcement is also used for penetration into a sloping rock surface. Proprietary pile points welded to pile toes are used commonly. H-piles are suitable for use as end bearing piles, and combinations of friction and end bearing. Since H-piles generally displace a minimum of material, they can be driven more easily through dense granular layers and very stiff clays. Using H-piles often reduces the problems associated with soil heave and ground vibrations during foundation installation. H-piles can be used for driving into soils containing obstructions such as boulders when properly protected at the toe. They are commonly used for any depth since splicing is relatively easy.

2.1.1.1.2. H-Piles as End-Bearing Piles

H-piles are most efficient when they can be driven to refusal or practical refusal on rock, or into dense materials overlying rock. The pile functions as a

Figure 2-1 Pile Classification

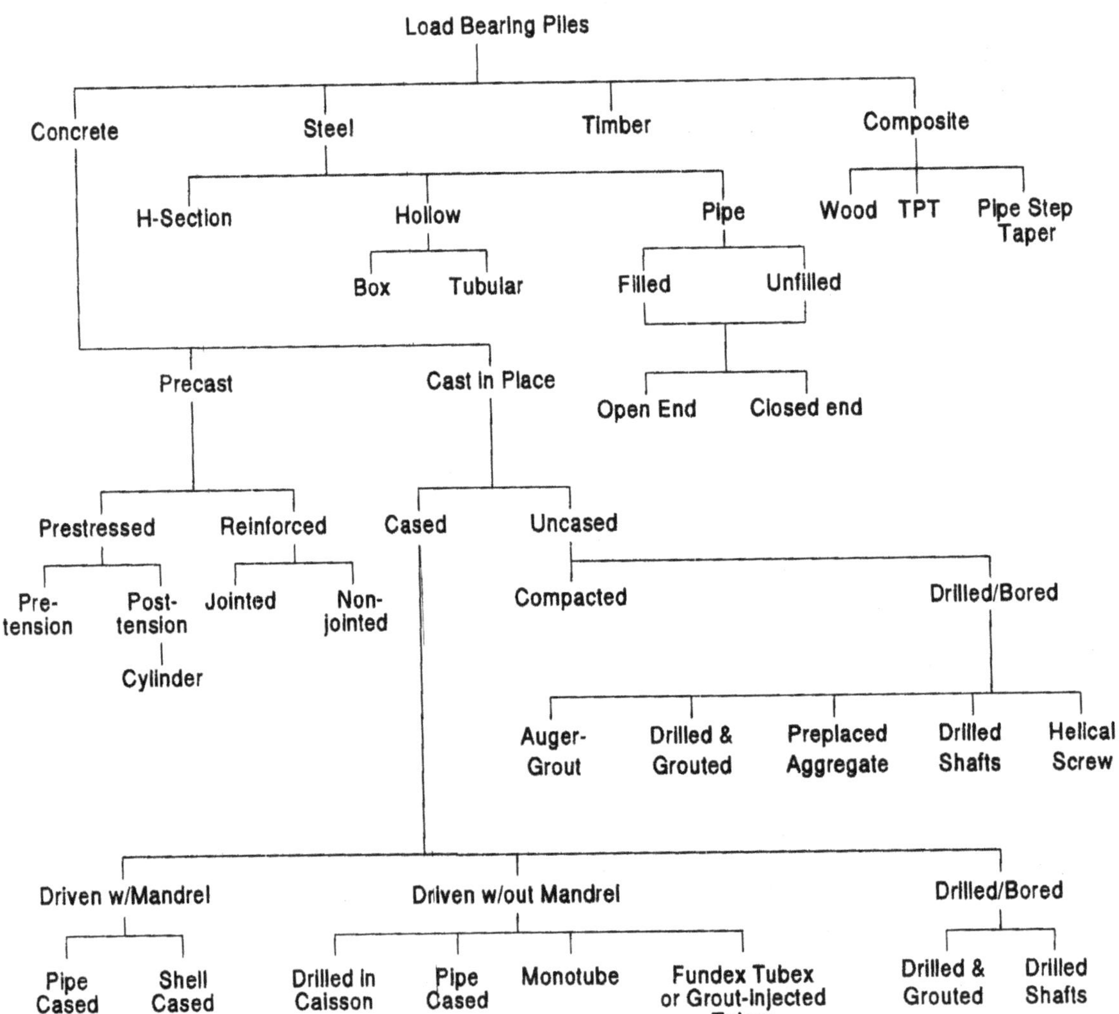

Table 2-1 Standard H-Pile Sections

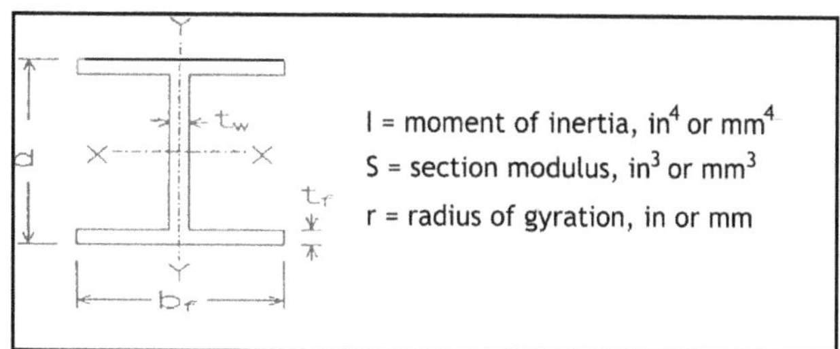

Designation	Cross-Sectional Area A	Depth d	Flange Width b_f	Flange Thickness t_f	Web Thickness t_w	Axis X-X I	Axis X-X S	Axis X-X r	Axis Y-Y I	Axis Y-Y S	Axis Y-Y r
English Units	in^2	in	in	in	in	in^4	in^3	in	in^4	in^3	in
HP14 x 117	34.4	14.21	14.885	0.805	0.805	1220	172.0	5.96	463.0	59.5	3.59
HP14 x 102	30.0	14.01	14.785	0.705	0.705	1050	150.0	5.92	380.0	51.4	3.56
HP14 x 89	26.1	13.83	14.695	0.615	0.615	904	131.0	5.88	326.0	44.3	3.53
HP14 x 73	21.4	13.61	14.585	0.505	0.505	729	107.0	5.84	261.0	35.8	3.49
HP13 x 100	29.4	13.15	13.205	0.765	0.765	886	135.0	5.49	294.0	44.5	3.16
HP13 x 87	25.5	12.95	13.105	0.665	0.665	775	117.0	5.45	250.0	38.1	3.13
HP13 x 73	21.6	12.75	13.005	0.565	0.565	630	98.8	5.40	207.0	31.9	3.10
HP13 x 60	17.5	12.54	12.900	0.460	0.460	503	80.3	5.36	165.0	25.5	3.07
HP12 x 84	24.6	12.28	12.295	0.685	0.685	650	106.0	5.14	213.0	34.6	2.94
HP12 x 74	21.8	12.13	12.215	0.610	0.610	569	93.8	5.11	186.0	30.4	2.92
HP12 x 63	18.4	11.94	12.125	0.515	0.515	472	79.1	5.06	153.0	25.3	2.88
HP12 x 53	15.5	11.78	12.045	0.435	0.435	393	66.8	5.03	127.0	21.1	2.86
HP10 x57	16.8	9.99	10.225	0.565	0.565	294	58.8	4.18	101.0	19.7	2.45
HP10 x42	12.4	9.70	10.075	0.420	0.420	210	43.4	4.13	71.7	14.2	2.41
HP8 x 36	10.6	8.02	8.155	0.445	0.445	119	29.8	3.36	40.3	9.88	1.95
SI Units	mm^2	mm	mm	mm	mm	mm^4	mm^3	mm	mm^4	mm^3	mm
HP360 x 174	22200	361	378	20.4	20.4	504	2810	151	184	974	91.0
HP360 x 152	19400	356	376	17.9	17.9	639	2470	150	159	846	90.5
HP360 x132	16900	351	373	15.6	15.6	375	2140	149	135	724	89.4
HP360 x 108	13800	346	370	12.8	12.8	303	1750	148	108	584	88.5
HP330 x 149	19000	334	335	19.4	19.4	368	2200	139	122	728	80.1
HP330 x 129	16400	329	333	16.9	16.9	315	1910	139	104	625	79.6
HP330 x 109	13900	324	330	14.4	14.4	263	1620	138	86.3	523	78.8
HP330 x 89	11300	319	328	11.7	11.7	211	1320	137	68.9	420	78.1
HP310 x 125	15900	312	312	17.4	17.4	270	1730	130	88.2	565	74.5
HP310 x 110	14100	308	310	15.5	15.4	237	1540	130	77.1	497	73.9
HP310 x 93	11900	303	308	13.1	13.1	196	1290	128	63.9	415	73.3
HP310 x 79	10000	299	306	11.0	11.0	163	1090	128	52.6	344	72.5
HP250 x 85	10800	254	260	14.4	14.4	123	969	107	42.3	325	62.6
HP250 x 62	7970	246	256	10.7	10.5	87.5	711	105	30.0	234	61.4
HP200 x 53	6820	204	207	11.3	11.3	49.8	488	85.5	16.7	161	49.5

short column, hence the rock may be stronger than the steel for the maximum design load that can be applied. It is generally accepted that competent rock in its natural bed can resist very high stress concentrations without crushing, except when the pile is installed through extremely soft soils. This combination therefore provides at least the potential for some high and therefore very efficient pile loadings. Other end-bearing materials include hardpan, marl, dense sand, and softer rock. Here the strength of the supporting material is often considerably less than that of the steel. H-piles driven into these materials obtain support by the development of an increased bulb of pressure around the toe and friction along the embedded development length.

Building codes set limits on loads and unit stresses as well as allowable settlements, which reflect the experience and recommendations of the related engineering community. Piles installed to end bearing are often permitted much higher loads than friction piles and those that are tested are allowed higher loads than those that are not.

Based on recorded experience, H-piles are probably most efficient when driven through relatively deep, soft to medium stiff clays to end bearing.

Fine grained soils (clays) are the most likely to exhibit large increases in capacity after installation. The extent to which soil freeze contributes to capacity of end-bearing piles is subject to much discussion. Soil freeze may be estimated using experience and static methods. With wave equation analysis, incorrect modeling of the setup (freeze) effects may lead to the discrepancies that are observed between estimated and actual driving performance.

2.1.1.1.3. H-Piles as Friction Piles

While many meters of H-piles have been driven for friction applications, they are non-displacement piles and tend to drive further in loose sands and silty sand. There may be good reasons however

Table 2-2 Advantages and Disadvantages of Steel H-Piles

Advantages	Disadvantages
High individual load capacity when driven to bear on or in hard or dense materials.	Relatively higher cost unless efficiently loaded.
Ready availability; can be installed with standard driving equipment; lengths can be easily extended or reduced to fit job requirements.	Inability to inspect the physical condition after driving (an advantage for closed ended tubular piles)
Compact shape with low displacement - minimum disturbance to adjacent piles or structures; able to penetrate where many other types could not.	Non-constant radius of gyration (an advantage of pipe piles for certain situations).
High bending strength for applications involving lateral loads. Readily driven on a batter if required.	Corrosion problems in certain environmental situations if unprotected.
Good tension piles for uplift - constant cross-section, plus steel is best material for tensile strength.	

to select H-piles for this use if, for example, a significant scour depth is computed for a bridge pier abutment.

When H-piles are driven into stiff clays with cohesion between 1 and 2 ksf, soil is usually trapped between the flange and the web and is compacted. In this case, the soil becomes a part of the pile and is carried down with it. An example of this (known as "H-pile plugging") is shown in Figure 2-2. This core aids in compressing the surrounding soil and building up resistance to further displacement. The principal load transfer is through frictional forces and not from end bearing. It is somewhat difficult to predict the capacity of any given length of pile driven in stiff clays and load testing is generally advisable. When soft or medium clays and silts are encountered, piles will develop shaft friction resistance almost equal to the surface area of the pile multiplied by the shear strength. These soils have high moisture content, which will make the pile resistance to driving seem lower than predicted. A time interval of a few hours to several weeks may be necessary to attain the true measure of the pile's long term geotechnical capacity.

Figure 2-2 H-pile Plug

2.1.1.1.4. H-Piles as Soldier Beams

One common application of H-piles is their use as soldier beams for retaining walls. These retaining walls can be either permanent or temporary for

Table 2-3 Mechanical Properties of Steel Grades for H-Piles

Specification and Grade	Yield Point (minimum)	Tensile Strength (minimum)	Minimum Elongation, 8" (203.2 mm), %	Minimum Elongation, 2" (50.8 mm), %
ASTM A36	36 ksi/248 MPa	58-80 ksi (406-552 MPa)	20	21
ASTM A572 Grade 42	42 ksi/290 MPa	60 ksi (414 MPa)	20	24
ASTM A572 Grade 50	50 ksi/345 MPa	65 ksi (448 MPa)	18	21
ASTM A572 Grade 60	60 ksi/414 MPa	75 ksi (517 MPa)	16	18

excavations and braced cuts. Typically, the H-piles are driven on 6' - 8' (1.8 - 2.4 m) centers in a row with the flanges facing each other. The lagging - either concrete or timber - is then stacked with the ends of the flanges facing the webs. The flanges of the H-piles thus retain the lagging. Cross bracing (in the case of braced cuts) or tieback systems can be used to provide additional lateral support for higher walls or loads.[1] H-piles also are used in conjunction with sheeting to form high-modulus walls; these are discussed in detain in the Pile Buck Sheet Piling Design Manual.

2.1.1.2. Steel Pipe Piles

Pipe piles usually consist of seamless, welded or spiral welded steel pipes of wall thickness in the range of 0.109" to 2.500" (2.8 - 63.5 mm). The piles are available in 8" (203.2 mm) to 48" (1219 mm) diameters. Typical pipe pile sizes are shown in Table 2-4. Much larger sizes and also used in special situations.

Common sizes of pipe piles can be considered for loads between 60 kips (267 kN) to over 400 kips (1779 kN). They are very competitive as combination end-bearing/friction piles for loads from 120 kips (534 kN) to 240 kips (1068 kN), driven closed-end and filled with concrete (particularly where pile lengths do not favor mandrel-driven piles.) Pipe also provides a strong casing for concrete fill where underground pressures are high. Pipe piles may be driven with an open end or a closed end.

Advantages and disadvantages of steel pipe pile are shown in Table 2-5.

2.1.1.2.1. Specifications

The basic specification for pipe piles is ASTM A-252, which covers welded and seamless product. There are three grades listed:

- Grade 1 has a minimum yield strength of 30 ksi (207 MPa);
- Grade 2 has a minimum yield of 35 ksi (241 MPa); and
- Grade 3 has a minimum yield of 45 ksi (310 MPa).

There are elongation requirements for ductility, but only minimum chemistry requirements.

2.1.1.2.2. Manufacture

Seamless pipe is rarely specified for pipe piles due to its relative cost. It generally comes on the market as surplus pipe. As the name implies, seamless pipe has no seams but is made in one piece from a hot steel billet by piercing the centre and expanding the steel to the shape and size desired.

Electric weld pipe is the most common type of pipe used for piles. Several manufacturing processes can be used:

- Electric resistance welding
- Fusion welding
- Flash welding.

The seams of these pipes may be straight, spiral-butt or spiral-lap construction. In each case, the manufacturer begins with hot rolled sheets or plates in either coils or flats. The production equipment determines how the pipe is assembled and the seams are welded. So-called "spiral mills" produce spiral pipe by butting or lapping the seams. This process accounts for a large share of the pipe pile produced. Larger and thicker pipe piles are generally made on the same mills which produce large diameter line pipe by the straight seam, electric butt-weld process. In some cases very heavy wall pipe piles are fabricated in specialty shops where penstocks or caissons might be fabricated. There seems to be no particular advantage of one process over the other as far as the pile foundation designer is concerned.

2.1.1.2.3. Closed Ended Pipe Piles

A closed ended pipe pile is shown in Figure 2-3. They may be filled with concrete or left unfilled. They may be filled with a structural shape such as an H-section in addition to the concrete and socketed into bedrock (rock socketed piles). If bearing capacity from the entire pile toe area is required, the pile toe should be closed with a plate or a conical tip. Mandrels are usually not used for driving pipe piles, which are generally driven from the pile head. When the end of a pipe pile is equipped with a closure device, the pile becomes a displacement pile and functions well as a friction pile particularly in loose sands. When driven open or closed end it can also function as a high capacity end-bearing pile.

Table 2-4 Typical Steel Pipe Pile Sizes

Designation and Outside Diameter, in.	Wall Thickness, in.	Area A, in.^{2}	Weight per foot, lb/ft	I, in.^{4}	S, in.^{3}	R, in.	Area of Exterior Surface, ft^2/ft	Inside Cross-Sectional Area, in.^{2}	Inside Volume, yd^3/ft	External Collapse Index
PP10	.109	3.39	11.51	41.4	8.28	3.50	2.62	75.2	.0193	62
	.120	3.72	12.66	45.5	9.09	3.49	2.62	74.8	.0192	83
	.134	4.15	14.12	50.5	10.1	3.49	2.62	74.4	.0191	116
	.141	4.37	14.85	53.1	10.6	3.49	2.62	74.2	.0191	135
	.150	4.64	15.78	56.3	11.3	3.48	2.62	73.9	.0190	163
	.164	5.07	17.23	61.3	12.3	3.48	2.62	73.5	.0189	214
	.172	5.31	18.05	64.1	12.8	3.48	2.62	73.2	.0188	247
	.179	5.52	18.78	66.6	13.3	3.47	2.62	73.0	.0188	279
	.188	5.80	19.70	69.8	14.0	3.47	2.62	72.7	.0187	324
	.203	6.25	21.24	75.0	15.0	3.46	2.62	72.3	.0186	409
	.219	6.73	22.88	80.5	16.1	3.46	2.62	71.8	.0185	515
	.230	7.06	24.00	84.3	16.9	3.46	2.62	71.5	.0184	588
	.250	7.66	26.03	91.1	18.2	3.45	2.62	70.9	.0182	719
PP10-3/4	.109	3.64	12.39	51.6	9.60	3.76	2.81	87.1	.0224	50
	.120	4.01	13.62	56.6	10.5	3.76	2.81	86.8	.0223	67
	.125	4.17	14.18	58.9	11.0	3.76	2.81	86.6	.0223	76
	.141	4.70	15.98	66.1	12.3	3.75	2.81	86.1	.0221	109
	.150	5.00	16.98	70.2	13.1	3.75	2.81	85.8	.0221	131
	.156	5.19	17.65	72.9	13.6	3.75	2.81	85.6	.0220	148
	.164	5.45	18.54	76.4	14.2	3.74	2.81	85.3	.0219	172
	.172	5.72	19.43	80.0	14.9	3.74	2.81	85.0	.0219	199
	.179	5.94	20.21	83.1	15.5	3.74	2.81	84.8	.0218	224
	.188	6.24	21.21	87.0	16.2	3.73	2.81	84.5	.0217	260
	.219	7.25	24.63	100	18.7	3.72	2.81	83.5	.0215	414
	.230	7.60	25.84	105	19.6	3.72	2.81	83.2	.0214	480
	.250	8.25	28.04	114	21.2	3.71	2.81	82.5	.0212	605
	.279	9.18	31.20	126	23.4	3.70	2.81	81.6	.0210	781
	.307	10.1	34.24	137	25.6	3.69	2.81	80.7	.0208	951
	.344	11.2	38.23	152	28.4	3.68	2.81	79.5	.0205	1,180
	.365	11.9	40.48	161	29.9	3.67	2.81	78.9	.0203	1,320
	.438	14.2	48.24	189	35.2	3.65	2.81	76.6	.0197	1,890
	.500	16.1	54.74	212	39.4	3.63	2.81	74.7	.0192	2,380
PP12	.134	5.00	16.98	87.9	14.7	4.20	3.14	108	.0278	67
	.141	5.25	17.86	92.4	15.4	4.19	3.14	108	.0277	78
	.150	5.58	18.98	98.0	16.3	4.19	3.14	108	.0277	94
	.172	6.39	21.73	112	18.6	4.18	3.14	107	.0274	142
	.179	6.65	22.60	116	19.4	4.18	3.14	106	.0274	161
	.188	6.98	23.72	122	20.3	4.18	3.14	106	.0273	186
	.203	7.52	25.58	131	21.8	4.17	3.14	106	.0272	235
	.219	8.11	27.55	141	23.4	4.17	3.14	105	.0270	296
	.230	8.50	28.91	147	24.6	4.16	3.14	105	.0269	344
	.250	9.23	31.37	159	26.6	4.16	3.14	104	.0267	443
	.281	10.3	35.17	178	29.6	4.14	3.14	103	.0264	616
	.312	11.5	38.95	196	32.6	4.13	3.14	102	.0261	784
PP12-3/4	.109	4.33	14.72	86.5	13.6	4.47	3.34	123	.0317	30
	.125	4.96	16.85	98.8	15.5	4.46	3.34	123	.0316	45
	.134	5.31	18.06	106	16.6	4.46	3.34	122	.0315	56
	.150	5.94	20.19	118	18.5	4.46	3.34	122	.0313	78
	.156	6.17	20.98	122	19.2	4.45	3.34	122	.0313	88
	.164	6.48	22.04	128	20.1	4.45	3.34	121	.0312	103
	.172	6.80	23.11	134	21.1	4.45	3.34	121	.0311	118
	.179	7.07	24.03	140	21.9	4.45	3.34	121	.0310	134
	.188	7.42	25.22	146	23.0	4.44	3.34	120	.0309	155
	.203	8.00	27.20	158	24.7	4.44	3.34	120	.0308	196
	.230	9.05	30.75	177	27.8	4.43	3.34	119	.0305	286
	.250	9.82	33.38	192	30.1	4.42	3.34	118	.0303	368
	.281	11.0	37.42	214	33.6	4.41	3.34	117	.0300	526
	.312	12.2	41.45	236	37.0	4.40	3.34	115	.0297	684
	.330	12.9	43.77	248	39.0	4.39	3.34	115	.0295	776
	.344	13.4	45.58	258	40.5	4.39	3.34	114	.0294	848

Table 2-4 Typical Steel Pipe Pile Sizes (continued)

Designation and Outside Diameter, in.	Wall Thickness, in.	Area A, in.^{2}	Weight per foot, lb/ft	I, in.^{4}	S, in.^{3}	R, in.	Area of Exterior Surface, ft^2/ft	Inside Cross-Sectional Area, in.^{2}	Inside Volume, yd^3/ft	External Collapse Index
	.375	14.6	49.56	279	43.8	4.38	3.34	113	.0291	1,010
	.406	15.7	53.52	300	47.1	4.37	3.34	112	.0288	1,170
	.438	16.9	57.59	321	50.4	4.36	3.34	111	.0285	1,350
	.500	19.2	65.42	362	56.7	4.33	3.34	108	.0279	1,760
PP14	.134	5.84	19.84	140	20.0	4.90	3.67	148	.0381	42
	.141	6.14	20.87	147	21.1	4.90	3.67	148	.0380	49
	.150	6.53	22.19	157	22.4	4.90	3.67	147	.0379	59
	.156	6.78	23.07	163	23.2	4.89	3.67	147	.0378	66
	.172	7.47	25.40	179	25.5	4.89	3.67	146	.0377	89
	.179	7.77	26.42	186	26.5	4.89	3.67	146	.0376	101
	.188	8.16	27.73	195	27.8	4.88	3.67	146	.0375	117
	.203	8.80	29.91	209	29.9	4.88	3.67	145	.0373	147
	.210	9.10	30.93	216	30.9	4.88	3.67	145	.0373	163
	.219	9.48	32.23	225	32.2	4.87	3.67	144	.0372	185
	.230	9.95	33.82	236	33.7	4.87	3.67	144	.0370	215
	.250	10.8	36.71	255	36.5	4.86	3.67	143	.0368	277
	.281	12.1	41.17	285	40.7	4.85	3.67	142	.0365	395
	.344	14.8	50.17	344	49.2	4.83	3.67	139	.0358	691
	.375	16.1	54.57	373	53.3	4.82	3.67	138	.0355	835
	.438	18.7	63.44	429	61.4	4.80	3.67	135	.0348	1,130
	.469	19.9	67.78	457	65.3	4.79	3.67	134	.0345	1,280
	.500	21.2	72.09	484	69.1	4.78	3.67	133	.0341	1,460
PP16	.134	6.68	22.71	210	26.3	5.61	4.19	194	.0500	28
	.141	7.02	23.88	221	27.6	5.61	4.19	194	.0499	33
	.150	7.47	25.39	235	29.3	5.60	4.19	194	.0498	39
	.164	8.16	27.74	256	32.0	5.60	4.19	193	.0496	52
	.172	8.55	29.08	268	33.5	5.60	4.19	193	.0495	60
	.179	8.90	30.25	278	34.8	5.59	4.19	192	.0494	67
	.188	9.34	31.75	292	36.5	5.59	4.19	192	.0493	78
	.203	10.1	34.25	314	39.3	5.59	4.19	191	.0491	98
	.219	10.9	36.91	338	42.3	5.58	4.19	190	.0489	124
	.230	11.4	38.74	354	44.3	5.58	4.19	190	.0488	144
	.250	12.4	42.05	384	48.0	5.57	4.19	189	.0485	185
	.281	13.9	47.17	429	53.6	5.56	4.19	187	.0481	264
	.312	15.4	52.27	473	59.2	5.55	4.19	186	.0478	362
	.344	16.9	57.52	519	64.8	5.54	4.19	184	.0474	487
	.375	18.4	62.58	562	70.3	5.53	4.19	183	.0470	617
	.438	21.4	72.80	649	81.1	5.50	4.19	180	.0462	874
	.469	22.9	77.79	691	86.3	5.49	4.19	178	.0458	1,000
	.500	24.3	82.77	732	91.5	5.48	4.19	177	.0455.	1,130

Figure 2-3 Closed End Pipe Piles

A flat plate of 1/2" to 2" (13-51 mm) thickness, or a conical point generally forms a closed ended pile. When pipe piles are driven to weathered rock or through boulders, a cruciform end plate or a conical point with rounded nose is often used to prevent distortion of the pile toe.

2.1.1.2.4. Open Ended Pipe Piles

Open-ended pipe piles are driven when hard driving, caused by the presence of debris, small boulders and the like is anticipated. The pipe can be fitted with a special driving shoe, which adds steel thickness at the toe to reduce stresses and damage. Periodically, the plugger materials are removed to aid in driving. Open-ended pipe piles may also be partially socketed into rock at site of steeply sloping bedrock or where pile fixity at the toe is a design requirement. Pipe piles driven open-end may be filled with concrete after cleaning out the plug, backfilled with sand, or the plug ignored. In the last two cases, the steel wall takes all stress and the pile would be treated

Table 2-5 Advantages and Disadvantages of Steel Pipe Pile

Advantages	Disadvantages
Wide selection of sizes and thicknesses available to choose from.	Open-end pipe piles are not as favorable as H-piles for non-displacement applications since the plug of soil inside the pipe also offers resistance to penetration.
Delivery is excellent since there are many manufacturers and distributors; popular sizes are stocked.	Closed-end they are full displacement piles with certain potential problems associated with displacement.
Standard sizes of pipe pile can be driven with conventional driving equipment. Light wall pipe makes an efficient shell for concrete fill when mandrel-driven.	They may not be price-competitive with other displacement piles.
Pipe piles driven open-end to rock, cleaned, inspected and filled with concrete can resist very high individual loads.	
Pipe piles with wall thickness over about 1/8" (3.2 mm) and filled with concrete are treated as a composite pile with both the steel and the concrete sharing the applied load. The advantages of both steel and concrete are enjoyed.	
Pipe piles can be inspected for material damage and curvature prior to acceptance.	
They can be readily spliced to extend lengths, resist hard driving, and drive straighter because of their constant radius of gyration. They make a more efficient column where unsupported length and large loads are design requirements.	

similar to an H-pile for design purposes.

In driving through dense materials, open-ended piles may form soil plugs. The plug may make the pile act like a closed end pile and increase the pile toe bearing capacity significantly. The plug should not be removed unless the pile is to be filled with concrete. For open-ended pipe piles not filled with concrete, the formation of soil plug should not be considered in computing pile end bearing capacity. Behavior of the plug during and after driving is a function of pile size and soil type and consistency. Only preliminary design guidance is available, which should always be confirmed by field observations and measurements.

This type of pile is also common in the installation of offshore oil platforms, whether driven from the surface or underwater. In these applications, they primarily are designed for uplift loads due to wave or wind action on the structure. Unless the pile plug generated during driving creates drivability problems, the plug is generally not removed. An illustration of this application is shown in Figure 2-4.

Figure 2-4 Pile Piles for an Offshore Oil Platform

Open ended pipe piles are recommended where the pile or pile group is to be subjected to horizontal loads and bending moments such as vessel impact and scour on large structures such as bridges. With a constant radius of gyration, pipe piles are also the most efficient columns and should be considered where freestanding columnar strength is important (such as open-platform marine piers and docks). This is advantageous in seismic situations, where liquefaction and several other factors have a strong influence on the design of deep foundations. In all of these applications, the pile size can range up to 3000 mm in diameter. Since such applications can require considerable length, this frequently requires adding-on to pipe piles, which is shown in Figure 2-5.

Figure 2-5 Welding an Add-On to Pipe Pile

An example of this kind of application was the Jamuna River Bridge in Bangladesh, built in the mid-1990's. The river is capable of scouring to a depth of 50 m, with a flood discharge of 100,000 m^3/sec. To deal with this, groups of high yield strength piles 2500 - 3150 mm in diameter up to 80 m long with a wall thickness of 45 - 60 mm were installed. After driving, the plug was removed and the toe of the pile was sealed with concrete, then the pile was grouted.[2]

2.1.1.2.5. RR Piles

A special type of pipe pile is the RR pile, manufactured by Makela Metals. These can be formed into sectional pile systems by mechanical joints. RR piles are type approved in Finland. They are used as toe-bearing piles in the repair of buildings, as supports under machine bases, and for house foundations. Light installation equipment, the cost-effective use of material and versatility of application are benefits offered by RR piles.

For the determination of bearing capacities and installation information, refer to the design instructions published by Makela Metals and approved by the Finnish Ministry of the Environment. For the technical properties and structure of RR piles, see Table 2-6.

Table 2-6 Technical Properties of RR Piles

DIMENSIONS						DETAIL DRAWING
Type of pile	Diameter D mm	Wall thickness t mm	Mass kg/m	Steel grade 1993	Yield strength R_{eH} N/mm^2	Pile cap Pile body
RR60	60,3	6,3	8,0	S355J2G3	355 [1)]	- steel pipe
RR75	76,1	6,3	10,4	S355J2G3	355 [1)]	
RR90	88,9	6,3	12,8	S355J2G3	355 [1)]	Splice
RR115	114,3	6,3	16,8	S355J2G3	355 [1)]	- rigid connection,
RR140/6	139,7	6,3	20,7	S355J2G3	355 [1)]	friction joint
RR140/8	139,7	8,0	26,0	S355J2G3	355 [1)]	
RR140/1	139,7	10,0	32,0	S355J2G3	355 [1)]	Point
RR170	168,3	10,0	39,0	S355J2G3	355 [1)]	- sand shoe or
RR220	219,1	10,0	51,6	S355J2G3	355	rock shoe
RR270	273,0	10,0	64,9	S355J2G3	355	
RR320	323,9	10,0	77,4	S355J2G3	355	

[1)] The load-bearing capacities have been calculated using the characteristic value 430 N/mm^2.
Piles RR60...RR170 are available in lengths of 1...6 m, piles RR140...RR320 in lengths of 1...16 m.
On piles RR60...RR170, the splice is fixed at the works. Spare splices are available for piles RR140...RR320.

RR piles are spliced using friction joints, so no welding is required for splicing. The Finnish Ministry of the Environment under decision number 282/533/87 and 10/5331/93 has approved the splices.

The pile toe is protected with either an RR sand shoe or RR rock shoe. The pile is furnished with an RR pile cap that is fixed to the pile body by means of sleeve connection.

The connections between the piles and superstructure are designed as flexible joints. However, the connections of piles shorter than 3 meters are preferably made rigid.

The permissible bearing capacity of an RR pile is determined by selecting the lowest of the following values:

- Maximum permissible centrally structural compression load;
- Permitted bearing capacity in respect of buckling;
- Geotechnical bearing capacity.

In accordance with the Pile Driving Instructions LPO-87 section 3.4231 of the Finnish Geotechnical Association, the maximum permissible structural compression load for RR piles is 33 to 58% of the yield point of the pile steel depending on the piling class. This has been considered in the permissible load ratings of Table 2-7.

Table 2-7 Permissible Bearing Capacities with Respect to Buckling, RR Piles

Type of pile	Piling class	Initial radius of curvature R m	Permissible bearing capacity with respect to buckling kN Closed shearing strength of the soil s (u) kN/m² 7	10	15	20
RR60	III	70	76	99	107	107
RR75	IB	150	150	187	205	216
	II	100	125	163	188	197
RR90	IB	200	208	237	256	267
	II	150	186	223	234	234
RR115	IB	200	296	320	343	256
	II	150	296	302	306	306
RR140/6	IB	200	330	354	377	391
	II	150	277	298	313	313
RR140/8	IB	200	402	437	471	491
	II	150	337	367	397	407
RR140/10	IB	200	439	571	637	645
	II	150	380	500	596	626
RR170	IB	300	680	738	789	789
	II	200	550	599	631	631
RR220	IB	350	740	785	828	854
	II	300	673	716	758	783
RR 270	IB	400	922	972	1020	1049
	II	350	845	893	939	968
RR320	IB	450	1095	1150	1202	1234
	II	400	1011	1063	1114	1145
R represents the initial radius of pile curvature after mounting, but prior to loading. Corrosion allowance is 1 mm.						

Where the pile is embedded in cohesionless soil layers, buckling must be taken into account as a possible determinant for its bearing capacity. Table 2-7 depicts the permissible load ratings of RR piles in respect of buckling. The initial radius of curvature R may be checked on the driven pile by illuminating it with a flashlight.

2.1.1.3. X-Piles

Makela Metals supplies the X piles presented in Table 2-8. X piles are mainly toe-bearing piles that are used for earth retaining applications. When used as driven piles, their X-shaped cross section minimizes soil disturbance and displacement. Their high rigidity allow X piles to be driven into hard-to-penetrate site conditions, including landfills.

Table 2-8 Dimensions and Installation, X-Piles

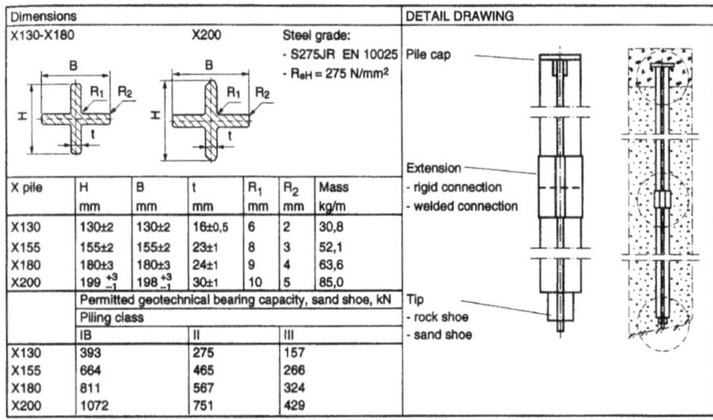

Dimensions

X130-X180 X200 Steel grade:
- S275JR EN 10025
- R_{eH} = 275 N/mm²

X pile	H mm	B mm	t mm	R_1 mm	R_2 mm	Mass kg/m
X130	130±2	130±2	16±0,5	6	2	30,8
X155	155±2	155±2	23±1	8	3	52,1
X180	180±3	180±3	24±1	9	4	63,6
X200	199^{+3}_{-1}	198^{+3}_{-1}	30±1	10	5	85,0

	Permitted geotechnical bearing capacity, sand shoe, kN Piling class IB	II	III
X130	393	275	157
X155	664	465	266
X180	811	567	324
X200	1072	751	429

X piles are predominantly used near existing structures in order to avoid damage and reduce ground vibrations. Typical applications include embankment foundation support, bridge structures, supports under machine bases and underpinning.

X piles have been approved in Finland and Sweden. In both cases the grantor's decision on each type approval is supplemented by design instructions, which are to be followed in the design and use of X piles.

The pile body functions satisfactorily for sand sites without special details. Rock shoes are attached at the fabrication location. The pile cap is fitted on the pile head by means of a sleeve connection or by welding. Piles can be spliced by welding or by means of extensions made of steel plate.

A corrosion allowance is taken into consideration in the design of piles. The average rate of corrosion in natural soil is 0.02 mm/year or 1 mm/50 years.

The permissible bearing capacity of an X pile is determined by selecting the lowest one of the following ratings:

- Permitted structural axial capacity;
- Permitted structural lateral capacity (especially buckling with very deep soft soil layers);
- Permitted geotechnical bearing capacity.

The specified geotechnical bearing capacity of the X pile is achieved by driving the pile to a point of sufficient resistance by soil. Steel piles usually function as retaining piles, which means that the geotechnical bearing capacity is equal to the toe-bearing capacity. The geotechnical bearing capacity of a pile with sand shoe is presented in Table 2-8.

The necessary soil analysis for the design of X piles includes the penetration depth and, in cohesionless soil, the shearing strength of the soil.

2.1.1.4. Corrosion of Steel Piles

Corrosion of steel[3] occurs where electric current leaves metal and enters a surrounding moist medium. Seawater at the splash line, where steel is alternately wet and dry is especially aggressive. Most fresh water corrodes steel only slowly unless there are pollutants. Piling driven into the ground so oxygen does not get to it suffers very little, unless there are stray currents. Piles generally are far enough apart that stray currents do not migrate between them.

Cathodic protection, use of a sacrificial anode, can be helpful but may not be effective in the splash zone where the steel is not always wet. Concrete jacketing may be needed in that area. Cathodic protection requires frequent maintenance.

Paint can provide protection if a proper material is selected; the steel is cleaned so paint will adhere and application is made with the steel dry and preferably warm. Cleaning means removal of all scale by mechanical means such as sandblasting. Be sure to follow the manufacturer's recommendations. For example, some bitumastics will not adhere to red oxide primer; they will stay on zinc chromate.

Metallized aluminum coatings have been helpful in protecting steel piling in seawater. About 6 to 12 mils of aluminum sprayed on steel cleaned to white metal and is then vinyl coated has been reported in excellent condition after 12 years of exposure.

Coatings can be applied commercially in specialty plants, usually near the rolling mills. Piles reach the job in good condition with only minor touch-up required.

Cast steel is naturally rust-resistive, so cast steel toe protection for H, sheet or pipe piles almost never rusts to any degree. Metal corrodes only when oxygen is present. Points of driven piles are usually well protected in the ground and protective coatings are unnecessary.

2.1.2. Concrete Piles

Concrete piles utilize concrete as the main structural material for compressive loads; however, concrete is deficient in resistance to tensile load. Therefore, when a concrete pile is subject to direct tension or bending, steel must be added to resist these stresses.

Concrete piles are classified as pre-cast or cast-in-place depending on the method of manufacture. Pre-cast piles are formed in a casting bed, cured, and then driven into place. There are several ways of manufacturing pre-cast piles:

1. Conventional steel reinforcing bars are used for tensile stresses and placed prior to casting the concrete;
2. High tensile rods or wires are pre-tensioned and the concrete shape cast around them (pre-stressed piles);
3. Hollow concrete cylinders are precast in manageable lengths, high strength rods or wire is strung through ducts in the several pre-cast sections and post-tensioned to form a complete pile.

Cast-in-place piles are, as the name implies, cast in a pre-formed excavation at the project site and hence the concrete is not subjected to driving forces.

2.1.2.1. Cast-In-Place Concrete Piles

In general, cast-in-place concrete piles are installed by placing concrete in an excavated hole in the ground. In some cases the hole is lined with a steel shell or casing which may be temporary or permanent. Steel pipe piles, when filled with concrete, can be classified under this category. Predetermination of pile lengths is not as critical as for precast concrete piling, since required pile lengths can be easily changed during installation.

Cast-in-place concrete piles can be installed with or without a mandrel[4], depending upon the wall thickness of the pile. Use of a mandrel allows piles with wall thicknesses of 0.02-0.13" (0.5-3.3 mm) thick, while those driven without a mandrel have wall thicknesses of 0.109-2.5" (2.8-63.5 mm) thick. With the latter, concrete is placed into the driven shell and is the primary basis for the structural strength of the completed pile. The steel in head-driven piles is generally of sufficient thickness to be included in the load capacity calculations. Contractors using this method avoid the expense of mandrels and long leads but pay more for the heavier shell. When the wall thickness exceeds about 0.1" (2.54 mm) (depending on building code requirements) the allowable load may be calculated in a way similar to closed-end pipe piles.

2.1.2.1.1. Raymond Step-Taper Piles

The best known of the mandrel-driven type piles is the Raymond Step-Taper® pile. In 1897, Alfred Raymond received a patent for a novel pile, which would consist of a tapered steel shell to be installed with the aid of an internal mandrel. After withdrawing the mandrel, the shell was filled with concrete to complete the pile. The maximum length was 37' (11.3 m). Until then, timber piles had dominated the market. This pile was the basis for forming the Raymond Concrete Pile Company, which fabricated and installed these piles as a general contractor. In the early 1930's, the modern Step-Taper® shape was introduced which was a marked improvement over the earlier design. The configuration for Raymond Step-Taper® piles is shown in Figure 2-6.

Figure 2-6 Raymond Step Taper Pile Configuration

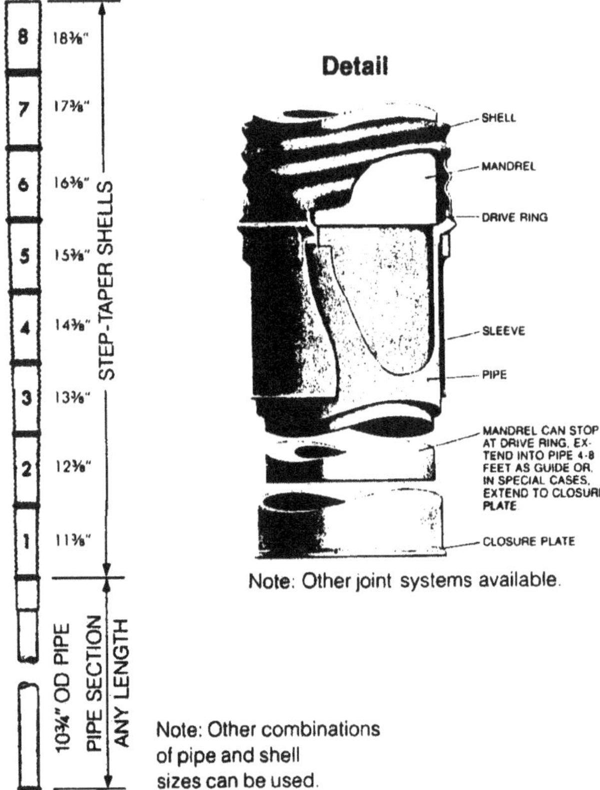

Where the original taper shell was a one-piece design, the Step-Taper® is assembled from short lengths of helically corrugated steel shells from 4' (1219mm) to 16' (4.88 m) in length. Nominal diameters of the sections range from 8" (203.2 mm) to 18" (457.2 mm); when assembled the pile diameter increases in increments of 1 inch per section and the rate of taper varies depending on the section lengths. Obviously if constant lengths are used, the rate of taper will be constant. The shells are made up from 10 to 18 gage steel with the heavier gages used in the lower portion of the pile assembly. At the project site, the combined shell lengths are pulled up over a heavy, steel, tapered mandrel. The mandrel is stepped to match the shell. Driving on the head of the mandrel drives both the mandrel and the shell into the ground. After driving is complete, the mandrel is withdrawn and the shell inspected and filled with concrete. The steel acts as a form for the concrete and is not assumed to carry any portion of the applied load. Reinforcing steel is added when necessary for lateral or tension loading. Lengths of 120' (36.6 m) are possible but not always practical since very long driving leads must be used.

Step-Taper® piles offer a wide selection of cross-sectional dimensions and length combinations for different loading and soil-bearing conditions. Being displacement piles, they function efficiently as friction piles and particularly in granular soils. By varying the toe diameter, or in combination with a pipe bottom, Step-Taper® piles can function as essentially end-bearing piles when driven to rock or compact strata overlying rock. Their efficiency as friction piles generally would result in a smaller transfer of load to the toe than a non-displacement type pile. The shape and drivability of this pile is especially effective for developing high individual capacities.

To install Raymond step-taper piles, metal shells are assembled at the project site by screwing lengths together. Joints are sealed with a neoprene ring and a bottom closure piece added. The problem of placing the shell over the mandrel can be easily addressed by dropping the shell into a previously driven shell and pulling it up onto the mandrel from the hole (see Figure 2-7). Driving on the mandrel transmits energy to the toe of the shell and the drive rings at the joints. Under certain conditions, such as stiff clays (soil heave) or dense overlying strata, predrilling is sometimes specified.

Figure 2-7 Shell-up Procedure for Step Taper Piles

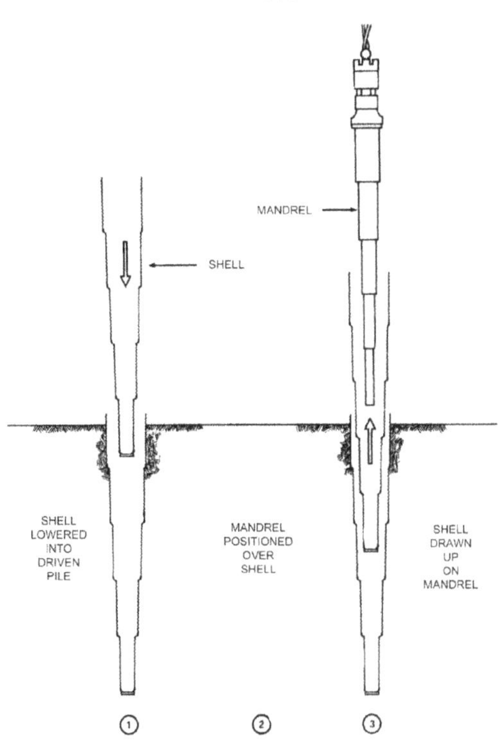

Continuing with Figure 2-8, after the shell has reached the required toe elevation, the mandrel is withdrawn and the shell inspected with a mirror or droplight. Water, should be removed at this time. Concrete is poured in a continuous operation through a drop chute. Generally, vibration is not required but Raymond suggested rodding the "top" 6' to 8' (1.8-2.44 m) of concrete. They also state that concrete may be placed in shells adjacent to driving since the driving has no detrimental effect on the new concrete. Raymond normally utilizes a concrete mix of their own formulation, chiefly based upon larger quantities of small sized coarse aggregate (a special mix to minimize aggregation.) If ground heave conditions are present, the levels of the pile heads should be monitored to see if they have risen. Step-Taper® piles that have heaved may be redriven if proper techniques are used.

Figure 2-8 Installation Sequence for Raymond Step-Taper® Piles

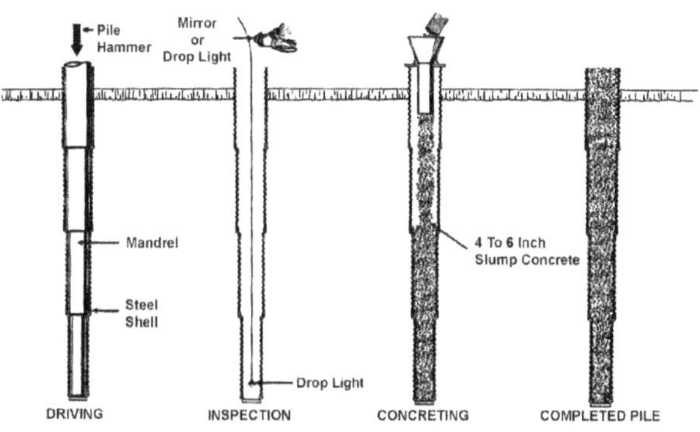

Table 2-9 Advantages and Disadvantages of Raymond Step Taper Piles

Advantages	Disadvantages
Versatility, a wide range of configurations and variations are possible to accommodate different loads and soil conditions.	Displacement piles are particularly vulnerable to pile heave in plastic soils. This condition should be monitored closely.
Drivability, the heavy mandrel permits the use of lighter hammers for more effective driving and development of the geotechnical capacity.	Thin gauge shells are vulnerable to damage where underground debris or boulders are encountered.
Internal inspection is possible after driving and before concreting.	Splicing to extend lengths is difficult.
Installation is made without damage to the working pile since driving is done on the mandrel and not on the concrete.	Shells are vulnerable to collapse from excessive earth or hydrostatic pressure, and special measures must be taken in those situations.
A range of pile capacities is possible from medium to very high.	
Shape characteristics: the configuration is that of a true displacement pile combined with the taper to develop capacity of the soil-pile system in shorter lengths than other types, particularly in loose granular soils.	
The pile shell insures that the hole is secure against soil intrusion.	

2.1.2.1.2. Other Types of Mandrel Driven Piles

In addition to the Raymond Step-Taper®, there are various straight sided shell piles. Steel companies Armco and Republic market helically corrugated shells, which are driven with proprietary or commercially available expanding mandrels. These grip the sides of the pile shell by pressure and friction in addition to bearing on the boot plate. Installation is somewhat like the Step-Taper® except long shells must be dropped down a specially drilled over-sized hole to be pulled up onto the mandrel. These piles can also be combined with a pipe tip or occasionally with a timber pile to create a composite pile. Other aspects are similar to the Step-Taper®. The Step-Taper® might be fundamentally more economical because of the tapering shape with less concrete and perhaps shorter lengths to develop the same capacity. These piles would have wall thicknesses less than 0.200" but heavier gauge piles might also be driven this way if especially hard driving and possible installation problems dictated this method. The steel in these pipe sections can be used in design calculations as an addition to concrete.

2.1.2.1.3. Monotube Piles

Monotube piles are a proprietary pile shell, which is rigid enough to be head driven. The rigidity is

Figure 2-9 Monotube Piles

TYPE	SIZE POINT DIAMETER x BUTT DIAMETER x LENGTH	Weight (N) per m 9 GA.	7 GA.	5 GA.	3 GA.	EST. CONC. VOL. m³
F Taper 3.6 mm per Meter	216 mm x 305 mm x 7.62 m	248	292	350	409	0.329
	203 mm x 305 mm x 9.14 m	233	292	336	394	0.420
	216 mm x 356 mm x 12.19 m	277	321	379	452	0.726
	203 mm x 406 mm x 18.29 m	292	350	409	482	1.284
	203 mm x 457 mm x 22.86 m	-	379	452	511	1.979
J Taper 6.4 mm per Meter	203 mm x 305 mm x 5.18 m	248	292	336	394	0.244
	203 mm x 356 mm x 7.62 m	263	321	379	438	0.443
	203 mm x 406 mm x 10.06 m	292	350	409	467	0.726
	203 mm x 457 mm x 12.19 m	-	379	438	511	1.047
Y Taper 10.2 mm per Meter	203 mm x 305 mm x 3.05 m	248	292	350	409	0.138
	203 mm x 356 mm x 4.57 m	277	321	379	438	0.260
	203 mm x 406 mm x 6.10 m	292	350	409	482	0.428
	203 mm x 457 mm x 7.62 m	-	379	452	511	0.657

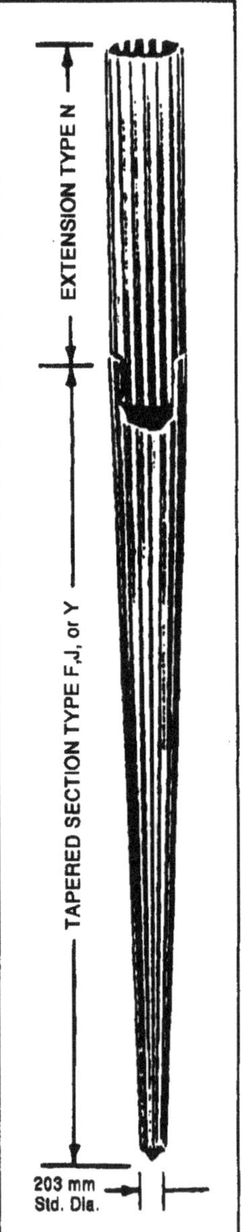

Extensions (Overall Length 0.305 m Greater than indicated)

TYPE	DIAMETER + LENGTH	9 GA.	7 GA.	5 GA.	3 GA.	m³ /m
N 12	305 mm x 305 mm x 6.10 / 12.19 m	292	350	409	482	0.065
N 14	356 mm x 356 mm x 6.10 m / 12.19 m	350	423	496	598	0.088
N 16	406 mm x 406 mm x 6.10 m / 12.19 m	409	482	569	671	0.113
N 18	457 mm x 457 mm x 6.10 m / 12.19 m	-	555	642	759	0.145

obtained by use of heavy gauge steel (3 to 9 gauge), which is longitudinally ribbed or "fluted" during the cold forming process. The basic shell is tapered with tips of about 8" (203.2 mm) diameter and butts 12" (304.8 mm) to 18" (457.2 mm). Lengths range from 10' (3.05 m) to 75' (22.9 m). Extensions to the tip sections are made with straight-sided tubes up to 40' (12 m) long. After installation, the shell is filled with concrete. Monotube pile configurations are shown in Figure 2-9.

Monotubes compete with lighter wall pipe piles, and mandrel-driven cast-in-place piles for both friction and end-bearing applications. They are designed assuming both the concrete and steel support the applied load.

2.1.2.1.4. Compacted Concrete Piles

This method was developed by the Franki Foundation Co. and was proprietary with them for many years. Recently the general process has become available from others, and there are variations on the Franki pile concept, such as the bulb piles that Raymond installed. The method utilizes a heavy, removable pipe shell and a charge of special mix concrete. Special equipment has been devised to handle the pipe and a heavy drop hammer, which rams the dry-mix concrete into the soil inside the pipe. As the mix descends it pulls the pipe with it. When the desired elevation is reached, the pipe is restrained and the concrete mix is pounded out the base where it forms a compact bulb. The pile shell is then rammed in on head of the bulb terminating at the surface.

This pile is most suited for granular soils and has developed working load capacities of over 300 kips (1334 kN). These piles experience the same general problems as augered piles and they generally are no longer than 40' (12 m).

2.1.2.1.5. Composite Piles

Piles, which combine two types of piles in a single length, are classified as composite piles. The most common of these have been briefly described in the cast-in-place section:

- A light metal shell filled with cast-in-place concrete is combined with a timber pile base providing the economy of timber below the ground water table and the durability of concrete above. These piles are generally utilized for light to low-medium loads.
- Shell pile head and pipe pile bottom, combining the length and penetrating ability of the pipe with the economy of the cast-in-place concrete head for medium to high capacity applications.

A very common type of composite pile is a prestressed concrete pile combined with an H-pile "stinger." This provides both toe protection and penetration assistance for the pile. Such a pile is shown in Figure 2-10.

Figure 2-10 H-pile "Stinger" on Concrete Pile

If required a very high capacity composite pile can be formed from a pipe pile that is driven or drilled to rock, cleaned out and socketed into the rock. A steel core section is added and the pipe filled with concrete. These piles are quite expensive but some building codes permit very high loads on this pile because of the controlled conditions under which it is installed. New York City, for example, has allowed a load up to 3000 kips (13.3 MN) on properly designed and installed piles of this type.

2.1.2.1.6. Drilled-In Caissons

Drilled-in Caissons are drilled shafts, which use a driven casing, either permanently or more typically temporarily. The caisson can be driven with an impact or vibratory hammer, depending upon the soil conditions. Use of a vibratory hammer simplifies removal of the casing. Design considerations are the same as those for drilled shafts.

2.1.2.2. Precast and Prestressed Concrete Piles

This general classification covers both conventional reinforced concrete and prestressed concrete piles. Both types can be manufactured by various methods and are available to a number of different cross sections. Frequently such piles are cast with a hollow core to reduce weight, in which case the head and toe of the pile are solid. The hollow core may be used for placing instrumentation during construction or for determining pile damage. Precast concrete piles are usually of constant cross section but may have a tapered tip. Concrete piles are considered non-corrosive but can be damaged by direct chemical attack (e.g., from organic soil, industrial wastes to organic fills), electrolytic action (chemical or stray direct currents), or oxidation. Concrete can be protected from chemical attack by use of special cements and by special coatings.

Requirements for precast concrete piles generally apply equally to prestressed units, except reinforcement. Such piles must be designed and installed in accordance with the general provisions for piling. Precast piles must be proportioned, reinforced, cast, cured, handled and driven to resist the stress induced by handling and driving as well as by structural loads. Design details should indicate suitable points of pickup and support for each length of pile. Handling equipment shall be constructed to equalize the reactions on multiple lines of pile pickups.

2.1.2.2.1. Reinforced Concrete Piles

These piles are manufactured from concrete and have reinforcement consisting of a steel rebar cage consisting of several longitudinal bars and lateral or tie steel in the form of individual hoops or a spiral. Reinforced concrete piles as compared to prestressed piles are more susceptible to damage during handling and driving because of tensile stresses. They are rarely used in current U.S. practice. These piles are easier to splice than the prestressed piles and are used where possibilities of variable pile lengths exist. These piles are best suited for friction piles in sand, gravel and clays. Typically, the maximum length allowed is 50'.

2.1.2.2.2. Prestressed Concrete Piles

This pile consists of a configuration similar to a conventional reinforced concrete pile except the prestressing steel replaces the longitudinal reinforcing steel. The prestressing steel may be in the form of strands or wires and is placed in tension. The prestressing steel is enclosed in a conventional steel spiral. Such piles can usually be made lighter and longer than normally reinforced concrete piles of the same rigidity.

Prestressed piles can either be pretensioned or post-tensioned. Pretensioned piles are usually cast full length in permanent casting beds. Post-tensioned piles are usually manufactured in sections and assembled and prestressed to the required pile lengths at the manufacturing plant or on the job site. Table 2-10 shows typical prestressed concrete piles, along with data for typical prestressed concrete pile sections. Figure 2-11 shows typical details of pile reinforcement.

The minimum lateral dimensions of precast concrete piles should be 10" (254 mm) except for taper at the toe. For piles designed with voids, the minimum wall thickness of concrete should be 4" (101.6 mm). The maximum departure of the pile axis from a straight line, measured before installation while the pile is not subject to bending forces, should not exceed 1/8" (3.2 mm) in 10' (3.05 m), 3/8" (9.5 mm) in 40' (12 m), or 0.1% of the pile length.

Concrete preferably should have a higher strength than the usually specified minimum of 5 ksi (34.5 MPa) at 7 days if steam cured or at 28 days if cured by other means. The mix should have a cement content of 6 to 8 bags per cubic yard. Type II cement may be used where moderate sulfate resistance is required. Air entraining cement or suitable admixtures may be used to increased workability of the concrete. Calcium chloride may not be used, except for a trace in formulations used to accelerate strength gain and reduce shrinkage.

The minimum amount of longitudinal reinforcement should be at least 1.5 to 2% of the concrete section, made up of a minimum of four bars symmetrically placed. For a length equal to at least three times the minimum lateral dimension at each end of the pile lateral tie reinforcement should be spaced 3" (76.2 mm) on centers, increased elsewhere to a maximum of 12" (304.8 mm). Lateral reinforcement should be No. 5 gauge or equivalent spiral. For piles designed with voids, the 3" (76.2 mm) spacing of ties or laterals can be extended for a distance of 12' (3.66 m) or one-third the length of the pile, whichever is smaller. Closely spaced ties or spirals improve the ductility of a pile. If hard driving is expected, spacing of 4" (101.6 mm) minimum is recommended.

Current practice is to cover longitudinal reinforcing

Table 2-10 Section Properties and Allowable Service Loads of Prestressed Concrete Piles

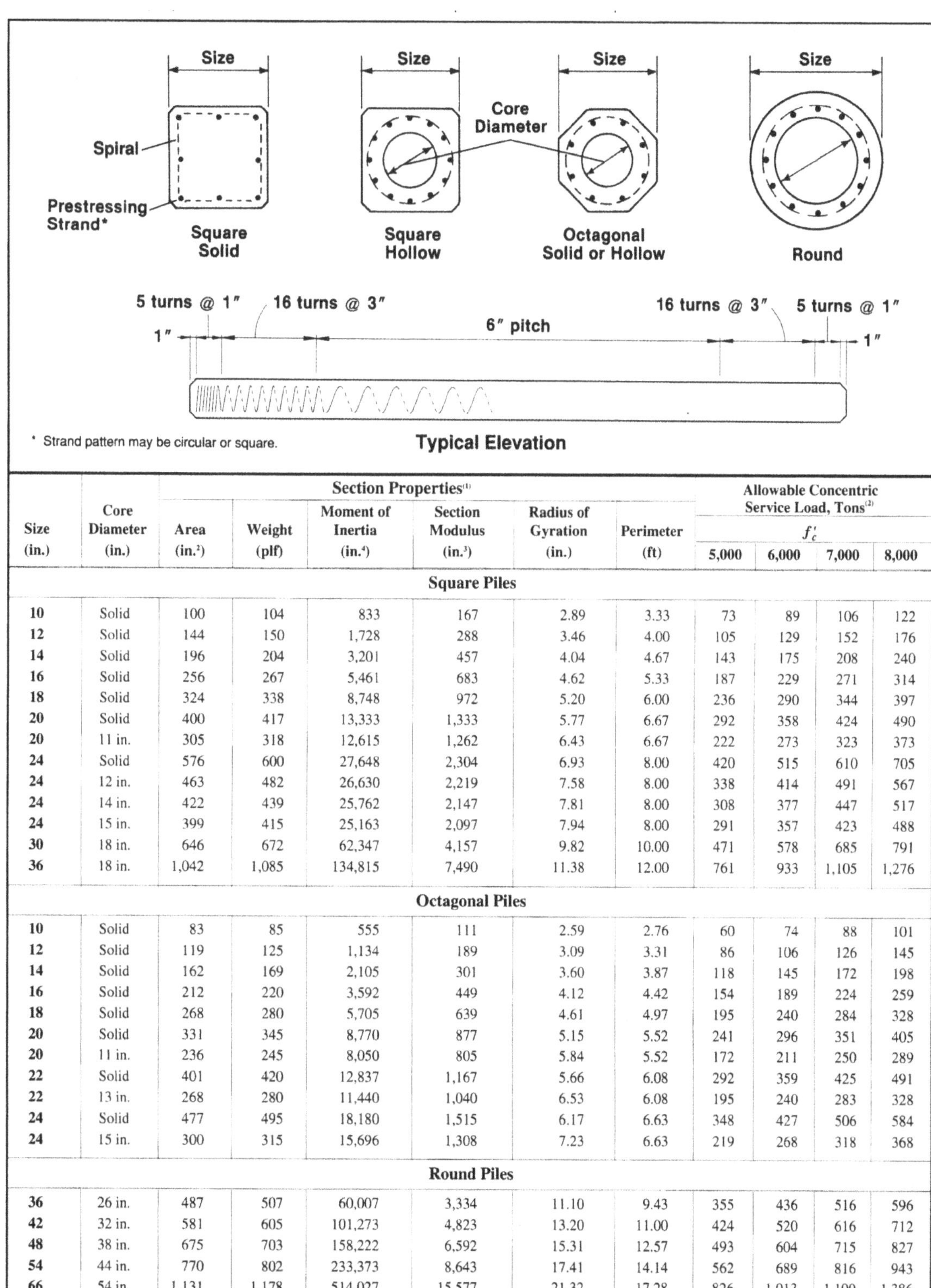

Size (in.)	Core Diameter (in.)	Section Properties[1] Area (in.²)	Weight (plf)	Moment of Inertia (in.⁴)	Section Modulus (in.³)	Radius of Gyration (in.)	Perimeter (ft)	Allowable Concentric Service Load, Tons[2] f'_c 5,000	6,000	7,000	8,000
					Square Piles						
10	Solid	100	104	833	167	2.89	3.33	73	89	106	122
12	Solid	144	150	1,728	288	3.46	4.00	105	129	152	176
14	Solid	196	204	3,201	457	4.04	4.67	143	175	208	240
16	Solid	256	267	5,461	683	4.62	5.33	187	229	271	314
18	Solid	324	338	8,748	972	5.20	6.00	236	290	344	397
20	Solid	400	417	13,333	1,333	5.77	6.67	292	358	424	490
20	11 in.	305	318	12,615	1,262	6.43	6.67	222	273	323	373
24	Solid	576	600	27,648	2,304	6.93	8.00	420	515	610	705
24	12 in.	463	482	26,630	2,219	7.58	8.00	338	414	491	567
24	14 in.	422	439	25,762	2,147	7.81	8.00	308	377	447	517
24	15 in.	399	415	25,163	2,097	7.94	8.00	291	357	423	488
30	18 in.	646	672	62,347	4,157	9.82	10.00	471	578	685	791
36	18 in.	1,042	1,085	134,815	7,490	11.38	12.00	761	933	1,105	1,276
					Octagonal Piles						
10	Solid	83	85	555	111	2.59	2.76	60	74	88	101
12	Solid	119	125	1,134	189	3.09	3.31	86	106	126	145
14	Solid	162	169	2,105	301	3.60	3.87	118	145	172	198
16	Solid	212	220	3,592	449	4.12	4.42	154	189	224	259
18	Solid	268	280	5,705	639	4.61	4.97	195	240	284	328
20	Solid	331	345	8,770	877	5.15	5.52	241	296	351	405
20	11 in.	236	245	8,050	805	5.84	5.52	172	211	250	289
22	Solid	401	420	12,837	1,167	5.66	6.08	292	359	425	491
22	13 in.	268	280	11,440	1,040	6.53	6.08	195	240	283	328
24	Solid	477	495	18,180	1,515	6.17	6.63	348	427	506	584
24	15 in.	300	315	15,696	1,308	7.23	6.63	219	268	318	368
					Round Piles						
36	26 in.	487	507	60,007	3,334	11.10	9.43	355	436	516	596
42	32 in.	581	605	101,273	4,823	13.20	11.00	424	520	616	712
48	38 in.	675	703	158,222	6,592	15.31	12.57	493	604	715	827
54	44 in.	770	802	233,373	8,643	17.41	14.14	562	689	816	943
66	54 in.	1,131	1,178	514,027	15,577	21.32	17.28	826	1,013	1,199	1,386

(1) Form dimensions may vary with producers, with corresponding variations in section properties.

(2) Allowable loads based on $N = A_c (0.33 f'_c - 0.27 f_{pc})$; f_{pc} = 700 psi. Check local producer for available concrete strengths.

Figure 2-11 Typical Details of Pile Reinforcement

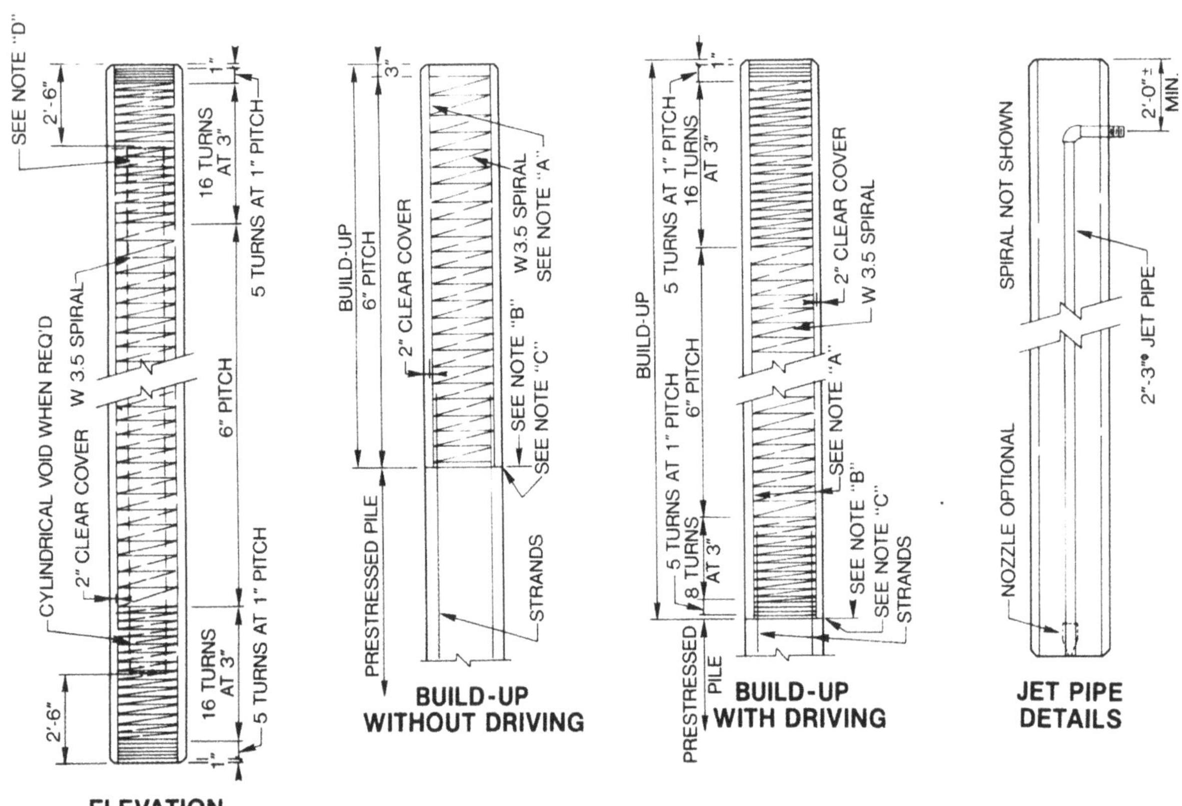

Notes on Build-up

Note A: The minimum area of reinforcing steel shall be 1½ percent of the gross cross-section of concrete. Placement of bars shall be in a symmetrical pattern of not less than four bars.

Note B: Method of attachment of pile to build-up may be by any of the methods given in the notes on alternate pile heads. If mild reinforcing steel is used for attachment, the area shall be no less than that used in the build-up.

Note C: Concrete around top portion of driven pile shall be bush-hammered to prevent feather edges in the cast-in-place build-up.

Note D: End fitting or form may be flat or tapered with proper taping to prevent leakage.

Note E: Additional mechanical and other types of splices are available to designers. Refer to PCI JOURNAL two-part series, "Splicing of Precast Prestressed Concrete Piles," September-October and November-December 1974.

steel with a minimum of 1-1/2" (38.1 mm) of concrete.[5] Voids, when used, shall be located within 3/8" (9.5 mm) of the position shown on plans. Voids may extend through either or both ends of the pile. If the void extends through the lower end of the pile, the pile head must be vented to prevent build-up of internal hydraulic pressure during driving. Paper or fiber used to form a void in the pile has been known to decompose and develop destructive gas pressures, so it should be removed or the pile permanently vented.

If prestressed piles are used, the minimum working net prestress in the pile should be 700 psi (4.83 MPa). Prestressing strands are of the ungalvanized seven-strand type conforming to the general requirements of ASTM designation A 416 and may be either regular or high-strength. Strand properties, manufacture and installation should conform to guidelines of the Prestressed Concrete Institute.[6]

The primary advantage of prestressed concrete piles versus conventional reinforced concrete piles is durability. Since the concrete is under continuous compression, hairline cracks are kept tightly closed and thus prestressed piles are usually more durable than conventionally reinforced piles. Another advantage of prestressing (compression) is that the tensile stresses, which can develop in the concrete under certain driving conditions, are less critical. These piles are best suited for friction piles in sand, gravel, and clays.

2.1.2.2.3. Prestressed Cylinder Piles

Prestressed cylinder piles, originally developed by Raymond, are post-tensioned piles that are spun cast in sections, bonded with a plastic joint compound, and then post-tensioned in lengths containing several segments. Special concrete is cast by a process unique to cylinder piles that achieves high density and low porosity. The pile is virtually impervious to moisture. Results of chloride ion penetration and permeability tests on prestressed cylinder piles indicate that the spun cylinder piles have excellent resistance to chloride intrusion. Table 2-11 shows the typical configuration of prestressed cylinder piles and provides appropriate engineering design data. Figure 2-12 shows typical details of cylinder pile reinforcement. Generally, cylinder piles are used for marine structures or dry land trestles. The piles typically extend above ground and are

Table 2-11 Dimensions and Properties of Prestressed Cylinder Piles

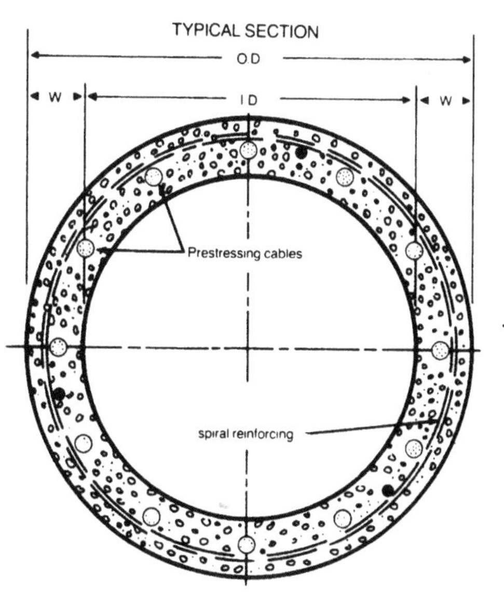

CYLINDER PILE DESIGN PROPERTIES

	SIZE										STRESS ON CONCRETE DUE TO EFFECTIVE STRESS PER CABLE (b)	
	O.D.	I.D.	W	A_c	I	S	r	CIRCUM-FERENCE	POINT AREA	WEIGHT PER FOOT (a)	Wire (c)	Strand (d)
	in.	in.	in.	in.²	in.⁴	in.³	in.	ft	ft²	lb	lb/in.²	lb/in.²
	36	27	4½	445	56,360	3,130	11.3	9.43	7.07	479	116.1	110.5
*	36	26	5	487	60,000	3,330	11.1	9.43	7.07	524	106.1	100.9
	42	32	5	581	101,300	4,820	13.2	11.00	9.61	625	89.0	84.6
	48	38	5	675	158,200	6,590	15.3	12.57	12.57	726	76.6	72.8
*	54	44	5	770	233,400	8,640	17.4	14.14	15.90	829	67.1	63.8
	54	42	6	905	264,600	9,800	17.1	14.14	15.90	973	57.1	54.3
	60	49	5½	942	353,200	11,770	19.4	15.71	19.63	1014	54.9	52.2
*	66	54	6	1131	514,000	15,580	21.3	17.28	23.76	1217	45.7	43.5
	72	60	6	1244	683,000	18,970	23.4	18.85	28.27	1339	41.5	39.5
	78	65	6½	1460	940,700	24,120	25.4	20.42	33.18	1572	35.4	33.7
	84	70	7	1693	1,265,300	30,130	27.3	21.99	38.48	1823	30.5	29.0
	90	76	7	1825	1,582,900	35,180	29.5	23.56	44.18	1964	28.3	26.9

*STANDARD SIZES

NOTES:

(a) The tabulated weights, intended for design purposes, are based on a unit weight of concrete of 155 lbs/ft³ and the nominal wall thicknesses. For handling purposes, allowance should be made for manufacturing tolerance in extra wall thickness with corresponding increase in weight per foot.

(b) Number of prestressing cables ranges from 8 to 16 for 36" piles, 12 to 24 for 54" piles, and 16 to 32 for 66" piles.

(c) Each prestressing cable consists of twelve 0.192"-diameter stess-relieved wires with initial stress of 175,000 psi (60,801 lbs) and effective stress of 148,750 psi (51,681 lbs).

(d) Each prestressing cable consists of two ½"-diameter 270 ksi 7-wire strands with initial tension of 57,820 lbs and effective tensionof 49,150 lbs.

designed to resist a combination of axial and lateral loads. They are available in diameters of 36" (914.4 mm) to 90" (2286 mm).

2.1.2.2.4. Pretensioned Spun Concrete Piles

A relatively new type of concrete pile is the pretensioned spun concrete pile, a cylindrical pile with a void. It is geometrically similar to cylinder piles but the manufacturing process has some differences.

These piles are manufactured as follows:

1. The strands are straightened from coil and cut to the desired length. Their ends are button headed for the prestressing machine. The spiral wire is automatically wound around and welded to the strands. The cage is then placed into the bottom half of the mould.
2. Concrete is fed into the bottom half mould, after which the top half of the mould is bolted to its mate.
3. The longitudinal strands are then prestressed against the mould through a central shaft. This operation insures uniformity of stress in all of the bars and contributes to the straightness of the pile.
4. The pile is then spun in the mould. While forcing the concrete to the sides of the mould and thus giving the pile its tubular shape, this also squeezes water out of the concrete,

Figure 2-12 Typical Details of Cylinder Pile Reinforcement

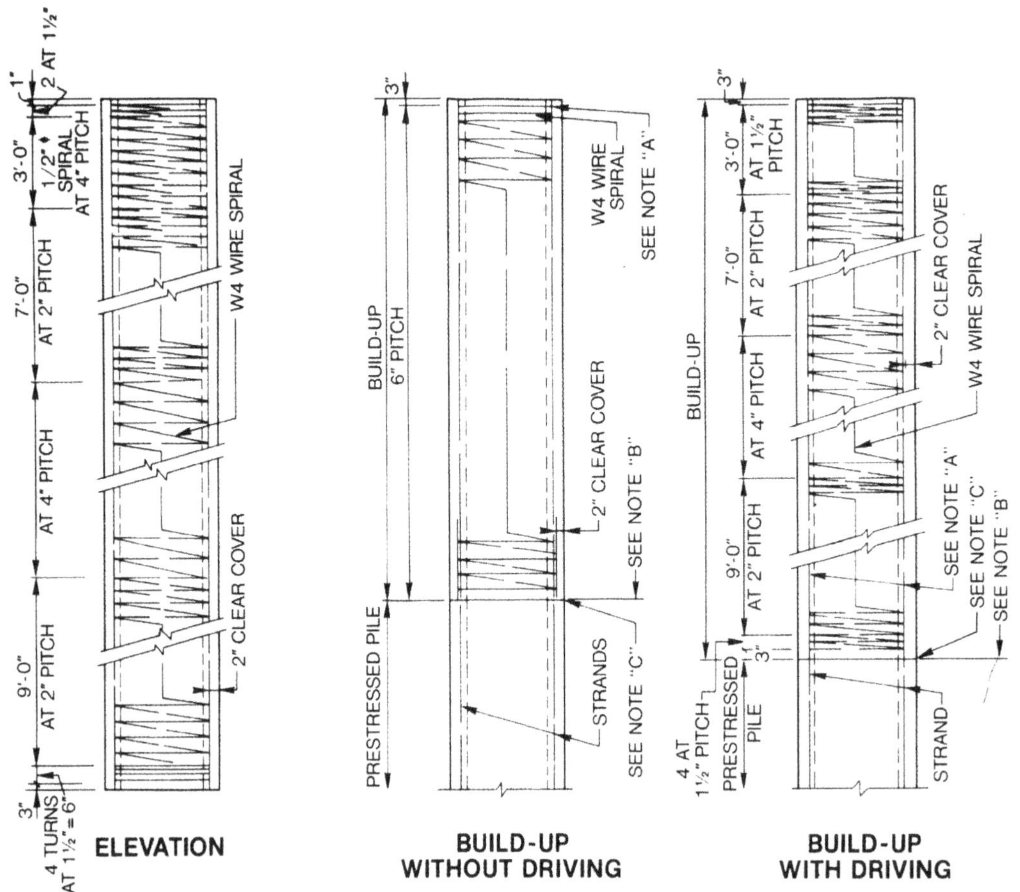

Notes on Build-up

Note A: The minimum area of reinforcing steel shall be 1½ percent of the gross cross-section of concrete. Placement of bars shall be in a symmetrical pattern of not less than eight bars.

Note B: Method of attachment of pile to build-up may be by any of the methods given in the notes on alternate pile heads. If mild reinforcing steel is used for attachment, the area shall be no less than that used in the build-up.

Note C: Concrete around top portion of driven pile shall be bush-hammered to prevent feather edges in the cast-in-place build-up.

increasing the strength of the pile.

5. The pile is removed from the mould using a vacuum lifter.
6. The pile is cured in an autoclave, which advances curing further and prevents the driving of uncured piles.

Pretensioned spun concrete piles can be obtained in diameters from 250 mm (9.8") to 1000 mm (39.4"), and in lengths up to 46 m (151'). They include a provision for splicing and can be cut off with the usual methods.

2.1.2.2.5. Material Specifications

2.1.2.2.5.1. Aggregates

Concrete aggregates should conform to "Specification for Concrete Aggregates" (ASTM C 33) or to "Specifications for Lightweight Aggregates for Structural Concrete:" (ASTM C 330); except that aggregates failing to meet these specifications, but which have been shown by special test or actual service to produce concrete of adequate strength and durability, may be used with the approval of the governing authority. [7]

2.1.2.2.5.2. Water

Water used in mixing concrete should be clean and free from injurious amounts of oils, acids, alkalis, salts, organic materials, or other substances that may be deleterious to concrete or steel. Mortar cubes made with nonpotable mixing water should have 7- and 28-day strengths equal to at least 90% of the strengths of similar specimens made with potable water.

2.1.2.2.5.3. Admixtures

Air-entraining admixtures should be considered where concrete piles are exposed to conditions of freezing and thawing. When used, air-entraining admixtures should conform to "Specification for Air-Entraining Admixtures for Concrete" (ASTM C 260).

The amount of air entrainment and its effectiveness depend on the admixture to be employed, the size and nature of the coarse aggregate, its moisture content, and other variables. Too much air will lower the strength of the concrete and too little will reduce its effectiveness. It is recommended that the air content of concrete is to be in the range of 4 to 7%, depending on the size of the coarse aggregate.

Air-entraining admixtures are less effective when used with low slump, high strength concrete. Furthermore, the need for air entrainment is reduced in high strength concrete because of its high density and low porosity. For this reason, the designer should carefully evaluate the site conditions compared to the pile quality specified before making a decision regarding air entrainment.

When used, water-reducing admixtures, retarding admixtures, accelerating admixtures, water reducing and retarding admixtures, and water reducing and accelerating admixtures should conform to "Specification for Chemical Admixtures for Concrete" (ASTM C 494). Calcium chloride or admixtures containing calcium chloride should not be used.

2.1.2.2.5.4. Concrete Quality

Concrete in precast prestressed piles and build-ups to be driven should preferably have a minimum compressive cylinder strength (f'_c) of 5000 psi at 28 days. Economy in handling and driving along with higher load capacity can be achieved with concrete strengths up to 8000 psi. Designers should check with local pile manufacturers to determine optimum strengths.

For acceptable durability, concrete piles should have at least six sacks of cement per cubic yard of concrete. The water-cement ratio (by weight) should correspond to the least water that will produce a plastic mix and provide the desired workability for the most effective placement of the concrete.

2.1.2.2.5.5. Reinforcement

All steel wires, prestressing strands and reinforcements, unless otherwise stipulated, shall conform to applicable ASTM standards.

2.1.2.2.5.6. Grout

Cement grout where used in prestressed piles should be of materials which conform to the requirements stipulated herein for cement, sand, admixtures, and water. Approved expanding admixtures or expansive cements may be used. Sand and cement grouts are generally used when grouting dowels into holes in heads of piles, sometimes with expanding admixtures.

Some expanding admixtures contain calcium chloride and should be avoided. Neat cement grout is frequently used to grout dowels in pile heads[8].

2.1.2.2.5.7. Anchorages

Anchorage fittings for post-tensioning assemblies should conform to the latest ACI 318, Building Code Requirements. ACI post-tensioning specifications

Figure 2-13 Tolerance Dimensions for Precast and Prestressed Concrete Pile

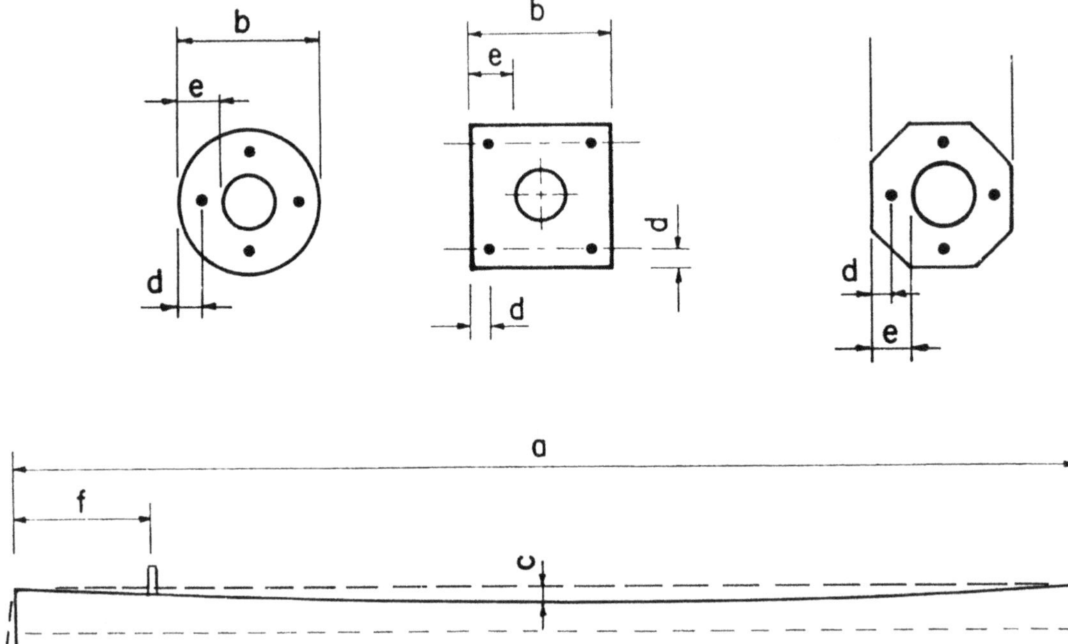

a = Length* .... ± 1 in.

b = Width or diameter .... ± ⅜ in.

c = Sweep (variation from straight line parallel to centerline of member) (considered to be a form tolerance) .... ± ⅛ in. per 10 ft

d = Position of tendons .... ± ¼ in.

e = Wall thickness .... – ¼ in., +½ in.

f = Position of handling devices .... ± 6 in.

g = Position of steel driving tips .... ± ½ in.

h = Variation from specified end squareness or skew .... ± ¼ in. per 12 in., ± ½ in. max.

i = Local smoothness any surface .... ± ¼ in. per 10 ft

j = Longitudinal spacing of spiral reinforcement .... ± ¾ in.

* In most cases, controlling pile length to + 6 in., – 2 in. is functionally acceptable.
Note: 1 ft = 0.305 in.; 1 in. = 25.4 mm.

may also be used for guidance.

2.1.2.2.6. Forms and Casting

Precast and prestressed concrete piles are usually cast at off site yards and hauled to the site by truck or barge. They are made in forms 400-600' (122-183 m) or more in length. This has the advantage of good manufacturing control; it has the problem of moving long, heavy units some distance through congested areas. Prestressing wires are pretensioned across this length by hydraulic jacks. Bulkheads are set in the form to make the desired length of pile; prestressing wires go through the bulkheads, which serve as spacers. Spiral reinforcing is slipped over the strands during stringing and distributed before placing the bulkheads. Forms for casting piles are preferably relatively permanent and made of steel or other reusable material. Provision should be made to chamfer edges and corners of the piles, except where reinforcement may protrude through an end. Reinforcement must be spaced away from the pile faces by small cement blocks or by metal chairs with plastic tips. Lateral bars must be effectively tied to the longitudinal steel. Figure 2-13 shows tolerance dimensions for concrete piles.

Hydraulic jacks must stress prestressing strands uniformly. An accurately calibrated gage incorporated in the system permits stress computation at any time. Elongation should be measured at completion of stressing and at time of placing concrete; it should conform to elongation tables furnished by the manufacturer of the strand. Strands must be kept free of oil and other substances harmful to the concrete bond. Some discoloring of the strand may not be harmful but corrosion must be prevented.

Concrete should be compacted by high frequency internal or external vibrators; contact of vibrators with prestressing strands should be kept to a minimum. Full tension on the strands shall be maintained until test cylinders, cast and cured under the same conditions as the piles, indicate strength of at least 4,000 psi. Pretension in the strands or wires must be released from the anchorage gradually and uniformly. Burning strands in alternate locations, rather than all strands across one side does this. When released from elongation strands expand and grip the concrete; full tension may not be achieved for the initial 50 diameters of the strand length at both pile ends.

Points at which piling are to be lifted or supported should be clearly apparent. When other picking methods are used (inserts, slings, vacuum pads) suitable markings to indicate correct support points should be provided. When not specified otherwise, the two pick-up points on a precast concrete pile should be 21% of the length from both head and toe.

For long columns for bridge piers and for offshore towers large diameter, open-centre piles are centrifugally cast in about 16' (4.88 m) lengths. Post-tensioning wires are inserted through openings formed in the 4 or 5" (127 mm) thick walls to develop the required lengths, which may be over 200'. Additional information is available from PCI and from the producers.

2.1.3. Timber Piles

Evidence of structures supported by timber piles can be traced back to Switzerland some six thousand years ago. Timber piles have been used in North America since the mid-eighteenth century and even to this day, are still very important to the foundation designer despite the inroads made by manufactured materials. An historical application of timber piling - that of major bridges - is shown in Figure 2-14.

Figure 2-14 Timber Piling Driven for a Bridge

Historically, almost all types of wood have been employed for piling purposes at one time or another. Two species however, now account for over 90% of the usage and are the basis for which most standards concerning timber piles are written. These species are Southern pine and Douglas fir.

Southern pine is grown mainly in the southern United States and consists of four sub-species: longleaf, loblolly, slash and short leaf. Douglas fir is a product of the Northwest Coast with the preferred product for piling identified as "Coastal" Douglas fir.

Some specialty timber is imported from the tropics for marine piling applications. Greenheart, imported from South America, is one such species. It features high strength and superior resistance to decay and to attack by marine borer organisms.

Timber piles are processed as clean-peeled (all outer bark and 80% of inner bark removed) rough peeled (all outer bark removed) and un-peeled (all bark retained). Piles that are to be further treated with preservatives must be clean-peeled.

Timber piles are frequently installed un-peeled and untreated. These are generally for use in temporary structures or installations with a planned short-service life. However, the majority of timber piles are now treated with wood preserving chemicals to extend their life.

Sawed timbers are very rarely used for piling, therefore timber piles are always round and tapered, which is an efficient shape for a pile.

2.1.3.1. Sizes of Timber Piles

Table 2-13 and Table 2-14 relate circumference at the toe of the pile to head circumference for ten-foot length increments up to 120' (36.6 m).

- Table 2-13 is to be used when the toe dimension is specified and the minimum circumference of the head is desired.
- Table 2-14 is used when the head dimension is specified and the corresponding toe dimensions are needed.

In the case of end-bearing timber piles, the toe dimension is more important since most of the load will be transferred to the toe. In the case of a true friction pile, the head dimensions are more critical since most of the load will theoretically be transferred to the soil long before reaching the toe. By utilizing the two charts, the timber industry can more efficiently match timber shapes to the requirements of the design.

Circumferential dimensions are used since it is difficult to accurately measure diameters of a tree trunk. Head dimensions listed are for a location 3' (914 mm) from the true head (since the butts are trimmed for driving and cut-off later.

2.1.3.2. Quality

Timber for piling should be of sound wood and free from decay and insect damage. Other possible defects are identified as follows:

- A check is a separation of the wood extending across the growth rings from the surface toward the centre but not completely across the section. A check should not extend any further than the pitch (centre core).
- A shake is a circumferential separation of

Table 2-12 Advantages and Disadvantages of Timber Piles

Advantages	Disadvantages
Low cost, per ton of capacity.	Cannot be spliced to extend lengths.
Dependable, renewable supply - available in a range of lengths and sizes.	More vulnerable to driving damage.
Long history of successful application to low and medium unit loads.	Vulnerable to deterioration from a number of natural sources unless effective protection is provided.
Easily handled and driven with conventional equipment.	Restrictive properties regarding strength, sizes and lengths.
Tapered shape and full displacement characteristics advantageous for developing soil capacity in shorter lengths.	
Strength in tension and bending applications.	

Table 2-13 Specified Butt Diameters with Minimum Tip Circumferences[9]

Specified Butt Diameters, in.	7	8	9	10	11	12	13	14	15	16	18
Required Minimum Circumference 3 ft from Butt	22	25	28	31	35	38	41	44	47	50	57
Length, ft	Minimum Tip Circumferences (in.) and Corresponding Diameter in Italics										
20	16.0	16.0	16.0	18 0	22.0	25.0	28.0				
	5.0	5.0	5.0	5.7	7.0	8.0	8.9				
30	16.0	16.0	16.0	16.0	19.0	22.0	25.0	28.0			
	5.0	5.0	5.0	5.0	6.0	7.0	8.0	8.9			
40				16.0	17.0	20.0	23.0	26.0	29.0		
				5.0	5.4	6.4	7.3	8.3	9.2		
50					16.0	17.0	19.0	22.0	25.0	28.0	
					5.0	5.5	6.0	7.0	8.0	8.9	
60						16.0	16.0	18.6	21.6	24.6	31.6
						5.0	5.0	5.9	6.9	7.8	10.0
70						16.0	16.0	16.0	16.2	19.2	26.2
						5.0	5.0	5.0	5.1	6.1	8.3
80							16.0	16.0	16.0	16.0	21.8
							5.0	5.0	5.0	5.0	6.9
90							16.0	16.0	16.0	16.0	19.5
							5.0	5.0	5.0	5.0	6.2
100							16.0	16.0	16.0	16.0	18.0
							5.0	5.0	5.0	5.0	5.8
110										16.0	16.0
										5.0	5.0
120											16.0
											5.0

Table 2-14 Specified Tip Diameters with Corresponding Minimum Butt Circumferences

Specified Tip Diameter in.	5	6	7	8	9	10	11	12
Tip Circumference, Required Minimum	16	19	22	25	28	31	35	38
Length, ft	Minimum Circumferences 3 ft from Butt in. with Diameter in Italics.							
20	22.0	24.0	27.0	30.0	33.0	36.0	40.0	43.0
	7.0	7.6	86	9.5	10.5	11.5	12.7	13.7
30	23.5	26.5	29.5	32.5	35.5	38.5	42.5	45.5
	7.5	8.4	9.4	10.3	11.3	12.2	13.5	14.5
40	26.0	29.0	32.0	35.0	38.0	41.0	45.0	48.0
	8.3	9.2	10.2	11.1	12.1	13.0	14.3	15.3
50	28.5	31.5	34.5	37.5	40.5	43.5	47.5	50.5
	9.0	10.0	11.0	11.9	12.9	13.8	15.1	16.0
60	31.0	34.0	37.0	40.0	43.0	46.0	50.0	53.0
	9.8	10.8	11.8	12.7	13.7	14.6	15.9	16.8
70	33.5	36.5	39.5	42.5	45.5	48.5	52.5	55.5
	10.6	11.6	12.6	13.5	14.4	15.4	16.7	17.7
80	36.0	39.0	42.0	45.0	48.0	51.0	55.0	58.0
	11.4	12.4	13.4	14.3	15.3	16.2	17.5	18.4
90	38.6	41.6	44.6	47.6	50.6	53.6	57.6	60.5
	12.2	13.2	14.2	15.1	16.0	17.0	18.3	19.2
100	41.0	44.0	47.0	50.0	53.0	56.0	60.0	
	13.0	14.0	15.0	15.9	16.8	17.8	19.0	
110	43.6	46.6	49.6	52.6	55.6	61.0		
	13.8	14.8	15.7	16.7	17.7	19.4		
120	46.0	49.0	52.0	55.0	58.0			
	14.6	15.6	16.6	17.5	18.4			

the rings of growth. The lengths of shakes in the head of the pile are limited.

- A split is a lengthwise separation of the wood across the growth rings but extending from one surface to the other. Splits may not be any longer than the diameter of the head.
- Knots are, of course, the source of limbs, which have been trimmed from the trunk. Restrictions are imposed on the sizes and depth of knots based on being classed as "sound" or "unsound."
- Straightness requires that a straight line from center of head to center of toe must lie entirely within the pile body.

2.1.3.3. Preservative Treatment

In the past, timber piles, which have remained continually wet due to their location below the ground water table, have proven to have a practically indefinite service life. Conversely, timber piles subject to a fluctuating water table or exposed to attack by insects, fungi or marine woodborers have shown rapid deterioration. The subject of proper and effective preservation treatment for timber piles is therefore very important to a potential owner or specifier of timber piles.

Attempts to prevent deterioration by surface treatment can be traced to ancient engineers who used various oils or pitch for surface treatment of some of their structures. Modern wood preserving began in England in 1832 with a process using mercuric chloride.

Pressure injection of creosote began in England in 1838. A plant was opened in Massachusetts, using imported creosote, in 1865. The usage of creosoted timbers for railroad ties and pile-supported structures grew rapidly with the expansion of the railroads. Thus, so did the number of creosote treating plants. In 1904, the American Wood Preservers Association was founded. This group, composed of timber growers, industry representatives, engineers, educators, and piling users has been responsible for developing wood preservation into an effective and reliable science.

2.1.3.4. Environmental Concerns

There are no materials that are immune to attack by at least some elements of the environment. Timber, in its natural state, is attacked by insect pests, fungi, decay and disease. Once trees are cut and converted to products for the building trades, wood is still vulnerable to most of the same problems in addition to some new ones. These problems have been identified and protection for each developed and improved over the years. An important task of the piling designer is to identify potential deterioration mechanisms and specify appropriate protection.

Timber foundation piles are used extensively on land to provide support for buildings, floor systems, machinery, equipment, retaining walls, storage tanks and bridge piers. This usage results in the pile being driven into the soil, with (the toe and part of the pile generally below the water table) and the head end "encased" with a concrete slab or footing. Piles for these applications are generally the least affected by the environment. A fluctuating water table however, could contribute to decay of untreated wood. Furthermore, the interface of pile to concrete and the disturbed area directly below is a potential site for fungi. It is now common practice to treat the entire pile with at least the minimum treatment shown in AWPA Standard C1 and C3 which lists creosote, creosote-tar solution, creosote-petroleum, pentachlorophenol, ammonical copper arsenate, and chromated copper arsenate as preservatives for selection. The purpose of this treatment is to prevent moisture from reaching the wood that would inhibit decay from wetting and drying cycles, and also provide an inhospitable environment for wood-destroying organisms.

Timber piles are used extensively in fresh water as well as on land for marina piers, railroad and highway trestle bents, supporting buildings above the flood plain, shear fences, and other applications. These piles are partially embedded in soil and partially exposed to water, air, or both, Timber piles in fresh water have no organic enemies as long as the pile is continuously wet. Splash and fluctuating water levels will result in wetting and drying periods that could result in deterioration at those locations. Untreated timber piles exposed to air are vulnerable to inspect attack; on land, the interface between ground and atmosphere is also a potential problem area. In any case, when one comes in contact with any of these situations, timber piles should probably be treated 100% (the full length) as spot treatment would prove impractical.

Land and fresh water piles should be treated in accordance with the requirements of AWPA Standard C1. Table 2-15 shows the difference between retention quantities intended for land and fresh water foundations and marine piling applications

Table 2-15 Minimum Retention, Pounds per Cubic Foot for Various Types of Timber Piling

	RETENTION PCF (MIN.) Southern Pine			Douglas Fir		
	Foundation	Land and Water	Fresh Marine	Foundation	Land and Water	Fresh Marine
Creosote and creosote solutions						
Creosote	12.0	12.0	20.0	17.0	17.0	20.0
Creosote-coal tar Creosote-petroleum	12.0	12.0	20.0	17.0	17.0	20.0
Oil-borne Preservatives						
Pentachlorophenol	0.60	0.60	N.A.	0.85	0.85	N.A.
Water-borne Preservatives						
ACA	0.80	0.80	2.50 (a) 1.50 (b)	1.00	1.00	2.50
CCA	0.80	0.80	2.50 (a) 1.50 (b)	1.00	1.00	
Penetration in inches of wood, min. and/or sapwood.	3.0 or 90	3.0 or 90	3.5 or 90	0.75 (c) and 85	0.75 (c) and 85	Variable see C3

(a) Zone 1
(b) Zone 2
(c) Up to Max of 1.60

2.1.3.5. Marine Applications

Timber piles are used extensively along the coasts in salt and brackish water for marine construction, commercial docks and piers, navigation devices, trestle bent bridge construction, mooring and turning posts, bridge protection and others. In addition, timber in marine environments are exposed to a variety of organisms which either feed directly on the wood or drill into it for nesting purposes. Woodborers are divided into two families, mollusks related to the clam, and crustaceans related to crabs or shrimp. Mollusks include teredinids (shipworms). These feed on and live in untreated wood. Within a year of infestation, they may grow to an inch in diameter and four feet long. Attack is generally heaviest near the mud line. Pholads, another mollusk, resemble minute clams and do not burrow as deep. Teredinids, including teredos and bankia, can be found along the entire coastline regardless of temperature or salinity. The crustacean branch includes Limnoria, of which there are many species and Sphaeroma, which use the wood only for shelter. Limnoria are responsible for the hourglass shape of piling as seen along the coast. Limnoria lignorum are colder water species found in coastal waters of the northeast and northwest. Limnoria tripunctata favors warmer water from Virginia southward on the East coast, San Francisco southward on the West Coast and all of the Gulf Coast. In the warmer water, this borer is very tolerant of creosote and therefore other means of discouragement have had to be employed. Sphaeroma is reported to be widespread on the Florida and Gulf coasts, particularly in brackish waters, however this borer is normally not the source of severe damage.

The AWPA has especially focused on piles for marine environments, having published their "Standard for Pressure-Treated Material in Marine Construction," C18 and also Standard C 2-85, "Lumber, Timbers, Bridge Ties and Mine Ties - Preservative Treatment by Pressure Processing," Table 2-16 summarizes the type treatment suggested for the four most common marine borers.

Table 2-16 Preservative Treatment for Lumber and Timber Exposed to Marine Borers

Marine Borer	Preservative Type and Retention (pcf) CREOSOTE	CCA-ACA-ACZA	DUAL
Teredinids	25	2.5	1.5 & 20
L. Quadripunctata or Lignorum	25	2.5	1.5 & 20
L. Tripunctata	NR	2.5	1.5 & 20
Pholads	25	NR	1.5 & 20
Spaeroma	25	NR	1.5 & 20

CCA = Chromated Copper Arsenate
ACA = Ammoniacal Copper Arsenate
ACZA = Ammoniacal Copper Zinc Arsenate
NR = Not Recommended
Duel = Combined Treatment

Creosote is and will probably remain the predominate treatment for wood piling. It is sometimes combined with coat-tar in solution. Pentachlorophenol (Penta) is combined with oil to provide an alternative to creosote. However, Penta cannot be used in salt-water applications. Metallic salts identified as CCA (chromated copper arsenate,) ACA (ammoniacal copper arsenate, and ACZA (ammoniacal copper zinc arsenate) which are water-borne preservatives and are commonly used for protection of marine piling against borers or in combination with creosote in a dual treatment process. Some embrittlement of the wood fiber is a disadvantage of the metallic salt treatment.

Timber piles to be treated are clean-peeled, then

dried and conditioned by steaming. Douglas fir is not steamed, but conditioned by another process. Preservative is forced into the wood under pressure. The retention of preservative is the key to adequate performance and the standards are very explicit about this quantity for various applications.

Dual treatment combines water-borne metallic salts (which deter certain specific marine borers that are immune to creosote) with creosote that deters other types.

It is obvious that unprotected timber piles are in jeopardy when installed in salt or brackish water. In addition, there have been effective means developed to prolong their life. The identification of the problems and the specification of protection is a part of the design process for these structures. Local experience and practice combined with the recommendations of those who know preservation treatment best is an appropriate method of handling the problem.

2.1.3.6. Specialty Woods

While the main theme of this chapter concerns foundation piling, timber has been used extensively for waterfront construction involving piles that function as foundation piles but also as part of the structure itself. Examples of this are timber piers and docks, trestle bents, shear fences, groins, jetties, dolphins, fenders and others. It was considered important to include some discussion of the specialty woods, particularly Greenheart. There are several origins for wood with this name.

Demerara Greenheart (Nectandra Rodioei) is so named because of the place where it is grown, which is the Demerara River area of Guyana, South America. The wood is characteristically very dense and consequently up to 3 times as strong as Douglas fir and Southern pine in bending and compression. The wood exhibits a very good resistance to marine borers and decay. This wood contains an alkaloid substance that deters marine organisms from attacking, but if attacked, the very dense heartwood is an additional deterrent to the borer. Demerara Greenheart piles are available in lengths up to 75' (22.9 m) Greenheart piles have not been commonly used for pure foundation pile because of cost, supply limitations and the fact that the longevity advantage is not generally a factor in underground installations.

It is said that the cost of Greenheart is about equal to properly creosoted conventional piling timbers. The resistance of Greenheart to attack by some of the more ferocious borers, particularly in warm or tropical waters is not well documented at this time.[10]

The service life of all wood piles can be expected to be curtailed in tropical waters, as is the case of most piling materials.

Other specialty woods, some bearing the general name "greenheart" are also imported for marine piling application and fendering systems. The designer who is contemplating the use of specialty piling woods should consult the supplier to insure that the wood he will get will fulfill the requirements of his design.

For reference, Table 2-17 shows test results of tests of Demerara Greenheart performed in 1987 in accordance with ASTM D-143.

Table 2-17 Test Results for Demerara Greenheart[11]

Property	Ultimate Strength, ksi	Allowable Strength, ksi
Bending Strength	8.3	3.943
Tension Parallel to Grain	7.55	3.586
Compression Parallel to Grain	10.75	5.403
Compression Perpendicular to Grain	8.85	2.159
Shear Parallel to Grain	0.95	0.634
Modulus of Elasticity	2990	2990
Unit Weight, kips/ft^3	0.0654	0.0654

2.1.4. Plastic Piling

Although, as shown above, both concrete and wood piling have features to prevent both decay and environmental degradation, both have limitations in this regard. To address these problems, especially for use with dock piling and marine fender systems, round recycled plastic piles have been developed. These range in diameter from 8" (203 mm) to 23 1/4" (590 mm), and can be made in lengths up to 120' (36.6 m). They are manufactured from recycled

Table 2-18 Pile Type Selection
Subsurface Conditions and Recommended Pile Characteristics

Typical Problem	Recommendations
Boulders overlying bearing stratum.	Use heavy nondisplacement pile with a point and include contingent pre-drilling item in contract.
Loose cohesionless soil.	Use tapered pile to develop maximum skin friction.
Negative skin friction.	Use smooth steel pile to minimize drag adhesion, avoid battered piles. Use bitumen coating for piles.
Deep soft clay.	Use rough concrete piles to increase adhesion and rate of pore water dissipation.
Artesian Pressure	Caution required for using mandrel driven thin-wall shells as generated hydrostatic pressure may cause shell collapse: pile heave common to closed-end pile.
Scour	Do not use tapered piles unless large part of taper extends well below scour depth; design permanent pile capacity to mobilize soil resistance below scour depth.
Coarse Gravel Deposits	Use prestressed concrete piles where hard driving is expected in coarse soils. Use of H-piles in these deposits often results in excessive pile lengths.

Table 2-19 Pile Type Selection
Pile Shape Effects

Shape Characteristics	Pile types	Placement Effect
Displacement	Closed end steel pipe	Increase lateral ground stresses.
	Precast concrete	Densify cohesionless soils, remolds and weakens cohesive soils temporarily. Setup time for large pile groups in sensitive clays may be up to six months.
Nondisplacement	Steel H	Minimal disturbance to soil.
	Open End Steel Pipe	Not suited for friction pile in coarse granular soils because piles have a tendency to "run"
Tapered	Timber Thin-Wall Shells Monotubes	Increased densification of soil, high capacity for short length in granular soils

plastic but, like concrete piles, have a reinforcing cage that can be steel, fiberglass, or a combination of the two. A pipe in the centre of the pile, in which case the pile is hollow, can also reinforce them. Their usual applications are marine piling, marine camels and marine lumber. Newer versions of these piles can also have a square cross section. They are designed to withstand both axial and lateral loads, including ship impact. They can be installed with any type of impact hammer.[12] An example of these piles in use is shown in Figure 2-15.

Figure 2-15 Plastic Composite Piling in Marine Application

2.1.5. Selection of Pile Type

General guidelines for the selection of a pile type are shown in Table 2-18, and Table 2-19. Table 2-18 provides pile type recommendations for various subsurface conditions. Table 2-19 shows the placement effects of pile shape characteristics. These are only general guidelines; specific project conditions and requirements may alter these substantially.

In addition to the considerations provided in the tables, the problem posed by the specific project location and topography must be considered in any pile selection process. The following are some of the problems usually encountered:

- Driven piles may cause vibration damage.
- Remote areas may restrict driving equipment size and, therefore, pile size.
- Local availability of certain materials may have decisive effects on pile selection.
- Waterborne operations may dictate use of shorter pile sections due to pile handling limitations.
- Steep terrain may make the use of certain pile equipment costly or impossible.

Although one pile type may emerge as the only logical choice for a given set of conditions, more often several different types may meet all the requirements for a particular structure. In such cases, the final choice should be made based on a cost analysis that assesses the over-all cost of alternatives. This would include uncertainties in execution, time delays, cost of load testing programs, as well as differences in the cost of pile caps and other elements of the structure that may differ among alternatives. For major projects, alternate foundation designs should be considered for inclusion in the contract documents if there is a potential for cost savings.

2.2. Sheet Piles

Sheet piles are by definition, structural units which when connected one to another, will form a continuous wall, generally for retaining earth or excluding water. Interlocking devices formed as part of the manufactured product provide the wall continuity. Sheet piling has been made of steel, concrete, wood, aluminum and other materials. Steel is by far the dominant choice, due to ready availability, relative strength, and ease of handling, storage and installation, although the use of other materials is increasing.

2.2.1. Steel Sheet Piling

Steel sheet piling is generally delivered to the project from the mill or from field-stocks in preordered lengths and stacked, ready for use. Individual pieces or pre-interlocked pairs are installed by driving using impact hammers, vibrators or by water jetting. Sheets are "threaded" one to another during the setting and driving operation so that a continuous, relatively soil and watertight wall is formed.

In functioning as a wall, the sheet piling acts as a beam under load and therefore must resist bending.

In certain applications, ability to resist bending is not as important but strength of the interlock is.

Sheet pile shapes have evolved over the years from simple channel sections with crude fabricated locks, to "U" shapes with integral, rolled-on interlocks, to today's wide range of high strength "Z" shapes.

A successful sheet piling may be ideally described as one having shape and strength to stand up under impact driving, containing free-sliding interlocks which permit one sheet to be continuously connected to its neighbor, durable in order to provide the desired life, and having the structural capacity to safely resist the service loads anticipated.

2.2.1.1. Applications

Steel sheet piling applications include the following:

- Artificial Islands
- Bulkheads-including dock walls for marine terminal facilities in water
- Cofferdams - temporary, in water or on land to permit excavation for and construction of permanent works
- Cut-off Walls - in connection with earth or concrete dams or dikes to retard seepage
- Dry Dock Walls
- Retaining Walls
- Seawalls, Flood Walls, Dikes, Jetties, Groins
- Navigation Lock Walls and other large navigation structures
- Mooring and Turning Cells, Dolphins.
- Barge Docks - Consisting of Individual Cellular Structures
- Bridge Protection Cells - to protect bridge piers from shipping collisions.

2.2.1.1. Types

Steel sheet piling is manufactured in three basic configurations - "Z", "U" and "straight" (flat). Historically such shapes have been "hot-rolled" products of structural mills. Like other shapes such as beams or channels, they are formed during a succession of passes through different roll stands of the mill. In the case of sheet piling, the rolled-on clutch or interlock is an additional special feature of sheet piling production.

Some producers use a cold-forming process in which hot-rolled sheet steel is fabricated into traditional sheet piling shapes manufactures some sheet piling shapes. These new additions to product availability contain interlocks that are considerably different from the hot-rolled products. Cold-formed sheeting is manufactured from a hot-rolled coil of steel, and then is slowly fed through a series of rollers, which gradually bends or forms the steel into its designated shape.

2.2.1.1.1. Z-Type Shapes

The Z type configuration for sheet piling is the strongest and most efficient. These shapes resemble wide-flange beams, having a web and two flanges. The interlocks are located out on the flanges at maximum distance from the neutral axis, and this provides a high section modulus for resisting bending moments. Because of this, Z-shapes have traditionally been used for heavier construction projects. However, they are now complemented by the arch or U shapes for lighter work, and lightweight Z-shapes have been also introduced into the marketplace.

Figure 2-16 Z-Type Sheet Piling

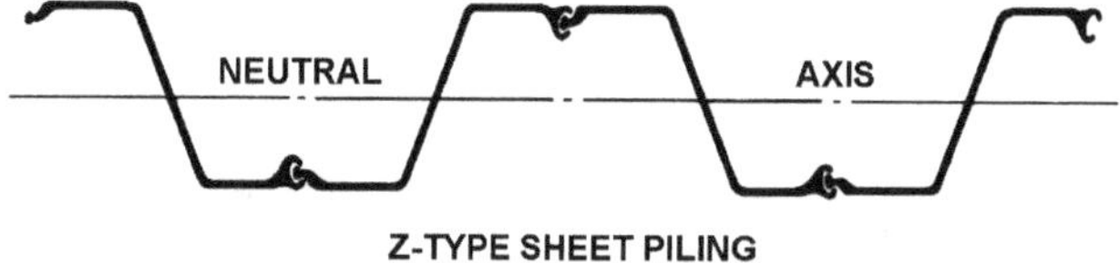

The interlocks of hot-rolled Z-type sheet piling are designed for free sliding and integrity during driving. Since most Z-piles are used to construct straight walls, there is generally no need for any guaranteed swing or deflection between sheets although there is almost always some attainable if needed, except perhaps in very long sheets. Sheet piling produced in the United States has been rolled with a "ball and socket" interlock design. Historically, it has been recommended that these be driven so that the ball of the interlock was leading. The socket then had a pre-cleared path into the ground. The importance of this recommendation has been properly questioned and many walls have been successfully installed without following this procedure. These inter locks are not designed for applications where resistance to tension is important. There are other shapes and interlocks specifically designed for tension applications, which will be discussed later in this chapter.

Although there are some general similarities in the styles, the interlocks provided on Z-piling available from European and Japanese producers vary with the producer. One manufacturer describes his lock as a "double jaw," each lock having a finger, a thumb and a socket formed by these elements. When interlocked, one thumb engages and is held in position by the fingers. Other manufacturers have designed "single-jawed" interlocks where one thumb engages the adjacent socket, and is restrained by the finger, similar to the ball and socket.

The interlocks of the cold-formed series of Z-Type shapes are best described pictorially in Figure 2-17. The forming process produces an interlocking structure on the end of each web, termed a "hook and grip."

Figure 2-17 Types of Interlocks for Sheet Piling

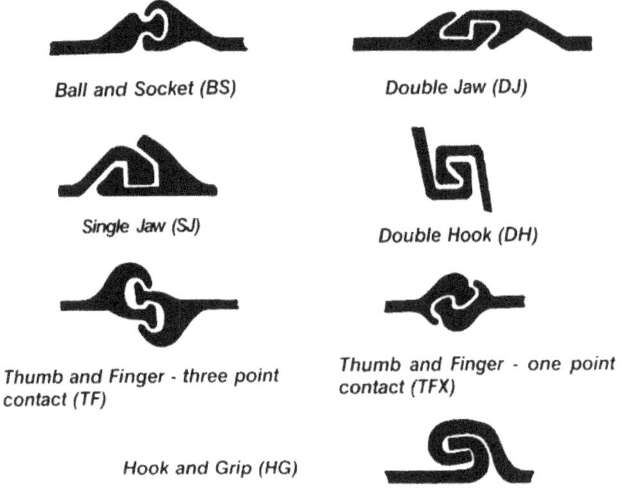

In general, whether the piling be foreign or domestic, the user should not assume sheets can be mixed on the job simply because the locks "look" the same. Tolerances may vary from producer to producer.

The interlocks of the heavier piling sections may not interlock with lighter ones from the same producer. The manufacturer should be consulted for accurate information. Specially fabricated pieces may be needed to change shapes or types in a run of wall.

Because of their inherent strength, Z-piles can be handled and shipped with less exposure to possible damage than some other style sheets. These interlocks do not lend themselves well to splicing and splices should be avoided if possible. Lengths up to 100' (30.5 m) have been produced and installed. However, these are difficult for the mills to process and ship and the manufacturer should be consulted regarding any requirement for lengths over approx. 65' (19.8 m)

Interlocks should be reasonably free sliding, that is, the pile should run to grade of its own weight when interlocked with its neighbor. The mills check interlock clearances and tolerances with templates and gauges as part of the inspection procedure prior to shipment. When "stickers" are encountered in the field, combinations of minor deficiencies such as camber or sweep in the piling length and interlocks, which may be on the edge of acceptable tolerance, could create a problem in installation.

Domestic sheet piles have been typically identified by their weight per square foot. For example, PZ-27 is a Z-type weighting twenty-seven pounds per square foot of projected wall. This is obtained by dividing the piling weight, which is 40.5 pounds per foot, by the width of each sheet, which in this case is 1.5'.

Z-piles are generally used in constructing straight wall structures - cofferdams, bulkheads and retaining walls. Sheet piling can be pulled after temporary use, the interlocks cleaned and reconditioned and the sheets used again for similar applications. There is a large business in rental sheet piling and such considerations as delivery requirements, project duration, number of uses, re-sale values must be evaluated in order to affect the decision to rent or purchase.

2.2.1.1.2. Arch Web and U-Shape Piling

These shapes resemble the hot-rolled channel sections produced on structural mills. The interlocks are formed on the web ends and interlock with their opposing mate along the centerline of the wall. These shapes are not nearly as efficient as the Z-type for the equivalent weight. For example a typical arch web piling wall constructed of PDA-27 sheets weighs the same (per square foot) as a PZ-27 wall but has only about one-third the strength based on section modulus per foot of wall. The reason for this is the location of the interlocks on the centerline of the wall. Since shear transfer across the interlocks cannot be guaranteed, the wall strength is generally based on the properties of a single sheet, which greatly reduces the strength available for design purposes.[13]

Figure 2-18 Arch Web Sheets

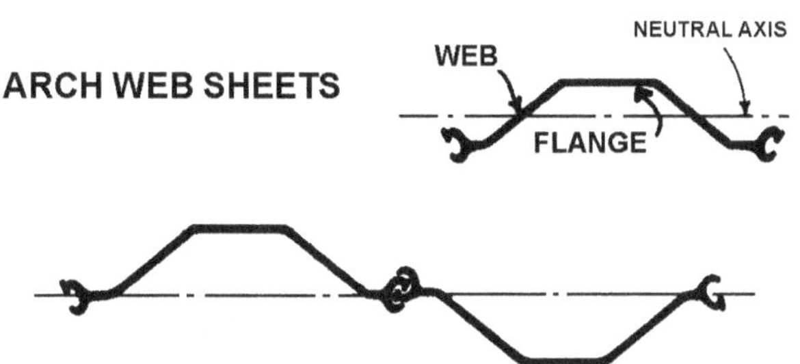

Except for very shallow-arch shapes used primarily in circular cells, hot-rolled arch web sheet piling is no longer produced in the U.S. Some arch web profiles are available from the cold-formed producers, and European and Japanese makers still offer a complete line of hot-rolled U-shapes. Despite their inefficiencies, these shapes have survived because they are somewhat easier to work with in the field than the Z-sheets. The interlocks are looser and more swing per lock can generally be obtained when needed.

Interlocks of domestic arch-web sheet piling were of the "thumb and finger" design. The thumbs of opposing sheets were threaded into the sockets formed by the thumb and curved finger providing a strong grip and one of relatively good water and soil tightness. The standard installation procedure calls for reversing every other sheet. Occasionally either accidentally or to save space, contractors have laid the sheets up in singles by using the fingers rather than the thumbs to run in the slots. It is almost impossible to hold a straight line in this manner and those who have attempted it have had mixed success.

The German engineer Larssen based interlocks of foreign U-type piling on an historical interlock design. This interlock is best described as a "double-clutch" design and is considerably less complex than the thumb and finger design of American manufacture. There has not been much recent experience with foreign U-type sheet piling in this country. This is probably attributed to the ready availability of the more efficient Z-type shapes and the refusal of domestic engineering firms to recognize shear transfer across the locks, and thus the higher published strengths claimed by some producers.

Figure 2-19 Larssen Type Interlock

Arch web and U-type sheets may be somewhat easier to splice for extending lengths than Z-sheets. In addition, since these locks are somewhat looser, there is a smaller chance of binding during the setting and driving operation when dealing with extremely long lengths.

The interlocks of the series of arch web shapes produced by cold forming are of the "hook and grip" type previously described.

Arch-web shapes have been used primarily for lighter construction, for example, trench shoring, shallow cofferdams in water or on land, light bulkheads for marinas or river port facilities, shallow retaining walls and cut-off walls (where strength is less important than interlock integrity). Large "U" shapes have traditionally been popular in other parts of the world for all classes of construction, probably because Z-type sheets are a more recent addition overseas.

2.2.1.1.3. Flat and Shallow Arch Web Types

Whereas the Z-type and Arch-web type piling are used in applications where their resistance to bending is the "primary" consideration, there is another series of piling shapes that find their application in circular, freestanding structures called Filled Cells.

The sheets used in these applications are subjected to hoop tension from internal pressure exerted by the retained soil, rather than bending. As a result, the ability to transfer this stress across the interlocks is most important and these sheet pilings have interlocks specifically designed for such loads. When used to build these large, barrel-like structures, the individual sheets strongly resemble barrel staves. They are purposely designed as flat profiles so that they will not elongate and flatten across the arch as would be the case with the arch-web or Z-sheets.

The interlocks of this group are of the thumb and finger type, as shown in Figure 2-20. Opposing thumbs are threaded into the slots formed by the thumb and finger and under hoop tension. The sheets then form a continuous circular wall that is earth tight and relatively watertight. (Hoop tension is a function of the internal soil pressure and the cell radius). The manufacturer guarantees the strength between interlocked sheets so that designs can be prepared which will provide adequate safety factors against failure. It has been found that a three-point contact interlock is somewhat stronger than the one-point contact produced by some European producers. This may be important only when striving for higher than normal interlock strengths to meet special design requirements.

Figure 2-20 Straight Web Sheet Piling Interlock Types

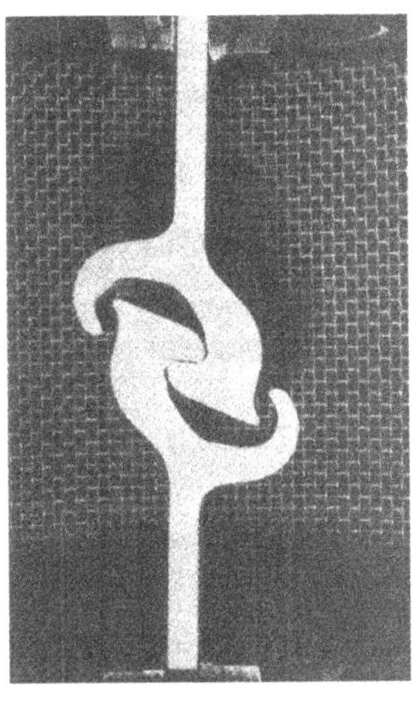

There are two basic profiles produced for these circular applications. A shape having a slight arch and identified as PSA23 by its domestic manufacturer is recommended for smaller diameter cells. The interlocks are the lightest produced for these applications but are said to provide an ultimate strength of at least 12 kips per inch. The manufacturer suggests that design stresses be held to 3 kips per inch since the sheet may stretch under additional loading.

This piling section weighs only 23 pounds per square foot and is extremely economical if it can be used. Since interlock pull is a function of cell diameter, this sheet finds application for diameters up to about 45' (13.7 m).

Several heavier sections are available with higher interlock strengths. A domestic manufacturer offers one weighing 27.5 pounds per square foot with ultimate interlock strength of 16 kips per inch of interlock. A similar shape but with slightly thicker web for corrosion allowance or hard driving is also offered at 31 pounds per square foot. An interlock with a pull of 28 kips per inch was developed several years ago to meet the demand for larger cell construction. Using high strength steel and thickening the restraining finger accomplished this. The limited market and the difficulty of manufacture apparently contributed to the decision to abandon production several years ago. Sheet piling of this grade is still available from at least one foreign producer. It should be noted that it is possible to special order interlock strengths over the standard 16 kips.

In order to turn arcs and close circles, flat or shallow arch piling is manufactured to provide a "swing" of approx. 10° between sheets. This available swing may be reduced when working with longer sheets, however, chances are if you're working with longer sheets, the cell is probably of a large diameter, and 2 or 3° may be all that is needed. Where swing between sheets will exceed the manufacturer's recommendation, bent sheets must be used to insure closing the arcs.

Filled cells built of sheet piling become large, free standing gravity structures capable of withstanding large overturning moments and sliding forces from external loads. An example of this is shown in Figure 2-21.

The flat sheets are threaded to continuously connect with each other around a circular guide template. This "barrel" is then filled with select material such as sand, sand-gravel or rock. Filled cells generally must be built on rock, hard clay or driven into sand or gravel. When used as single cells in diameters from about 15' (4.6 m) to 65' (19.8 m) they can function as artificial islands, dolphins, mooring structures and navigational lock walls.

Their primary application however is for deep-water cofferdam construction and in bulkheads, piers or other waterfront construction where the water is deep and the loads are high. In this case, individual cells are built and connected to each other with intermediate connector cells. The result is a continuous wall of steel and fill. Temporary cofferdams can be dismantled and the piling reused many times. It should be pointed out that manufacturer warranties extend to the first use only. The sheets of one manufacturer should not be randomly interlocked with another's even though they thread easily. Interlock strength is a function of the interlock dimensions. Tolerance gauges are used by the mill in their inspection procedures and are backed by laboratory pull tests on representative samples. If any sheets seem loose or sloppy with any tendency to separate, even for a short length, then all sheets should be closely inspected before use.

Figure 2-21 Cellular Cofferdam

Figure 2-24 Various Anchorage Systems for Sheet Pile Walls

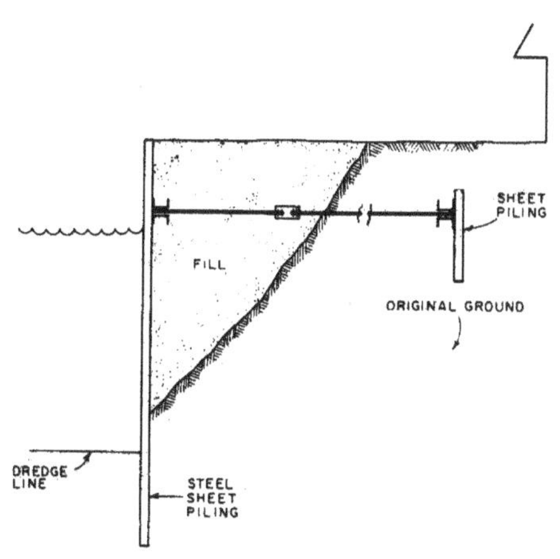

TIE RODS & DEAD MAN

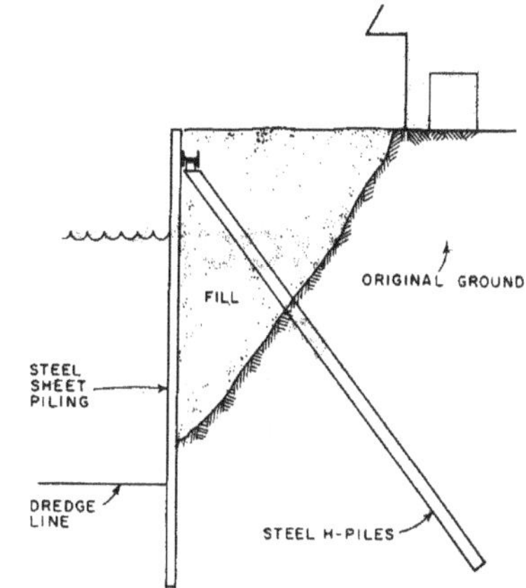

STEEL H-PILES TENSION ANCHORS

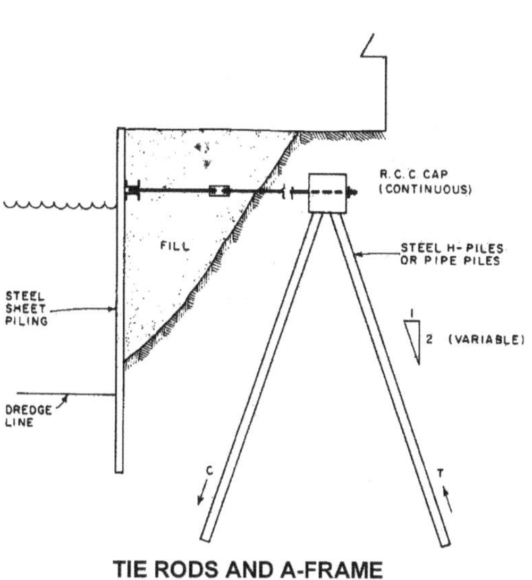

TIE RODS AND A-FRAME

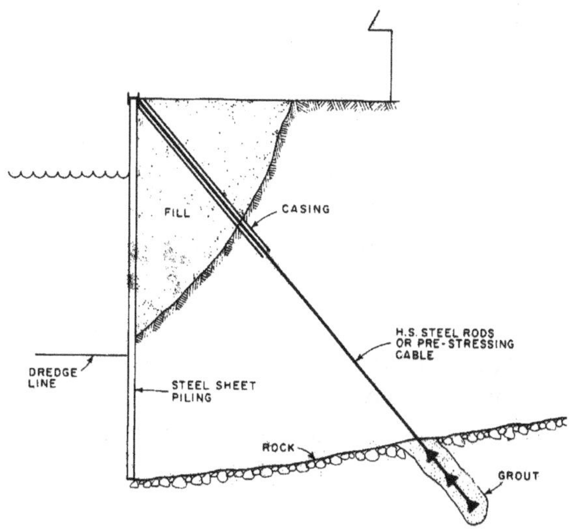

ROCK ANCHORS (OR EARTH ANCHORS)

The cold-formed process does not produce flat sheets since the required strength in tension cannot be attained with the typical "hook and grip" interlock.

2.2.1.2. Grades of Steel Sheet Piling

In the United States, sheet piling is specified by reference to the ASTM standard. The basic grade is A-328. This requires that the yield strength be a minimum of 38.5 ksi. Higher strength steels are available in ASTM Grades 50 and 60 which have minimum yields of 50 ksi (345 MPa) and 60 ksi respectively. This permits possible economies in meeting calculated bending moments by combining lighter sections with higher strength steel.

An additional grade of steel for piling has been developed for possible use where corrosion from salt water is a consideration. This grade is ASTM A-690 which is a 50 ksi (345 MPa) yield steel and also of a formulation which has demonstrated increased life in salt-water splash zones compared to regular grades.

Allowable stresses for design of sheet pile structures are generally based on about 65% yield (25 ksi for the standard A 328 grade). Proportionately higher allowable stresses of 32 ksi and 38 ksi respectively are allowed for the higher strength grades. Some increase is generally allowed for temporary stresses.

2.2.1.3. Other Wall Systems

Economic considerations as well as mill facility limitations dictate just how big a sheet piling shape can be produced and marketed, regardless of short-term demand. The mills have done a good job of keeping up with the need for sheet piling for deeper port facilities - as well as larger and deeper cofferdams. The strongest Z-type shape offered by any manufacturer (European) currently has a published section modulus of 78 cubic inches per foot of wall.

One domestic manufacturer offers a Z-type having a section modulus of 60.7 cubic inches per foot of wall. A pile this large combined with high strength steel could be expected to handle all but the most severe design requirements. To address exceptional situations, one European producer has developed a wall system employing wide-flange beams with interlock devices rolled into the flange edges. These are combined with Z-shapes to form a wall, which benefits from the use of a very efficient wide-flange beam to provide higher section moduli at less weight per foot of wall. Various combinations are offered which give a wide selection of solutions.

Traditionally, producers of U-shaped sheets have offered fabricated master piles to stiffen deep bulkhead walls and provide very large moments of inertia at the expense of weight and fabrication costs. In the United States, deep construction was generally designed around wide flange beams with the wall continuity maintained by welding pieces of piling interlocks to the beams. These structures were sometimes difficult to build although with more experience, some recent structures have turned out very well for both owner and contractor.

2.2.1.4. HZ Walls

Wide flange beams are the most efficient structural units for handling loads that result in bending moments developing. A wall constructed entirely of interlocking wide-flange beams would be very difficult to construct and would be inefficient from the standpoint of wall face coverage. A system has been developed which combines wide flange beams with Z-type sheet piling to provide a wider range of section moduli. The basic layout combines one wide flange with two light Z-shapes as the basic unit as shown in Figure 2-22 and Figure 2-23. By including wide flange in the unit, elevated section moduli, well above anything attainable from plain sheets, can be realized. The system is not quite as efficient as a conventional wall when bending moments are less than about 2 million inch-pounds, since these can be handled with "as-rolled sheets" and high strength steel. An alternate to the proprietary HZ system for deep-water design is to utilize wide flange beams to stiffen a conventional Z-pile wall. Either the beams are welded to pairs of Z-sheets or portions of the Z sheet are slit and welded to the beam flanges to provide an interlock. Very high section modulus can be obtained using these methods.

Figure 2-22 Master Pile Section

Figure 2-23 Master Pile Wall

2.2.1.5. Anchor Systems

The top support for the sheet pile wall is provided by the anchor system. Typical anchor systems are shown in Figure 2-24. The reaction obtained during the design is transferred from the wall to the anchors by structural steel wales generally fastened to the rear of the wall by bolting. Occasionally, wales are located on the outboard face of the wall but for working bulkheads, they are best placed on the inside face. Wales generally consist of two channels, back to back, with spacers.

Tie rods, spaced at regular intervals, extend from the face wall through the wales to an anchor wall or anchor piles to the rear. Depending on the loads per rod, tie rod spacing is generally an even multiple of the driving distance (width) of a pair of sheet piles to facilitate installation. Tie rods are fabricated from steel bars, the ends upset to provide additional metal at the threads, and assembled on the job with holding plates, washers, nuts and turnbuckles. Tie rods should be located as close to the low water elevation as possible to reduce the span between supports, but should be above water to facilitate installation. The dead man or rear anchor wall should be located well outside the influence of the active failure zone of the soil against the front wall. Tie rods should be coated and wrapped to protect them from corrosion losses and protected from overstressing due to ground settlement.

Other methods of anchoring bulkheads include rock or earth anchor systems and H-pile tension piles. Earth anchors are useful where there is not sufficient space behind to install a more conventional tie rod system. Earth anchors utilize slant-drilling techniques to install high strength steel rods or cable between the wall and rock or stable soil at some lower elevation. The anchor is grouted in place and a pre-test of its capacity is possible during the prestressing or post-tensioning phase. Steel H-piles may be driven on a batter and fastened to the wale system. They function as tension piles.

2.2.1.6. Temporary Cofferdams

Steel sheet piling is virtually the only means for a contractor to keep water out of a construction site while the permanent structure is built in the dry. On land, other methods such as slurry walls and steel soldier beam walls have replaced sheet piling for temporary construction. The exception would be where ground water must be retained or to comply with legal requirements for safety.

Straight wall cofferdams are designed using techniques similar to bulkheads and retaining walls. The cofferdam consists of a closed square or rectangular wall of sheet piling. An internal system of wales and struts provide support. On land, earth or rock anchors keep the inside of the cofferdam clear for working ease. Out in the water however, conventional bracing systems of walers and struts are still necessary.

Figure 2-25 Sheet Piling Box Cofferdam

In river cofferdams, external water pressure is the dominant load and removes a lot of uncertainty from the load assumptions since the loading is triangular at 62.4 pounds per foot of depth. Below excavated elevation, the pressure consists of submerged soil and water, which increases the pressure. The sheeting is designed as a beam on multiple supports. Loading conditions both during construction of the cofferdam and after dewatering must be considered. Stability of soil under and in front of the wall should also be examined for seepage effects. Internal bracing is spaced so that the sheet piling will not be overstressed. Moment calculations can be based on continuity over the supports. Reactions from the sheeting provide loads on the wales, which are designed as beams on supports. Loads on wales should include end thrust from the other members in the tier. Cross struts are designed as columns and located so that they do not seriously interfere with the work inside.

Circular cofferdams with internal ring wales and no cross struts have sometimes been constructed where it has been necessary to keep the interior clear. Wales have been fabricated of steel or have been of cast in-place reinforced concrete.

Several serious collapses of steel ring-wale braced cofferdams in water have occurred due to incorrect assumptions regarding the nature of stress distribution in the wales. Experienced designers should work on these applications.

2.2.2. Aluminum Sheet Piling

2.2.2.1. Introduction

Aluminum sheet piling has been available since 1969 in various forms and has had an excellent success rate during this period in both salt and fresh water environments. As many as seven different companies have produced piling sections in various sizes, shapes and thicknesses. Examples of aluminum "Z" sections are shown in Figure 2-26.

Figure 2-26 Aluminum "Z" Sections

There are several important questions to be addressed when the choice of an aluminum sheet pile wall is being considered:

- It is strong enough?
- How long will it last?
- Will it look good?
- Is it functional?
- What are its initial costs and total costs over its life?
- Is it acceptable to the owner?

One of the most significant advantages of aluminum sheet piling systems is its light weight. It has one of the most efficient strength to weight ratio of any type of building material. The ease of handling the relatively lightweight sheets, cap and hardware is a pleasant surprise to most contractors who are new to the use of aluminum piling. It allows the installer to work in tight spots that otherwise might be impractical from a cost standpoint with other types of piling materials.

Historically, almost 90% of all aluminum piling applications have been in a saltwater environment, and generally without protective coatings. If care is taken in properly applying the material to the site, protective coatings will normally not be required.

There are five important areas, which should be thoroughly addressed when considering aluminum sheet piling.

1. Material Specification,
2. Corrosion,
3. Construction Suggestions,
4. Design Principles, and
5. Engineering Data.

2.2.2.1. Material Specifications

The following material specification covers the mechanical properties of the aluminum allows used in the sheeting, wale, cap, tie rods, anchors, corner extrusions and fasteners in marine retaining wall applications.

2.2.2.1.1. General

Tolerances should conform to the specifications listed in "Aluminum Standards and Data", Fifth Edition. Typical safety factors, except for anchor rods and clips, as recommended in the Aluminum Association's "Specifications for Aluminum Structures - Section 1" are 1.95 for ultimate strength or 1.65 for yield strength.

The welding filler used on all wall system welds shall be alloy 5356 in conformance with the American Welding Society's Specification A5.10 and with chemical composition in accordance with "Aluminum Standards and Data".

2.2.2.1.2. Sheeting, Bracing Assemblies, Wales, Cap, Backing Beams, Shims, Tie Rods, Rod Shims, Wale Clips

All material shall be made from aluminum alloy 6061-T6. The chemical composition shall conform to American Society for Testing Materials, ASTM, designation B 221 alloy 6061-T6, shown in Table 2-23. The mechanical properties as given in Table 2-20 shall be met.

Table 2-20 Aluminum Alloy ASTM B221 6061-T6

Thickness, Inches	Minimum Ultimate Tensile Strength, ksi	Minimum Yield Tensile Strength, ksi	Minimum Elongation in 2 inches, percent
< 0.125	38	35	8
0.125 - 1.000	38	35	10

The sheeting shall be furnished in standard sizes to permit assembly in uniform increments as shown on the plans. The sheeting shall have a minimum section modulus of _____in^3/LF of wall and shall have a minimum constant thickness of _____ inches.

2.2.2.1.3. Tie Rods, Bolts and Nuts for Connections

Tie rods and bolts for connections shall be of the diameter specified by the Engineer. Threads shall be American Standard Course Thread Series, Class 2, Free Fit.

Aluminum bolts and nuts material shall conform to either the chemical requirements of the ASTM designation B 221 alloy 6061-T6, as provided in Table 2-20, or to the ASTM designation B 211 alloy 6061-T6, as provided in Table 2-21. The allowable design tensile stress shall be 18,000 psi on the root area. The bolts may be sampled and tested before erection or may be accepted on the manufacturer's certification.

Stainless steel type 18-8 (300 series) bolts and nuts, of the same diameter as the aluminum bolts and nuts may be substituted in lieu of aluminum bolts and nuts. This material shall meet ASTM specification A 193B8.

Table 2-21 Aluminum Alloy ASTM B211 6061-T6

Thickness, Inches	Minimum Ultimate Tensile Strength, ksi	Minimum Yield Tensile Strength, ksi	Minimum Elongation in 2 inches, percent
0.125 - 8.000	42	--	10

2.2.2.1.4. Alternate Sheeting, Cap Insert, Corner Extrusions

Alternate anchor sheeting sections and any alternate sheet piling sections specified by the engineer, cap inserts and corner joints shall be furnished in aluminum alloy 6063-T6. The chemical composition shall conform to ASTM designation B 221 alloy 6063-T6, shown in Table 2-23. The mechanical properties as given in Table 2-22 shall be met.

Table 2-22 Aluminum Alloy ASTM B221 6063-T6

Thickness, Inches	Minimum Ultimate Tensile Strength, ksi	Minimum Yield Tensile Strength, ksi	Minimum Elongation in 2 inches, percent
< 0.125	30	25	8
0.125 - 1.000	30	25	10

2.2.2.1.5. Anchor Plates

The anchor plates shall be fabricated from sheeting and backing beams as required in the plans.

2.2.2.1.6. Field Inspection and Acceptance of Parts

The field inspection shall be made by the Engineer, who shall be furnished by the manufacturer of all the wall parts, an itemized statement of the number and size of the parts in each shipment. Each part included in a shipment shall meet fully the requirements of these specifications.

2.2.2.1.7. Methods of Testing

Unless otherwise provided, chemical analysis, when required, shall be in accordance with Standard Method E34 of the ASTM except when suitable spectrographic analysis may be employed.

Table 2-23 Chemical Composition Limits of Aluminum Materials

Chemical Element	Alloy 6061-T6, Percent	Alloy 6063-T6, Percent
Si	0.40 - 0.80	0.20 - 0.60
Fe	0.70	0.35
Cu	0.15 - 0.40	0.10
Mn	0.15	0.10
Mg	0.80 - 1.20	0.45 - 0.90
Cr	0.25	0.10
Zn	0.15	0.10
Ti	0.05	0.05
Other Alloys	Each 0.05	Total 0.15

2.2.2.2. Corrosion

Probably the most common question asked about aluminum sheet piling is "will it work in this environment?" The following information regarding aluminum's corrosion resistance is provided below and should be helpful in determining if your site is suitable for the use of aluminum piling.

2.2.2.2.1. Introduction to Marine Aluminum

Aluminum alloys have been selected as materials of construction in many fields because of their ability to resist corrosion. Aluminum's ability to resist corrosion by atmospheric weathering has been well demonstrated by its application in agriculture, industrial and residential roofing, siding and other building materials for many years. The use of aluminum for storage tanks, tank cars, heat exchangers and other process equipment is ample evidence of its resistance to corrosion by chemicals and food products. Aluminum's resistance to corrosion both by fresh and salt waters can be shown by its many applications in ships, pleasure boats, irrigation pipe, heat exchangers, sewage disposal plants, rain carrying equipment, etc. Experience has also been gained over the years from installations of culvert sheeting and buried pipelines, which indicate that aluminum will perform satisfactorily in contact with many soils.

2.2.2.2.2. Why Is Aluminum Corrosion Resistant?

In order to have knowledge of and thoroughly understand proper installation and maintenance suggestions it is first necessary to know the mechanism by which aluminum derives it resistance to corrosion.

Although aluminum is an active metal, its behavior is stable because of the protective, tightly adherent, invisible oxide film on its surface. Even when disrupted, this film begins to re-form immediately in most environments when oxygen or air is present. The oxide is present on the surface of the cast ingot and continually reforms after being disrupted by rolling, forging, drawing, extruding or other fabricating processes.

As long as this oxide film is intact and continuous or can reform, if damaged, the aluminum metal will maintain its high resistance to corrosion. The oxide film is tenacious, hard and relatively insoluble and is therefore able to endure under a wide variety of environmental conditions. There are, however, some conditions that can lead to a breakdown or dissolution of the oxide film. Many years of study by the aluminum industry have been devoted to defining these conditions and developing means of minimizing their effect.

2.2.2.2.3. Causes of Corrosion

In most environments, the corrosion of aluminum (like that of other common structural metals) is associated with the flow of electric current between various anodic and cathodic regions. The electrochemical corrosion produced depends on the electrical potentials of these regions.

In order to investigate the electrochemistry of aluminum corrosion compared to other metals, scientists have developed a test solution, which can be used to establish the potential difference between aluminum, its alloys and dissimilar metals. Table 2-24 presents the potential difference of aluminum, its alloys and other metals.

Table 2-24 Electrode Potential of Several Metals Measured under Different Conditions

Metal	Standard Electromotive Series (volts)	Static Seawater (volts)	Flowing Seawater 13 ft. per sec. (volts)
Magnesium	– 2.34	– 1.45	
Aluminum	– 1.67	– 0.74	
Zinc	– 0.76	– 0.80	– 1.03
Iron	– 0.44	– 0.5	0.61
Hydrogen/Platinum	0.00	0.00	
Copper	+ 0.35	+ 0.80	– 0.36
Silver	+ 0.80	+ 0.12	– 0.13
Stainless Steel Type 316			– 0.05
Platinum	+ 1.20	+ 0.4	+ 0.15

Although the potential differences shown are useful in predicting the possibility of galvanic corrosion, they are only a guide. To establish the actual potential difference between aluminum and some dissimilar metals under actual project conditions, a potential measurement in the solution actually used in the intended application must be made because the potential difference depends upon the electrolyte, as the table indicates. Furthermore, the amount of galvanic corrosion is determined, not only by the potential difference, but also by the overall electrical resistance in the galvanic circuit. Special resistances to current flow, called polarization, can exist at the metal-liquid interfaces that are relatively large compared to the resistance of the solution.

The phenomenon of polarization accounts for the fact that even though an aluminum-stainless steel couple has a greater potential difference that does an aluminum-copper couple, the resistance at the metal-liquid interface on stainless steel is greater than on copper. Hence, the stainless steel causes less galvanic current flow from anodic regions on the aluminum than does copper. In actual practice this means that aluminum is quite compatible with stainless steel but problems can and do arise when aluminum is coupled to copper or copper-bearing alloys in certain electrolytes.

Corrosion of aluminum (as well as other structural metals) is electrochemical in nature and involves the flow of electric current between various anodic and cathodic regions. Several major factors basic in determining this flow of current and the resulting corrosion are:

- Alloy constituents
- Metallurgical and thermal treatments
- Effect of pH
- Galvanic corrosion (dissimilar metals)
- Stray currents
- Soil resistivity

2.2.2.2.3.1. Alloy Constituents

Virtually all of the aluminum used commercially today are alloys in which the primary ingredient is aluminum metal but include additions of other metals, usually for the purpose of increasing strength and/or workability without sacrificing corrosion resistance.

The variables that influence the amount and distribution of corrosion are:

2) Composition of the micro-constituents and their location;
3) Quantity of the micro-constituents and their location;
4) Continuity of the micro-constituents and their location; and
5) Electrical potential relative to the aluminum solid solutions.

Table 2-25 shows the electrode potentials of aluminum solid solutions and constituents. Note that iron (Fe), for example, forms constituents that are cathodic to aluminum. These constituents, because they form cathodic points over which the oxide film is weak, may promote electrochemical attack of the surrounding aluminum. The same analogy may also be drawn in the case of alloys containing various amounts of copper constituents. For this reason, alloys containing these metallic additions are rarely used when resistance to corrosion is of paramount importance.

Table 2-25 Electrode Potential of Aluminum Solid Solutions and Constituents

Solid solution or constituent	Potential, v(a)	Solid solution or constituent	Potential, v(a)
Mg_2Al_3	−1.24	99.95 Al	−0.85
Al + 4 $MgZn_2$(b)	−1.07	Al + 1 Mg_2Si(b)	−0.83
Al + 4 Zn(b)	−1.05	Al + 1 Si(b)	−0.81
$MgZn_2$	−1.05	Al + 2 Cu(b)	−0.75
$CuMgAl_2$	−1.00	$CuAl_2$	−0.73
Al + 1 Zn(b)	−0.96	Al + 4 Cu(b)	−0.69
Al + 7 Mg(b)	−0.89	$FeAl_3$	−0.56
Al + 5 Mg(b)	−0.88	$NiAl_3$	−0.52
Al + 3 Mg(b)	−0.87	Si	−0.26
$MnAl_6$	−0.85		

(a) 0.1*N* calomel scale, measured in an aqueous solution of 53 g per liter NaCl + 3 g per liter H_2O_2 at 25 C. (b) Solid solution.

From these data it would appear that silicon additions could produce alloys, which would be very cathodic and cause corrosion. Conversely, when silicon and magnesium are both used as additions in the proper amounts, magnesium silicide (Mg_2Si) forms as a constituent, which, in solid solution, has very little effect on the electrode potential. Alloys 6061 and 6063, which are used exclusively in the major manufacturers' wall systems, are alloys of this type and are well known for their corrosion resistance in seawater.

Aluminum alloys containing magnesium in amounts up to about 5 % are also known to have good corrosion resistance in marine environments. Alloys 5052, 5083, 5086, and 5154 are examples of this alloy type.

2.2.2.2.3.2. Metallurgical and Thermal Treatments

Metallurgical treatments of aluminum alloys that can be used to develop desired mechanical properties also could influence resistance to corrosion. Thermal treatment and cold work processing influence the quantity and distribution of the constituents and the magnitude of residual stresses. Thus, these factors are very important influences on the type and rate of corrosion.

Commercial treatments used in aluminum producing plants assure that the alloy specified will exhibit the properties attributed to that alloy. However, subsequent working or thermal treatments applied by the customer can and often do alter the properties of the alloy. One result can be a lowering of the alloy's natural corrosion resistance. If one portion of an alloy surface receives a thermal or mechanical treatment different from the remainder of the alloy, differences in potential between these regions can result and resistance to corrosion lowered (For example, the heat due to welding). Careful selection of welding filler material must be made in order to avoid or minimize corrosion problems that might result from the heat of welding.

2.2.2.2.3.3. Effect Of pH

Generally, the protective oxide film is stable in aqueous solutions in the pH range of 4.5 to 8.5. Usually the oxide film is readily soluble in strongly acid or alkaline solutions; consequently, such solutions may attack aluminum. However, as with all general rules there are exceptions.

Aluminum alloys are used in environments such as ocean, lake, river and municipal waters. No significant correlation is known between the corrosiveness of waters on aluminum and such factors as chloride content, sulfate content, total solids, total hardness, or total alkalinity. Some generalizations can be made but a sufficient number of exceptions are found to necessitate caution in applying them.

In water with a pH of 8.5 or more, the resistance of aluminum depends primarily on the nature of the compounds causing the high pH. Service experience has demonstrated that many natural alkaline waters are compatible with aluminum.

Similarly, the extent of corrosion of aluminum in acid water depends to a large degree on the nature of the compounds causing the low pH. Acid waters containing chlorides are particularly corrosive to aluminum. Sulfate waters of low pH are also corrosive, but less severe than chloride waters of the same pH.

These general statements are not valid if the waters contain traces of heavy metals. Copper, lead, tin, nickel, mercury and cobalt compounds generally referred to as heavy-metal compounds, promote localized corrosion attack.

2.2.2.2.3.4. Galvanic Corrosion - (Contact With Dissimilar Metals)

Corrosion currents of substantial magnitude may be caused by contact between different metals in the presence of an electrolyte. In general, the behavior of the various metals can be predicted from their electrode potentials as shown in Table 2-24. The less negative metal is likely to stimulate attack of the more negative.

Although Table 2-24 can be used to predict which metal or alloy in a couple would suffer galvanic attack, the extent of this special attack cannot be predicted from the table. For instance, as cited previously, although the difference in potential between an aluminum alloy and stainless steel is greater than that between the aluminum alloy and copper, the current with the former couple is less than the latter because of polarization of the stainless steel. The table does suggest that unless the materials are plated or coated in some manner, contact of aluminum with mild steel, copper or copper bearing alloys should be avoided where possible. When it is not possible to avoid the use of these metals, they should be electrically separated from the aluminum by the use of non-metallic materials.

2.2.2.2.3.5. Stray Current Corrosion

Electric currents (either AC or DC) caused by externally generated potentials can be responsible for severe corrosion, particularly in marine and underground structures. Such stray currents can be associated with the track-return of street railways, grounding of electric generators and welding equipment or buried pipelines having induced cathodic protection. The attack occurs at the point where the stray current leaves the aluminum retaining wall to enter the soil or electrolyte. The magnitude of such stray currents and possibilities of encountering them are subjects of speculation because few factual data are available. Such corrosion is becoming less prevalent because of improved design and installation practices.

Aluminum piling has been used successfully in many marinas. Proper design of the electrical system of course is the key to prevent stray current corrosion from affecting any metals at marine installations.

2.2.2.2.3.6. Soils

The corrosion performance of unprotected, buried aluminum alloys varies considerably with the type of soil. No satisfactory classification of soils with regards to their corrosive action on aluminum has been developed. It has been assumed that the "safe" range of pH values for soil is the same as for aqueous solutions, pH 4.5 to pH 8.5, but this has not been adequately substantiated. As in aqueous solutions, the particular compounds in the soil that are causing the high or low pH is undoubtedly a factor. Some data indicate that soils in the "safe" pH range, which have a resistivity greater than 500 ohm-cm, have proven to be compatible with aluminum. Testing of soil samples and resistivity determinations can be only used as guides. Further studies are underway by various companies to further address this issue.

2.2.2.2.3.7. Clay Soils

In general, clay and organic soils are corrosive to aluminum and should be avoided. Protective coatings or cathodic protection should be added to the aluminum to provide a longer service life, if aluminum is to be used in these types of soils. The least expensive and most commonly used method is to coat the material that is in contact with the clay soils with a coal tar epoxy coating. The decision to use aluminum or not is generally left to the user when dealing with these types of soils.

2.2.2.2.4. Types of Corrosive Attack

The aluminum alloys used for all components of the system are highly corrosion resistant. It is possible, however, that under extremely adverse environmental conditions some corrosion could occur. If corrosion does occur, it most probably would be either the uniform or pitting type. These types of corrosion will be discussed briefly to aid in recognition.

2.2.2.2.4.1. Uniform Attack (Etching)

During uniform attack, the metal corrodes evenly. Such attack usually occurs in the presence of strongly acid or strongly alkaline electrolytes that simply dissolve the oxide film and prevent its reformation. The appearance of the metal being uniformly attacked may range from superficial etching and staining to rapid dissolution of the metal. Uniform attack is easy to evaluate by a measurement of weight loss or decrease in thickness. The rate of attack usually is expressed in mils per year (mpy). Etching may be a serious problem if it continues at a lineal rate.

2.2.2.2.4.2. Pitting Attack

Pits, the most common form of corrosive attack on aluminum, may form at localized discontinuities in the oxide film when aluminum is exposed to weather, fresh or salt water, or other neutral electrolytes. Depending upon the alloy composition, the quality of the oxide film and the nature of the corrodent, the pits may be minute and concentrated or can vary in size and be widely scattered. Pitting type corrosion often appears to be more severe than it actually is because the build-up of corrosion product occupies many times the volume of the metal from which it was formed. Removal of this corrosion product will often reveal corrosion of only minor significance.

The evaluation of pitting corrosion is difficult. Weight losses are of little value and tension tests can be misleading. Measurements of depth and distribution of pits made at several time intervals provide a means of determining whether the rate of penetration changes with time.

2.2.2.2.5. General Resistance To Corrosion

Since laboratory exposure tests, such as salt spray or immersion in electrolytes, are only useful for comparative information and do not necessarily predict actual service performance, actual long-term atmospheric exposure and weathering tests have been necessary.

In the past 30 to 35 years, thousands of specimens have been exposed throughout the U.S. and elsewhere. Test reports published in the literature demonstrate convincingly the excellent atmospheric weathering characteristics of aluminum alloy products in industrial, chemical, seacoast, tropical and many other environments.

One obvious phenomenon that has emerged from these long-term exposure tests is that corrosion of aluminum in these environments is "self-limiting." Whereas corrosion during the early months and years of exposure may appear to be severe, this rate of weathering decreases with time. The tendency is for the attack to proceed laterally along the surface rather than to become progressively deeper. The decreased rate of attack, as evaluated by losses in tensile strength as well as depth of attack measurements, indicates that corrosion diminishes with time over the entire surface to a very low rate. The curves in Table 2-26 show test data from many exposure sites, which demonstrate this effect.

Table 2-26 Comparison of test data and service for atmospheric exposure at seacoast and industrial sites

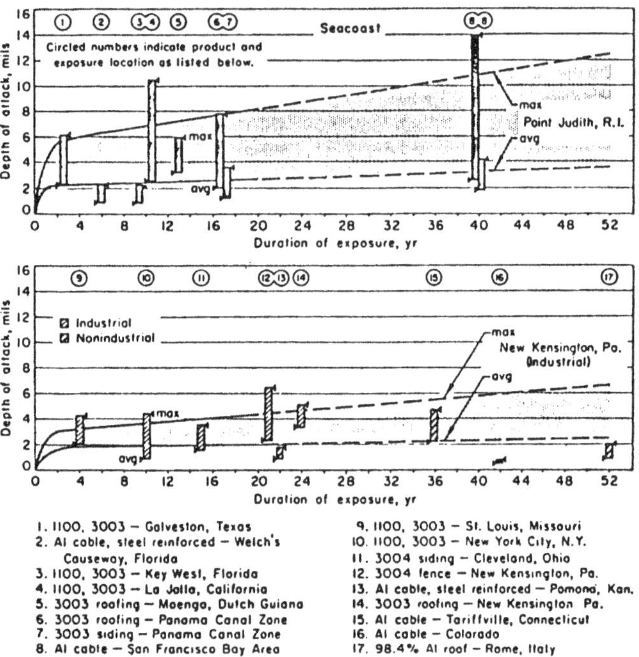

Curves for Point Judith and New Kensington are based on data obtained on aluminum alloys 1100, 3003, and 3004, extrapolated to 52 years. Data obtained on test specimens at other exposure stations and on related aluminum alloys from a variety of service conditions are shown as vertical bars on the charts. Comparison shows that performance of aluminum alloys at Point Judith and New Kensington can be used with confidence to predict performance in most seacoast and industrial regions.

2.2.2.2.6. General Suggestions

There is no absolute "safe" method for testing soils. On a "normal" homeowner lot, one can easily miss an "unsafe" area of bad soil. The best suggestion is to take soil samples at the left, centre and right sides along the intended bulkhead installation line. One sample in the centre back from the wall line should also be taken. Two water samples should be taken also at the one-third points. These suggestions assume the natural ground to be homogeneous throughout the property. If the natural soils are not homogeneous, the same procedures should be followed for each type of soil present. Heavy metals should be tested for if you suspect their presence.

If the test results show either the soil or water to be

Table 2-27 Typical Material Specifications for Pultruded Sheet Piling

Mechanical Property	ASTM Test	Property, Parallel to Fibers	Property, Perpendicular to Fibers
Tensile Strength	D638	30 ksi	7 ksi
Tensile Modulus	D638	2,500 ksi	800 ksi
Compressive Strength	D695	30 ksi	15 ksi
Compressive Modulus	D695	2,500 ksi	1,000 ksi
Flexural Strength	D790	30 ksi	10 ksi
Flexural Modulus	D790	1,800 ksi	800 ksi
Full Section Modulus of Elasticity		2,800 ksi	
Notched Izod Impact	D256	25 ft-lbs/in	4 ft-lbs/in

outside the "safe" ranges, as discussed previously, the decision is then left to the designer to either use protected aluminum, non-protected aluminum or no aluminum.

If time permits while applying for a Corps of Engineers permit, install a section of sheeting in the ground along the intended bulkhead line and remove the sample in 2 to 3 months time. This should provide sufficient evidence of corrosion for one to make a reasonable conclusion as to the suitability of aluminum for that site. Aluminum, if unsuitable for a site, will normally react adversely in that 2 to 3 month period.

It is unnerving to some potential users but there is really no safe method of testing for soils. Potentially corrosive areas can be missed during the normal testing procedures. Nothing is 100% sure in any occupation but when pH and resistivity are checked and are found within the suggested general safe guidelines, the user will find the odds in favor of a successful application to be very good.[14]

2.2.3. Vinyl Sheet Piling

2.2.3.1. Introduction

Vinyl sheet piling is a relatively new type of sheeting that can be applied in a wide variety of applications. It is lightweight, simple to set and install, resistant to environmental attack and can be configured in various colors. It was designed to be:

- Weatherable
- Corrosion Resistant
- Unaffected by Fungi and Marine Borers
- Environmentally Friendly
- Easy to Install
- Cost Effective

It is ideal for the lighter bulkheads that are typical with residential, recreational and marina construction. Vinyl sheeting can also be used for Navy walls as well.

2.2.3.2. Configuration

Vinyl sheet piling is generally manufactured by continuous extrusion. This is the same process used to form PVC pipe. The raw material, plastic resin compound, is melted and pushed through a die. This die shapes the plastic into the desired cross section. The shape is then cooled and cut to the desired length.

The individual sheets have interlocking male and female edges. The interlocking edges are extruded as part of the sheet to insure continuity.

Vinyl sheeting is available in a number of configurations. The most common configuration is a Z-sheet type of configuration, which is similar to steel piling. Such sheets are made to be driven two at a time and, as is the case with their steel counterparts, offer a high section modulus. Since it is extruded, vinyl sheet piling can have a wider variety of sections than rolled steel sheeting. Vinyl sheet pile manufacturers take advantage of this advantage and frequently include stiffening ribs and/or thickened corners. The configuration of these stiffening elements varies with the manufacturer.

Another configuration that is common with vinyl

sheet piling is an "AWL" configuration as shown in Figure 2-27. This configuration combines the high section modulus of Z-sheeting with the ease of interlock alignment of flat and U-type sheeting - in fact, it is in effect two U-sections put together. AWL sections also eliminate interlocks and thus potential leakage. When installed by vibratory drivers, however, setting the clamp arrangement for multiple sheet driving can be difficult due to the physical layout of the sheeting.

Figure 2-27 "AWL" type sheeting

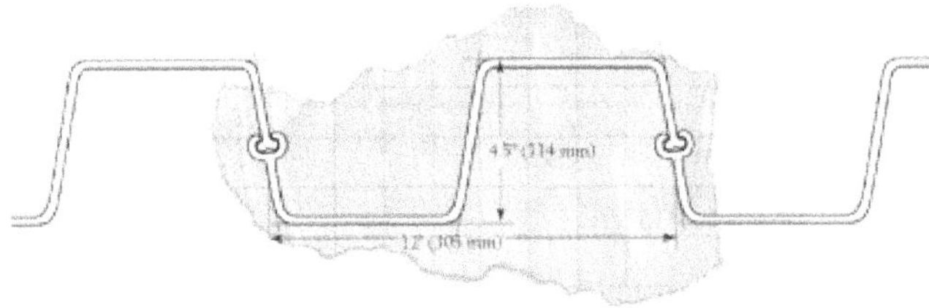

As is the case with other sheeting, vinyl sheeting requires transition pieces such as corners and intersections. These are customized for the other sheeting the manufacturer makes.

2.2.3.3. Material

Vinyl sheeting is made of a modified polyvinyl chloride (PVC), which makes it suitable for most marine environments and not subject to leaching, corrosion or similar deterioration mechanisms. The technology that has brought us vinyl siding for homes, plastic automotive parts such as bumpers and dashboards, and durable home appliances, is now being utilized to produce a sheet piling for marine retaining walls, sea walls or bulkheads. The vinyl also includes a UV stabilizer to reduce deterioration due to sunlight. Tensile strength for this material is approximately 6-7 ksi[15] (41.3-48.3 MPa), with an initial modulus of elasticity of 350-400 ksi (2.41-2.76 GPa.) It cannot be overemphasized, however, that with plastics of any kind material properties cannot be applied as simplistically as with metals. Tensile strength of the material, for instance, can vary both with the way the sheeting is loaded (purely tensile vs. flexural loads) and with time. Time variation is also significant with the modulus of elasticity. It is important to follow the manufacturer's recommendations on loading, and in many cases the manufacturer will specify a maximum moment per foot or meter of sheeting wall as opposed to allowing the designer to compute the maximum fiber stress in flexure.

The relatively low values of strength and modulus of elasticity given above are very important to consider in the design of vinyl sheet walls. Not only is it important to consider the strength of the material but also that it is subject to greater deflection than other types of sheeting as well. Designers will realize that, as support spans increase, wall deflection will become the controlling aspect of the design.

2.2.3.4. Design

The design of vinyl sheet piling is similar to that of other types of sheet piling. Since vinyl sheeting is light and commonly used in residential or other light applications it is tempting to assume that one can shortcut the design as well. Vinyl sheet piling, however, is an engineered marine product and should be treated as such. Most manufacturers of vinyl sheet piling have engineering specifications and recommendations that should be followed during the application.

The following presents several key points related to the design of vinyl sheeting:

- The type and compaction of backfill is very critical to the success of a vinyl sheet pile wall. Ideally, backfill should be free draining cohesionless soil, compacted in layers or "lifts," with an angle of internal friction of 34° or more. Cohesive soils should be avoided as backfill to avoid rupturing the sheets as they both expand with changes in water content and have very low permeability. Expansive soils or soils with high plasticity indices (such as fat clays) should be avoided as backfill.
- Weep holes should be considered to allow for drainage behind the sheets during rapid changes in the groundwater level. The success of these weep holes depends on the permeability of the soil.
- Vinyl sheet walls are almost always tied back; they are seldom cantilevered. Tiebacks must be sufficiently long to carry the load. Make sure the tieback spacing is not excessive or the purpose of the tiebacks will be defeated.
- Consider the long-term properties of the material, as they will change with time. A

factor of safety of at least 1.5 should be used when designing vinyl sheet pile walls, taking creep effects into consideration.

- Walls are generally capped with wood or concrete, although occasionally steel caps are used.
- Although sheet piling does not generally work against rock, it is especially important to avoid rock with vinyl sheet piling. As with any vinyl product, it is subject to rupture and will damage severely during installation if rock is hit.
- Use of a proper waling system is very important for a successful vinyl sheet pile wall. Single waling is suitable for lower walls while multiple waling is necessary for taller ones.
- Design should take into consideration wave and ice loading.
- The layout of vinyl sheeting is also similar to other types of sheeting. Careful consideration should be made to the sheet geometry and corner details.

2.2.3.5. Installation

Vinyl sheet piling can be installed using a variety of equipment types, which include:

- Vibratory hammers, either excavator or crane mounted. Vibratory plate compactors are also used, but these are exclusively excavator mounted. It is important when using a plate compactor to keep a steady, downward force on the pile and vibrator during driving.
- A portable air-compressor or hydraulic jackhammer with a sheet shoe. This is only suitable for short sheets and is generally used by occasional installers.
- A drop impact hammer, either land-based or barge-mounted.
- A water jet fed by a high output pump, either held by hand or suspended from a crane.

As with other types of sheet piling, vinyl sheet piling is best set before being driven. Because it is lightweight, when safety conditions permit it can be set by hand. It can also be installed with a crane or excavator if the conditions require.

The selection of an installation method is a matter of both jobsite conditions and contractor preference. However, as with any driven pile, the preparation before driving is frequently as important as the driving itself. The following is an outline of a technique for installing vinyl sheet piling, in this case a single wale design.

The optimum method of installation is to set the sheets first, then driven them. Manufacturers' recommendations for sheet pile wall installation in terms of sheet alignment, panel driving and other techniques apply to vinyl sheeting as well.

2.2.4. Pultruded Fiberglass Sheeting

2.2.4.1. Introduction

Pultruded sheet piling is another relatively new product. It has two distinct characteristics:

- It consists of very strong fiberglass roving (long extended fibers) and mat that are cured into a high performance resin. The roving imparts most of the longitudinal strength with tensile strength in excess of 500 ksi and an elastic modulus of 10,000 ksi. The tensile strength of the roving is greater than most steel. The continuous strand mat provides the material its transverse (longitudinal) strength.
- It is pulled through the die that forms its shape, not pushed through like extruded aluminum and vinyl profiles. The continuous manufacturing process makes the parts very consistent in tolerances and properties from one piece to the next.

The pultrusion manufacturing process allows for producing continuous lengths of reinforced plastic structural shapes with constant cross-sections. The process involves pulling the raw materials through a heated steel forming die using a continuous pulling device. The reinforcement materials are in continuous forms. As the reinforcements are saturated with the resin mixture ("wet-out") in the resin bath and pulled through the die, heat from the die initiates hardening of the resin. A rigid, cured profile is then formed that corresponds to the shape of the die. In the forming and curing die, the thermosetting reaction is heat activated and the composite is cured (or hardened.)

On exiting the die, the hot product must be cured to prevent cracking and deformation before the pull blocks grip it. The final product can then be cut to any length.

Other properties of Pultruded sheet piling include the following:

- Resistant to corrosion.
- Resists attack of marine borers and other destructive elements in the marine environment.
- Resists UV degradation.
- Lightweight - allows for easier installation, even setting by hand in many cases.

2.2.4.2. Material Specifications

Typical material specifications for pultruded sheet piling are shown in Table 2-27.

2.2.4.3. Sheeting Profiles

Pultruded sheeting comes in a wide variety of profiles and shapes. A typical profile is shown in Figure 2-28.

Figure 2-28 Typical Sheeting Profile for Pultruded Fiberglass Sheeting

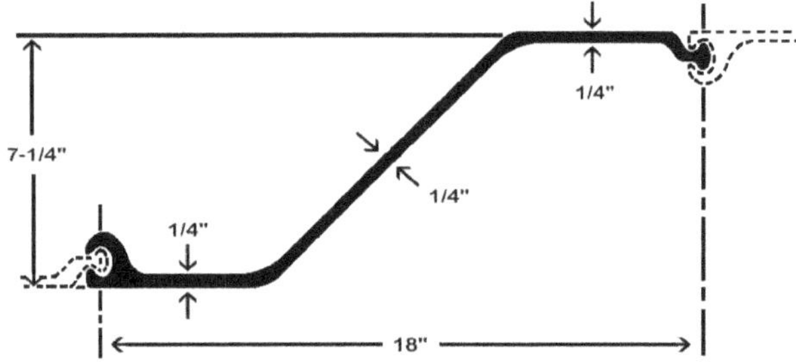

2.2.4.4. Design of Pultruded Sheet Walls

Pultruded sheet pile walls are designed using the same principles as other sheet piling materials. There are some special design considerations of these walls that need to be noted:

- Deflection. Although stiffer than other non-metallic sheeting, Pultruded sheeting is still more susceptible to deflection than steel sheeting. Designers should insure that excessive deflections do not occur. Also, the deflections shown here - and those of most methods used to compute deflections of sheet piling - do not take into consideration shear deflection, which is significant in short sections and more important in fiberglass than in materials such as steel.
- Local Buckling and Transverse Bending. In addition to the flexural stresses computed by conventional analysis methods, Pultruded fiberglass shapes - both the sheeting and the H-piles used in the wales - are subject to both compression flange buckling and lateral-torsional buckling. The former is a function of the shape geometry and the properties of the material; the latter includes the unsupported length of the sheeting. Although both of these can generally be avoided through the limitation of conventional deflection and bending stresses, the designer should check these conditions. If H-piles are used in the wales, these elements can be included in the design by use of the property tables available from the manufacturers of Pultruded H-piles, along in some cases with discussions of these conditions in general.
- Interlock strength. The transverse strength of the material is considerably less than the longitudinal strength. Applications such as cofferdams where interlock strength is critical should be avoid with Pultruded sheeting.
- Compaction. The type and compaction of backfill is very critical to the success of a Pultruded sheet pile wall. Backfill should be free draining cohesionless soil, compacted in layers. Cohesive soils should be avoided as backfill to avoid rupturing the sheets when they expand with changes in water content.
- Weep holes should be drilled into Pultruded sheeting to allow for drainage behind the sheets during rapid changes in the groundwater level. Weep holes are more effective if the soil is cohesionless and has high permeability.
- Factors of safety should be similar to those used with other sheet pile materials.
- Although no sheet piling does really well against rock, it is especially important to

avoid rock with Pultruded sheet piling.

The layout of Pultruded sheeting is also similar to other types of sheeting. Careful consideration needs to be made to the sheet geometry and how it comes out in corners.

2.2.4.5. Modifying Pultruded Sheeting

Pultruded sheeting is furnished cut to specified length. In most cases, however, it will be necessary to cut, drill or tap the material to complete the installation.

2.2.4.5.1. Important Notes when Cutting on Pultruded Sheeting

1. Observe common safety precautions. For example, the operator of a circular power saw should wear safety glasses to protect his eyes as well as an OSHA approved dust mask.
2. A coverall or long sleeved shirt will add to the operator's comfort during sawing, machining or sanding operations. Although the dust created is non-toxic and presents no serious health hazard, it can cause skin irritation. This can be reduced or eliminated by use of a protective cream and/or the wearing of proper attire when cutting.
3. Always provide adequate support to keep the material from shifting when making a cut. Without adequate support fiberglass reinforced profiles can shift and may cause chipping at the cut edges. Proper support will also prevent any warping.
4. When cutting and drilling, use light evenly applied pressure. (Avoid excessive pressure!) Heavy pressure tends to clog the blade teeth with dust particles shortening the cutting life of the blade. In addition, cutting speed is a critical variable. If the edges begin to fray, slow the cutting speed. In addition, too much force can rapidly dull the tool. Diamond or carbide grit edge saw blades, carbide tip drill bits and carbide router bits are recommended.
5. Water-cooling is desirable when cutting numerous pieces or when thick cross sections are being sawed. With cooling, cutting speeds increase, smoother cuts result, and dust is often eliminated.
6. Do not generate excessive heat in any machining operation. Excessive heat softens the bonding resin in the fiberglass - resulting in a ragged rather than a clean-cut edge. Excessive heat can also burn resin and glass.
7. Shearing and/or punching are not recommended.

2.2.4.5.2. Sawing or Cutting

The cutting or sawing of pultruded sheet piling, cap or walers can be accomplished quickly and accurately with a circular power saw. A table or radial model is better than a portable hand model because of the built-in rigidity and guides, which insure accurate cuts. However, a hand model utilized by a capable individual is effective.

For infrequent cutting with a circular power saw, a metal blade with coarse, offset teeth can be used satisfactorily. For frequent cutting, a masonry saw blade - preferably carbide tipped - will give accurate cuts and reasonably long blade life. For production cutting, use a 60 to 80-grit diamond tipped blade for best results.

When sawing relatively few pieces, a disposable-blade hacksaw (24 to 32 teeth per inch) is suitable. Although an ordinary carpenter's saw can be used, frequent resharpening makes this tool less desirable.

One problem that may be encountered with a circular power saw is that larger pultruded sections (e.g. walers/H-piles) cannot be cut in one pass due to the blade vs. wale size. This problem does not exist with Composite Z™ Sheet Piling. However, larger wale sections can be sawed in two passes by cutting halfway through from one side, inverting the material or saw and cutting from the opposite side. Obviously this would be best performed prior to placement of the waler however can be accomplished in place by a skilled carpenter.

If the cross section is too large for the circular saw "two-pass method," or if large sections are being sawed in quantity, use a power band saw with a carbide or diamond tipped blade - preferably a machine with automatic feed to insure a light, even pressure on the blade.

2.2.4.5.3. Drilling

Any standard twist bit is a good tool for drilling pultruded sheet piling, caps and walers. Carbide tipped drills are recommended when cutting large quantities. Drill speeds should be approximately equivalent to those used for drilling hardwood. When drilling large holes, a backup plate of wood will prevent the hole from breaking out on the backside.[16]

2.2.4.5.4. Threading and Tapping

Threading and Tapping of fiberglass reinforced material is not recommended as a means of fastening when high strength is required, and should be avoided in the design of fabricated components whenever possible. The threading operation cuts the continuity of the glass fibers and leaves only the sheer strength of the resin component to provide the strength of the thread. Although threaded connections have been used satisfactorily where strength is not an important consideration (e.g. mounting of "signs" or lightweight "lighting components", etc.) it generally requires a bonding of the threaded connection with an epoxy or urethane in to improve the strength of the connection.

2.2.4.5.5. Sanding

Open grit sandpaper on a high speed sanding wheel gives best results. Use very light pressure – do not force the sander against the fiberglass surface because heavy pressure may heat up and soften the resin. Wet sandpaper applied by hand or with an orbital sander will produce a high gloss finish if desired.

2.2.4.5.6. Grinding

Grinding is generally not recommended on composite shapes. In ordinary grinding operations, the dust tends to load the stone and stop the grinding action. If grinding is required, use a coarse grit wheel and water as a coolant.

2.2.4.6. Mechanical Fastening

2.2.4.6.1. Bolted Connections

Using standard bolts, nuts, and washers can make satisfactory connection between pultruded fiberglass sheet piling and composite cap and wale components[17]. In marine applications these are generally galvanized or stainless. Since fiberglass materials can fail under high, localized stress conditions, such as those encountered around a bolt, it is important to use the proper washer. Remember, the strongest joint between pieces of composite shapes is obtained by using properly fitted bolts and a larger diameter washer to help distribute the load. This is extremely important particularly with regard to the wale and tie-rod installation[18].

2.2.4.6.2. Nailed Connections

Nailing is a satisfactory way of fastening Pultruded shapes to wood and to other materials that provide enough grip to hold the nail. Although, common nails can be driven through 1/16" thick composite shapes without re-drilling holes – tempered nails are generally required when penetrating 5/16" thick material. Fiberglass heavier than 5/16" requires pre-drilled holes, slightly oversize, to admit the nail and to allow for expansion and contraction between the fiberglass and the material. It is also advisable to pre-drill slightly oversized holes before nailing long lengths of lighter fiberglass sections. One important note, never nail fiberglass to fiberglass.

2.2.4.6.3. Screwed Connections

Self-tapping screws have been used successfully in many applications involving mechanical connection when high strength fasteners are not required. A better use of self-tapping screws is in combination with adhesives. In this application the screws can serve to hold the adhesive bonded surface of the two parts together while the adhesive cures in addition to contributing limited mechanical strength to the connection. Appropriately sized pilot holes should be provided in the Pultruded shape for the screws. In corrosive environments, galvanized, stainless steel or Monel screws should be used. If corrosion is still a concern, a suitable coating of epoxy or urethane can be applied to the exposed screw heads to aid in the prevention of rusting. Lag screws are not recommended because they do not bite well in the fiberglass.

2.2.4.6.4. Bolting Into Tapped Holes

Mechanical fastening can be accomplished with screws into tapped holes, however, the properties of tapped holes are not good nor will the connection be strong. For removable cover plates, sheet metal screws can be used. The strength of the connection can be improved by use of threaded inserts bonded into place with suitable adhesives.

When removable bolts are required: threaded metal inserts of fasteners should be installed in the fiberglass and preferably bonded in place with a suitable adhesive. Fiberglass threads can wear out quickly and may not provide sufficient strength. Many types of metal inserts and fasteners are commercially available to eliminate this concern.

2.2.4.6.5. Riveted Connections

"POP" Rivets are very effective in joining pultruded fiberglass sections or attaching certain accessories. Rivets are available in various sizes and head styles in aluminum, steel, Monel, copper, and stainless steel. Those materials subject to corrosion should not be utilized in a marine environment. Other types of rivets, such as Drive Rivets, those formed by a rivet gun or the conventional rivet formed with a ball peen hammer, can produce an effective mechanical connection. The strength of the connection can also be improved with suitable adhesives. The riveting of any accessory to a pultruded shape is subject to weight, load and/or strength limitations. As a result only lightweight accessories such as signs, lightweight lighting components or conduit should be attached using a riveted method. Backup washers are recommended for distributing load stresses. As in drilling operations, it is necessary to use a slightly larger drill than the exact diameter of the rivet. For a 1/8" rivet, use a No. 30 drill rather than a 1/8" drill.

2.2.4.7. Wales, Tiebacks and Caps

2.2.4.7.1. Caps

All permanent sheet pile walls should be capped when installation is complete. Although this can be done with wood or concrete caps, a more sensible solution is the use of pultruded sheeting cap that is customized to fit the sheeting profile. A drawing for this is shown in Figure 2-29.

Figure 2-29 Pultruded Sheet Pile Cap

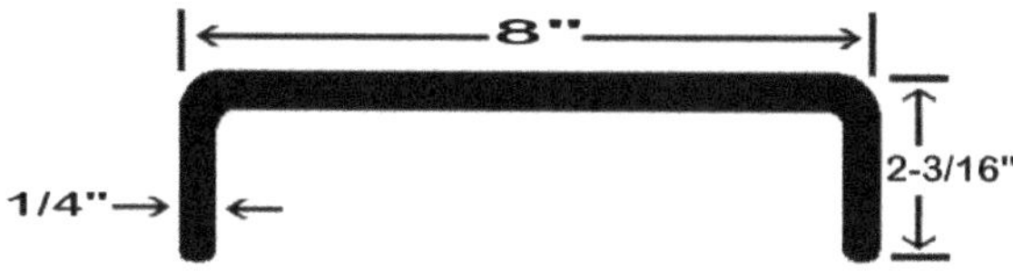

2.2.4.7.2. Wales

Most installations of Pultruded sheet piling will include some type of additional support for the wall. Although this support usually includes a tieback system, it always will include wales. Proper waling is very important in successful use of pultruded sheet pile walls. Low wall heights can be strengthened with single waling but taller walls require multiple waling.

The most suitable waling for this application is pultruded H-piles, which avoid the deterioration and environmental difficulties of wood wales. These H-piles come in several sizes and are configured according to the design requirements. Specifications are shown in Figure 2-30.

Figure 2-30 Pultruded Wale Specifications

SIZE	A, in.	B, in.	C, in.	I_{xx}, in^4	I_{yy}, in^4
12" x 1/2"	12	1/2	1/2	452.45	144.11
10" x 1/2"	10	1/2	1/2	256.20	83.42
10" x 3/8"	10	3/8	3/8	198.53	62.54
8" x 1/2"	8	1/2	1/2	126.96	42.74
8" x 3/8"	8	3/8	3/8	99.19	32.03
6" x 3/8"	6	3/8	3/8	40.17	13.52

Although single wales are sometimes acceptable, in some cases double waling - or more -- is required.

2.2.4.7.3. Tiebacks

Tiebacks are essential to resist wall overturning when the wall is sufficiently tall. There are three recommended options to tie back Pultruded sheet pile walls:

- Conventional galvanized or stainless steel tie-rod, washer and bolt systems anchored into a properly engineered "deadmen", tieback wall, pile or other anchored device when properly engineered will be suitable. These consist of three basic components;

- o Tieback rod, which is generally a threaded, galvanized stainless steel rod cut to suitable length. For maximum corrosion resistance, an 18-8 stainless steel (300 series) should be used for these rods. Remember that these rods are completely buried so they are subject to deterioration due to groundwater, soil pH and other environmental agents.
- o Washers, bolts and other spacers to connect the tiebacks to the wales and thus the wall. These are threaded onto the tieback rods and tightened after suitable holes are drilled in the wall and wales.
- o Anchor for the tiebacks. This can be an anchor plate, pile or other type of anchor, depending upon the soil conditions and tieback pull loads. The tie rods are connected to the anchors.

- Manta Ray Anchor system, manufactured by Foresight Products, LLC: This anchoring system (similar to a "toggle bolt") has a "hydraulic/load locker" device, which can give an immediate proof test of the installed anchor.
- A.B. Chance Company, a Hubbell Company: The A.B. Chance helical tieback anchor (screw anchor) for tieback applications has been used throughout the industry in retaining wall, seawall, and bulkhead applications. It has a long history of successful applications. As with the Manta Ray system, it also can provide an immediate true load test of the anchor.

The design and spacing of the tiebacks depends upon the loading requirements. Spacing of the tiebacks is also influenced by the rigidity of the wales or top cap. If tiebacks are spaced too far apart, the wales and thus the wall will excessively deflect.

Any of the above three options can provide for a structurally sound tied back anchoring system when properly engineered. A retaining wall, seawall, or bulkhead is only as strong as its properly engineered wale and tieback system.

2.2.4.8. Installation of Pultruded Sheet Pile Walls

Installation of Pultruded sheet piling is similar to that of other types of sheet piling. Pultruded sheet piling can be installed using a variety of equipment types, which include:

- Vibratory hammers, either excavator or crane mounted. Vibratory plate compactors can also be used, but these are exclusively excavator mounted. It is important when using a plate compactor to maintain a steady, downward force on the pile and vibrator during driving.
- A portable air-compressor or hydraulic jackhammer with a sheet shoe.
- A drop impact hammer, either land-based or barge-mounted.
- A water jet driven by a high output pump, either manually held or suspended from a crane.

As with other types of sheet piling, Pultruded sheet piling is best set before being driven. Because it is lightweight, when safety conditions permit it can be set in place by hand. It can also be installed with a crane or excavator if the conditions require.

The selection of an installation method is a matter of both project conditions and contractor preference. However, as with any driven pile, the preparation before driving is frequently as important as the driving itself.

After the sheeting is driven, the wales, tiebacks, caps, etc. are to be installed. Make certain all recommendations in this guide concerning cutting, drilling, tapping and mechanical connections are followed during any or all of these operations.

2.2.5. Wood Sheeting

2.2.5.1. Types of Wood Sheeting

In its simplest form, wood sheet piling can consist of a single line of boards or "single-sheet piling" but it is suitable for only small excavations where there is no serious ground water problem.

Figure 2-31 Single Sheet Piling

In saturated soils, particularly in sands and gravels, it is necessary to use a more elaborate form of sheet piling which can be made reasonably watertight with overlapping boards spiked or bolted together, such as the "lapped-sheet piling" or "Wakefield" system. Wakefield sheeting consists of three planks 2" (50.8 mm), 3" (76.2 mm) or 4" (101.6 mm) thick and 12" (304.8 mm) or more wide, bolted and/or spiked together with the centre piece an inch or more ahead of the others to form a tongue and groove. For most Wakefield assemblies, timbers should be surfaced on all four faces.

Figure 2-32 Lapped and Wakefield Sheet Piling

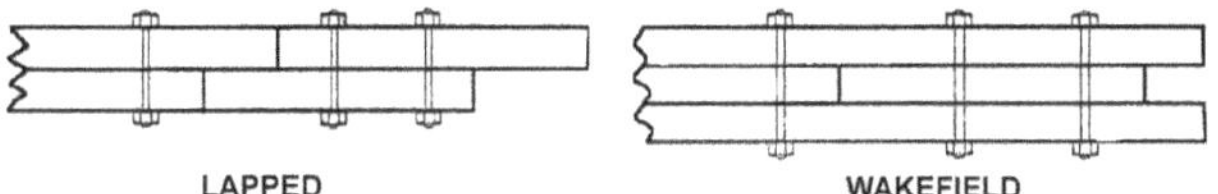

In areas where steel sheet piling may not be available, Wakefield Piles are useful; this system was in wide use before the advent of steel sheet piles. For the driving case, the forward bottom edge of the wood can be beveled to force the sheet being added tightly against the in-place units. Wakefield or similar piles may be driven with a small conventional pile hammer or a heavy, hand-held air hammer or "paving breaker." A vibratory earth compactor has been successfully used for vibrating lightweight sheet piling, as have been excavator mounted vibratory hammers. "Tongue and groove" sheet piling is also used as shown in Figure 2-33. This is made from a single piece of timber that is cut at the mill with a tongue and groove shape.

Figure 2-33 Tongue and Groove Wood Sheet Piling

2.2.5.2. Typical Timber Bulkheads

Figure 2-34 shows a conservatively designed, low height bulkhead where the existing grade along the sheet piles is somewhat higher than the low water level. Figure 2-35 shows an intermediate height bulkhead suitable for retaining fill at the site of a marina or for providing a finished waterfront in a housing development. If these bulkheads are located inshore, or if the outside water level variations are less than shown in the figures, the heights of bulkheads and lengths of sheet piles may be reduced based on design computations.

Figure 2-34 Bulkhead Design, Zero-Foot Water Depth

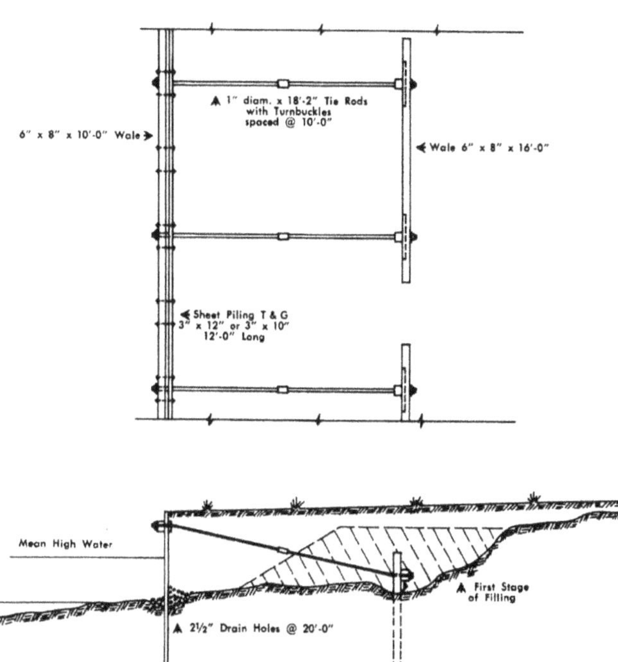

Figure 2-35 Bulkhead Design, Four-Foot Water Depth

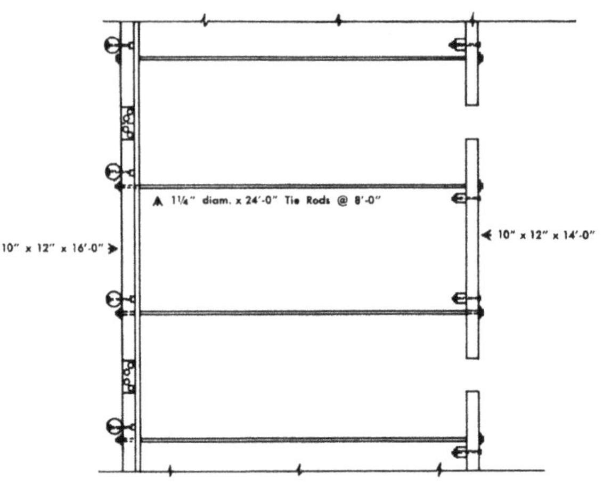

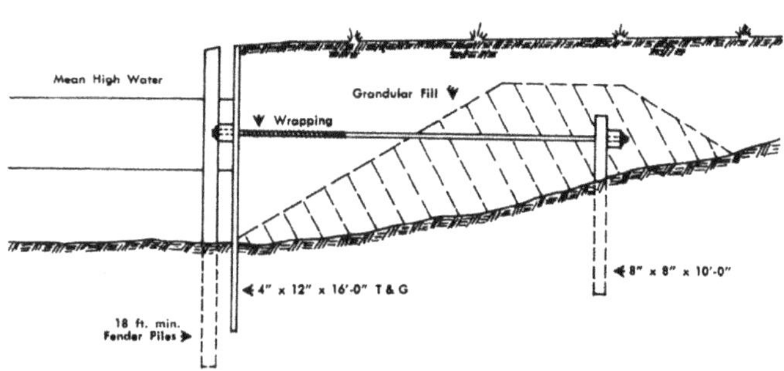

Figure 2-36 Bulkhead Design, Eight-Foot Water Depth

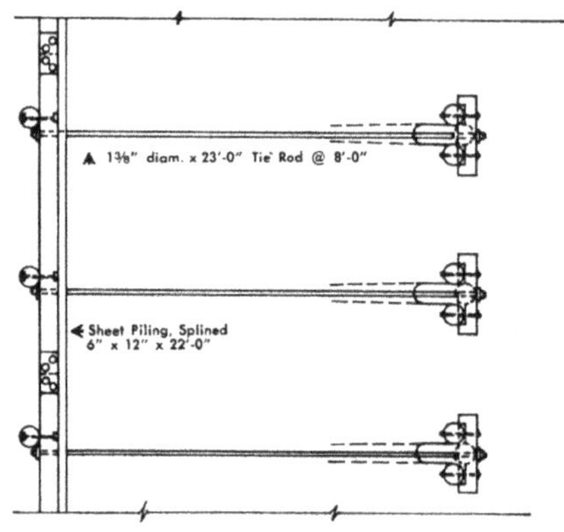

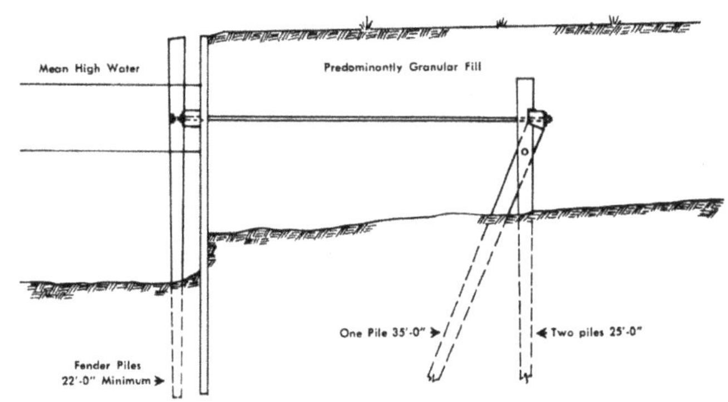

The anchorage systems (Figure 2-34 and Figure 2-35) depend upon the passive resistance of the earth immediately around the anchor post and wales. The theoretical mound required to develop this passive resistance is shown by a dotted line. If backfill is placed directly against the sheet-pile bulkhead before this mound of earth is placed around the anchor system, the resulting forces may displace or fail the bulkhead, resulting in a costly and disastrous failure. After the mound is placed over the anchorage system, backfill can be deposited against the sheet piles by suitable methods.

Figure 2-36 shows a bulkhead suitable for the deepwater marinas, or for locations where the existing water depths are 6 to 8 (1.8-2.4 m) and extensive landfills are desirable. The anchorage system (Figure 2-36) is a self-supporting A-frame that does not depend on passive earth resistance. This anchor system is particularly adaptable to filling by the hydraulic method because the backfill can be raised behind the sheet piles without regard for the placement of backfill at the anchorage location.

To install the A-frame anchorage, a pile-driving rig is required to drive the piles to specified bearing capacity and length. At locations with less water level variation that the 4 (1.2 m) shown, the height of finished grade may be lowered proportionally. For an increase in water level variation, a similar increase in height of finished grade can be made with a corresponding reduction in water depth. The 5 (1.52 m) vertical distance from finished grade to tie rod level should be maintained.

A typical completed timber bulkhead is shown in Figure 2-37.

Figure 2-37 Completed Timber Bulkhead

2.2.6. Concrete Sheet Piles

For sea walls, groins and other waterfront structures, precast concrete sheet piles are often economical. They are normally used in situations where the precast members will be incorporated into the final structure or are going to remain in place after they fulfill their purpose. Precast concrete sheet piling is usually made in the form of a tongue and groove section. The individual pieces vary in width from 18" (457.2 mm) to 24" (609.6 mm) and in thickness from 8" (203.2 mm) to 24" (609.6 mm). They are reinforced with vertical bars and hoops in much the same way as precast concrete bearing piles. For seawater contact especially, prestressing is most desirable, as the minor cracks are kept closed so salt does not get to the reinforcing. This type of sheeting is not watertight, but the spaces between the piles can be grouted.

Figure 2-38 Concrete Sheet Piling

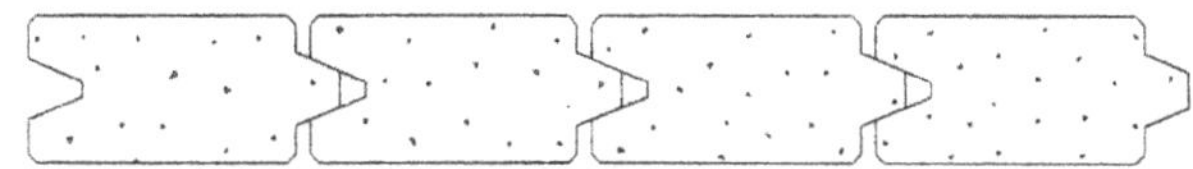

Tongue and groove design, or a slot for grouting, will help to make the wall watertight. The slot would have to be washed out after pile installation and grout forced into the opening. For driving, a bevel of about 30° at the bottom on the leading edge of the pile will force the section being added tightly against the wall already in place.

In order to provide a watertight precast concrete sheet pile, two halves of a straight steel web sheet pile, which has been split in half longitudinally, are embedded in the pile, as shown in Figure 2-39.

Figure 2-39 Concrete Sheet Piling with Steel Interlocks

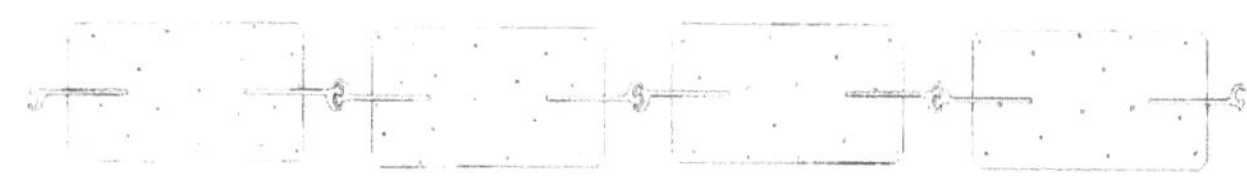

Jetting is frequently used to install concrete sheet piles in sand. The pile must be installed while jetting, as the jetted hole may not stay open.

The Portland Cement Association and the Prestressed Concrete Institute have data on manufacturing and installing concrete sheet piles.

Photo and Figure Credits

- GRL and Associates/Pile Dynamics, Inc.: Figure 2-3, Figure 2-10.
- Plastic Pilings, Inc.: Figure 2-15
- J.A. Rauch Construction Co., Inc.: Figure 2-37.
- Prestressed Concrete Institute: Figure 2-11, Figure 2-12, Figure 2-13, Table 2-10.
- Don C. Warrington, P.E.: Figure 2-2, Figure 2-3, Figure 2-4, Figure 2-5, Figure 2-14, Figure 2-26.

Footnotes

1 Bethlehem Steel Piles, Bethlehem Steel Corporation, July 1979

2 Gerwick, B.C., Jr., Bach, P.B., and Fotnos, G. "In the Wet." Civil Engineering, May 1995, pp. 46-48.

3 A thorough discussion of this topic can be found in *Sheet Pile Design by Pile Buck*.

4 Mandrels are discussed in the chapter on installation equipment.

5 Bengt Fellenius, who has had long experience with precast piles, suggests less cover for greater pile strength. This requires that close positioning tolerance be strictly adhered to. Placing the longitudinal reinforcing accurately so it has a dependable cover of just 1" (25.4 mm) to the outside of the steel will result in a stiffer pile that is better able to withstand driving stress. This will provide greater strength and minimize risk of cracks in the pile.

6 For some years this was a standard of the American Association of State Highway Officials and PCI. AASHTO has discontinued participation in such joint activities.

[7]ACI recommends a limit of 8% C_3A for sulphate concentrations between 0.1 and 0.2% and a limit of 5% for concentrations over 0.2%. However, for the higher strength concretes (8000 psi and over) employed in the manufacture of prestressed piles, the 8% limit on C_3A is considered adequate for sulfate concentrations over 0.2% (see ACI Journal, August 1973).

[8]Neat cement grouts used in bonding post-tensioned tendons should follow the "Recommended Practice for Grouting Post-Tensioned Prestressed Concrete" (PCI JOURNAL, Nov.-Dec., 1972).

[9] Where the taper applied to the butt circumferences calculate to a circumference at the tip of less than 16 in., the individual values have been increased to 16 in. to assure a minimum of 5-in. tip for purposes of driving.

Diameters are approximate. Circumferences are the specified requirements.

[10]J. Much (U.S. Coast Guard's "The Engineer's Digest" March-April, 1958) reported greater resistance of Greenheart piling to marine borer attack compared to creosoted yellow pine at the San Juan, P.R. Base. Both species were included in test Dolphins built at the base. After three years, the Greenheart was still in good condition while the pine had revealed some attack. Presumably, these tests were continued and information compiled beyond 1957

[11] The tests were conducted by Shimel and Sor, T.L., for Greenheart Associates-Demerara Inc.

[12] Heinz, R., "Plastic Piling." Civil Engineering, pp. 63-65, April 1993.

[13] This statement reflects design practice in the U.S. In the E.U., the interlocks are frequently ignored as they are along the neutral axis, and two sheets are treated as one section.

[14] An example of a real down-to-earth rule of thumb corrosion check is this: if grass is not growing where you want to place the wall, there probably is a problem with pH. Grass will normally grow in the 4.5 to 8.5 range.

[15] Tensile strength figures such as these do not take into consideration that vinyl is subject to creep. A reasonable figure for creep limited tensile stress is 4,000 psi. With a factor of safety of 2, this would mean an allowable stress of 2,000 psi. A factor of safety of 1.5 (and a corresponding allowable stress of 2,667 psi) can be used only with certain types of materials and with the recommendation of the manufacturer. The industry standard for determining the creep limited stress is ASTM D5262 for a minimum duration of 10,000 hours. This test is done on samples collected from the finished product. Therefore, it not only verifies the performance of a particular vinyl compound, but also the method of manufacture as well.

[16] Important Note for Close Tolerance work: Holes drilled in composite structurals are generally .002" to .004" undersize. Thus, a 1/8" drill will not produce a hole large enough to admit a 1/8" expanding rivet. Instead, a No. 30 drill must be used. This should be noted when drilling for wale and/or tie rod hardware placement. Remember, just as in cutting operations, drilling should be done in a light, evenly applied pressure.

[17] Consider carefully the use and design of fastening devices for mechanical connections of any composite structure.

[18] The proper washer will aid in distributing "load stresses" which is critical in the design of any composite structure.

Chapter 3. Installation Equipment

3.1. Overview

Over the years, technology has advanced toward the development of larger and faster hammers. These developments included rams that were raised by ropes through human or horse muscle, those hoisted by steam winches or powered by steam, air, or hydraulic pressure, and finally the diesel hammer. Double acting hammers were developed in order to speed up operations. Hammers from all stages of this technological history are still encountered on construction sites. This section describes these various types of pile hammers.

If the only purpose of a hammer were to advance the pile into the ground, then the force-time characteristics of the hammer blow would not be so important. However, the vertical advance of a pile under a given hammer blow can be used as a measure of the pile's bearing capacity. Thus, the hammer takes on a second function as a piece of testing equipment. The hammer's interaction with the pile-soil system can be both modeled before driving (wave equation analysis) and monitored during pile installation (pile driving analyzer.) These methods are used to both control driving stresses and estimate the load capacity the pile. From this, field inspection has two important purposes:

- Ascertain that the hammer conforms to certain minimal standards.
- Record observations on hammer and driving system performance.

These two functions will enable the engineer to reduce the uncertainty in his or her interpretation of pile behavior. Furthermore, better productivity plus a reduced likelihood of construction claims can be expected.

3.1.1. Overview of the Driving System

The driving system consists of the following components (see Figure 3-1):

- The lead, also called the set of leads, sometimes including a brace that allows for the adjustment of the lead angle.
- The hammer cushion, which protects both the hammer and the pile from excessive stress. Hammer cushion materials are relatively soft compared to steel. Usually, a striker plate protects the cushion.
- The helmet, which aligns the pile head and hammer. It is usually a steel casting.
- The pile cushion. This is only needed on concrete piles, and usually consists of several layers of plywood.

All components of the driving system have some effect on the performance of a hammer and/or how the hammer transfers its energy to a pile.

Figure 3-1 Components of a Pile Driving System

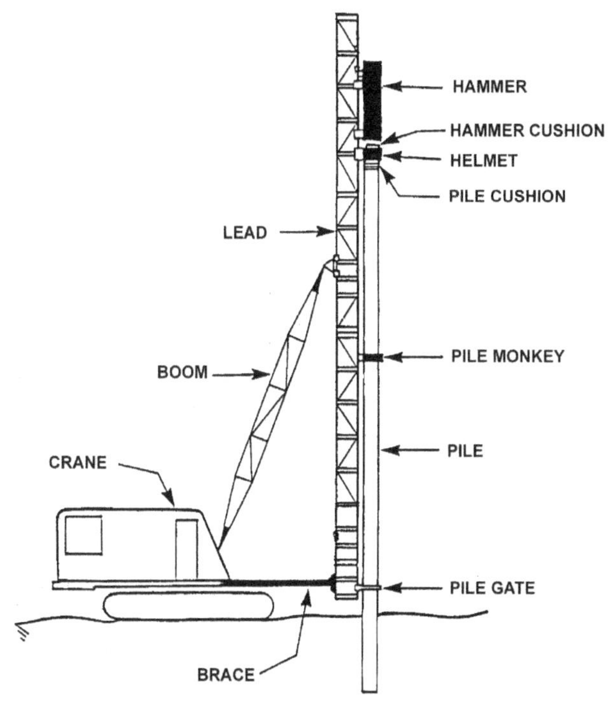

3.1.2. Equipment Selection

Hammer selection may be the most important aspect of pile installation. In some installations only one hammer type may be applicable for the pile-soil combination, while for others several types may be suitable. Evaluation must consider the need to use pile penetration rate as the means to end driving, the ability to drive the pile without structural damage or reducing soil capacity, the ability to obtain penetration rates within the desired band, and the realization that some hammer types may cause reduced capacities for identical pile lengths.

In general, wave equation analysis supplemented by construction experience and engineering judgment should be the basis for hammer approval and criteria such as allowable driving stresses, desired penetration rates, and any other data used as a basis for approval that are clearly defined in the specifications.

Selection of a particular hammer should consider the anticipated driving resistance, ultimate capacity, pile stresses expected during driving, and soil set-up. The hammer type and size used for production installation should always match that used in the test program because a different hammer would likely result in a different capacity. The designer or contractor may designate a number of hammers for the test program when warranted. Any changes in hammer type or size will usually require additional testing. Methods for determining the drivability of a specific hammer-pile-soil system in advance of driving are discussed in Chapter 4.

One factor that designers overlook is that of availability of various types and sizes of pile driving equipment. Designers should always strive to specify the most broadly available hammers possible. Failure to do so will result in a narrowed bid list of contractors and indirectly equipment manufacturers, which will adversely affect the installation costs.

3.1.3. Penetration Limitations

For impact hammers the rate of penetration is customarily defined as the blow count per unit length of pile penetration. Blow counts are typically recorded in the field on a per-foot basis until the pile approaches a designated tip elevation or the end of driving. At that point the blow count is usually recorded for each inch of penetration. Limiting penetration rates are designated to prevent overdriving, which may cause structural damage to the pile, and to provide guidance for determining the relative capacity attained during driving. Pile tip damage due to very difficult driving (commonly referred to as refusal) is not readily detectable when the pile encounters an obstruction or a hard bearing stratum prior to reaching the indicated tip elevation. Therefore, the limiting penetration rates, or the criteria necessary to determine limiting rates, or the criteria necessary to determine limiting rates, should be specified.

3.2. Impact Hammers

Impact pile driving hammers consist of a ram and an apparatus that allows the ram to move quickly upwards and then fall onto the driving system and pile. The ram must have a mass and impact velocity that is sufficiently large to move the pile as shown in Figure 3-2.

Figure 3-2 Impact Hammer System Schematic

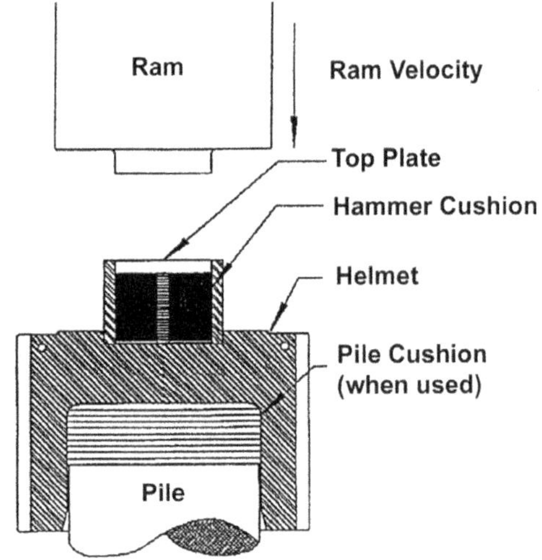

Upon impact with the pile accessory, the ram creates a force far larger than its weight, which, if sufficiently large, then moves the pile an increment of penetration into the ground. The mass falls a certain distance called the stroke. If the stroke is too low, the pile will not move into the ground. If the stroke is too high, the pile may be damaged. The stroke of a pile-driving hammer is usually between three and ten feet (900 to 3000 mm).

A properly functioning hammer strikes the pile in quick succession. It transfers a large portion of the kinetic energy of the ram into the pile.

Figure 3-3 shows various types of pile hammers. These will be discussed in the following sections.

3.2.1. External Combustion Hammers

External combustion hammers are hammers that burn the fuel that provides the energy for the operation of the hammer outside of the hammer itself. These hammers have external power sources such as the crane itself, steam boilers, air compressors, and/or hydraulic power packs to provide the energy to move the ram upward, and in some hammers, downward

Figure 3-3 Impact Pile Driving Hammers

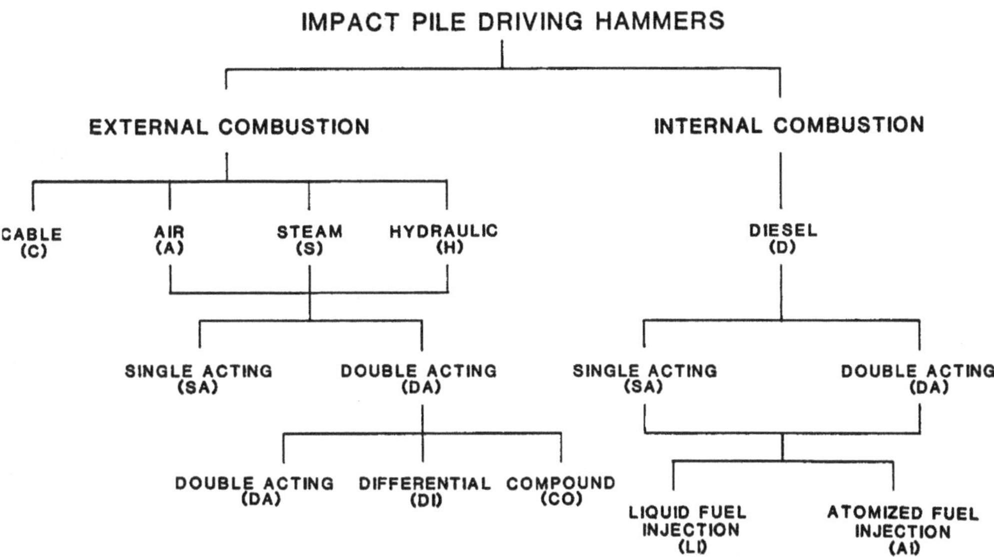

as well. The various types of external combustion hammers are detailed below. Figure 3-4 shows the basic components of an external combustion hammer.

Figure 3-4 Basic Components of an External Combustion Hammer

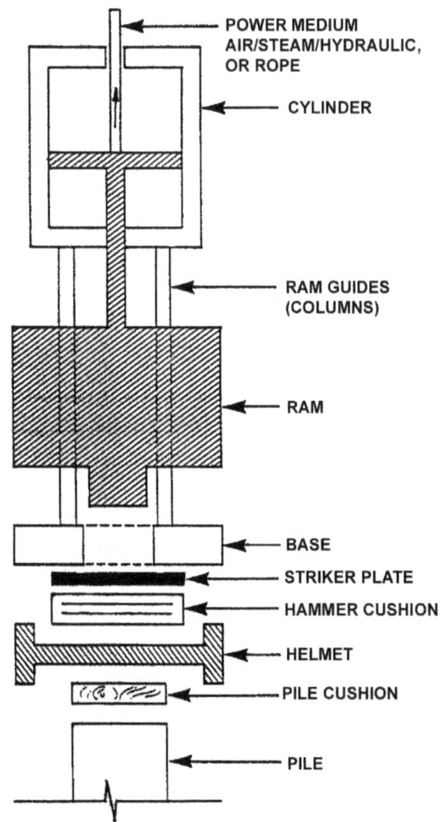

3.2.1.1. Drop Hammers

The drop hammer is the oldest type of pile driving hammer in existence. A typical drop hammer is shown in Figure 3-5. The hammer is connected to a cable that is attached to a winch on the crane. The hammer is raised to the desired stroke. The winch has a clutch on it that then allows the operator to release the hammer, which falls by its own weight and strikes a pile cap and the pile. A hoist from the same crane, which supports the pile and the leads, usually lifts it. After lifting the ram to the desired stroke, it is dropped by either a tripping mechanism or by releasing the hoisting drum, as shown in Figure 3-6. In the former case, all the developed energy remains in the ram; in the latter case, some energy is needed to unwind the cable from the drum. The available energy per blow of a drop hammer is the product of the ram weight times the drop height. Standard ram weights range from 500 to 10,000 lbs (2.5 to 50 kN). Typical drop heights are approximately 4’ (1.2 m). For a given hammer, varying the drop height can vary the energy per blow. Drop hammers are typically used on very small projects and for small piling. The advantages and disadvantages of drop hammers are shown in Table 3-1. It should be noted that a pile cap and leads should always be part of a driving system to ensure that the ram hits the pile axially, and to protect the pile top from damage.

Figure 3-5 Typical Drop Hammer

Figure 3-6 Drop Hammer in Use

Table 3-1 Advantages and Disadvantages of Drop Hammers

Advantages	Disadvantages
Simplicity of operation.	Relatively slow rate of operation.
Ease of mobilization and demobilization.	Driving efficiency depends upon operator's skills.
Low investment and maintenance costs.	Variations in drop height.
Drop height can be easily changed.	

One variation of the drop hammer currently is one that requires only a minimal amount of headroom. The idea utilizes a pipe pile with a large enough diameter to allow the pile hammer to move up and down inside the pipe's walls. The hammer impacts onto a "stop" built into the bottom, inside of the pipe pile. As the pile is driven, the impact occurs near the toe of the pile. The pile is actually pulled down into position in lieu of being pushed. This configuration minimizes the need for the additional overhead clearance (leads, crane, etc.).

For drop hammer arrangements that require the ram unwind the cable from the drum during its fall, the effective energy of the hammer is reduced because of the friction loss and the inertia of the drum and the cable assembly. Just before impact, the operator sometimes "catches" the ram, thereby reducing the impact velocity. The operator often does not achieve constant strokes, and variable energy outputs result from to blow. An additional source of energy losses is the friction between the ram and the leads. Going through the checklist that follows can minimize these factors.

3.2.1.1.1. Checklist for Drop Hammer Inspection

3.2.1.1.1.1. Before Driving Begins

- Check ram weight.
- Check material, size, and the condition of the hammer cushion.
- Check the pile cushion (if present), size and condition.
- Check that helmet and pile are well matched.
- Check helmet weight.
- Check straightness of leads and connections of lead parts for tightness.
- Make sure that the portion of the leads that guide the ram, as well as all pulleys used are well greased.
- Clearly mark the leads as a reference for drop height observations.
- Check the alignment between the leads and the pile.
- Make sure that the cable and the drum components are well lubricated.

- Check the condition of the hoisting cable.

3.2.1.1.1.2. During Driving

If a trip mechanism is not used, be certain that the hoisting cable is spooling freely, i.e., that there is no tension in the cable just before and during impact.

- Make sure that the desired hammer stroke is consistently maintained.
- Check that ram, helmet, and pile maintain alignment during driving (no excessive swaying due to the lifting and releasing of the ram).

3.2.1.2. Single Acting Air/Steam Hammers

Single acting air/steam hammers are essentially drop hammers, except that the hoisting cable is replaced by pressurized air or steam (motive fluid). The ram is usually a short, stocky block of steel that is connected at its top to a piston. It is guided by columns or inside guiding enclosures, depending on the particular design[1]. Vulcan, Conmaco, MKT, Menck, and Raymond manufacture single acting air/steam hammers most commonly encountered in the United States. Vulcan, Raymond and Conmaco have very similar designs. Single acting air/steam units from Vulcan and Menck are commonly used to drive piles offshore. A typical single acting air/steam hammer is shown in Figure 3-7.

Figure 3-7 Typical Single-Acting Air/Steam Hammer

The maximum available energy per blow for a single acting air/steam hammer is expressed as the product of ram weight times stroke. Ram weights for current production models are between 3 and 300 kips (14 and 1,500 kN). Strokes vary from 2 to 5 ft (.6 to 1.5 m). Thus, energy ratings may be between 6 and 1,500 kip-ft (10 and 2,250 kNm).

The operating rate of single acting air/steam hammers range between 35 and 60 blows per minute. This rate is comparable to most other hammer types that lack downward assist.

Most single acting air/steam hammers have a fixed stroke. A slide bar with cams (Figure 3-8) is attached to the ram for the tripping of valves on the ram cylinder. However, the stroke of some hammers can be varied if the slide bar is equipped with a dual set of cams and with a valve trip which can be moved to either set of cams as shown in Figure 3-9. This adjustment is usually done remotely by means of a hydraulic line. The primary advantage of variable stroke hammers is their ability to reduce damaging driving stresses during the early stages of driving when lower soil resistance is encountered.

Figure 3-8 Slide Bar with Cams

Figure 3-9 Stroke Changing Device for Air/Steam Hammers.

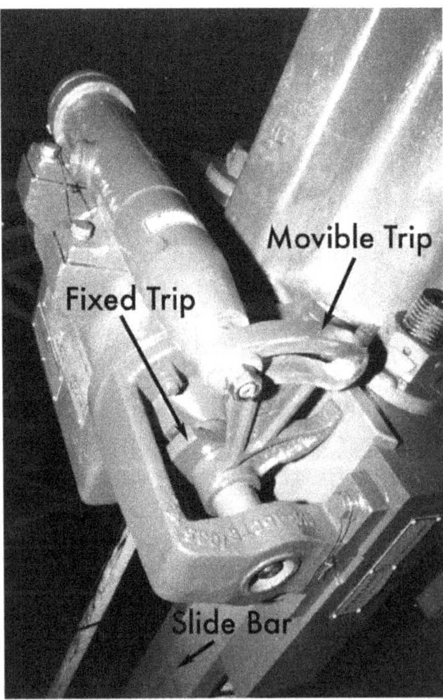

Advantages and disadvantages for single acting air-steam hammers are shown in Table 3-2.

Table 3-2 Advantages and Disadvantages of Single Acting Air/Steam Hammers:

Advantages	Disadvantages
Higher rate of blows per minute than drop hammers.	Additional equipment needed (boiler, compressor, hoses, etc.)
Relatively consistent operation.	Relatively heavy hammer; requires higher crane and handling equipment capacities.
Simple hammer design (compared to all except drop hammers)	

3.2.1.2.1. Operation

The operating cycle of a single-acting air/steam hammer is shown in Figure 3-10. The hammer operates in a two-step cycle as follows:

1. A blow of an air/steam single-acting hammer is initiated by introducing the motive fluid at constant pressure in a cylinder under the piston. The pressure increases in the cylinder and causes the ram to move upward. At a certain height (usually 50-75% of the maximum fall height), the pressure inlet valve is closed and the pressure chamber is opened to the atmosphere.
2. The ram will coast upwards under its momentum for an additional distance. Towards the top of a full stroke, the piston penetrates a safety chamber formed by trapped air above the so-called relief ports, which assists in the deceleration of the piston. Upon its descent, the ram reaches a position - usually immediately preceding impact - where the valve opens to allow motive fluid to enter the cylinder. The exhaust is closed at the same time.

Figure 3-10 Single-Acting Air/Steam Hammer Operating Cycle

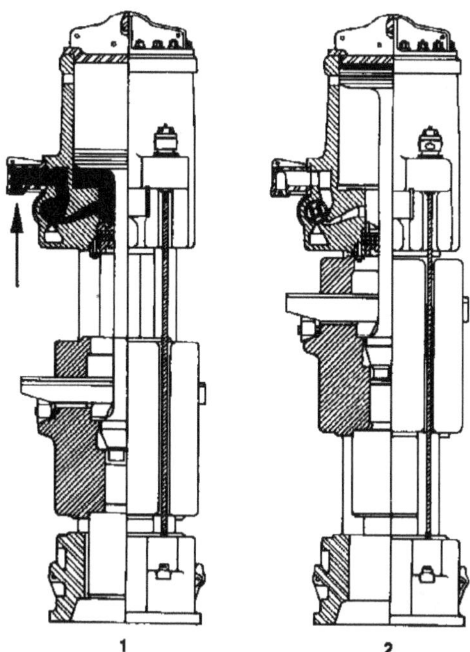

Obviously, if pressurized motive fluid enters the cylinder too soon before impact occurs, this pressure will tend to slow the fall of the ram. This effect is called pre-admission; it may have a significant effect on the striking velocity. An incorrect hammer cushion thickness, poorly adjusted valves, or an improper slide bar may cause this problem.

During impact, the ram pushes the helmet and pile downwards, allowing the hammer base, columns, and cylinder (these components are also called the assembly) to fall under the action of both gravity and the reaction forces of the pressurized fluid. When the hammer base meets the helmet, a so-called "assembly impact" occurs.

Since the air/steam hammer is an external combustion hammer, its performance depends on the boiler (steam generator) or air compressor. Either unit must be of sufficient capacity to provide the operating pressures at the necessary fluid flow rate. The hammer specifications therefore require an operating pressure and an operating volume (e.g., cubic feet per minute or cubic meters per minute).

The specified pressures must be available at the hammer, and the system losses must be minimized. Pressure losses may result from leaky or worn hoses, or in hoses of insufficient number of size or excessive length.

Single acting air/steam hammers should be used with a pile helmet inserted between the ram and the pile top. The helmet usually supports the hammer assembly, and houses the striker plate and the hammer cushion, if so required by the manufacturer. The helmet must be well matched to both pile type and hammer model. For high hammer cushion stacks (such as those commonly used with micarta and aluminum stacks), a follower or shield may be inserted between hammer and helmet; its effect is to extend the upper portion of the helmet. Such a follower or shield should only be used after obtaining approval from the engineer.

A correct cushion thickness is important for the proper performance of a hammer. Thinner than specified cushions may cause pre-admission, which cushions the blow. A taller cushion stack than recommended shortens the ram stroke. The shorter stroke will, of course, produce blows with lower energy. If the cushion thickness increases even further, then the inlet valve may not open at all, and the hammer will not run. It is very important to start with and maintain a hammer cushion thickness as specified by each manufacturer for each hammer model.

3.2.1.2.2. Checklist for Single Acting Air/Steam Hammers

3.2.1.2.2.1. Before Driving Begins

- Check hammer model and serial number to ensure that the hammer has the proper energy rating.
- Check material, size, and the condition of the hammer cushion. When measuring the hammer cushion thickness, the thickness of the striker plate should also be measured and then added.
- Check the pile cushion (if present) for size and condition.
- Check that the helmet and pile are well matched.
- Check helmet weight; if in doubt have it weighed.
- Check straightness of leads and connections of lead parts for tightness.
- Make sure that the portions of the leads that guide the hammer do not offer excessive

friction.

- Check the alignment between the leads and the pile.
- Mark ram guides or check hammer for reference points as a reference for stroke observations.
- Check number, size, and length of steam or air hoses.
- Make sure that hoses are cleared of obstructions before they are connected to the hammer.
- Check that the motive fluid lubricator is filled with an appropriate lubricant.
- Inspect the slide bar and its cams for excessive wear.
- Check the columns (ram guides), piston rod and slide bar are well lubricated.
- Make sure that the ram keys and column keys are all tight.

3.2.1.2.2.2. During Driving:

- Make sure that the desired hammer stroke is maintained.
- Check that the ram, helmet, and pile maintain alignment during driving.
- Check that pressure at compressor or boiler is equal to the rated pressure plus hose losses. Pressure should not vary significantly during driving.
- Measure the rate of hammer operation in blows per minute. A stopwatch may accomplish this.
- Make sure that the motive fluid is well lubricated. After a short period of pile driving, lubricant should drip out of ports.
- Continuously check the quality of the motive fluid; condensation or icing of air, and water in steam are all signs of a motive fluid in a poor condition.
- Check that the hammer hoist line is always slack.
- Record when hammer or pile cushioning material is replaced or added to the driving system. New cushion material should not be added towards the end of driving; a minimum of 100 blows should be applied before final blow counts for capacity determination are taken.

3.2.1.3. Double, Differential and Compound Acting Air/Steam Hammers

The desire to create a hammer that applies blows in rather quick succession for increased productivity led to the development of double acting air/steam hammers. For an increase in blow rate, the stroke was shortened, and in order to maintain a comparable energy rating, the ram was accelerated during its down stroke by means of active pressure. Advantages and disadvantages of these hammers are shown in Table 3-3.

Table 3-3 Advantages and Disadvantages of Double, Differential and Compound Acting Air/Steam Hammers

Advantages	Disadvantages
The speed of operation of a double acting hammer is approximately twice that of a single acting hammer, i.e., 120 instead of 60 blows per minute.	Energy output is extremely sensitive to proper valve timing, and to pressure and volume of the motive fluid.
	The available energy of each blow is more difficult to inspect and to verify.
	The applied hammer energy output is sensitive to soil resistance (rebound in hard driving).

3.2.1.3.1. Double Acting Hammers

The term "double acting" is frequently used to refer to any hammer that uses the air, steam or hydraulic pressure to accelerate the ram in the down stroke along with gravity. This definition is misleading because it does not take into consideration the different methods by which this is accomplished, which can be significant both in the operation of the equipment and in its maintenance. A more accurate term to generally refer to the application of downward and down stroke pressure on the ram is "downward assist," which has come into common usage with hydraulic hammers.

In the U.S., the best-known example of a double acting air/steam hammer is the MKT "B" series, which include the 9B3, 10B3 and 11B3. Double-acting hammers are especially popular in driving sheet piling where vibratory hammers cannot penetrate the soil or where they are favorable economically. Such an application (with sheet piling) is shown in Figure 3-11. The hammer is guided by pants, which eliminate the need for leaders.

Figure 3-11 MKT "B" Series Hammer Driving Sheet Pile

The operating cycle of these hammers is shown in Figure 3-12. The ram carries a rigidly connected cam throw (some double acting hammers use a fluid valve) that engages a cam rod suspended in the intermediate head of the hammer. In operation, the motive fluid first enters the inlet port and flows through the lower opening of the valve to the underside of the piston. The top opening of the valve completes a path from the topside of the piston to the exhaust port. As the fluid lifts the piston, thus lifting the ram, the lugs of the cam throw slide past the edges of the cam rod until, at the top of the stroke, they engage a spiral portion of the cam rod, causing it to rotate. The valve connected to the top of the cam rod also rotates, allowing the inlet motive fluid to enter the top of the cylinder, and permitting the exhaust fluid on the underside of the piston to escape through the exhaust port. The ram then falls, its velocity increased by the fluid pressure on the top of the piston. The cam throw lugs slide down where another spiral portion of the cam rod is engaged. The cam rod rotates; rotating the valve to the original position, and the motive fluid path reverses.

Figure 3-12 Operating Cycle of a Double Acting Air/Steam Hammer

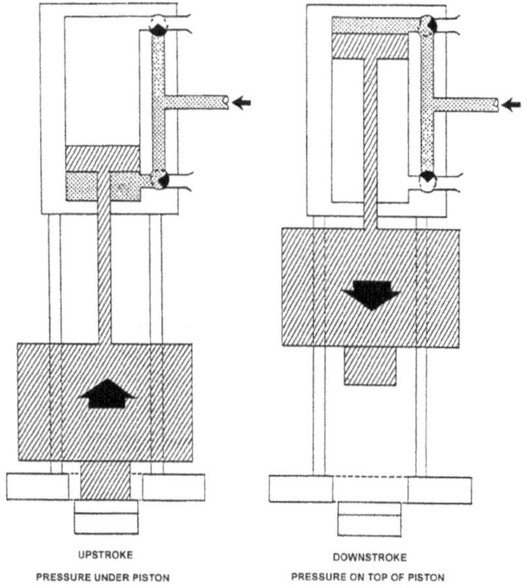

The MKT double acting hammers are designed for a metal-to-metal impact between the ram and the impact block also called the anvil. The impact block may be seated directly on the pile without cushioning. The ram is guided by the hammer housing, also called the cylinder. This housing completely encloses the ram and other moving parts, and an inspection is therefore difficult.

Since hammer and pile experience ram impact at the same time, no discernable assembly drop is present for this hammer type. On the other hand, there may be several separations between the impact block and the pile top.

3.2.1.3.2. Differential Acting Hammers

These hammers are similar to double-acting models except that the air or steam is constantly pressurized under the piston. This allows for a simpler valve configuration than with a double-acting hammer with similar operating characteristics. A typical differential-acting hammer is shown in Figure 3-13. Its operating cycle and characteristics are shown in Figure 3-14. The best examples of this type of hammer are the Vulcan and Raymond "C" series hammers.

Figure 3-13 Typical Differential-Acting Air/Steam Hammer.

Figure 3-14 Operating Cycle for Differential-Acting Air/Steam Hammers.

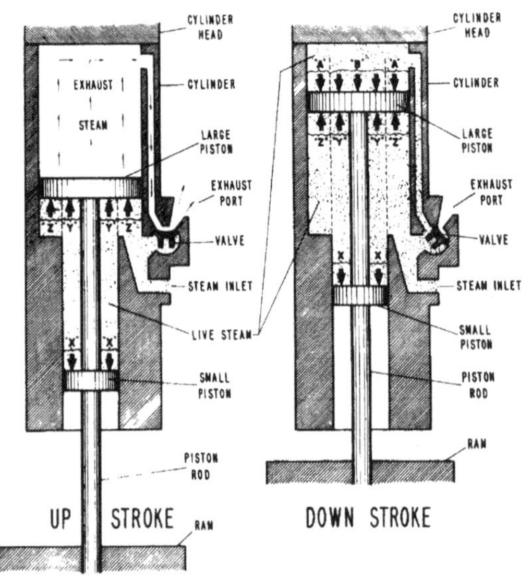

Referring to Figure 3-14, the cycle begins at impact, the valve rotated so that the area above the large piston is open to the atmosphere and exhausting the compressed air or steam from the previous stroke. The area in the cylinder between the large and small pistons is always pressurized, and, as in the beginning of the cycle, when there is only atmospheric on the top of the large piston, this creates an unbalanced force on the piston and the ram accelerates upward. As the ram moves upward the intake wedge actuates the trip, rotating the valve and admitting steam to the cylinder above the large piston. This produces an unbalanced force downward on the ram, bringing the ram to a halt at the top of the stroke. The ram is then forced downward, gaining kinetic energy from both gravity and the downward acting steam or air force, to impact. Just before impact the exhaust wedge rotates the valve once again to exhaust the compressed air or steam above the large piston and the cycle starts once again.

3.2.1.3.3. Compound Acting Hammers

A compound acting hammer is any air/steam hammer that a) uses air or steam expansively[2] during the operating cycle and b) applies downward assist to the ram. Thus the Menck air/steam hammers, which use the motive fluid expansively, are not really "compound hammers" as they are single acting.

The first compound hammer to gain wide acceptance was the "California" series of hammers, invented by James N. Warrington and produced by Vulcan between the World Wars. An example of this is shown in Figure 3-15. After World War II, MKT developed the "C" series hammers; this is described in more detail below.

Figure 3-15 Vulcan California "E" Hammer

Figure 3-16 Compound Air/Steam Hammer

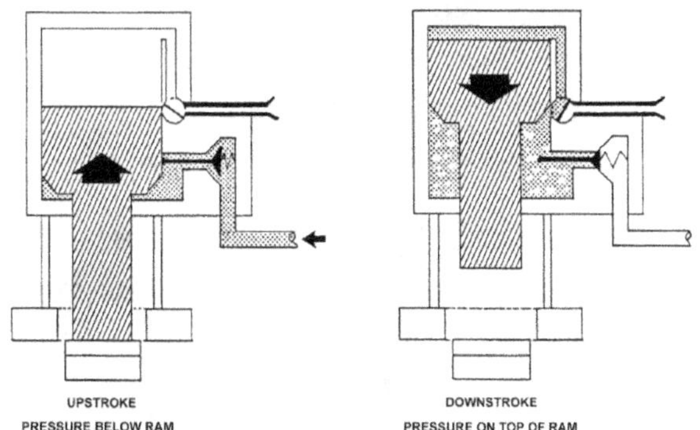

A schematic of the operation of a compound hammer is presented in Figure 3-16. The ram of a compound hammer is raised under full pressure and with the upper chamber vented. After reaching the top of the stroke, the upper chamber that is containing full pressure in the beginning of the downward stroke. At the same time, the inlet valve is closed and as the ram descends, the motive fluid expands in both upper and lower chambers with decreasing pressures. Since the top ram area is significantly larger than the bottom area, a net downward force results, which accelerates the ram at a rate greater than gravity.

Compound mechanisms, similar to differential ones, utilize the motive fluid more economically than double acting mechanisms. Again, for full efficiency, this hammer must be supplied with as much pressure as specified.

3.2.1.3.4. Checklist For Double Acting And Differential Acting Air/Steam Hammers

3.2.1.3.4.1. Before Driving Starts:

- Check hammer model and serial number to ensure that the hammer has the proper energy rating.
- Check material, size, and the condition of the hammer cushion. When measuring the hammer cushion thickness, the thickness of the striker plate should also be checked and then added (R(aymond), C(onmaco), V(Vulcan) hammers).
- Check the pile cushion (if present), for size and condition.
- Check that helmet and pile are well matched. If the hammer type is such that no helmet is necessary, check that the hammer and pile are well matched.
- Check helmet weight; if in doubt have it weighed.
- Check straightness of leads and connections of lead parts for tightness.
- Make sure that the portions of the leads that guide the hammer do not offer excessive friction.
- Check the alignment between the leads and the pile.
- Mark ram guides or check hammer for reference points as a reference for stroke observations (R, C, V).
- Check number, size, and length of steam or

air hoses.

- Make sure that hoses are blown down before they are connected to the hammer.
- Check that the motive fluid lubricator is filled with appropriate lubricant.
- Inspect the slide rod and its cams for excessive wear. (R, C, V).
- Check that columns (ram guides), piston rod, and slide bar are well lubricated.
- Make sure that ram keys and column keys are all tight.

3.2.1.3.4.2. During Driving

- Make sure that the desired hammer stroke is maintained. (R, C, V)
- Check that ram, helmet, and pile stay in alignment during driving.
- Check that the pressure at the compressor or boiler is equal to the rated pressure plus hose losses. Pressure should not vary significantly during driving.
- Record when and to what level pressure needs to be reduced because of uplift.
- Measure the rate of operation of the hammer in blows per minute. This may be accomplished by using a stopwatch.
- Make sure that the motive fluid is well lubricated. After a short period of pile driving, lubricant should drip out of ports.
- Continuously check the quality of the motive fluid; condensation or icing of air, and water in steam are all signs of a motive fluid in bad condition.
- Check that the hammer hoist line is always slack.
- Record when hammer or pile cushioning material is replaced or added to the driving system. New cushioning should not be added towards the end of driving; at least 100 blows should be applied before final blow counts are taken.

3.2.1.4. Hydraulic Impact Hammers

A hydraulic hammer is one that incorporates the use of an external energy source to lift the hammer to the top of its stroke. For the single acting hydraulic hammer, the free-falling piston develops the actual energy induced into the pile, much the same power stroke as a drop hammer or a single acting air/steam hammer. A hydraulic hammer is shown in Figure 3-17.

These hammers substitute hydraulic fluid for air or steam, and it is applied to the piston to move the ram. A hydraulic power pack, such as one shown in Figure 3-18, provides the pressurized fluid to operate the hammer. Hydraulic impact hammers can be single acting, double acting, differential acting, or other variations. Most but not all hydraulic hammers employ the use of an electric valve operated with a variable timer. The timer allows for very flexible control of the output energy. Others use a purely hydraulic system to control the valve and thus the cycling of the ram, which obviates the need for electronics on the hammer. Depending upon the flow and pressure requirements of the hammer and the capabilities of the crane or excavator, some hydraulic hammers can be powered using the hydraulic power units on the crane or excavator, dispensing the power pack altogether.

Figure 3-17 Hydraulic Impact Hammer Driving Steel Pipe Pile

Figure 3-18 Hydraulic Power Pack

The hydraulic hammer has a variable stroke, which is readily controlled from a control box. The stroke can be varied infinitely so that the stroke can be optimized to the point of matching the dynamic spring constant of the hammer and pile. Many hydraulic hammers have the ability to induce a single blow, which is very useful in dynamic testing or in cases of severe pile run. The control systems can also output a wide variety of information, such as kinetic energy output of the hammer and diagnostic information.

Most hydraulic hammer manufacturers claim high efficiencies for their hammers. Although there are many improvements in hydraulic hammers that enable a more efficient drop, the main reason for the higher efficiencies is that most hydraulic hammers - even those that are "single acting" - have some kind of downward assist to equalize the hydraulic flow during the hammer cycle. This enables the apparent efficiency of the hammer to increase.

3.2.1.4.1. Checklist For Hydraulic Impact Hammers

- Determine/confirm the ram weight. There may also be identifying labels as to hammer make, model, and serial number that should be recorded.
- Check the power supply and confirm it has adequate capacity to provide the required pressure and flow volume. Also, check the number, length, diameter, and condition of the hoses (no leaks in hoses or connections). Manufacturers provide guidelines for power supplies and supply hoses. Hoses bent to a radius less than recommended could adversely affect hammer operation or cause hose failure.
- Hydraulic hammers must be kept clean and free from dirt and water. Check the hydraulic filter for blocked elements. Most units have a built in warning or diagnostic system.
- Check that the hydraulic power supply is operating at the correct speed and pressure. Check and record the pre-charge pressures or accumulators for double acting hammers. Allow the hammer to warm up before operation, and do not turn off power pack immediately after driving.
- Most hydraulic hammers have built in sensors to determine the ram velocity just prior to impact. This result may be converted to kinetic energy or equivalent stroke. The inspector should verify that the correct ram weight is entered in the hammer's "computer". This monitored velocity, stroke, or energy result should be constantly monitored and recorded. Some hammers have, or can be equipped with, a printout device to record that particular hammer's performance information with pile penetration depth and/or pile driving resistance. This is the most important hammer check that the inspector can and should make for these hammers.
- For hydraulic hammers with observable rams, measure the stroke being attained and confirm that it meets specification. For hammers with enclosed rams, it is impossible to observe the ram and estimate the stroke.
- Check that the ram guides and piston rod are well greased.
- Where applicable, the total thickness of hammer cushion and striker plate must be maintained to match the manufacturer's recommendation for proper valve timing and hammer operation.
- Make sure the helmet stays properly seated on the pile and that the hammer and pile maintain alignment during operation.
- The hammer hoist line should always be slack, with the hammer's weight fully carried by the pile. Excessive tension in the hammer hoist line is a safety hazard and will reduce energy to the pile. Leads should always be used.
- Compare the observed hammer speed in blows per minute from near end of driving with the

manufacturer's specifications. Blows per minute can be timed with a stopwatch or a Saximeter. Slower operating rates at full stroke may imply excessive friction, or incorrect hydraulic power supply.

- As the driving resistance increases, the ram stroke may also increase, causing the ram to strike the upper hammer assembly and lifting the hammer from the pile (racking). If this behavior is detected, the pressure flow should be reduced gradually until racking stops. Many of these hammers have sensors, and if they detect this condition, the hammer will automatically shut down. The flow should not be overly restricted so that the correct stroke is maintained.
- Some manufacturers void their warranty if the hammer is consistently operated above 100 blows per 250 mm of penetration beyond short periods such as those required when toe bearing piles are driven to rock. Therefore, in prolonged hard driving situations, it may be more desirable to use a larger hammer or stiffer pile section.
- Common problems and problem indicators for hydraulic hammers are summarized in Table 3-4.

Table 3-4 Common Problems And Problem Indicators For Hydraulic Hammers

Common Problems	Indicators
Hoses getting caught in leads.	Visually evident.
Fittings leaking.	Hydraulic fluid dripping.
Electrical connections.	Erratic performance.
Sensors.	Erratic performance.

3.2.1.5. Jacking

Pile jacking machines are not impact hammers but advance the pile by simply pushing it into the ground. Pile jacking machines are most effective when the soil resistance is lower than the maximum ram force and when there are piles for reaction (such as with sheet piling).

The operating cycle is as follows:

1. The jack is set on the reaction stand for the installation of the first two piles.
2. The jack moves by elevating its travel carriage while supporting itself on the previously installed pile.
3. The travel carriage then slides forward.
4. The travel carriage lowers itself, drops onto the installed piles, and continues its hydraulic installation process.

After the third or fourth pile is driven, the jack moves off the reaction stand and travels independently on the piles.

Pile jacking devices can be used for both sheet piles (Figure 3-19) and tubular piles (Figure 3-20.)

Figure 3-19 Pile Jacking Device for Sheet Piling

Figure 3-20 Pile Jacking Device for Tubular Piles

Pile jacking is becoming more important as a method of installing piling. This is because of noise and vibration difficulties, especially around sensitive structures and areas where noise attenuation requirements are especially restrictive.

3.2.2. Internal Combustion Hammers

These hammers burn the fuel that powers them inside of the hammer itself. The diesel hammers are the only constituent of this class, although other types of fuel are being used.

3.2.2.1. Single Acting (Open End) Diesel Hammers

3.2.2.1.1. General Description

An open-end diesel hammer consists of a long slender piston (the ram), which moves inside a cylinder. The cylinder is open at its upper end, thus allowing the ram to partially emerge from the cylinder. Since the ram falls only under gravity, the OED is also called single acting.

There are many manufacturers of open-end diesel hammers throughout the world. The hammer ratings vary from 5 to more than 300 kip-ft with ram sizes between 1 and 35 kips.

Many of these hammers are built using the same fuel combustion principle: impact atomization. This is also referred to as liquid fuel injection. The Delmag "02" models (e.g., D30-02) use a medium pressure injection, which may be classified as impact atomization. A typical open-end, single-acting diesel hammer is shown in Figure 3-21.

Figure 3-21 Open-End, Single-Acting Diesel Hammer

A diagram of an open end diesel hammer in various stages of its operating cycle is shown in Figure 3-22. This also illustrates the major parts of the machine.

The drive cap - anvil interface is illustrated in more detail in Figure 3-23.

Figure 3-22 Major Parts and Operating Cycle of a Diesel Hammer

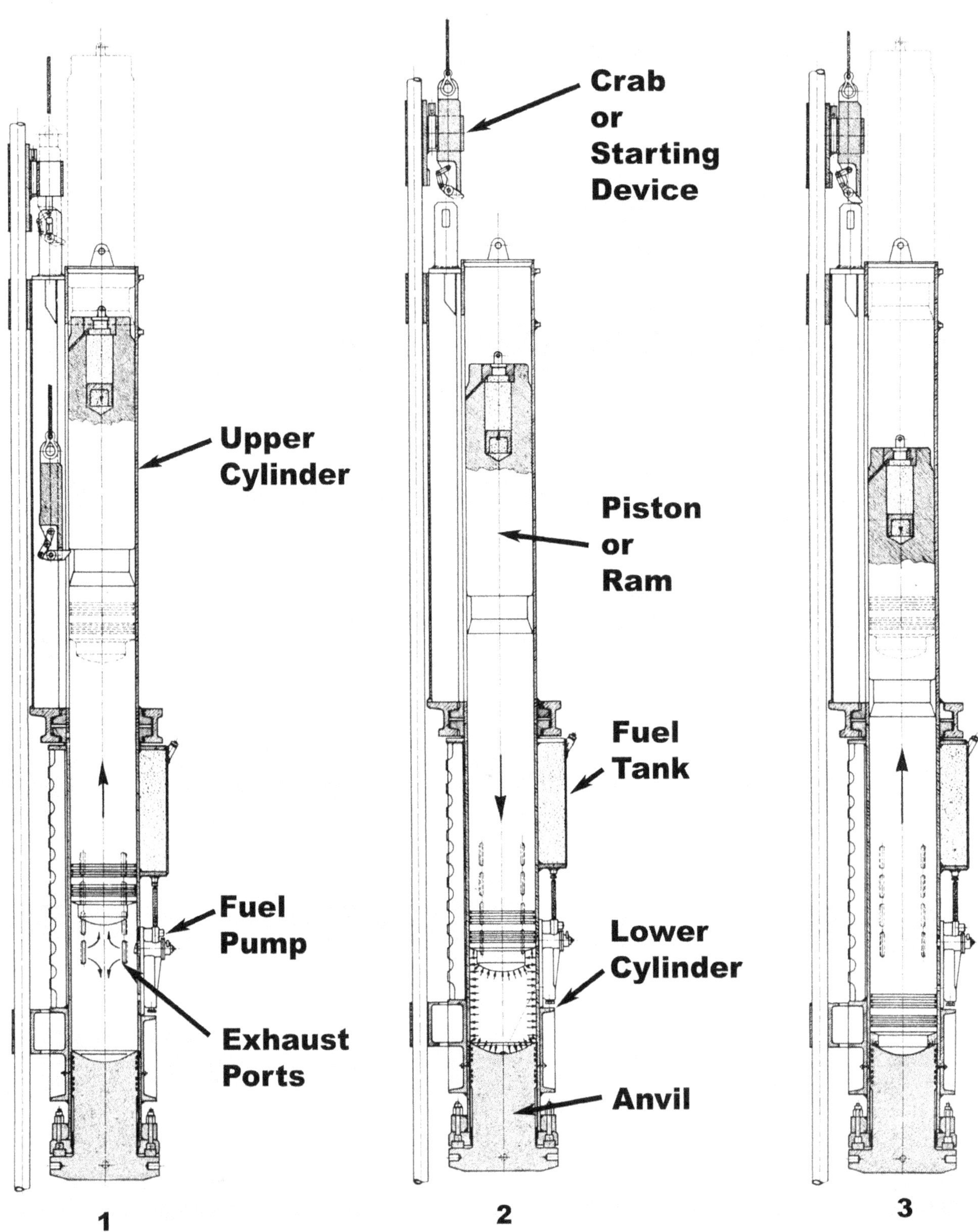

Figure 3-23 Drive Cap - Anvil Interface Details

The impact block usually has enlarged top and bottom diameters. The cylinder bottom through which the centre of the impact block passes is narrowed to a diameter just large enough to allow free movement. Thus, the impact block can move up and down a few inches, but is prevented from falling out of the cylinder when the entire hammer is raised. The exhaust ports are merely openings in the cylinder wall. The recoil dampener is located between the cylinder bottom and the impact block bottom. Its purpose is to cushion the cylinder against a strong upward impact block movement, due to pile rebound.

The open-end diesel hammer operates as is shown in Figure 3-22. The hammer operates in a three-stage cycle as follows:

1. The piston, with the assistance of the starting device (crab) driven from the winch of the pile driving rigs or hydraulically, is raised to an upper position. At this point, it is released by the starting device and falls down under its own weight.
2. Before the bottom of the ram passes the exhaust ports, the piston pushes the fuel pump lever, and fuel from the pump is supplied to the spherical recess of the anvil (some models directly inject atomized fuel into the combustion chamber). As the ram passes the exhaust ports, it closes them, beginning the compression of the air used for combustion. At the bottom of the stroke, the piston impacts the anvil.

   The energy of impact is divided between fuel vaporization and its mixing with heated air and driving of the pile. Depending on the pile length, after impact the pile starts to rebound a certain time, thereby pushing the impact block upwards. The impact block then collides with the cylinder, which is softened by the recoil dampener. For short piles, the rebound happens quickly and then a second impact between impact block and ram may occur.
3. After a short period, the air-fuel mixture is ignited, and because of the pressure of both the expanding exhaust gases and the rebound of the pile, the piston is raised up and additional driving impulse is transmitted to the pile. The upward moving ram eventually clears the exhaust ports and excess pressure is blown off. The ram continues upwards, now subject only to the slowing force of gravity and friction of the ram on the cylinder wall. Naturally, the larger the ram's upward velocity at the time it clears the ports, the higher the height it reaches before starting to descend for the next blow.

The magnitude of the upward ram velocity depends on:

- The hammer combustion pressure
- The pile mass and stiffness
- The stiffness of the hammer and pile cushions
- The soil stiffness and/or resistance.

Practically all components of the hammer-pile-soil system have an effect on the diesel hammer's stroke. Under normal conditions, the ram stroke increases as pile driving becomes harder. In soils with high resistance but low stiffness, a "spongy" driving situation results and the stroke will be relatively low. A similar situation exists with very long and flexible piles.

Advantages and disadvantages of single acting (open end) diesel hammers are shown in Table 3-5.

Table 3-5 Advantages and Disadvantages of Single Acting Diesel Hammers

Advantages	Disadvantages
Self-contained unit does not require additional equipment, thus enhancing economy of operation.	Stroke dependent on hammer-pile-soil system.
Increasing strokes in hard driving provide increasing energies; low strokes in easy driving protect concrete piles.	Relatively low blow rates (blows per minute) with high strokes.
Relatively low ram weight compared to energy (high strokes).	Potential for environmental problems (very dependent upon how hammer is outfitted with fuel and lubricants.)

Upon impact, the ram pushes the impact block, hammer cushion, helmet, and pile head rapidly downward, allowing the cylinder to fall under gravity. The impact block separates from the ram within a very short time and the pressure of the combusting air-fuel mixture will cause further separation as the ram is forced upward.

3.2.2.1.2. Performance Characteristics

The inspection of a diesel hammer should focus on the following two questions:

- Is the stroke of the hammer adequate?
- Is the energy output at a given stroke adequate?

3.2.2.1.2.1. Lack Of Stroke

There are a number of ways in which the stroke of an open-end diesel hammer can be measured. The most basic approach is to attach a so-called "jump stick" with footmarks to the cylinder. As the top of the ram emerges from the cylinder, the stroke can be read. Binoculars are recommended equipment to assist accuracy. Jump sticks are not a very good means of measuring hammer strokes, particularly since they often break off, thereby causing a safety hazard.

A better solution is the use of a Saximeter™. This unit measures the period between blows and, from this information, calculates the ram stroke. The Saximeter™ operates on the following principle: Since the ram falls freely until it is even with the exhaust ports, and since it falls only a short distance against an appreciable pressure, the assumption of a free-fall condition is accurate. If the time of fall is t_d, and if only the gravitational force exists, the height of the fall is

Equation 3-1:

$$h = \frac{g t_d^2}{2}$$

Where

- h = height of stroke, ft or m
- g = acceleration due to gravity = 32.2 ft/sec^2 or 9.8 m/sec^2
- t_d = descent time of ram, sec.

Since the ram has to fall and rise, the time of one complete cycle t_p is

Equation 3-2:

$$t_p = 2t_d$$

The stroke may thus be computed from

Equation 3-3:

$$h = \frac{g t_p^2}{8}$$

However, corrections of strokes measured with a jump stick and those computed according to this equation indicated that the computed value would usually be high by 0.3' (100 mm). For this reason a correction was made. Substituting for g the Saximeter™ formula becomes

Equation 3-4: (U.S. Units)

$$h = 4.02 t_p^2 - 0.3$$

Equation 3-5: (SI Units)

$$h = 1.23 t_p^2 - 0.1$$

The Saximeter™ formula may be used with a stopwatch. In that case, for accuracy the stopwatch should be started with blow number "1" and stopped with blow number "11". Thus, ten times periods have elapsed and the time t_p is one tenth of the measured time.

If a diesel hammer does not reach 80% of its full stroke with a blow count of six blows/inch or more, then the following potential causes for poor hammer performance should be investigated:

1. Soft driving system, long flexible piles, spongy soils. These conditions can be identified by closely observing the hammer cylinder. If the cylinder experiences a high recoil, say more than one inch, then a "bouncy" driving situation exists. The cylinder motion can most easily be observed by its movement relative to the impact block.
2. Lack of fuel or improper fuel. Lack of fuel often causes low strokes. Most commonly, the first check on a diesel hammer concerns the fuel pump. Routine inspections are difficult, however, as they involve the removal of the fuel line from the injector and a manual activation of the pump. With a measuring glass, the amount of fuel injected per blow can be determined. If the fuel pump is working properly, then a clogged injector may be the cause for low fuel injection. Improper fuel should be a rare occurrence. Improper fuel may cause low strokes through early or late ignition. Of course, if water or dirt entered the fuel system, poor hammer performance would result. Draining of the fuel through a clear cloth may indicate the presence of contaminants.
3. Excessive friction. Excessive friction not only reduces the stroke but also causes a low energy transfer to the pile. Although other hammer components also need attention, for a maximum stroke the ram lubrication is most important. Fortunately, open-end diesels show the upper ram during driving and it is easily ascertained whether the ram has a wet, shiny, lubricated surface. Scratches or dried patches of residue on the ram indicate that lubrication is insufficient. Note that too much lubrication may result in burning of grease and oil and thereby possibly cause pre-ignition. This condition would be indicated by black smoke.
4. Lack of compression. If either piston or impact block rings allow pressure to leak off, then full compression and thus full combustion pressure will not develop. Furthermore, if the recoil dampener is too thin, then the hammer's starting volume will be small, and full compression will not be developed. In bad situations, the leakage of pressure past the impact block can be noticed during pile driving when smoke appears between the recoil dampener and the impact block bottom. Compression is easily checked under a "cold blow" on a bearing pile. To perform this check, the ram is picked up as if the hammer was to be started, but the fuel pump line is kept stretched such that no fuel is pumped. The ram is then released and, after impact, it should continue to bounce on the air trapped in the chamber. Each bounce can be easily heard. At least ten bounced should occur if the hammer has good compression.

3.2.2.1.2.2. Pre-ignition

Pre-ignition means that the fuel combusts before impact occurs. Thus, pre-ignition reduces the ram impact velocity and cushions the impact.

When a hammer pre-ignites, the ram energy is not transmitted to the pile, but rather returned to the ram, which therefore strokes very high. The low energy in the pile results in a higher blow count. Thus, the pre-ignition situation has all the symptoms of a hard driving condition with a potentially low soil resistance.

The following are signs of pre-ignition in hard driving (when it may be particularly detrimental to a successful pile installation):

- Black smoke while strokes are high.
- Flames in exhaust ports.
- Blistering paint (due to excessive heat).
- No obvious metal-to-metal impact sound.

If pre-ignition is suspected, then the hammer should be stopped, allowed to cool for half an hour, and then restarted. Stroke and blow count should then be accurately monitored. If strokes are lower and blow counts are lower during the first two minutes of driving, then chances are that the hammer pre-ignited before the cooling period.

Most atomized fuel injection hammers have some pre-ignition by design. The fuel usually starts to burn when the ram is a small distance above the impact block. If the ram descends slowly, the pressure has more time to act on the ram than in the case of a high stroke, when the ram reaches the impact block within a short time. Thus, in hard driving, with high strokes and therefore high ram velocities, such as "design pre-ignition" is of little consequence. For easy driving, it is often beneficial in keeping the hammer running.

3.2.2.1.2.3. Friction

Friction, of course, is one of the most serious reasons for low energy outputs. Fortunately, during driving, the ram of open-end diesels is usually visible above the cylinder, and it can be easily checked whether the ram is shiny from lubrication. If the ram appears dry or scratched, then friction losses may be serious.

Not only the ram, but also the impact block must be well greased. The impact block moves a certain distance relative to the cylinder during impact. Most manufacturers require that the impact block be greased at half hour intervals. Some diesel hammers have a lubrication system that obviates the greasing; in these cases, the lubrication system must be functioning and have oil at all times during operation.

3.2.2.1.2.4. Overheating without pre-ignition

Overheating of a hammer may not only cause the hammer to pre-ignite, it may also lead to problems with the lubrication and fuel systems. The diesel fuel line may develop a vapor lock and reduced strokes could result. Burning of the lubricants may reduce combustion efficiency and increase friction. Thus, overheating should be avoided, particularly in a hot climate, by ensuring continuous maintenance of the hammer.

3.2.2.1.2. Checklist for Single Acting Diesel Hammers:

3.2.2.1.2.1. Before Driving Begins

- Check and record hammer model and serial numbers.
- Check the material, size, and the condition of the hammer cushion. When measuring the hammer cushion thickness, the thickness of the striker plate should also be measured and recorded.
- Check the pile cushion (if present).
- Check that helmet and pile are well matched. If the hammer type is such that no helmet is necessary, check that the hammer and pile are well matched.
- Check helmet weight; if in doubt, have it weighed.
- Check straightness of leads, and check lead connections for tightness.
- Make sure that the portions of the leads that guide the hammer do not offer excessive friction.
- Check the alignment between leads and pile.
- Make sure that all exhaust ports are open.
- Inspect recoil dampener for damage.
- Fill the fuel tank.
- Check that lubrication of all hammer grease nipples is regularly done.

3.2.2.1.2.2. During General Driving

- Use a Saximeter™ or stopwatch to calculate the stroke of the open-end diesel hammer and record these results along with the blow count.
- Check that the ram, helmet, and pile stay in alignment during driving.
- As the ram moves out of the cylinder, check its surface for signs of uniform lubrication and ram rotation.
- Ascertain that the cylinder rests on the recoil dampener between blows.

3.2.2.1.2.3. During Driving as Resistance Develops

- Observe and record color of smoke.
- Check whether smoke escapes between cylinder and impact block bottom.
- Check for any signs of hammer overheating, such as blistering paint.
- Check and record whether cylinder rebounds excessively.
- Listen as ram impacts. There should be a clear metallic sound.

3.2.3.1. Double Acting (Closed End)

Diesel Hammers

3.2.3.1.1. General Description

A closed end diesel hammer consists of a long slender piston (the ram), which moves inside a cylinder. The cylinder is closed at its upper end, thus causing the ram to compress the air trapped between ram and cylinder top. When the ram falls, it is subject to both gravity and the pressure in the "bounce chamber." For this reason the CED is also called double acting. An example of a closed end, double-acting diesel hammer is shown in Figure 3-24.

Figure 3-24 Closed End, Double-Acting Diesel Hammer

There are three manufacturers of closed end diesel hammers distributing their products in North America: Bermingham, ICE and MKT Geotechnical systems. The hammer ratings vary from 5 to not more than 80-kip-ft, with ram sizes between 1 and 10 kips. The MKT hammers are built using liquid fuel injection. ICE and Bermingham's closed end models utilize atomized fuel injection.

As shown in Figure 3-25, a closed end or double acting diesel hammer consists of eight major parts:

- A piston or ram with piston rings.
- An impact block also with rings, onto which the ram impacts.
- A cylinder in which ram and impact block move.
- One or more fuel pumps.
- Fuel injector(s).
- Exhaust ports.
- Recoil dampener.
- A bounce chamber including bounce chamber ports which allow for a regular venting of the bounce chamber.

Figure 3-25 Double Acting Diesel Hammer

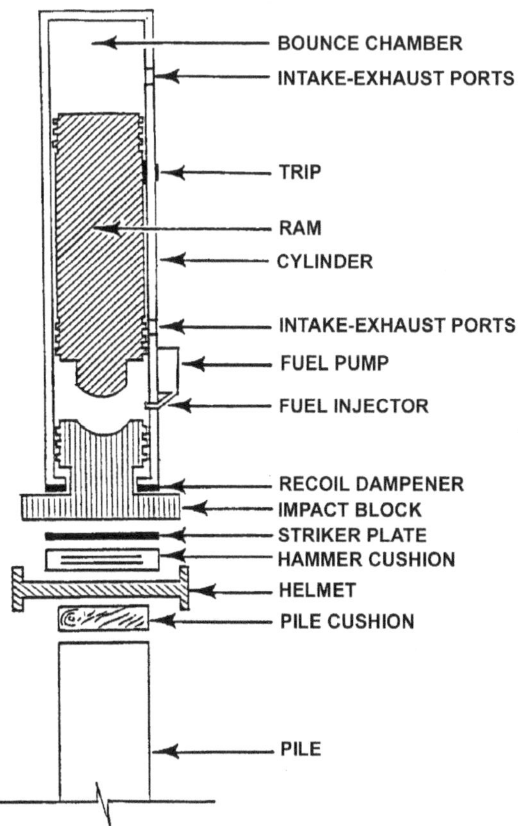

The exhaust ports are merely openings in the cylinder wall. The impact block usually has enlarged top and bottom diameter. The cylinder bottom, through which the centre of the impact block passes, has a diameter just large enough to allow free movement. Thus, the impact block can move up and down a few inches, but is prevented from falling out of the cylinder when the entire hammer is raised. The recoil dampener is located between the cylinder bottom and the impact block bottom. Its purpose is to cushion the cylinder against a strong upward impact block movement, due to the pile rebound.

Advantages and disadvantages are shown in Table 3-6.

Table 3-6 Advantages and Disadvantages of Double Acting Diesel Hammers

Advantages	Disadvantages
Self contained unit not requiring additional equipment.	Uplift in hard driving.
High blow rate (blows per minute) compared to open-end diesels.	Uncertain energy when combustion prevents ram-anvil impact.
	Complex maintenance.
	Stroke is not easily determined.

3.2.3.1.2. Operation

The ram is initially lifted a certain distance above the exhaust ports to start the hammer. This is accomplished by means of a tripping device and either a hoist or a hydraulic jack. As the ram rises, it closes the bounce chamber ports and air is compressed in the upper (bounce) chamber (see Figure 3-26).

Figure 3-26 Operating Cycle of A Double Acting Diesel Hammer

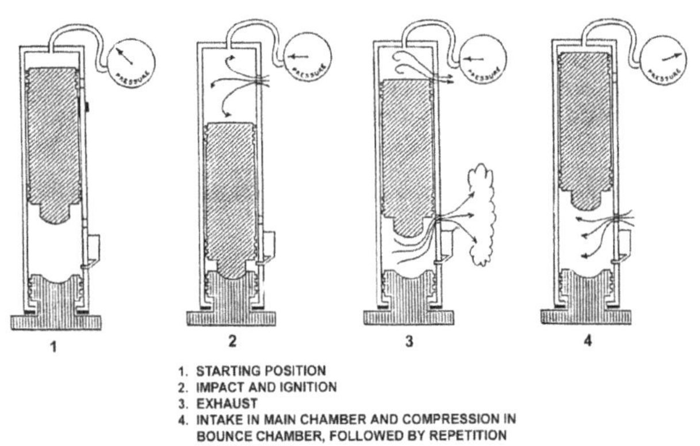

When the ram has reached the starting height, the trip is released and the ram begins to fall under the action of both gravity and the pressure of the air in the bounce chamber.

When the bottom of the ram passes the exhaust ports, it closes them and a certain volume of air is trapped inside the chamber formed by the cylinder, ram, and impact block. As the ram descends further, it compresses the trapped air, which becomes hot.

After the ram has reached a certain position between the ports and the impact block, it pushes a lever and plunger system into the fuel pump, which in turn injects a certain quantity of fuel into the chamber.

At this point, the difference between impact atomization and atomized fuel injection becomes important. For most hammers, the fuel is injected into the chamber at a low pressure and therefore in liquid forms. Even though the air in the chamber is hot, combustion of the liquid fuel will not occur before the ram impacts against the impact block, thereby atomizing the fuel.

For atomized fuel injection, combustion will take place shortly after injection starts (assuming that the air in the cylinder has become sufficiently hot) and will last as long as the injection.

Upon impact, the ram forces the impact block, hammer cushion, helmet, and pile head rapidly downward, allowing the cylinder to fall under gravity. The impact block separates from the ram within a very short time and the pressure of the combusting air-fuel mixture will cause further separation.

Depending on the pile length, the pile begins to rebound a certain time after impact, thereby pushing the impact block upward. The impact block then collides with the cylinder, which is cushioned by the recoil dampener. For short piles, the rebound occurs so quickly that a second impact between impact between impact block and ram may occur. A rebounding pile will always return energy to the ram since the gas pressure increases as the impact block moves upward. The upward moving ram eventually clears the exhaust ports, and excess pressure is blown off. The ram continues upward, now fighting against the forces of both the bounce chamber pressure and gravity. Naturally, the higher the ram's upward velocity at the time it clears the ports, the higher the height it reaches before it starts to descend for the next blow.

The magnitude of the upward ram velocity depends on:

1) The hammer combustion pressure.
2) The pile mass and stiffness.
3) The stiffness of the hammer and pile

cushions.

4) The soil stiffness and/or resistance.

5) The rate at which the pressure in the bounce chamber increases.

All components of the hammer-pile-soil system have an effect on the diesel hammer's stroke. Under normal conditions, the ram stroke increases as pile driving becomes harder. In soils with high resistance but low stiffness, a "spongy" driving situation results and the stroke will be relatively low. A similar situation exists with very long and flexible piles.

3.2.3.1.3. Performance Characteristics

The inspection of a closed end diesel hammer should focus on the following two questions:

- Is the stroke of the hammer adequate?
- Is the energy output at a given stroke adequate?

3.2.3.1.3.1. Lack of Stroke

The stroke of a closed end diesel hammer can be indirectly measured by means of a bounce chamber pressure gauge (BCPG). This pneumatic gauge is commonly connected to the bounce chamber by means of a hose. There are a few MKT hammer models with a BCPG directly attached to the hammer cylinder. Such a gauge type is difficult to read due to its location.

For each blow, the dial of the BCPG registers the peak bounce chamber pressure. The hammer manufacturer usually provides a chart that allows the conversion from bounce chamber pressure to equivalent stroke (equivalent to the free fall stroke) or the corresponding energy (the product of ram weight and equivalent stroke). Note that it is not important to know the actual stroke. An example of such a chart is shown in Figure 3-27.

Figure 3-27 Sample Energy Chart for Closed Ended Diesel Hammer

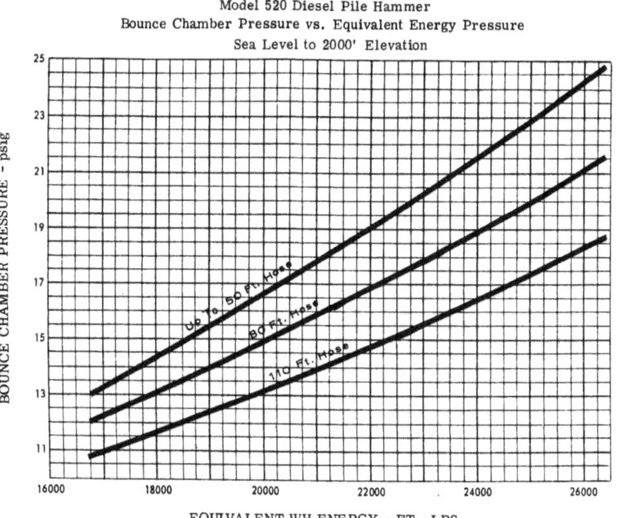

Of course, the bounce chamber pressure is an indirect indicator of the ram stroke. The measurement is sufficiently accurate if the manufacturer's calibrated hoses are used and if the bounce chamber volume is not changed by lubrication oil. Lubrication oil may enter the compression tank (ICE hammers) or stay with unvented bounce chambers. Thus, if the hammer manufacturer recommends draining the bounce chamber, then this should be done daily or the measured bounce chamber pressures may be meaningless.

It is possible for the bounce chamber pressure to reach such a magnitude that the cylinder weight cannot balance it. Then the cylinder lifts up, together with the ram, and they settle back onto the recoil dampener when the ram falls. Such a driving condition is highly undesirable since it is rather unstable. Thus, when lift-off occurs, the hammer operator must reduce the amount of fuel injected into the hammer. From that point on, the energy output of the hammer will be smaller than at the time just before lift-off became imminent. This is because there is no longer as much combustion pressure acting on the pile head. When observed this condition should be recorded and the bounce chamber pressures reread after a fuel pump adjustment has been made.

In addition to measuring the bounce chamber pressure, the hammer rate should be checked by means of a stopwatch or a Saximeter™. This unit measures the time between hammer blows and prints "blows per minute." Although not as important for closed end as for open-end diesels, the hammer speed is related to the stroke and therefore the Saximeter provides an independent check. The causes for low strokes are the same as those for open-end diesel hammers.

3.2.3.1.3.2. Other Hammer Problems

These are the same as those for open-end diesel hammers.

3.2.3.1.4. Checklist for Closed End Diesel Hammers:

3.2.3.1.4.1. Before Driving Begins:

- Check and record hammer model and serial numbers.
- Check the material, size, and the condition of the hammer cushion. When measuring the hammer cushion thickness, the thickness of the striker plate should also be measured and recorded.
- Check the pile cushion (if present).
- Check that helmet and pile are well matched. If the hammer type is such that no helmet is necessary, check that the hammer and pile are well matched.
- Check helmet weight; if in doubt, have it weighed.
- Check straightness of leads, and check lead connections for tightness.
- Check that the leads do not offer excessive friction.
- Check the alignment between leads and pile.
- Check that all exhaust ports are open.
- Inspect recoil dampener for damage.
- Fill the fuel tank.
- Check that lubrication of all hammer grease nipples is regularly done.

3.2.3.1.4.2. During General Driving:

- Using the manufacturer's gauge and the correct connection hose, measure the bounce chamber pressure and record.
- Using a stopwatch or Saximeter ™, measure blows per minute and record this, together with the blow counts.
- Check that the ram, helmet, and pile stay in alignment during driving.
- Ascertain that the cylinder rests on the recoil dampener between blows.

3.2.3.1.4.3. During Driving as Resistance Increases:

- Observe and record color of smoke.
- Check whether smoke escapes between cylinder and impact block bottom.
- Check for any signs of hammer overheating, such as blistering paint.
- Check and record whether the cylinder rebounds excessively.
- Listen as ram impacts. There should be a clear metallic sound.
- Observe and record whether lift-off occurs, and when fuel pump adjustments are made.

3.2.3.3. Diesel Hammer Fuel Systems

3.2.3.3.1. Liquid Fuel Injection

Liquid fuel injection, low-pressure injection, and impact atomization are all names for the same process, which is unique to pile driving hammers. In this process, the falling ram compresses the air trapped in the combustion chamber, and fuel is then injected such that it collects in liquid form on top of the impact block. The impacting ram then disperses this fuel, which is thereby mixed with the hot, compressed air. Combustion takes place soon after.

For a cool hammer, it may be two milliseconds after impact before combustion begins. As the hammer heats up, the fuel will begin to evaporate as soon as it is injected, and combustion may commence without delay at the time of impact. The hotter the hammer, the more likely there will be an early combustion. An overheated hammer is likely to preignite, i.e., the combustion will be well advanced by the time of impact.

Diesel hammer manufacturers have focused their attention to the fuel injection process and the shape of the impacting ram and impact block surfaces. One reason for experimenting has been the sometimes-poor starting behavior of impact atomization hammers. At the beginning of driving, pile rebound is usually so weak that the stroke tends to remain small. Thus, in easy driving, high hammer energy is required to maintain a good stroke. At the same time, however, the hammer is cold and the relative speed of ram and impact block is low. Thus, the fuel neither evaporates nor does it become as finely atomized as in hard driving. A relatively inefficient combustion results.

3.2.3.3.2. Atomized Fuel Injection

Using the same principles as seen in other types of diesel engines, atomized fuel injection diesel hammers compress an air charge in the cylinder, and once the air is hot and compressed, fuel is injected under high pressure. The injection pressure may be 1000 psi (6895 kPa) or more.

High-pressure injection causes the fuel to enter the chamber as a finely distributed mist. Thus fuel begins to burn as soon as this fuel mist enters the hot air. The initial injection time corresponds (with only slight variations) to a specific distance between the ram and the impact block, and therefore to a certain volume. Similarly, injection and/or combustion end after the ram has reached a certain point during rebound.

In easy driving, combustion will begin independently of the relative speed between the ram and impact block. However, since the ram is falling slowly (short stroke), the time between the start of combustion and impact will be longer. This situation allows the combustion pressure to develop to the point where it stops the ram before impact. Thus, little energy is transferred into the pile.

Atomized fuel injection hammers work very well if they are maintained perfectly. It is important that the fuel injection timing is accurate down to the millisecond. Worn injectors or fuel pump parts, or damaged fuel lines between the pump and injector can all greatly affect performance.

3.2.4. Driving Accessories

It is not possible for the striking end of the ram of an impact hammer to directly adapt itself to all shapes of piles. It is necessary to have driving accessories of various types and sizes to be inserted between the bottom of the hammer and the pile to both mate the two geometrically and transmit the impact hammer force to the pile.

3.2.4.1. Hammer Cushion

Most impact hammers have some type of cushion under the end of the ram, which receives first the striking energy of the hammer. This cushion is necessary to protect the hammer's striking parts from damage. Cushion material, in conjunction with the hammer's ram and the pile, also affects the impact characteristics of the hammer; a softer cushion, for example, will produce an impulse that is longer and has a lower maximum force. At the top of the helmet there is usually a pot shaped recess that contains the hammer cushion. A striker plate sits atop and protects the hammer cushion, such that the material is compressed as uniformly as possible. The actual cushion material and its configuration will vary, depending upon the hammer configuration and the cushion material being used. Any hammer cushion should be used in accordance with the recommendations of the hammer manufacturer. Table 3-7 presents a summary of cushion materials and their characteristics.

Most hammer cushions have a limited life. They compress, and material may need to be added to maintain a certain thickness. Badly broken or burnt hammer cushions are inefficient. Hammer cushions with poor properties or improper thickness lead to inefficient hammer operation and/or hammer failure.

Figure 3-28 shows typical cushion configurations for both air/steam and diesel hammers. Pile cushion is only included in concrete and plastic piling; with steel and wood piling, no pile cushion is normally needed.

In some instances, no hammer cushion is required. Some diesel and hydraulic hammers have no hammer cushion. With wood piling, some air/steam hammers can be equipped with special bases to drive wood piling without hammer cushion material.

For air/steam hammers, Figure 3-28(A) shows an integrally cast hammer cushion pot. Most air/steam hammers can use as an alternative a capblock follower or shield, where micarta and aluminum cushion material is stacked into a piece separate from the pile cap. Such a configuration is shown in Figure 3-28(C).

3.2.4.2. Anvil

The rams of most external combustion hammers strike the hammer cushion or top plate directly. With

Table 3-7 Summary of Cushion Material Characteristics

Name of Material	Description	Advantages	Disadvantages
Wire Rope Biscuits	Wire rope coiled into flat biscuits, then placed into cushion receptacle.	Inexpensive material plentiful on project site.	Material hardens rapidly into steel mass extensive hammer and pile damage possible.
Force Ten	Wires braided into steel cloth plates.	Long life	These cushions must be carefully monitored since hard driving tends to fuse the strands into a solid steel plate. Before these strands have fused, these materials must be replaced. A daily inspection is necessary. Not allowed on may state transportation projects.
Micarta and Aluminum (Conbest)	Phenolic plates alternated with aluminum plates. Aluminum plates used for heat dissipation; these are frequently omitted with diesel hammers.	Long life versatile, can be used with many hammers.	Material too hard to be used with some hammers. It must be replaced when it starts to pulverize or when it disintegrates into various layers.
Hamortex	Aluminum foil bonded with various plastic and paper materials and spirally wound into disks.	Excellent range of elasticity moduli. Can be used as concrete pile cushion as well as hammer cushion.	Inconsistent life varies widely between projects.
MC-904/Blue Nylon	Type of nylon, cast into discs. Sometimes alternated with aluminum for heat dissipation. Usually supplied in 2" thick disks.	Excellent range of elasticity moduli.	Material breaks up and melts easily under hard conditions. Best suited for diesel hammers. Occasional vertical cracking is not detrimental. However, after the cushion develops horizontal cracks, it should be replaced.
Plywood	Plywood sheets stacked and mounted onto concrete pile used almost exclusively as a pile cushion.	Inexpensive and soft.	Poor coefficient of restitution; absorbs much impact energy. Not acceptable as a hammer cushion, either in plank or chip formats.
End grain hardwood (oak, hickory, bongossi, etc.)	Wood cut to use in cushion receptacle as single block or into blocks that are then fit into receptacle. For a hammer cushion, hardwood is only acceptable if it is well fitted into the helmet and if it is placed with the wood grain parallel to the hammer-pile axis.	Good range of modulus of elasticity.	Low coefficient of restitution sometimes expensive and hard to find. Burns during use. Where the wood shows signs of burning or brooming, it should be replaced. An inspection should be conducted after 2000 blows or daily, whichever comes first.

Figure 3-28 Typical Cushion Configurations

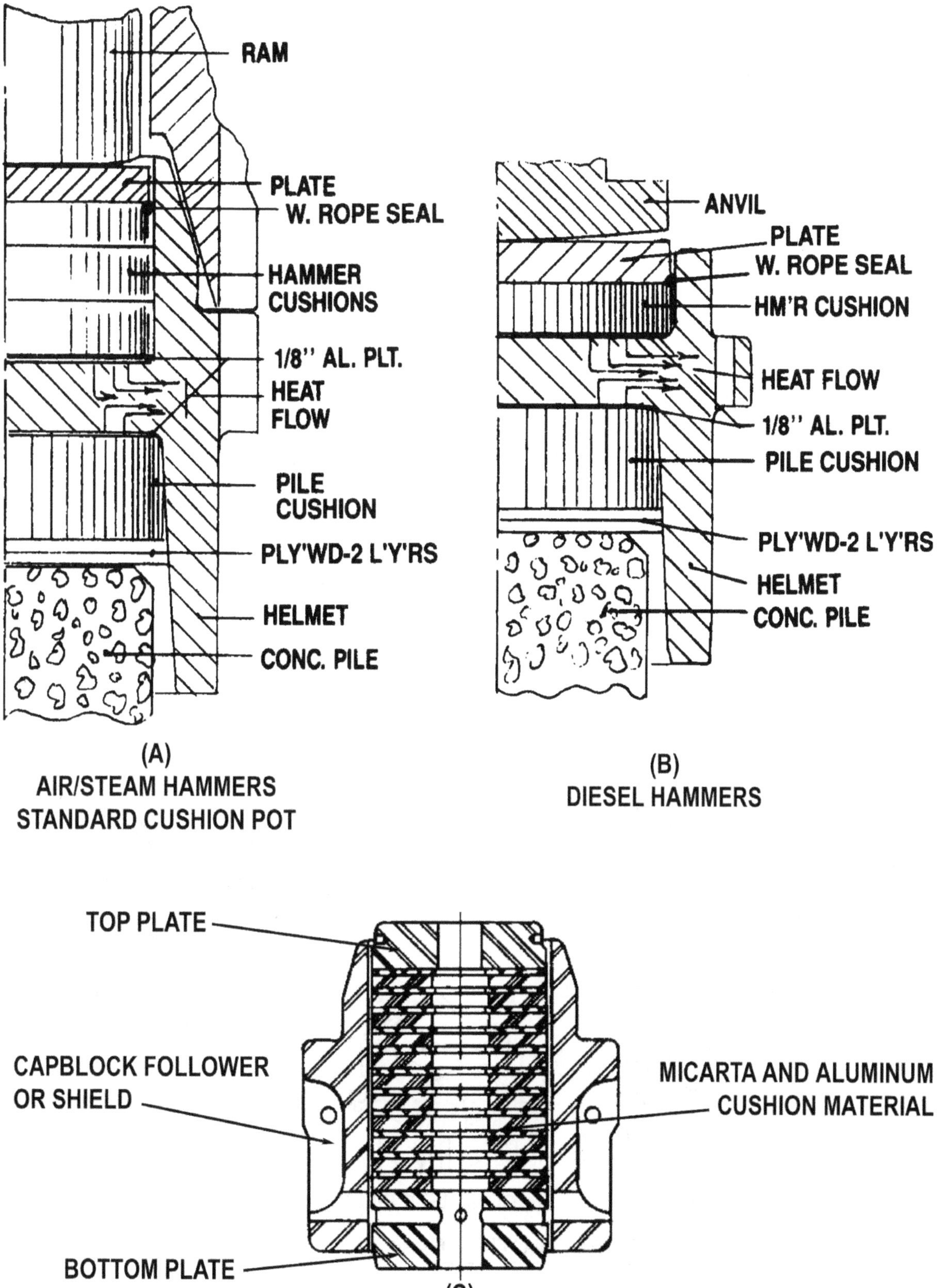

RAM
PLATE
W. ROPE SEAL
HAMMER CUSHIONS
1/8'' AL. PLT.
HEAT FLOW
PILE CUSHION
PLY'WD-2 L'Y'RS
HELMET
CONC. PILE

(A)
AIR/STEAM HAMMERS
STANDARD CUSHION POT

ANVIL
PLATE
W. ROPE SEAL
HM'R CUSHION
HEAT FLOW
1/8'' AL. PLT.
PILE CUSHION
PLY'WD-2 L'Y'RS
HELMET
CONC. PILE

(B)
DIESEL HAMMERS

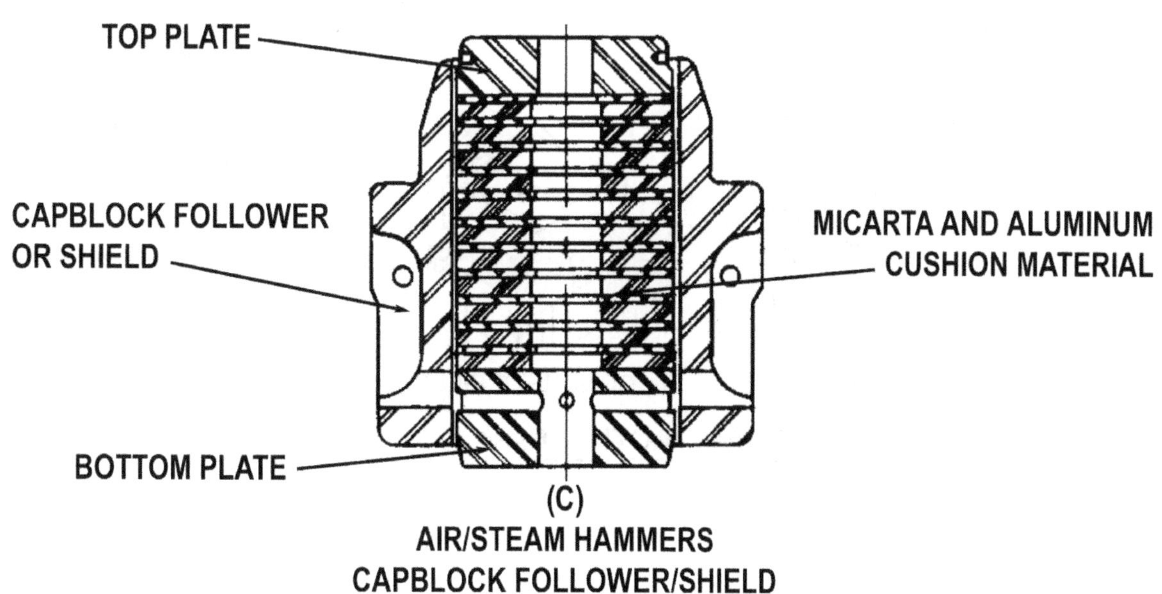

(C)
AIR/STEAM HAMMERS
CAPBLOCK FOLLOWER/SHIELD

Figure 3-30 Typical Driving Helmets

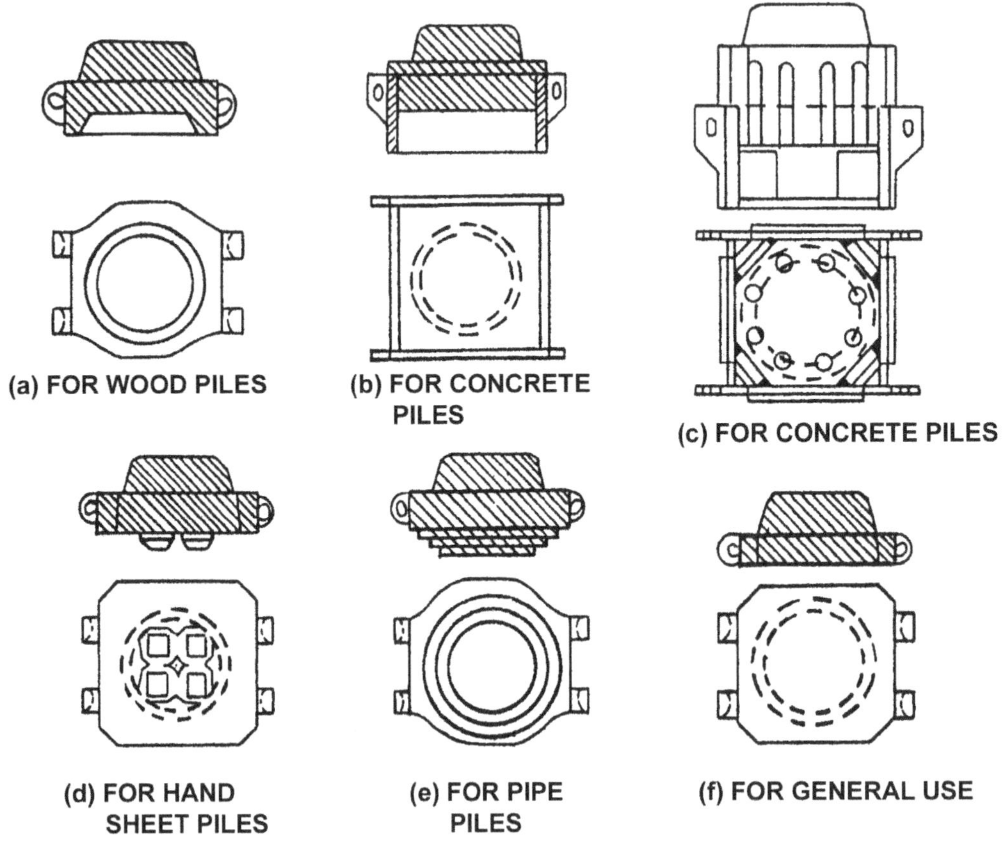

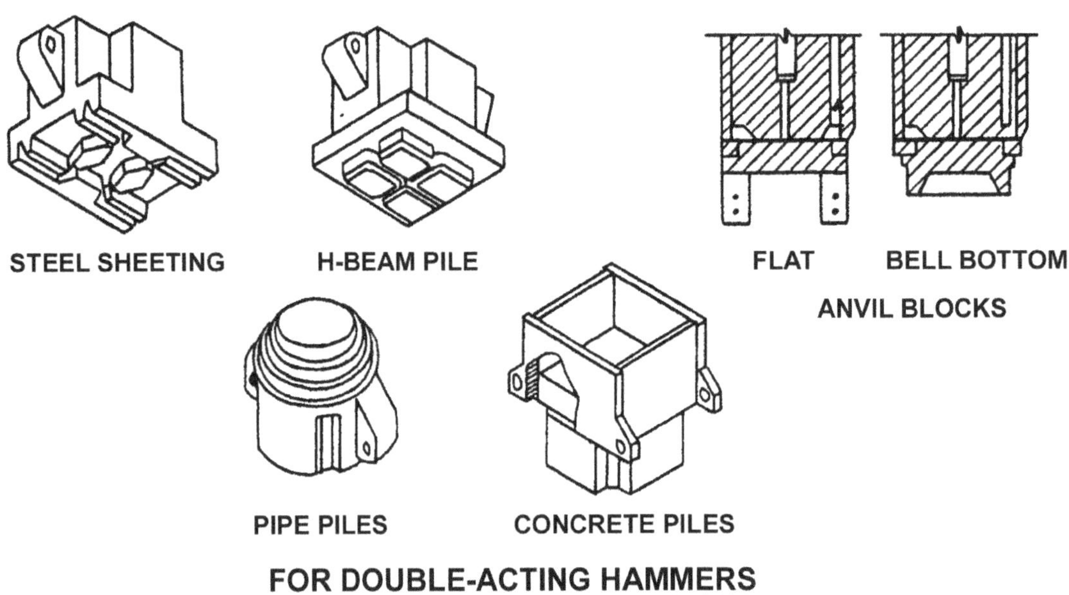

internal combustion hammers, an anvil is necessary to trap the combustible mixture and allow it to build pressure. Figure 3-29 shows a diesel anvil out of the cylinder. The term "anvil" is also sometimes used to describe the drive cap or helmet (see next paragraph).

Figure 3-29 Diesel Hammer Anvil

3.2.4.3. Helmet

The helmet is the heavy and rigid steel block between hammer and pile. Often the hammer manufacturer supplies helmet inserts or adapters. The helmet distributes the blow from the hammer more uniformly to the head of the pile to minimize pile damage. It is therefore important that the helmet's striking surface is smooth and contacts the pile top evenly. It should not allow more than approximately 2 inches (50 mm) of lateral movement. The helmet weight (more properly mass) must be known to evaluate the drivability of a pile using the wave equation.

A poorly seated helmet will cause high-localized stresses at the pile top. In addition, poor seating may cause the exposed portion of a long pile to buckle elastically under each hammer blow (this is known as "pile whipping"). Figure 3-30 shows some typical driving helmets.

Because diesel hammers do not impact the pile with the intensity of air/steam hammers, it is possible to use a two-piece driving helmet. Figure 3-31 shows a universal driving helmet and Figure 3-32 shows the filler helmet (in this case for pipe piles) for diesel hammers. The top of the universal helmet (to the left in the figure) is filled with the hammer cushion and top plate (if desired.) The bottom is a square adapter that mates with the filler; the two are cabled together through the cabling holes.

Figure 3-31 Universal Driving Helmet for Diesel Pile Hammer

Figure 3-32 Filler Helmet for Diesel Pile Hammer

3.2.4.4. Pile Cushion

When driving concrete piles, it is necessary to use a pile cushion between the hammer and the pile. In the United States, plywood is the most common pile cushion material. Hardwood boards may also be used with the grain perpendicular to the pile axis. If the pile is in danger of being cracked in tension during driving, then pile cushions may be very thick (up to 18" (457 mm)). For the protection of the pile top against a compressive damage, thinner cushions are usually sufficient. The depth of these cushions can vary from 6" (152 mm) to 18" (457 mm).

The pile cushion material should be dry and unburned. Once the wood cushion begins to burn, it should be replaced. Also, after approximately 1000 hammer blows the cushion may become compressed and excessively hard. Recommendations must be obtained for each individual concrete pile job as to the required cushion properties and frequency of replacement.

3.2.4.5. Mandrel

A mandrel is used to install thin-wall shell piles, which are subsequently filled with concrete. The mandrel is necessary because the pile wall is too thin to withstand the stresses of driving. Figure 3-33 shows diagrams for several mandrels. Figure 3-34 shows a mandrel in use during driving.

Figure 3-33 Mandrel for Shell Piles

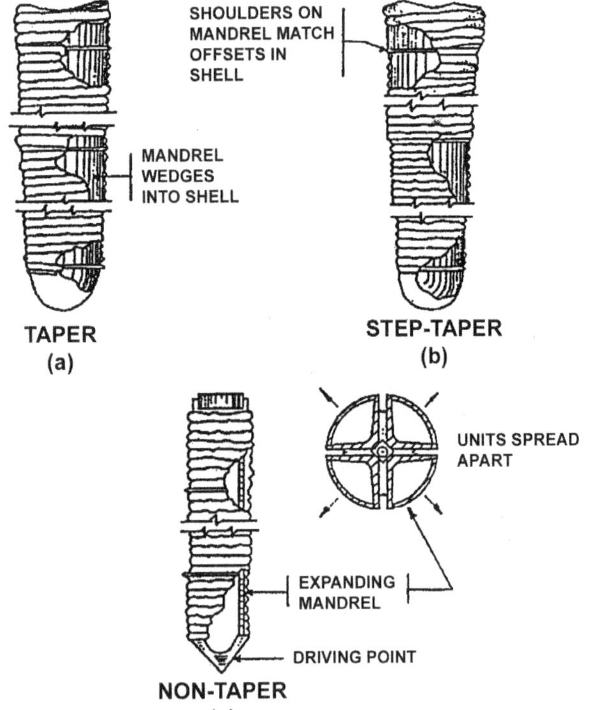

Figure 3-34 Mandrel in Use During Driving

Mandrels may be tapered, stepped or expanding, as the configuration of the pile requires. Where used with uniform diameter shell they should be expandable to tightly grip the pile for driving and to prevent collapse during installation. Mandrels must resist hammer blows and transmit energy to the bottom of the pile as well as distributing forces along the corrugations.

Uniform diameter shell piles are installed with a variety of expandable mandrels. Typical shell piles are 12 to 18 gage with 1/2 x 2 in. corrugations formed as the steel passes through a corrugator. The shell is then spirally rolled (12-1/4 in. O.D. is a popular nominal size) and fusion welded or crimped to make the pile watertight.

Most mandrels are the segmental leaf type. They have corrugations to match those in the shell and are expanded by mechanical, pneumatic or hydraulic means to grip the shell tightly. They also support the shell during installation to protect against deformation and tearing.

In a mandrel such as the Guild Mandrel, there are groupings or sets of internal sheaves, spaced at predetermined intervals. These sheaves are separated by a cable, which is anchored at the mandrel tip, threaded alternately over and under the sheaves and tied off to a hydraulic ram in the head section of the mandrel.

As the ram is retracted, the cable pull forces the alternating sheaves apart. This separates the mandrel leaves and, in turn, causes the helix bars on its exterior to mesh with the corresponding corrugations in the shell. Because of the high, positive expansion pressure developed in this system, the shell and mandrel operate as a single unit.

Steel shells are slipped on the mandrel for driving. This process is simpler with tapered piles than for uniform diameter ones.

3.2.4.6. Followers

A follower is a steel member placed between the pile hammer and pile that allows the pile to be driven below the reach of the leads. The most common uses are to drive a pile below the top of an existing structure or for driving piles over water. Although followers can make driving less difficult, there are several problems associated with their use. Experience shows it to be quite difficult to maintain alignment between the pile and follower, especially for battered piles. Additionally, erratic energy losses due to poor connection between the pile and follower, frequent misalignment, and follower flexibility make it nearly impossible to equate blow count with pile capacity. For these reasons, most specifications exclude the use of followers. If a follower must be used, it should be selected so that its impedance is approximately the same as the pile impedance[3]. If concrete piles are being driven, then some cushion must be used between the follower and the pile.

In the case of driving piles underwater, one way to eliminate the need for a follower is to use a pile hammer that can be driven underwater. This is discussed in 3.5.2.

3.3. Vibratory Drivers

A vibratory pile driver is a machine that installs piling into the ground by applying a rapidly alternating force to the pile. This is generally accomplished by rotating eccentric weights about shafts. Each rotating eccentric produces forces acting in a single plane and directed toward the centerline of the shaft. Figure 3-35 shows the basic set-up for the rotating eccentric weights used in most current vibratory pile driving/extracting equipment, and Figure 3-36 shows actual eccentrics and gears in a vibratory hammer. The weights are set off centre of the axis of rotation by the eccentric arm. If only one eccentric is used, in one revolution a force will be exerted in all directions, and a lateral whipping effect would result. To avoid this problem, the eccentrics are paired so the lateral forces cancel each other, leaving only axial force for the pile. Some machines have several pairs of smaller, identical eccentrics synchronized and obtain the same effect as with one larger pair[4].

Figure 3-35 Eccentrics for Vibratory Hammers

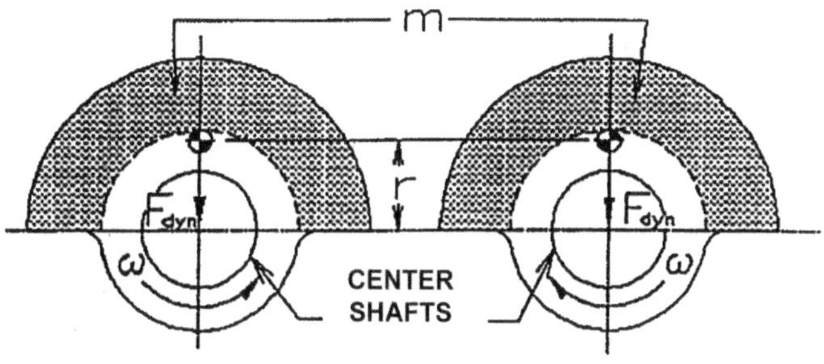

Figure 3-36 Eccentrics and Gears in a Vibratory Hammer

The force generated by these rotating eccentrics is transmitted to the gearbox – and ultimately to the pile – by bearings. A typical eccentric bearing is shown in Figure 3-37.

Figure 3-37 Eccentric Bearing

Figure 3-38 Elastomers or Springs Between the Vibrating and Non-Vibrating Portions of the Vibratory Hammer

The rotation of the eccentrics produces a sinusoidal motion of the pile, clamp and gearbox of the vibratory hammer. This is discussed in more detail in the next section.

In addition to the vibrating mass, most vibratory hammers have mass that does not vibrate when the eccentrics rotate (suspension housing, half of the elastomers or springs, and the part of the hoses or cables suspended by the suspension housing). This mass can be likened to a downward pull from a crane and is usually beneficial when driving. Some models have additional weights that can be added to the suspension system. The suspension system does affect the amplitude of the system; in fact, excessive suspension weight and/or down crowding (as is the case with excavator mounted units) can actually attenuate the vibrations of the system. A good suspension system minimizes boom vibration, gives the manufacturers a shock free location to attach hoses or cables, and when leads are required, is the only place to locate guide channels. Figure 3-38 shows elastomers used in vibratory hammers. The bracket shown is then mounted on the gearbox.

3.3.1. Performance of Vibratory Systems

Vibratory hammers[5] are able to install piles into the ground because they loosen the soil surrounding the pile sufficiently to enable the pile to penetrate into the ground by the combination of its own weight and whatever static force is exerted by or through the suspension system. With cohesionless soils, this involves the upward force in the vibration cycle overcoming both gravity and the overburden pressure and essentially rendering the surrounding sand "weightless," creating a "fluid" effect on the soil (even though water does not have to be present for this to take place.) With cohesive soils, this effect is enhanced and in some cases supplemented by thixotropy, where the soil is transformed in to a gel by a reversible chemical process.[6] In both cases, the elevation of pore water pressures during vibration also assist the penetration of vibrated piles.

The result of these phenomena is an effective system for installing piles. The main obstacle to successful vibratory driving is the toe resistance of the pile, especially in cohesive soils. This is one reason why piles such as sheet and H-piles, with their small toe areas, are the most widely and successfully vibrated type of piles.

Consider the system shown in Figure 3-39, where counter rotating eccentric masses similar to those found on a vibratory hammer excite a mass. As the rotating masses complete one revolution, they create vertical motion in the pile.

Figure 3-39 Schematic of Vibratory System

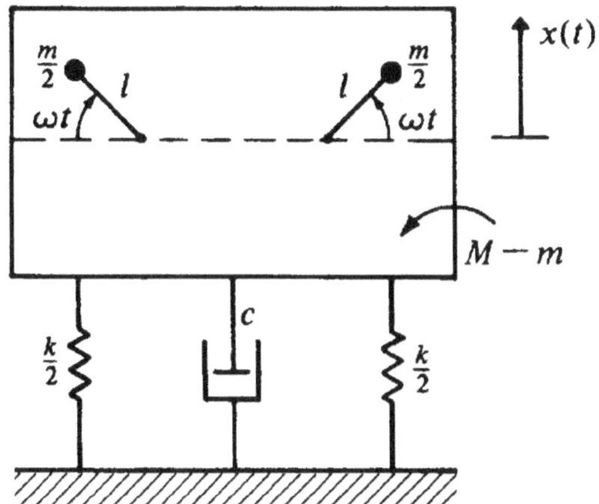

The system is connected to an infinite mass by a spring and dashpot system. The equation for the motion of the system is

Equation 3-6:

$$Mx''(t) + cx'(t) + kx(t) = F(t) = K\omega^2 \sin(\omega t)$$

Where

- M = total vibrating mass of system, kg
- K = eccentric moment of the rotating masses, kg-m
- ω = rotational speed (frequency) of the rotating masses, rad/sec
- c = viscous dampening of system, N-sec/m
- k = spring constant of system, N/m
- t = time, sec
- x = displacement, sec
- x' = instantaneous velocity of the system, m/sec
- x" = instantaneous acceleration of system, m/sec^2

The first variable we need to discuss is the frequency ω. The frequency range of a vibratory hammer is usually selected for the widest range of soils but since this factor is a function of horsepower, a compromise is sometimes reached., the frequency should be varied during driving since the soil varies and different soils react favorably to frequencies near the natural frequency of that particular soil.

Vibratory frequency is seldom quoted in radians per second. There are two more common ways of expressing this. The first is in cycles/second, or Hertz, which is computed by the equation

Equation 3-7:

$$\theta = \frac{\omega}{2\pi}$$

Where

- θ = frequency of vibrations, Hz = (Eccentric RPM)/60

The second (and in the U.S. the most common) is in revolutions per minute (RPM), calculated using the equation

Equation 3-8:

$$RPM = 60\theta = \frac{30\omega}{\pi}$$

One revolution per minute (RPM) of the eccentrics equals one vibration per minute (VPM) of the pile.

Eccentric moment is the most commonly specification used by both manufacturers and users to rate vibratory hammers. In calculating the eccentric moment, the only cross-sectional area is included in the eccentric moment is the shaded area shown in Figure 3-35. This area, multiplied by both the thickness of the eccentric and the density of the material, will yield the eccentric mass m. The eccentric moment can then be computed by the equation

Equation 3-9:

$$K = mr$$

Where

- m = eccentric mass, kg
- r = eccentric radius of the mass, m

Some eccentrics have a smaller eccentric radius than others so they must be heavier to have the same moment. Generally speaking a smaller eccentric radius has a negative affect for two reasons: it increases the vibrating mass, and it increases the starting torque the motors must apply to the eccentrics to reach their operating rotational speed.

Let us now consider the "free-hanging[7]" case, i.e., only considering the vibrating mass of the system, thus c=k=0. Equation 3-6 thus reduces to

Equation 3-10:

$$x'' = \frac{K\omega^2 \sin(\omega t)}{M}$$

Where

- M = total vibrating mass of hammer, kg

The solution of this equation is

Equation 3-11:

$$x = \frac{K \sin(\omega t)}{M}$$

Where

- x = system displacement, meters

This sinusoidal motion (in this case with an impact-vibration hammer) is shown by the chalk lines on the front the hammer shown in Figure 3-40. This also illustrates a quick method of measuring the amplitude of a vibratory hammer. A piece of chalk or other is passed across a vertical surface of the machine; the result produces a rough "oscillograph" of the hammer motion. The relation between this motion and the position of the eccentrics during rotation is shown in Figure 3-41.

Figure 3-40 Sinusoidal Motion of the Eccentrics

Figure 3-41 Position of the Eccentrics During Rotation

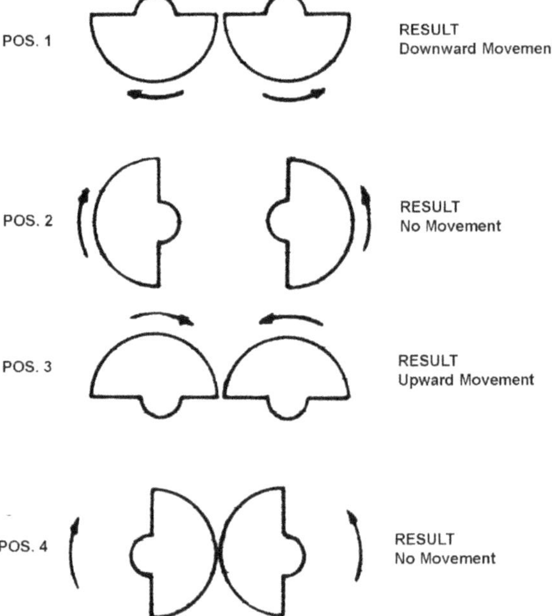

In the process of integrating Equation 3-10, we can derive three very important quantities. The first is the ratio of accelerations, or the peak acceleration during a vibratory cycle it is

Equation 3-12:

$$n = \frac{K\omega^2}{gM}$$

Where

- n = ratio of maximum acceleration of system to acceleration due to gravity, g's
- g = acceleration of gravity = 9.8 m/sec^2

The second is the peak velocity, which is

Equation 3-13:

$$v_{dyn} = \frac{gn}{\omega} = \frac{K\omega}{M}$$

Where

- v_{dyn} = peak dynamic velocity during the cycle, m/sec

These quantities are important because the power transmitted to the soil must be done in an efficient manner from a high-energy source through the pile-soil interface to a low one in the soil. Minimum values for n have been established from 1.5 to 9, but there is no consensus on this.

Finally, the maximum displacement is

Equation 3-14:

$$x_{max} = \frac{K}{M}$$

Where

- x_{max} = maximum displacement of system (zero to peak), meters

Since the acceleration, velocity, and displacement of the system as determined from Equation 3-10 are all sinusoidal with respect to time, these quantities are measured from the zero line of the sine wave. Customarily, the maximum cycle displacement of the vibrator, called the amplitude (sometimes called the double amplitude), is measured from peak to peak and is expressed as

Equation 3-15:

$$A = 2x_{max} = 2\frac{K}{M}$$

Where

- A = amplitude of system (peak to peak), m[8]

Equation 3-15 is valid for free hanging vibratory hammers. However, when the vibratory hammer is clamped to the pile and pile weight is added to soil friction, amplitude is reduced, so it is important to have as much free hanging amplitude as possible. Increasing eccentric moment or decreasing vibratory mass, if possible, can increase amplitude. The fallacy that the heavier vibratory hammer drives well is therefore not true.

For a rotating body, the force exerted on the centre shaft is given by the equation

Equation 3-16:

$$F_{dyn} = \frac{K\omega^2}{1000}$$

- F_{dyn} = dynamic force of eccentrics, kN

This force is referred to as the dynamic force. In U.S. units, the formula for Dynamic Force is:

Equation 3-17:

$$F_{dyn_{U.S.\ Tons}} = \frac{K_{in.-lbs.}(RPM)^2}{70,423,000}$$

It is interesting to note that, for optimum performance,

Equation 3-18:

$$F_{stat} = W_{nv} + F_{down} + gM = \frac{F_{dyn}}{2}$$

Where

- F_{stat} = total downward static force, kN
- W_{nv} = Non-vibrating (suspension) weight, kN
- F_{down} = down crowd or down pull force, kN

The "free hanging" model can be used to rate and compare vibratory hammers. However, these are based on a vibratory hammer with no load. In practice vibrators a) operate with a load to do useful work and b) require a certain amount of power to do this work. The following considers the matter of vibratory power consumption from a theoretical standpoint, although it has been applied to real vibratory hammers with success.

A theoretical power requirement could be derived from Equation 3-6. The difficulty in doing so concerns the spring constant k, which is unrealistic as it stands for the following reasons:

- A purely linear k would preclude penetration into the soil and movement of the pile. Plastic deformation is necessary for pile movement. This is also true with impact driving as well.
- The existence of k without c will result in no net power consumption. Although k can certainly modulate power consumption in the system power dissipation can only take place when c > 0.
- The general object of vibratory driving (especially when overcoming shaft resistance) is to liquefy the surrounding soil and reduce the resistance to pile penetration. Given this, a more appropriate model of the system would be to assume that k =0.

Making this assumption reduces Equation 3-6 to

Equation 3-19:

$$Mx''(t) + cx'(t) = F(t) = K\omega^2 \sin(\omega t)$$

It can be shown[9] that the continuous vibratory power consumption of the system is given by the equation

Equation 3-20:

$$N = \alpha N_o$$

Where

- N = power requirements for the vibratory hammer, kW
- $\alpha = \frac{\sin(2\phi)}{4}$
- $N_o = \frac{K^2\omega^3}{M}$
- φ = phase angle of the vibrations, radians

The maximum power takes place when φ = π/4 = 45°, thus α_{max} = 1/4. Substituting this into Equation 3-20, the maximum possible power is

Equation 3-21:

$$N_{max} = \frac{K^2\omega^3}{4M}$$

This model has the possibilities of some degree of realism[10] since a) the liquefaction of the soil is an essential part of vibratory driving, and a possible source of a fluid type dampener, and b) the possibility to model energy radiation into the soil as a dashpot, as is done in the wave equation. Equation 3-21, however, should be understood as a theoretical minimum power requirement when designing and specifying vibratory pile driving equipment. Values of α for actual use should include energy losses in the hydraulic or electric system driving the eccentrics, losses in the gearbox itself, losses in the clamp to the pile, and effects of the soil. With this theory, however, these values can be derived experimentally as well. Moreover, the model above is a linear one; research continues on the behavior of vibratory hammers relative to the capacity of the piles they are driving, with which considerations such as these are relevant.

The "bottom line" of these discussions is simple: enough horsepower to drive the eccentrics at their maximum frequency is what is required. For a hydraulic powered vibratory hammer,

Equation 3-22:

$$\text{Horsepower} = \frac{(\text{Oil Flow in GPM})(\text{Pressure in psi})(\text{Volumetric Efficiency})}{1714}$$

Any shortfall in the power requirements of a vibratory hammer will result in both slower and uneven rotation of the eccentrics, which will result in degradation of the dynamic force and of the driving capabilities of the hammer.

3.3.2. Types of Vibratory Hammers

3.3.2.1. Low-Frequency Hammers

These are vibratory hammers with a vibratory frequency of 5 to 10 Hz, used primarily with piles which have high mass and toe resistance such as concrete and large steel pipe piles. These hammers typically have large eccentric moments to achieve their dynamic force with high resultant amplitudes. An example is shown in Figure 3-42.

Figure 3-42 Low frequency Vibratory Hammer

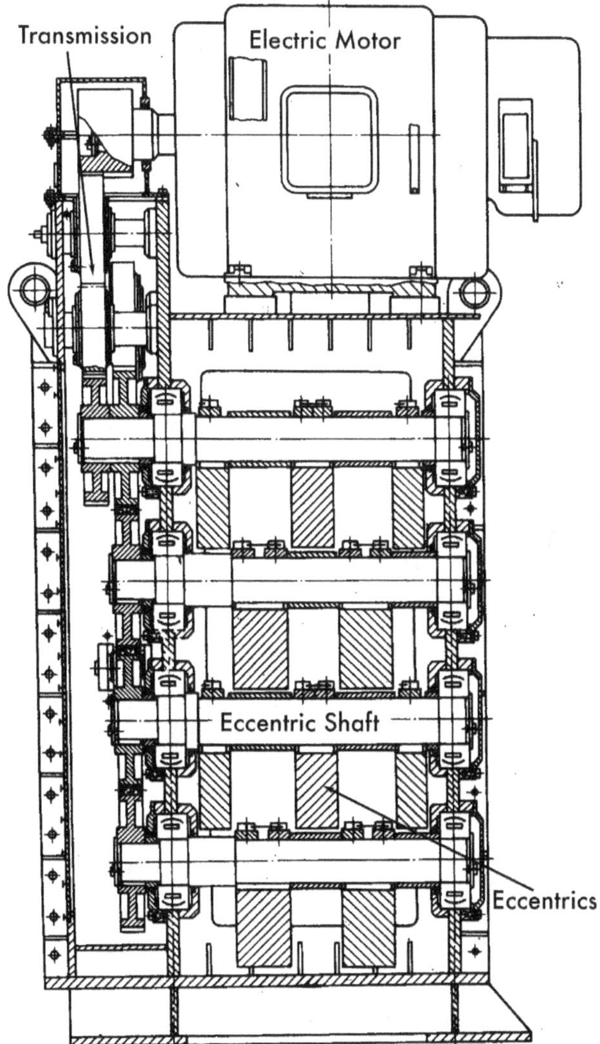

3.3.2.2. Medium-Frequency Hammers

These are hammers with a vibratory frequency of 10 to 30 Hz, used for piling such as sheet piles and small pipe piles. An example is shown in Figure 3-43. These hammers are the majority of vibratory pile drivers in use today, since they combine the dynamic force necessary to excite the soil, the correct frequency to properly interact with most soils, and the sufficient amplitude to penetrate hard subsurface layers.

Figure 3-43 Medium-Frequency Vibratory Hammer

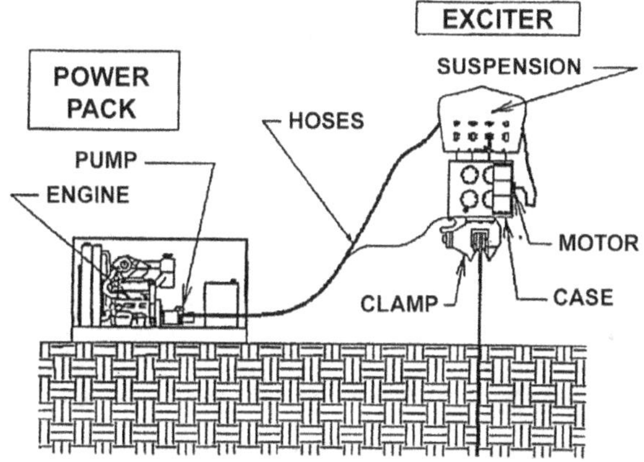

3.3.2.3. High-Frequency Hammers

These consist of all vibratory hammers that vibrate at frequencies greater than 30 Hz. There are of two basic types. The first machines are in the 30 to 40 Hz range that are designed primarily to minimize vibration of neighboring structures. These have been developed simultaneously both in Europe and in the United States, and they are similar in construction to the medium-frequency machines. The primary advantage of these machines is their lowered transmission of ground excitation to neighboring structures. These machines may have problems in overcoming toe resistance. Figure 3-44 shows an example of a high frequency vibratory hammer.

Figure 3-44 High Frequency Vibrator Near Sensitive Structure

3.3.2.4. Sonic or Resonant Hammers

In a class by itself is the resonant pile driver, first introduced in the early 1960's. The central principle of the resonant driver is to induce resonant response in the pile, thus facilitating driving and extracting. The resonant driver operates at frequencies in the range of 90 to 120 Hz. In early tests, much of the driving took place at the half-wave frequency of the pile. The ability to achieve this response was dependent on properly matching the frequency range of the machine to the length of the pile. In cases where this was not possible in a normal hammer/ pile set-up, a heavy wall follower connected the pile with the hammer. When the pile is exceptionally long, second and third overtones can be achieved. In principle, this concept has held great potential, but the mechanical complexity of this machine has withheld limited its acceptance. An example is shown in Figure 3-45.

Figure 3-45 Resonant Vibratory Hammer

3.3.1.5. Impact-Vibration Hammer

The term "impact-vibration hammer" refers to a vibratory pile driver that imparts both vibrations and impacts to the pile during operation. Such a machine is shown in Figure 3-46. In common with conventional vibratory hammers, it contains counter-rotating eccentrics that impart vertical vibrations; however, these are contained in a head that is free of the pile. Its motion is regulated by a set of springs, which link it to the frame. The frame can be connected to the pile in numerous ways. These springs transmit their compression force, which is of a vibratory nature, to the pile. In addition, depending upon the position of the head relative to the impact point and the effect of the springs on the vertical motion of the head, (at or near the bottom of the vibratory cycle), the head strikes the anvil and produces an impact similar to traditional impact hammers, but at a higher blow rate. Although this can produce variations in the eccentric rotational speed of up to 40% (as opposed to the 5% or so normal for vibratory hammers), this variation generally does not impede the continuous, stable operation of the equipment. These hammers have very limited usage and are not manufactured in the United States.

Figure 3-46 Impact-Vibration Hammer

3.3.3. Types of Pile Driven with Vibratory Hammers

Vibratory hammers can be used to drive a large variety of piles. Methods used to determine drivability are discussed in Chapter 4. Some of the general benefits of using a vibratory hammer are as follows:

- Use As an Extractor. One of the best applications for the vibratory hammer (see Chapter 5 for details.)
- Use to Limit Noise. In general, the vibratory hammer is much less noisy than impact hammer. The principal source of noise is the generator or hydraulic power pack, which can be muffled.
- Use as a Probe. The vibratory hammer, with its facility for extraction without uncoupling from the pile, can be used to probe for obstructions.
- Increased Mobility. The vibratory hammer often is used without leads. Power equipment is limited to a hydraulic power pack.
- Use to Reduce Pile Damage. Impact damage to the pile (crushing of head or tip, splits, etc.) may be avoided by use of a vibratory hammer. However, if the pile does not penetrate continuously, it is possible to build up enough heat in the pile to melt the pile or to fuse the interlocks in sheet piling.
- Miscellaneous
    - Can be used to move piles laterally by exerting a lateral strain as vibration is application.
    - Can be used to compact sandy soil by successive insertions.
    - Hydraulic vibratories can be used for underwater driving or extraction. These machines must be modified with special “seals” etc. Contact the manufacturer for specific requirement.

3.3.3.1. Sheet Piling

Vibratory hammers perform best with nondisplacement piles such as H Beam, Open-End Pipe, and Sheet Piling.

To drive steel sheet piling, the sheeting wall is set in place using a template and then the piles are driven to the desired depth. This practice requires that the vibratory hammer be no wider at the throat than about 14” as the hammer must clear the adjacent piles, as shown in Figure 3-47. In driving sheeting in this way, it is also common to drive the sheets two at a time, using a jaw with two sets of teeth and a recess between them large enough to accommodate the interlock. An alternate method of driving sheeting with a vibratory hammer is to set the sheet piling as they are driven. As a rule, in this case the sheets are driven one at a time. More information on driving and extracting sheet piles with a vibratory hammer can be found in Chapter 5.

Figure 3-47 Narrow Throat Width of Vibratory Hammer

Figure 3-48 Vibratory Hammer Installing H-Piles

3.3.3.2. H-Piles

The conditions to drive H-piles are similar to driving sheeting; however, when the pile's batter is large, the vibratory hammer can be mounted in a set of leaders as is done with impact hammers. In addition to a bearing application (where the pile might be driven to refusal), vibrated H-piles are used for soldier beams and in slurry wall construction.

3.3.3.3. Caissons and Pipe Pile

Caissons are a versatile item, extensively used with drilled shafts. Application of the vibratory hammer revolutionized caisson installation in the 1970's. Vibratory hammers are very successful when used to install large diameter caissons (up to 15 feet diameter) and, in general, open end pipes of sufficient size that a plug does not build up in the end of the pile. For this application, a special device called a caisson beam is employed. This is a horizontal slide with a set of two clamps attached to it. The clamps affix the pile to the hammer on opposite sides of the caisson. The clamps are locked to the slide during use but can be moved along the slide to enable a caisson beam set-up to drive a variety of pile sizes. A vibratory hammer, equipped with a caisson beam and clamps while driving a caisson, is shown in Figure 3-49.

Figure 3-49 Vibratory Hammer Installing Caissons

Figure 3-50 Caisson Clamp Arrangement for Large Caissons

Once the caisson is installed, the material is removed from the inside of the caisson, the reinforcement cage is placed, and then the concrete is poured to completely fill the pipe. Here a vibratory hammer saves its cost as the pipe can be extracted leaving a concrete pile the shape and size of the hole. In the conventional method of installation, an oversize hole is drilled and the caisson installed with drilling fluid. When the concrete was poured and the pipe removed, the concrete displaces the drilling mud, creating a much larger pile than required. The overage on concrete can be as much as 25%.

The equipment set-up for pipe piles is the same as with caissons. It is more common to vibrate open-end pipes; vibrating closed-end piles is less effective. An example application of driving pipe pile is the installation of pipe piles for offshore structures such as petroleum production platforms. For some of these, two caissons beams where two sets of clamps are used, the beams are being configured in an "x" arrangement. An example is shown in Figure 3-50.

3.3.3.4. Concrete Piles

Displacement piles such as timber, precast concrete, closed-end pipe, or concrete sheet piling can be installed with a vibratory hammer (depending largely on the size - the larger the displaced volume, the less effective the vibratory hammer is), although they are not the most effective with these types of piles. Concrete pile installation with vibratory hammers is rare in the United States but more common elsewhere. It is done with both prismatic (square and octagonal) and cylinder piles. As concrete piles are always displacement type piles, the vibratory hammer must develop some toe impact by raising and lowering the pile during the vibration cycle, thus allowing penetration. This is generally accomplished using low-frequency vibrators with high amplitudes. An alternative is to use an impact-vibration hammer, which can more effectively overcome with high toe resistance than can a vibratory hammer. Indeed, the need to drive concrete piles has been one of the most important factors in the development of these hammers. Extraction of concrete piles is more common in the

U.S., using specially designed clamps.

3.3.3.5. Wood Piles

Wood piles are rarely installed with a vibratory hammer in the United States. Extraction of wood piling, however, is common, and the vibratory hammer is an effective tool for this purpose. The wood piling can be extracted intact by the use of special wood clamps. An example of this is shown in Figure 3-51.

Figure 3-51 Wood Pile Extracted Using Wood Pile Clamp

3.3.4. Excavator Mounted Vibratory Hammers

An important type of vibratory hammer is the excavator mounted vibratory pile driver. Although these units are similar to crane mounted units, there are some important differences that the user needs to be aware of when selecting or applying such a machine to a particular project.

Excavator mounted vibratory hammers have the following advantages:

- Eliminate the need for a separate power pack.
- Easier to use in low headroom or indoor situations.
- More precise positioning of the vibrator for both installation and extraction of the pile.
- Lower expense of excavator vs. crane rental or purchase.
- Very portable; can be mounted onto the excavator before the job begins.
- Down crowding improves drivability without heavy bias weights on the suspension.

3.3.4.1. Overview

Figure 3-52 Excavator Mounted Vibratory Hammer

A typical excavator mounted unit is shown in Figure 3-52. As is the case with crane mounted units, the vibratory driver has a vibrating gearbox and clamp and a non-vibrating suspension unit to allow proper interface with the excavator.

Figure 3-53 Typical Mechanical Set-up of Excavator Mounted Vibratory Hammer

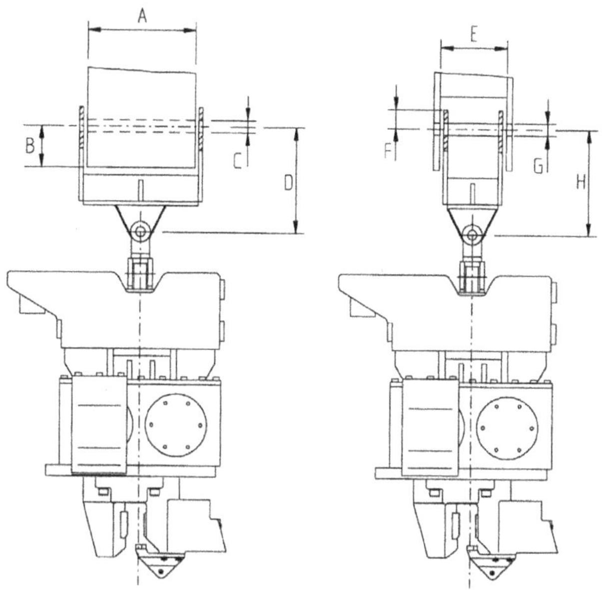

Figure 3-54 Mounting of Hydraulic Hoses on Excavator Boom.

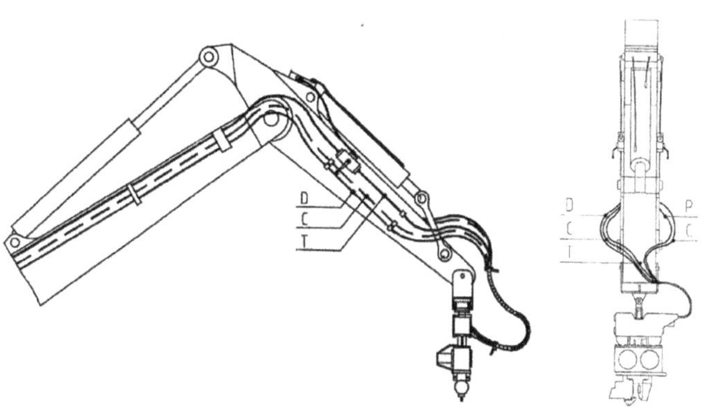

The hammer is connected to the excavator boom with a swivel connection as shown in Figure 3-52. The bucket is removed from the boom and the vibrator is mounted in its place. When mounting a vibrator to an excavator boom the following considerations are important:

- Each excavator is different.
- Mounting must be matched properly. This is both a product of advance planning for the specific excavator and some careful fitting during installation. The beginning of this is to ascertain the dimensions "A" through "H" as shown in Figure 3-53.
- Check that all connections are secure.
- Make sure suspension capacity is adequate for the boom pulling capacity.

Hose Nomenclature:

- C - Clamp Lines
- D - Drain Line
- P - Pressure Line
- T - Return Line

Figure 3-54 shows the excavator hoses on the boom. When rigging these hoses remember that:

- Hydraulic hoses must be of proper length and mounting.
- Dragging hoses must be avoided to prevent to premature wear.
- Use quick disconnects when desirable
- Avoid backpressure, especially on case drain

The manufacturer's instructions must be followed explicitly, along with general hydraulic and mechanical safety procedures.

One major difference between excavator-mounted units and their crane-mounted counterparts concerns the power pack. Crane mounted units generally use a separate power pack to power both the eccentrics and the clamp, using separate pumps for each. An excavator mounted unit uses the bucket circuit on the excavator unit to both turn the eccentrics and to operate the clamp. This eliminates the need for a separate power pack, which enhances the economy of the excavator-mounted unit. Remember to consider the following:

- Check that hydraulic flow and pressure are adequate to operate the unit according to the manufacturer's specification. An underpowered unit will not perform according to specifications and generally will not meet the user's requirements.

- Check the hydraulic system on the excavator completely.

Although excavator mounted units are ideal for driving aluminum, vinyl and fiberglass sheet piles, special are should be taken when handing these materials during both setting and driving to minimize damage. Excavator mounted vibratory hammers can also be used for wood piles as shown in Figure 3-55.

Figure 3-55 Excavator Mounted Vibratory Hammer with Wood Pile Clamp

3.3.4.2. Special Configurations

3.3.4.2.1. Swiveled Hammers

Figure 3-56 Swiveled Excavator Mounted Hammer

Figure 3-56 shows a swiveled vibratory hammer. Certain types of steel sheet piling (especially U-shaped sections such as Larssen sheets) are picked up, set and driven by the vibratory hammer in one operation. The swivel mount allows the hammer to rotate 90° and clamp on to the sheets without having to maneuver the boom to pick up and again to drive. Such a hammer is unnecessary when sheets are set before driving and certainly should never be used with aluminum, vinyl or pultruded fiberglass sheets.

3.3.4.2.2. Goosenecks

A gooseneck effectively lengthens the boom on the excavator and enables the hammer to be used in a wider variety of positions. As is the case with any boom extension, a gooseneck requires greater skill from the excavator operator, as the risk of turning over the excavator is greater with the load further from the centre of gravity.

3.3.4.3. Other Considerations

Excavator mounted units are generally part of the small range of vibratory hammers, so although the types of projects they can complete is broad they are limited in size. Many excavator-mounted units are high frequency units, which make them lighter but also make them less effective in cohesive soils. Finally, the performance of excavator-mounted units is dependent upon the size and condition of the bucket circuit, which can limit the performance of the machine even if the vibrator is properly sized for the job. Consult the manufacturer for more information on proper matching of hammer and excavator.

3.4. Rigs

All pile drivers require some kind of rig to lift hammer and pile and to guide the system as the pile is driven. The typical modern pile driver consists of a crawler crane to which pile driving components have been attached - leaders, hammer, spotter, boiler, moonbeam, etc. Although there are many different configurations used for this purpose, for impact hammers the most common is shown in Figure 3-1. This is a commercial crane adapted for pile driving. There are many variations of this configuration, depending upon the application. However, the ultimate choice of a rig configuration is very dependent upon its availability. Most of these rigs can be adapted to a floating, barge-type configuration as well.

3.4.1. Cranes for Impact Hammers

The crane is the single most important part of a pile driving system other than the hammer. Its proper selection is essential for correct, safe, and economical installation of the piling.

3.4.1.1. Basic Crane

The basic crawler crane consists of two crawlers, which distribute the weight of the driver to the ground; a car body (also known as "mounting" or "truck base") to which the crawlers are attached; and a cab (also called "upper works," "house," "turntable" or "machinery deck") containing the hoisting machinery. The crawlers and car body are frequently referred to as the "lower works."

Figure 3-57 Driver

Each crawler consists of a belt of pads connected by pins, rollers that transfer the weight of the rig to the pads, and chain driven sprockets that engage lugs on the pads to propel the machine. On most modern machines, the crawler assemblies are bolted or keyed to the car body so that they can be removed for shipping. The operator can apply power to either crawler, or lock either crawler, to steer the machine. On some cranes, the crawlers can be extended for increased width to provide stability, and retracted for shipping as shown in Figure 3-58.

Figure 3-58 Extendible Crawlers

The car body is a structural steel weldment (casting in older machines) to which the crawlers are attached, a roller path to support the cab, a kingpin about which the cab rotates and a ring gear which can be engaged to swing the machine. Mechanically driven machines have a vertical propel shaft extending down through the king pin, with bevel gears at the bottom, driving a horizontal propel shaft connected through clutches to each crawler. In addition, the king pin there is usually a shaft or air lines used to operate the travel clutches and travel locks.

Figure 3-59 Lower Works

In the cab are located the diesel engine, the hoisting machinery (also called the "draw works"), the operator's controls, the boom seat to take the boom, a gantry to provide an elevated position for the boom lifting rigging, and means for attaching one or more counterweights. In a basic machine, there are usually two drive shafts. One shaft can be driven in either direction by means of reversing clutches and is used to swing and propel the machine as well as raise and lower the boom. Shifting gears to engage the reversing shaft can perform any one of these functions. The second main shaft carries the hoisting drums, each of which can be engaged by means of its clutch. Each function is normally provided with a brake and lock. Controls vary somewhat and reference should be made to the operator's manual with each crane. On most large modern machines, clutches and brakes are air or hydraulically operated or assisted.

Figure 3-60 Machinery Deck

On most cranes, several "house rollers" in contact with the carbody roller path carry the weight of the upper works. Hook rollers prevent separation of the cab from the lower works. Some machines have a ball bearing mounting, with one race attached to the carbody and the other to the cab.

Figure 3-61 Hook and House Rollers

3.4.1.2. Optional Features

Crane manufacturers offer a number of optional

features on most models. In order to make a crane more suitable for pile driving, certain of these features are desirable. While not all may be essential, the lack of one or more may significantly reduce production.

1. Independent swing and travel. By the addition of another reversing shaft and appropriate gearing, the operator is able to swing and travel simultaneously. This facilitates spotting the machine to reach a pile location and avoids delay in shifting the machine from swing to travel and back again.
2. Independent boom hoist. While this is not standard, the addition of another reversing shaft permits the boom to be powered up and down, independently of swing or travel.
3. One or two additional drums. Additional drum shafts are installed on which are mounted hoist drums with clutch and brake. These are frequently needed for handling a drill, jet, pile guides, pipe, and shells. Also used are tugger hoists, particularly for handling shells.
4. Wide crawler pads. Many rigs travel heavily loaded and wider pads, when available, reduce ground pressure and minimize the need for mats. Wider pads do not provide any significant additional lifting capacity. In certain areas, the use of wide pads may require the removal of the crawlers to transport the rig between job sites, due to width or weight laws.
5. Gantry lifting and counterweight removal devices. These speed set-up and dismantling.

3.4.1.3. Modification for Pile Driving

To convert a crane to a pile driver, certain modifications are necessary:

- Boom. If several types of boom are available, the heaviest should be selected. A Raymond-type boom is more suitable for pile driving than commercial booms, due to its ability to resist some torque and lateral load, and it is essential for side batters. If a Raymond-type boom is not used, a suitable link must be fitted to the commercial boom to accept the leaders.
- Spotter. Plates are welded to the front of the turntable and suitable spotter installed to hold the bottom of the leaders at varying distances from the crane.
- Boiler or air compressor. One must remove sufficient counterweight to maintain balance and attach a power unit for the pile hammer. All hoses and piping need to be mounted as well.
- Hydraulic power pack. If a hydraulic hammer, drill, spotter or moonbeam is to be used, one must mount a hydraulic power pack or make mechanical modifications to provide for the mounting of one or several hydraulic pumps. Control valves must be installed at the operator's station and necessary piping and hose added to connect all components. It may be possible to use the crane's own power system to operate the hammer; both hammer and crane manufacturer should be consulted to insure proper mating of the two machines.
- Other pile driving accessories may be added, as the project requires.
- The modification of a rented crane to make it suitable for pile driving will require one to several shifts at the time of set-up.

3.4.1.4. Crane Capacity

Crane manufacturers normally rate crane capacities at 75% of the load required to tip the machine. The nominal rating is usually the maximum load that the crane can pick at a 12' (3.66 m) radius with the shortest boom, and this of little practical value in comparing cranes for pile driving.

Capacity charts for a number of pile drivers have been prepared and are included in subsequent sections. If a chart is not available for a particular machine, a qualified engineer as required by the Federal Safety and Health Regulations must determine the capacity of any crane.

For preliminary estimating and planning purposes, the capacity of various commercial cranes is shown in Table 3-8.

Because pile drivers usually move fully loaded, with their centre of gravity high in the air, more conservative limitations should be imposed than for ordinary lift crane work. This requires that the location of the vertical centre of gravity be determined.

3.4.1.5. Use of Capacity Charts

Crane capacity charts should be used as follows:

1. Determine the length of leaders required to handle the desired hammer and pile.
2. Calculate the weights to be hung from the leaders, including:
    a. Core;
    b. Shell;
    c. Followers and driving accessories;
    d. Pile;
    e. Drills, drill guides, fairleaders, augers, pile driving equipment, etc.;
    f. Pile lifting brackets, jets, etc.
3. Determine the appropriate crane to be used based on lifting capacity, boom length and boom type. Capacity charts are based on the machine being operated in a safe manner by a competent operator under adequate supervision. The maximum safe radius is often not the most productive radius and experienced supervisors will usually work at a shorter radius to provide greater stability and faster operation.

Ground stability is the most important consideration in driver operation, particularly when operating at or near capacity. The charts assume that the rig will be on mats except when the general superintendent authorizes you to work without them.

The capacity charts also assume that driving will be done with the rig level. Except when shelling-up, the bottom of the core should be no higher than the bottom of the leaders. When traveling with a core or pile, it should be blocked or tied to prevent it from swinging out.

Batter pile capacity is beyond the scope of the charts and must be computed separately for each situation.

3.4.1.6. Truck Cranes

Truck cranes have limited application as pile drivers. The nominal rating of a truck crane is calculated at 85% of tipping at 10' (3.05 m) radius with outriggers set. The rig cannot move in this condition so its capacity as a pile driver must be based on 65% of tipping in the least stable position, with outriggers raised. With outriggers raised, it will usually have very poor capacity for its weight.

It will help to lower the hammer and bring leaders in to the minimum radius before raising outriggers to move, but this slows production. Under no circumstances should it be assumed that any additional capacity could be safely secured by moving with the outriggers "just clear of the ground."

The cab of the prime mover will interfere with swinging a full circle. This will make it impractical to operate on mats that are frequently necessary when piles are to be driven. Except in unusual circumstances, a light crawler crane will usually prove more economical.

3.4.1.7. Maintenance

The superintendent should familiarize himself with the lubrication and maintenance instructions that are with each crane.

So that the user will be generally aware of lubrication requirements, the major items requiring daily lubrication with a grease gun (number of places to be lubricated shown in parenthesis) are:

Main Drive Shaft

- Rear Drum Shaft
- Front Drum Shaft
- Boom Hoist
- Crawler Sprockets
- Track Rollers
- Transmission
- Hydraulic Pump Bearing
- Centre Pin Bushing
- Boom Seat Pins
- House Rollers
- Hook Rollers

Track roller lubrication is a difficult job and one that is frequently overlooked, particularly when the rig is working in the mud. However, the rigs rock back and forth with every hammer blow, heavily

Table 3-8 Crane Capacity

Manufacturer	Model	Rating, U.S. Tons	Capacity, 90' Boom, 20' Radius, kips	Capacity, 150' Boom, 30' Radius, kips
American Hoist and Derrick	5299	50	50.5	N/A
	5300	70	64.5	34.5
	7250	60	67.0	36.8
	7260	100	108.2	57.3
	999	110	N/A	91.9
	9260	125	N/A	112.5
	9270	150	N/A	120.3
	9299	185	N/A	128.9
	9310	225	N/A	171.9
American Terex	HC-60	60	58.3	31.4
	HC-80	80	87.3	36.5
	HC-110	110	122.9	65.8
	HC-150	150	165.1	96.9
	HC-210	210	264.7	145.8
Kobelco	CK-550	55	50.1	N/A
	CK-800	80	72.9	N/A
	CK-1000	100	103.1	51.1
	CK-1750	175	266.7 @ 21'	142.7 @ 32'
	CK-2000	200	247.0 @ 21'	161.6 @ 31'
	CK-2500	250	326.7 @ 22'	224.9 @ 32'
Liebherr	HS-853-HD	100	113.2	54.0
	HS-883-HD	130	160.5 @ 85'	77.0
Link Belt FMC	LS-108	45	34.8	N/A
	LS-110	50	51.0	N/A
	LS-118	60	61.2	26.7 @ 35'
	LS-318	82	67.8	36.3
	LS-338	100	111.4	60.1
	LS-418	110	143.7	72.5
	LS-518	150	219.0	114.0
Link Belt Sumitomo	LS-108H	50	53.2	N/A
	LS-138H	75	72.2	39.3
	LS-208H	75	70.4	39.8
	LS-218H	100	110.8	60.2
	LS-238H	150	162.6	87.9
	LS-248H	200	223.4	138.1
Manitowoc VICON Series	2900WC	65	88.6	42.5 @ 32'
	3900V	100	136.6	73.4
	3900WV	140	176.0	105.0
	4000WV	132.5	177.4	114.4
	4000WVS2	175	250.5	127.5
	4100WV	200	319.2	162.4
	4100WVS2	230	336.9	210.3
Manitowoc EPIC Series	M-50W	60	53.3	28.8
	M-65W	75	83.0	40.4
	M-80W	88	96.1	51.5
	M-85W	95	100.4	53.2
	111 B	75	85.6	46.3
	222 B	100	93.1	50.2
	555 S2	150	220.4	128.6
	777 S2	175	230.7	135.6
	888 S2	220	348.6	230.5
	999 S2	250	396.4	255.6

loaded. Neglect of this important lubrication too often results in the need for a major overhaul long before it should be necessary.

The following other items require daily attention:

- Check crankcase oil level
- Inspect engine for oil or fuel leaks
- Check torque converter oil level
- Check radiator coolant level
- Check air cleaner oil level
- Check chain case oil level
- Check hydraulic tank oil level
- Open gear lubricant on main driver shaft bevel pinions
- Oil front and rear drum shaft chain sprockets
- Clean and oil roller path
- Observe engine oil pressure and engine temperature

Weekly lubrication covers many more points to be greased:

- Main Driver Shaft
- Rear Drum Shaft
- Front Drum Shaft
- Boom Hoist
- Crawler Sprockets
- Crawler Drive Shaft
- Track Rollers
- Transmission
- Hydraulic Pump Bearing
- Centre Pin Bushing
- Drum Brake Shafts (front)
- Drum Brake Shafts (rear)
- Swing and Slide Pinion Shafts
- Boom Seat Pins
- Gantry and Equalizer Sheaves

Weekly attention should be given to the following:

- Brush, drum gear and pinion, boom hoist bevel gears, swing bevel, slide pinion, swing gear, swing pinion and ring gear with open gear lubricant
- Oil operating linkage
- Oil engine accessories
- Drain water and sediment from filters and tanks
- Check engine fan belt
- Service air cleaner
- Clean crankcase and air compressor breathers
- Check all filters
- Check clutch and brake adjustment
- Check crawler chain adjustment

Periodic attention must also be given to the following:

- Drain engine crankcase
- Clean or replace fuel and lube filters
- Drain boom hoist gear case
- Drain transmission chain case
- Drain converter housing
- Drain crawler transmission cases
- Replace hydraulic oil filter
- Drain hydraulic oil tank

3.4.2. Booms for Impact Hammers

In this section, emphasis will be on Raymond-type booms, which were developed with the vibrational and other loads of pile driving in mind. These recommendations may be helpful with other booms in conjunction with the recommendations of the crane manufacturer.

3.4.2.1. General

Raymond-type booms are designed to resist the torsion and side thrust that are introduced in pile driving operations, particularly for side batter driving and drilling. All Raymond-type booms have a basic length of 40’ (12 m) consisting of a lower section designed to fit a particular crane and an upper section that may be the same for several different machines. To lengthen the boom, one inserts sections in multiples of 8’ (2.44 m).

Raymond-type booms have a built-in steam or hydraulic line and are provided with racks to store the boom connection bolts and, in some cases, pendants. They also have a fairleader at the boom tip.

While Raymond-type booms are stronger in some respects than the equivalent commercial booms, they are not better (or even as strong) under all circumstances.

In selecting a boom for pile work, one normally uses a boom length equal to the leader height below the boom tip, a minimum of one-half total leader height. When operating at the maximum radius or when driving in-batter piles, the leaders may hit the ground. In such cases, it may be desirable to lengthen the boom or to replace an 8' (2.44 m) leader section with a special 4' (1219 mm).

Commercial booms may be used for driving plumb and in- and out- batter piles. They should never be used for driving side batter piles.

3.4.2.2. Medium Duty Booms

The most common Raymond-type boom is the Raymond Medium Duty Boom illustrated in Figure 3-62. This boom is about 4-1/2' wide and 2-1/2' deep at the joints. Lower sections are available for many different commercial cranes. They can be adapted to other machines of similar size.

The basic 40' (12 m) boom consists of a 20' (6.1 m) lower section (suitable for the particular crane) and a 20' (6.1 m) upper section (the upper sections with 4" (101.6 mm) x 4" (101.6 mm) main angles should be used only on SUD, steam 703 and Hilland Rigs). Inserts are available in 8' (2.44 m), 16' (4.88 m) and 24' (7.32 m) lengths and the boom can be extended to 80' (24 m) under normal circumstances. With a 72' boom, a 24' (7.32 m) insert should be used; for an 80' (24 m) boom, the 24' (7.32 m) insert must be used. The 24' (7.32 m) section provides the extra depth in the centre required for the 80' (24 m) length boom.

The boom sections are connected with one 1 1/2" diameter high strength alloy bolt in each corner.

3.4.2.3. Heavy Duty Booms

For the largest machines, a "Heavy Duty Boom" illustrated in Figure 3-63 can be used. This boom has a section roughly 4' (1219 mm) deep and 6-1/2' (1981 mm) wide. Lower sections are available for Manitowoc 3900 (100 ton) or 4000W and adaptations can be made for other cranes of similar size.

The basic 40' (12 m) boom consists of a 24' (7.32 m) lower section and 16' (4.88 m) upper section. Inserts are available in 16' (4.88 m), 24' (7.32 m) and 32' (9.76 m) lengths. These inserts can be used in any combination to provide a boom of 56' (17.1 m) to 120' (36.6 m) in length in 8' (2.44 m) steps.

The Heavy Duty Boom uses the same 1-1/2" (38 mm) diameter connecting bolts and the same sheaves as the Medium Duty Boom, but the boom sections cannot be mixed.

3.4.2.4. Assembly

1. Inspect boom sections; and significant damage or buckling of any main angle or any brace must be repaired or the section replaced. Visually check for racking; check diagonal distance between corners across sides and ends; the two diagonals should not differ by more than 1/8" (3.2 mm). Check boom carefully for broken or cracked welds.
2. When sections are assembled on ground, boom must be straight. A line from the boom tip to the centre of the top should not vary more than 1" (25.4 mm) from the centerline of any section.
3. All joints must be tight. The ends of all four main angles must bear against the corresponding angles of the next section. Shims (up to 1/8" (3.2 mm)) may be placed between joints to straighten booms or get bearing between angles. Shims should have a hole for the boom bolt and be large enough to cover the entire end of the main angle and bolt block. Tighten boom bolts as much as needed to fit sections snugly. After rig is erected, bolts should be visually observed for any sign of looseness.
4. Select proper pendants and topping lift adapter. Inspect topping lift adapter for structural damage and cracks. Inspect pendants for worn or broken wires, corrosion, abrasion, crushing or other physical damage. Give special attention to points where wire rope enters end fittings. Replace the pendants if there is more than one broken wire near a fitting or more than two broken

Figure 3-62 Medium Duty Boom

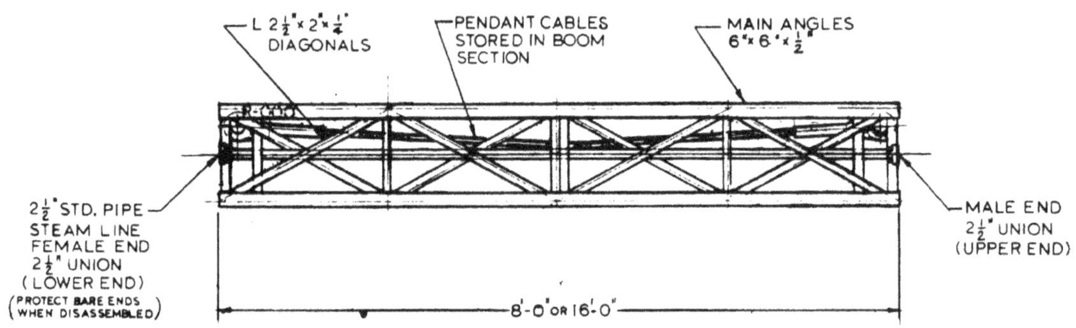

8'-0" & 16'-0" INTERMEDIATE SECTIONS
(SIDE VIEW)

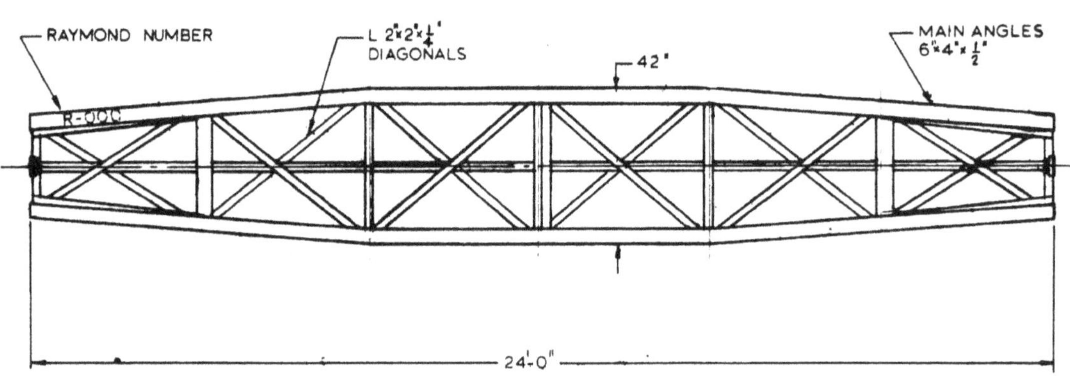

24'-0" INTERMEDIATE SECTION
(SIDE VIEW)

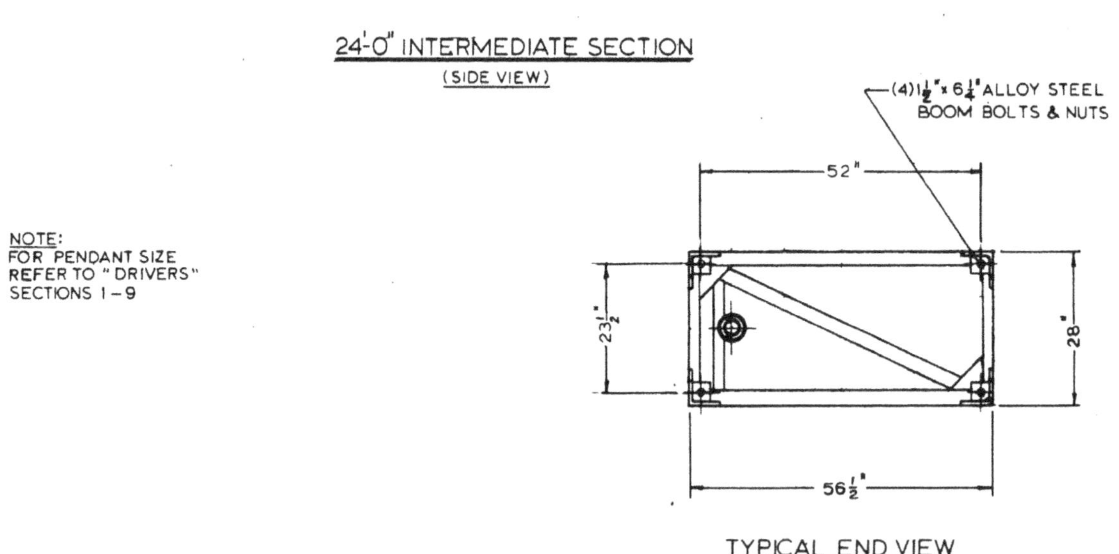

TYPICAL END VIEW

Figure 3-62 Medium Duty Boom (continued)

END VIEW (SEE PAGE 3)
MAIN ANGLE 6" × 4" × $\frac{1}{2}$
BOOM WITH LIGHTER 4" × 4" MAIN ANGLE USED ONLY ON SUD, HILLAND & STEAM 703
28"
$2\frac{1}{2}$" GOOSENECK WITH $2\frac{1}{2}$" BOSS HOSE SPUD
$13\frac{1}{2}$"
19'-10"
20'-$2\frac{1}{2}$"

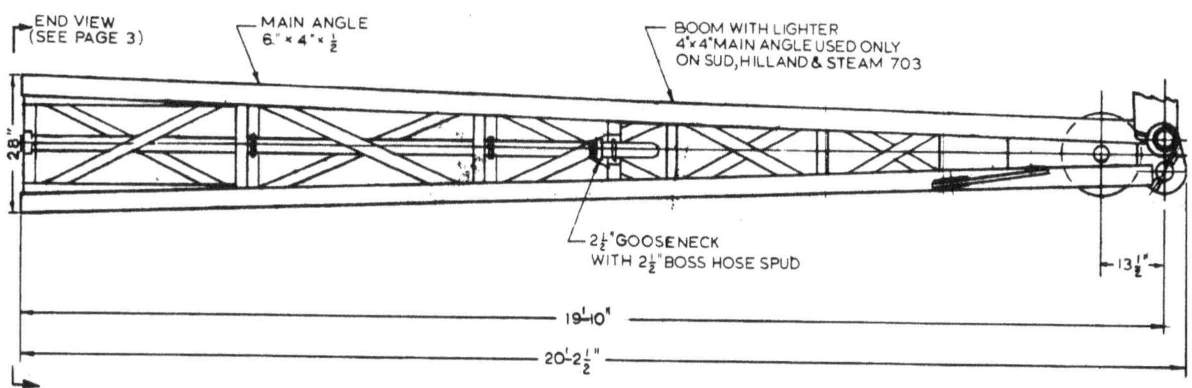

20'-0" UPPER SECTION

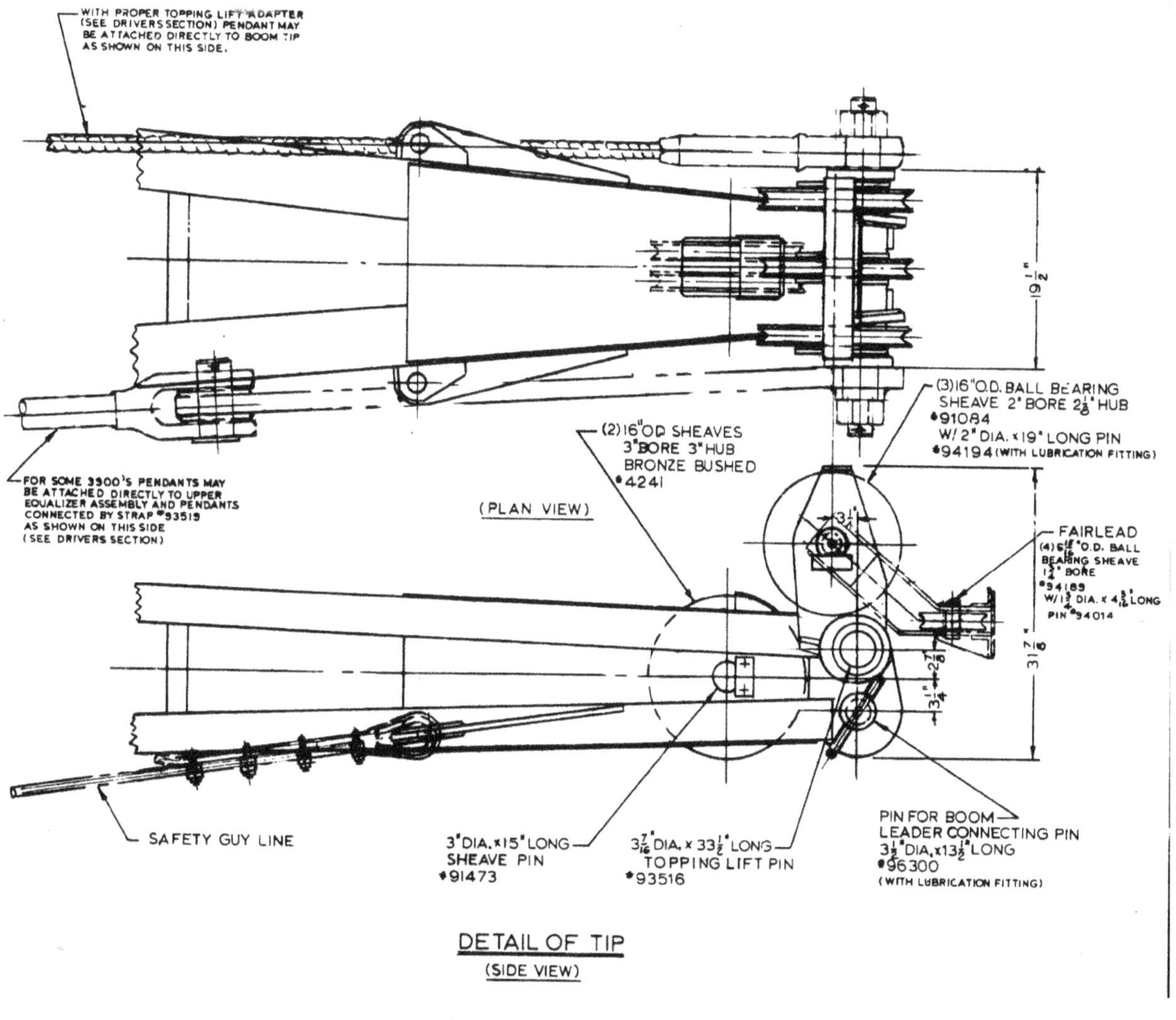

WITH PROPER TOPPING LIFT ADAPTER (SEE DRIVERS SECTION) PENDANT MAY BE ATTACHED DIRECTLY TO BOOM TIP AS SHOWN ON THIS SIDE.
$19\frac{1}{2}$"
(3) 16" O.D. BALL BEARING SHEAVE 2" BORE $2\frac{1}{8}$" HUB #91084 W/ 2" DIA. × 19" LONG PIN #94194 (WITH LUBRICATION FITTING)
(2) 16" OD SHEAVES 3" BORE 3" HUB BRONZE BUSHED #4241
FOR SOME 3900'S PENDANTS MAY BE ATTACHED DIRECTLY TO UPPER EQUALIZER ASSEMBLY AND PENDANTS CONNECTED BY STRAP #93519 AS SHOWN ON THIS SIDE (SEE DRIVERS SECTION)
(PLAN VIEW)
$3\frac{1}{4}$"
FAIRLEAD (4) $6\frac{15}{16}$" O.D. BALL BEARING SHEAVE $1\frac{3}{4}$" BORE #94189 W/ $1\frac{3}{4}$" DIA. × $4\frac{3}{16}$" LONG PIN #94014
$31\frac{7}{8}$"
$2\frac{7}{8}$
$3\frac{1}{4}$"
SAFETY GUY LINE
3" DIA. × 15" LONG SHEAVE PIN #91473
$3\frac{7}{16}$" DIA. × $33\frac{1}{2}$" LONG TOPPING LIFT PIN #93516
PIN FOR BOOM LEADER CONNECTING PIN $3\frac{1}{2}$" DIA. × $13\frac{1}{2}$" LONG #96300 (WITH LUBRICATION FITTING)

DETAIL OF TIP
(SIDE VIEW)

Figure 3-63 Heavy Duty Boom

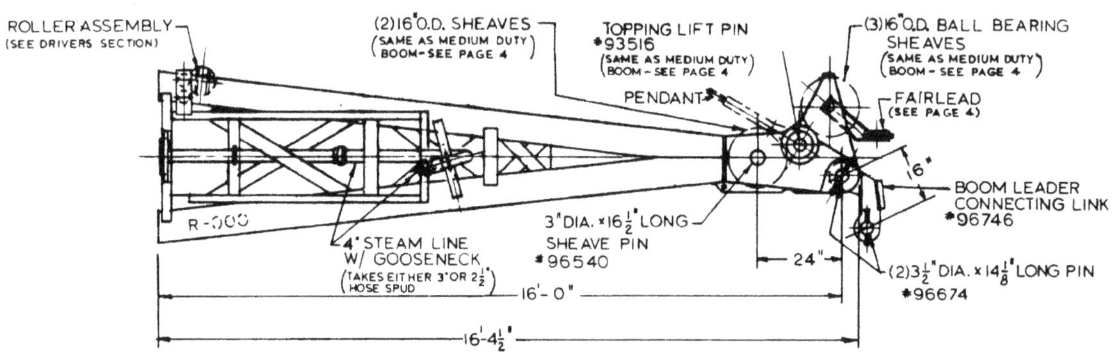

16'-0" UPPER SECTION

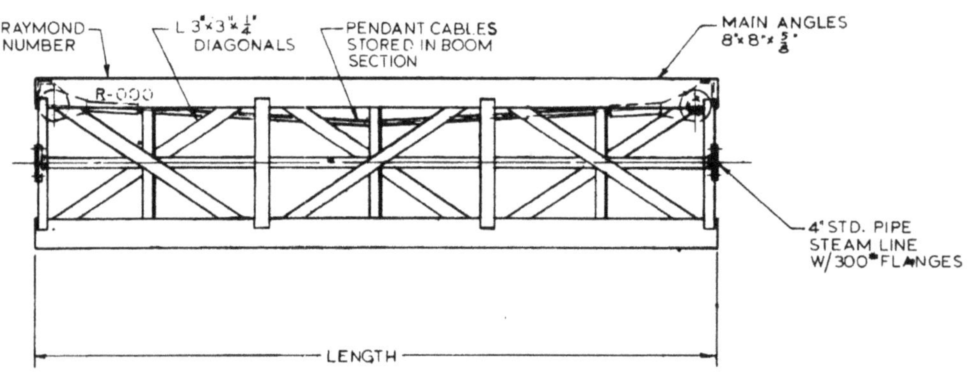

INTERMEDIATE SECTIONS

(SIDE VIEW)

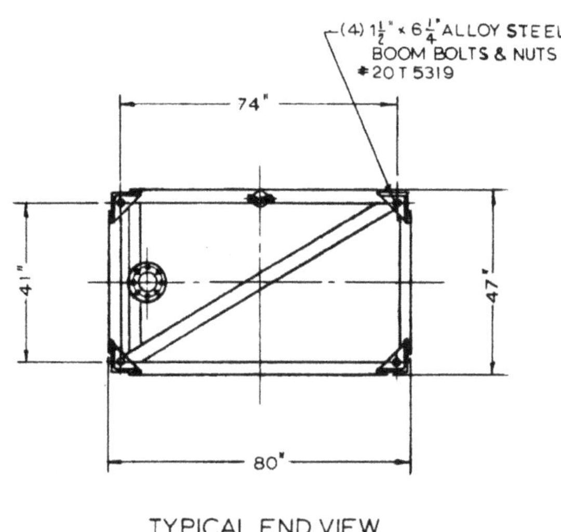

TYPICAL END VIEW

wires at any other point.

5. Compare length of each pendant in a pair; lengths should agree within 1/2" (12.7 mm). Pendants are sent out initially in matched pairs and should be kept together. Pendant pins should be kept with their pendant.
6. Check safety cables or other standing rigging for condition, that length is proper, and that cable clips are correctly applied.
7. Lubricate all wire rope; the best method is to immerse it in hot wire rope lubricant. Alternately, the wire rope should be wire brushed, if rusty, and thoroughly coated with a penetrating type lubricant.
8. Inspect sheaves, rollers and sheave pins. Any that show significant looseness, binding or wear should be replaced, particularly sheaves which have the wire rope pattern worn into them. Inspect steam, hydraulic and jet line fittings and supports, tighten, repair or secure as needed.
9. Inspect boom seat pins and pinholes for signs of wear or cracking; check that pins and boom butt match crane boom seat. Make certain that washers, spacers, sleeves and keepers are available to insure a snug fit of boom-to-boom seat with no play.

3.4.2.6. Precautions in Use

1. No attempt should be made to pull a stuck mandrel core on a side batter with any type boom. Modern cranes can exert such a great line pull that a boom can be buckled without warning.
2. Booms should not be used to drag loads sideways; loads should be picked up and swing.
3. Most booms are provided with safety cables to prevent the boom pin wheeling back over the cab. They are far more reliable than boom stops. Adjust safety cables so they are just tight when boom is in the highest safe working position.

3.4.2.7. Maintenance

1. Boom and boom rigging should be visually inspected daily, particularly for loose fittings.
2. Boom and boom rigging should be thoroughly inspected at assembly, as previously detailed, and similarly at disassembly.
3. Heating and straightening may repair minor localized damage to structural angles.
4. Damage or buckling of an angle should be repaired by doubling with another angle welded continuously along both sides and at both ends.
5. Any severe damage or racking of a boom section is impractical to repair in the field and the section should be replaced with a new section.
6. Sheaves and pins should be lubricated regularly; loose or worn sheaves and pins should be replaced.
7. Dirt and mud should be removed; rust spots should be scraped and given a coat of black paint. Sign should be kept clean; when it becomes chipped or worn, it should be replaced.

3.4.2.8. Handling

1. Lift boom sections by a sling around the main angles at each end or near a panel point; never lift by the bracing.
2. Store on blocking placed at end of sections or under panel points. When unbolting, be certain that boom is blocked at each side of the joint.
3. Lower sections should be stored or shipped flat or braced to prevent their falling over. Other sections may be stored or shipped flat or on their sides, whichever is convenient.
4. Do not ship or store heavy items on top of boom sections. When necessary, boom sections may be stored or shipped on top of one another.
5. When disassembling sections, cap hydraulic lines and protect fittings; remove hoses and store in protected location; lubricate boom bolts and pendants and store in racks; replace boom-seat pins in crane boom seat.

3.4.2.9. Boom Point Connection

The boom point connection is important with both under-hung and fixed leaders. Inadequate or poorly designed and manufactured boom point connections can lead to accidents and job delays. Typical boom point connections are shown in Figure 3-64.

Figure 3-64 Boom Point Connections

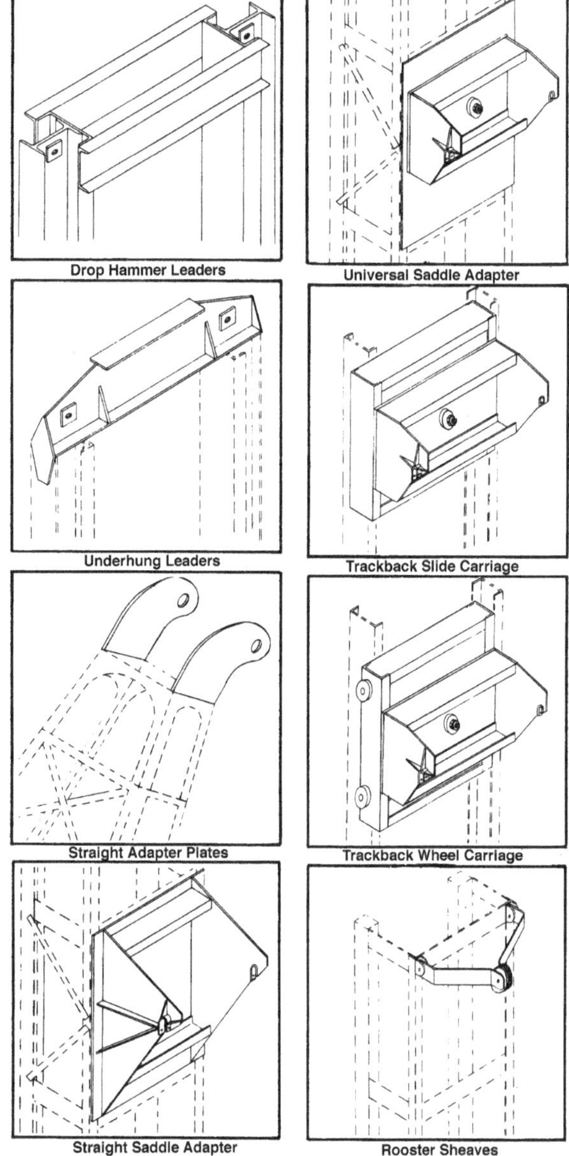

3.4.3. Cranes for Vibratory Hammers

Vibratory hammers are generally not operated in leaders. They are free hanging from the crane. It is important that cranes for vibratory hammers have adequate capacity to both lift the hammer and pile for driving and any possible extraction load that might be needed to extract piles. The boom must also be able to withstand any residual vibration transmitted to it by the suspension.

3.4.4. Power Systems

All external combustion hammers and the vast majority of vibratory hammers require some type of power source. This power source depends upon the nature of the hammer.

3.4.4.1. Air/Steam Hammers

Except for the largest marine rigs, air compressors power most air/steam hammers. These can be mounted either on the ground or, more commonly, on the back of the crane as a substitute for the counterweight. When used, boilers are either of the fire tube Scotch marine type or the vertical water tube Raymond type. Both compressors and boilers should be configured in accordance with the manufacturer's recommendations. Also, essential for air/steam hammers is a three-way shut-off valve that can blow pressure out of the lines after the hammer is stopped or in an emergency, and a line oiler of the sight feed or pump type to provide a continuous stream of atomized oil to the cylinder.

3.4.4.2. Hydraulic Systems

For a variety of reasons, hydraulic systems have become dominant for a wide variety of equipment used on pile driving projects such as:

- Vibratory hammers
- Hydraulic impact hammers
- Augers and drills
- Spotters and pile monkeys

A basic hydraulic power pack (in this case for a vibratory hammer) contains all of the components necessary to both power and to control the hydraulic equipment being used. Views of a hydraulic power pack (in this case for a vibratory hammer) are shown in and .

Figure 3-65 Internal View of a Power Pack

Figure 3-66 External View of a Power Pack

In some cases, the crane's hydraulic system can be employed to power either a vibratory hammer or a hydraulic impact hammer, generally the smaller models. This eliminates the need for both an external power pack and diesel engine. However, all other comments on control systems and operation apply.

3.4.4.2.1. Hydraulic Oil

Since most hydraulic problems are due to improper condition of the hydraulic oil, it is essential that one start with the proper type and grade of oil.

While a wide variety of fluids (even water) will transmit power in a hydraulic system, a good hydraulic fluid should also be anti-corrosive, stable at high temperatures, fluid at low temperatures and adequately lubricate all parts of the system.

Engine crankcase oils are used in some hydraulic systems (particularly in truck mounted systems) because of their availability at service stations. However, they have detergents and dispersant additives, which make them less desirable as hydraulic oils. In addition, they are more expensive than hydraulic oils.

Turbine grade, straight mineral oils, while suitable, do not have the additives that are needed in severe operating conditions. Aircraft type, fire-resistant hydraulic fluids are not compatible with the seals and gaskets and should not be used. Neither should one use hydraulic brake fluid.

The hydraulic systems should be filled with a premium grade, anti-wear type hydraulic oil. These oils contain oxidation inhibitors that permit them to be used at higher temperatures without breakdown. As temperatures rise, "varnish" is deposited and sludge is formed. Even with premium oils, oxidation becomes a problem at temperatures above 150°F (66° C), and will occur very rapidly if hydraulic oil exceeds 200°F (93° C) in any part of the system.

These premium grade oils also contain rust inhibitors, anti-foam agents and anti-wear additives. Typical hydraulic oils should meet the following standards:

Viscosity at starting	3000 SSU max.
Viscosity at 210°F (99° C)	40 SSU min.
Viscosity index	90 or higher

When start-up temperatures are below 10°F (-12° C), it is desirable that some provision be made for keeping the system warm.

3.4.4.2.2. Reservoirs

The oil reservoir not only supplies oil to the system but also allows the oil to cool and impurities to settle. Because of space and weight limitations, the reservoirs are sometimes smaller than the desired 2 to 3 times pump capacity. Reservoirs are either pressurized or elevated to provide gravity flow to the pump. They are provided with magnets to attract steel particles and have special interior preparation to minimize oil contamination, the worst enemy of hydraulic systems.

Reservoirs should normally be kept full, with the system at rest, and large cylinders retracted. The suction line is provided with a strainer to remove impurities large enough to be seen by eye. A suction hose or pipe leads from the reservoir to the pump; all joints must be tight and not leak air.

3.4.4.2.3. Hydraulic Pumps and Drives

The hydraulic pump is connected to the engine through a pump drive. Sometimes this pump drive is a gearbox acting as a speed changer to optimize the pump, but in others, a direct drive is employed, eliminating gear losses.

Pumps are usually driven by a diesel engine. Most of the pumps are either the gear type which one calls "low pressure" (2500 psi (172 bar)) or the piston type that one refers to as "high pressure" (5000 psi (345 bar)). One also uses a few vane type pumps operating at 2000 psi (138 bar).

Hydraulic pumps are precision made units with very small clearances; their efficiency depends on absolute cleanliness. Like a ball bearing, they can be damaged by a small quantity of dirt or dust. They should never be disassembled outside under job conditions. Pump alignment is also very critical. Pumps should never be run, even for short periods, without oil. Newly installed pumps should be run in for 30 minutes at zero pressure; failure to do so may result in immediate failure of the new pump.

Most units have separate pumps for the hydraulic clamps, but some integrate these into the main power source.

Hydraulic pumps can be used either in open or closed loop mode. Both appear on power packs in this application. Closed-loop systems have better control of starting, running, and stopping of the machines, but have traditionally been more complicated. Open-loop systems are more adaptable for powering other equipment.

With low-pressure pumps, as the pump rotates, oil is trapped between the gear teeth and the pump case and discharged on the pressure side. One normally limits these pumps to a pressure of 2500 psi (172 bar). They are used to power hydraulic spotters, moonbeams, pile guides, low-pressure hydraulic drills and hydraulic hoists. They are also used to supercharge high-pressure pumps; for supercharging, they are operated at not over 40 psi. Some units may have two or three pumping units mounted in the same housing. Capacity depends on gear width.

High pressure (5000 psi (345 bar)) pumps are used to operate hydraulic hammers and high-pressure drills. They consist of a number of pistons that are pushed by a rotating inclined plate and returned by springs. Although older hydraulic units used for pile driving equipment employed fixed displacement pumps, most of the newer ones - especially for drills and augers - use variable displacement pumps, whose reliability is now suitable for the application. These frequently are equipped with load sensing devices to properly adjust the flow to the load. Variable displacement pumps can have very sophisticated flow control mechanisms.

3.4.4.2.4. Filters

Most new hydraulic systems are provided with 5000-psi (345 bar) filters that can be used in either high or low pressure systems. Their purpose is to remove dust-like particles that cannot be seen, but will ruin hydraulic equipment.

Filters should be changed once a month. If the element is not renewed they will by-pass and allow dirt to pass into the system. Some filters are equipped with indicators to slow when they are dirty but these indicators are not completely reliable.

3.4.4.2.5. Pressure Gauges and Relief Valves

Each hydraulic circuit should be provided with at least one pressure gauge. Each circuit also may have a relief valve to protect pump and other elements from damage. Relief valve settings should be checked at the start of each job. Any attempt to exceed recommended pressure limits will result in abnormally short pump life. Many newer systems with variable displacement pumps and load sensing dispense with the relief valve altogether.

3.4.4.2.6. Motors and Cylinders

Hydraulic motors are similar in construction and appearance to equivalent model pumps; the differences are often not visible. All motors are provided with a drain to dispose of leakage. If the leakage is not drained off, it will build up pressure and blow the seals.

Hydraulic cylinders are used in several different sizes to power spotters, moonbeams and pile guides. They are double acting and have counterbalanced lock valves so that positive pressure is required to operate them in either direction. Lock valves insure that external forces cannot move the cylinder.

3.4.4.2.7. Controls

These units can employ air, electric, or manual controls for the hydraulic circuitry. Manual controls are the simplest; however, they confine the operator of the unit to the power pack's location, which, depending upon visibility and other factors may not be the most convenient place from whence to operate the machine. Remote controls allow more flexibility for the operator but are an added expense and source of trouble for the machine.

The hydraulic systems for impact hammers make accommodations to the intermittent flow characteristics of these hammers. Hydraulic impact hammer power packs also include the stroke and blow rate control features of the hammer; in some cases the hammer can be made to execute only one blow at a time.

3.4.4.2.8. Hydraulic Lines

Most hydraulic lines on mobile equipment are high-pressure hydraulic hose. Tubing and pipe are used in rare conditions but are not practical not only because of mobility problems but also due to vibrations. Teflon tape or "Loctite" is often used on screwed joints to prevent leakage.

In order to avoid mixing high and low pressure hose, one should have a standardized system of hoses for high pressure (5000 psi (345 bar)) service and for low pressure (2500 psi (172 bar)) service.

3.4.4.2.9. Enclosure

Most power packs have a sheet metal enclosure, but some do not. The principal advantage of an enclosed power pack is protection from weather and criminal activity. Enclosures are also helpful if they provide sound deadening, although many do not. Open power packs are more economical and there is better access to the parts for service.

3.4.4.3. Electric Vibratory Hammers

The exciters for these units usually employ three-phase induction motors driven at a single frequency, which has encouraged the development of many systems to vary the eccentric moment and thus the driving force. In some cases, electric vibratory hammers can be driven from nearby three-phase outlets, obviating the need for a generator set. The hammer thus only requires a switch box to control it. A separate, small power pack, driven with an electric motor, is required to operate the hydraulic clamp, if there is one. This either can be on the ground or mounted on the static overweight. Electric systems are less and less popular because of maintenance and reliability considerations.

3.4.5. Leader Systems

Leader systems are essential for properly aligning the impact hammer and pile during driving. Their configuration is dependent upon the application.

3.4.5.1. Leader Sections

Although most of the attention on leaders is given to the type of connection with the crane and method of operation, the leader section is essential to successful leader operation.

Leader sections can be divided into two broad categories: leaders that run the hammer inside of the leaders (U-sections, such as shown in Figure 3-67) and those which operate the hammer in front of the leaders (spud type leaders, such as is shown in Figure 3-68.) U-sections are preferred by U.S. contractors and are capable of more precise hammer positioning and alignment and can have a higher section modulus than other sections. Sections that place the hammer in front of the leaders are more advantageous with sheet piling and are favored by European and other contractors.

Figure 3-67 U-Type or Box Leaders

Figure 3-68 European or Spud Type Leaders with Hammer

3.4.5.2. Fixed Leaders

Fixed leaders are attached by a mechanical joint at the boom point and at the bottom of the leaders. The boom point connection should always enable the leaders to rotate about the boom point, and should be below the top of the leaders. A typical fixed leader set-up is shown in Figure 3-69. Advantages and disadvantages of fixed leaders are described in Table 3-9. The spotters, as described below, generally differentiate fixed leaders.

Figure 3-69 Fixed Leaders

Table 3-9 Advantages and Disadvantages of Fixed Leaders

Advantages	Disadvantages
Requires only a two-drum crane.	Heaviest and most expensive leader type.
Superior accuracy when locating leader vertically and all batter positions.	More troublesome to assemble and maintain.
Rigid control of leader during positioning operation.	Cannot reach pile when pile head is below ground surface (excavation, below water)
Compound batter angles can be set and accurately maintained.	Less mobile than other leader systems.
Boom can be lowered and leaders folded under (for short haul over the road and railroad travel) when crane of adequate capacity is used. This depends on the length of leader and boom and the configurations of the crane.	

Typical fixed leaders consist of a main section attached to the boom point, a top section, a bottom section and sufficient intermediate sections to obtain the required height. There should always be at least an equal, but preferably a greater length of leaders below the boom tip than there is above.

The top section has a headblock and fairleaders for handling the hammer, piles, drills, shells, jets, etc. For handling very heavy piles or piles that do not fit in the leaders, a pile lifting bracket may be added.

The main section has a special link that connects the boom tip with the leaders. Intermediate sections come in various lengths. In theory, the entire lead assembly could be one piece but it is broken up into sections for transportation convenience. Since each joint adds weight and is a potential weak point, when available longer sections should be used rather than shorter sections. The bottom section is similar to intermediate sections.

When driving steep out-batter piles, or in deep excavations, the bottom of the leaders may be far above the ground. It may be necessary to use special longer and stronger hammer extensions and sliding frame.

3.4.5.2.1. Spotters for Fixed Leaders

3.4.5.2.1.1. Manual Spotters

These fixed leaders use either a fixed length spotter (useful only when all of the pile on a particular project has the same batter) or a spotter with a system of holes through which pins can be run. The batter of the leaders can thus be changed by telescoping the spotter to the desired length, lining up the holes in both inner and outer tubes, and inserting the pins.

3.4.5.2.1.2. Moonbeam Spotters

These are fixed in the usual way to the boom; however, a curved beam at the bottom of leaders allows rotation about the boom point. A wheeled carriage on the leaders connects the leaders to the moonbeam as shown in Figure 3-70. Moonbeam spotters are generally obsolete, having been replaced by the hydraulic spotters; however, there are hydraulic moonbeams, as shown in Figure 3-71.

Figure 3-70 Moonbeam Spotter Leader

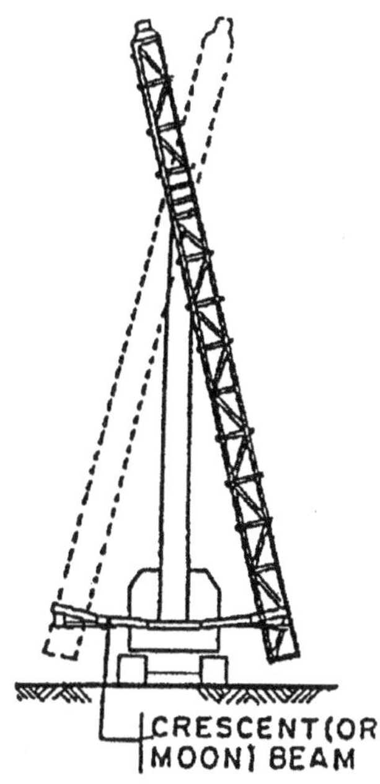

Figure 3-71 Hydraulic Moonbeam Spotter

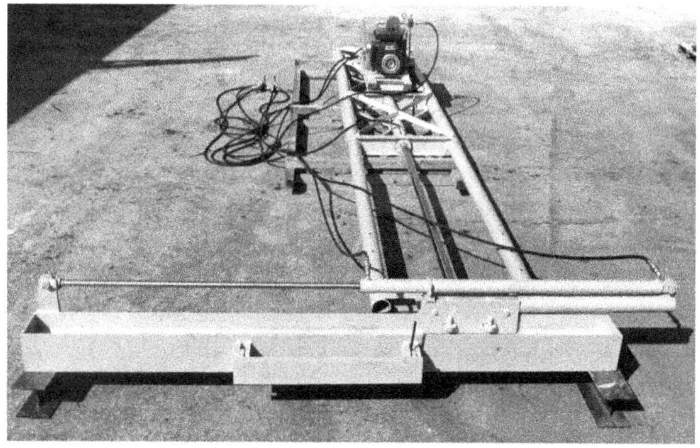

3.4.5.2.1.3. Hydraulic Spotters

These spotters use hydraulic cylinders to control the movement. Hydraulic cylinders are mounted to telescope the overall length in and out, and to pivot the spotter ends about their connection either with the crane cab or with the leaders, or both. These spotters are very versatile. Within their travel range, they can be adjusted in an infinitely variable manner from the crane. They can be powered either by their own hydraulic power system or from the crane's own hydraulic system. An example is shown in Figure 3-72.

Figure 3-72 Hydraulic Spotter

3.4.5.2.1.4. Maintenance of Spotters

- Spotters and moonbeams should be inspected frequently for structural damage, for wear and cracking at pinholes, and for breaks and cracks. Welding should repair them, if needed. Sliding surfaces should be lubricated at the beginning of each project and at least once a month thereafter.
- Care must be taken when the spotter is disconnected from the rig. It should never be done with the hammer above the boom tip, as the leaders will pinwheel about the boom leader-connecting pin. The leaders should be hanging plumb and, unless there is more length of leaders below the boom tip than above, the leaders must be tied off or counterweighted.
- Hoses of hydraulic spotters should be protected and monitored for signs of physical damage. Hoses that have wire braid showing or which have been crushed should be replaced.
- Dirt is the greatest enemy of hydraulic systems and particular care must be taken to keep dirt out when coupling and uncoupling hoses. When moving or storing, all hydraulic fittings should be capped and covered to protect them from damage and to keep out dirt and moisture.
- Some spotters are provided with wear plates on the sliding surfaces that can be replaced or built up when the spotters become loose.
- Before spotters are stored, rust spots should be scraped and touched up with paint. The sheaves, cables and sliding surfaces should be lubricated. They should be stored on blocked to keep them off the ground. Spotter pins should be replaced in their holes for storage.

3.4.5.2.2. Assembly of Fixed Leaders

- Leaders should be assembled on the ground alongside the rig.
- Sections should be inspected; cracked or broken welds, or any damage repaired or the section replaced.
- Check ladder, straighten if bent; replace broken rungs.
- Use drift-pin to line up holes; if holes cannot be aligned, check location with diagram and ream incorrectly located holes. Do not burn.
- Tighten bolts with the proper torque on each bolt. Be certain you have full thread on each nut.
- Check that leaders are straight. A line from the headblock to bottom of the leaders should not vary more than 2" (50.8 mm) from the centre of any section. Shims may be added at the joints to get bearing area and to be retained by the leader bolts.
- Inspect and lubricate connecting pin, headblock and fairleaders; replace worn or loose sheaves.
- Install drill guide frames or other hardware that is required for the job; elastic stop nuts must be used for all bolting aloft. All standard hardware is designed to fit existing holes drilled in the leaders at the time they are built; if holes are not there, they must be drilled and not burned as burning can severely weaken the leaders.

- Clean all platforms; remove all unneeded hardware; remove or secure all loose fittings and bolts.
- If working near an airport, leaders may have to be painted and lighted in accordance with FAA regulations.

3.4.5.2.3. Maintenance of Fixed Leaders

- Boom leader connecting pin, all sheaves and rollers should be lubricated at the beginning and end of each project and weekly while in active use.
- Visually inspect leaders daily for damage and loose fittings. Check leader bolts for tightness after first day's operation and regularly thereafter.
- Minor localized damage to diagonals can be repaired by heating and straightening. Damaged or buckled diagonals should be cut out and replaced; doubling may later interfere with installing drill guides but may be done in an emergency. Welds should be chipped out and ground smooth (not burned), so as not to weaken adjacent members.
- Leaders that are significantly damaged cannot be economically repaired.
- Ladders must be kept in good repair by straightening, welding when loose, or replacing. In interests of safety, platforms should be kept clean and in good repair.
- Rust spots should be scraped and painted black; signs should be cleaned or replaced when they become chipped or damaged.

3.4.5.2.4. Handling and Storage

- Leaders should be lifted with hoods or slings at individual back angles at four locations near the yokes; never pick up with a choker, by the front columns or by a diagonal.
- Leader sections should be stored or shipped resting on both front columns on blocking directly under the yokes; never store or ship leaders on their sides.
- When unbolting, make certain that both sides of the joint are supported on blocks (under a yoke).
- Inspect hammer extensions; inspect and lubricate tapered boom-leader connecting pin; both should be left in the main leader section. The sliding frame is stored and shipped with the bottom section.
- Don't store or ship heavy items (like tuggers) on top of leader sections. Leaders may be stacked, when necessary, but a 4" (101.6 mm) x 6" (152.4 mm) timber placed near the yokes should separate them; this will prevent local bending of angles.
- Remove pile lifting brackets and similar hardware on front of the leaders before shipping or storing. Drill and jet guides add weight and may be damaged; unless they are to be used on the next job, it is preferable to remove them.
- Store leader sections so they are accessible for inspection, inventory and repair and so serial numbers are visible.

3.4.5.3. Vertical Travel or Semi-Fixed Leaders

Vertical travel leaders (referred to also as semi-fixed leaders) are fixed leaders with hydraulic spotters that can also move the leaders up and down relative to the spotter and the boom point. They are most advantageous when positioning of the leaders is exceptionally difficult, such as with railroad construction. Figure 3-73 shows an example of the lower parts of vertical travel leaders, illustrating their versatility in positioning the hammer and pile.

Figure 3-73 Vertical Travel or Semi-Fixed Leaders

3.4.5.4. Swinging Leaders

Swinging leaders are suspended from the crane using a wire rope. They are by far the most common leaders in use. Advantages and disadvantages of swinging leaders are shown in Table 3-10. A typical swinging leader is shown in Figure 3-74. Swinging leaders are generally used with plumb piles; however, under certain conditions, and with the proper equipment and crane operator, they can be used on batter piles, as shown in Figure 3-75. It is important when swinging leaders are used to avoid supporting weight of hanging leaders on the pile to prevent buckling.

Figure 3-74 Swinging Leader

Figure 3-75 Swinging Leaders Used on Batter Piles

Table 3-10 Advantages and Disadvantages of Swinging Leaders

Advantages	Disadvantages
Lightest, simplest, and least expensive.	Requires a 3-drum crane (one for leader, one for hammer, one for pile) or a 2- drum crane with lead hung on sling from boom point.
With stabbing points secured in the ground, this leader is free to rotate to align hammer with pile without precise alignment of crane and the pile.	Difficult control twist of leader if pile stabbing points are not secured into the ground.
Leaders are generally 15 to 20 feet shorter than the boom; crane can reach out further, assuming crane capacity is sufficient.	Crane positioning is more difficult than with any other type of leader. Operator must rely on balance while centre of gravity continues to move.
Can drive in an excavation or over the edge of an excavation.	
For long leader and boom requirement, the leader weight can be supported on the ground while the pile is lifted into place without excessively increasing the working load.	

One issue that must be dealt with swinging leaders is that of proper pile alignment. Since the leaders have some freedom of lateral movement, additional steps should be taken to assure that the pile is aligned properly when driven. One method - and in some cases a required one - is to use a template such as shown in Figure 3-76.

Figure 3-76 Template for Pile Alignment

3.4.5.5. Underhung Leaders

Underhung leaders are similar to fixed leaders, except that the boom point connection is made at the top of the leaders, and generally the leader can move only fore and aft from the crane. These leaders can be used with or without a spotter. Table 3-11 lists the advantages and disadvantages of these leaders, and typical underhung leader set-ups are shown in Figure 3-77.

Figure 3-77 Underhung Leader

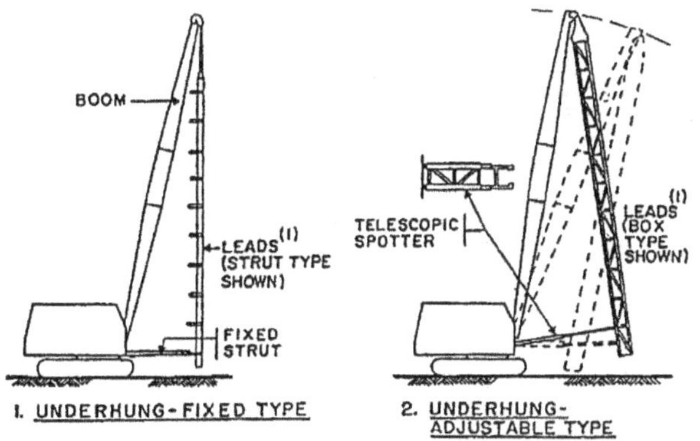

Table 3-11 Advantages and Disadvantages of Underhung Leaders

Advantages	Disadvantages
Lighter and generally less expensive than extended type leader.	Cannot be used for side-to-side batter driving.
Requires only a two-drum crane.	Length of pile limited by boom length, as this type of leader cannot extend upon the boom point.
Accurate in locating leader in vertical or fore and aft batter positions.	When long leaders require a long boom, working radius of crane is reduced, which reduces the effective capacity of the crane.
Relatively short rigging time in setting up and breaking down. Utilizes sheave head in crane boom.	

3.4.5.5. Hammers without Leaders

Most impact hammers require leaders to operate; however, some hammers can be fitted with pants as shown in Figure 3-78. Not all hammers can be fitted with pants. These should only be used in accordance with the manufacturer's recommendations. Impact hammers should never drive piles without leaders, pants or other form of hammer guidance under any circumstances.

Figure 3-78 Impact Hammer with Pants

3.4.5.6. Offshore Type Leaders

A very special type of leader - a type of "flying

extension" – is the leader used for installation of offshore piles in conventional platforms. These leaders are freely suspended from the crane and use the pile to properly align the hammer. This set-up is also used for large pipe and concrete cylinder piles in onshore construction, as shown in Figure 3-79. Care should be exercised when using this type of leaders with batter (inclined) piles. The pile has to have sufficient bending strength to carry the additional weight of the hammer and the leaders.

Figure 3-79 Offshore Type Leader

3.4.5.7. Leader Accessories

Leaders are used with a wide variety of accessories.

3.4.5.7.1. Cradle or Extension

A cradle or extension is used when the hammer is either too small or large for the leaders used. Hammer extensions are of three general types:

1. Inboard extensions hold the hammer, hammer cushion and driving accessory in the leaders and support the hammer when driving below the leaders, as shown in Figure 3-80. The extension wraps around the hammer and enables the hammer to be used in a set of leaders normally unsuitable for the hammer's jaw size. Inboard extensions permit easy insertion and removal of the hammer from the leaders, and also increase the flexibility of a contractor's hammer and leader collection.
2. Outboard extensions are used when it is desired to support the hammer in front of the leaders, usually when driving sheet piles or when the pile cannot fit between the leader columns. An outboard extension is shown in Figure 3-81
3. Flying extensions are used to guide the hammer, hammer cushion and follower when leaders are not used. Offshore leaders are an example of flying extensions.

Figure 3-80 Inboard Extension

Figure 3-81 Outboard Extension

The weight of an inboard extension is carried on top of the hammer. For many years, hammers were provided with heavy coil steel springs to support the extensions. Newer extensions are provided with rubber bumper assemblies, which rest directly on the hammer.

Outboard extensions may have sheaves for reeving the hammer line. The hammer is then secured to the outboard crossbeam with wire rope.

The following should be noted and followed during the use of an extension:

- The loads imposed on extensions may bend them out of shape (particularly when they extend below the leaders) causing them to bind in the leaders.
- The flanges at the lower end of extensions tend to be distorted by the core head ears trying to twist out. They should be repaired and reinforced before they are too badly sprung.
- Extensions should always be secured to the hammer with wire rope. If this is not done, the extensions may bind in the leaders and allow the hammer to escape and possibly cause an accident. When it is necessary to drive below the leaders, the hammer may be released and the extensions held temporarily with the #1 line. As soon as the hammer is raised, the extensions should be resecured.
- When driving in-batter piles, the hammer must be watched carefully for any tendency to come out of the extensions. In this position, there is not only the usual twisting action but the weight of the hammer is being supporting by the extensions. If the hammer comes out of the extensions in an in-batter position, the results could be disastrous.
- Hammer extensions should be inspected daily with particular attention given to the rubber bumper assemblies, the flanges at the lower end and to welds where the yokes are joined.
- When the rubber bumpers or the bumper attachment bolts become worn, they should be replaced. Broken or cracked welds should be repaired and bent flanges straightened and reinforced.
- While localized damage can be repaired, any severe twisting or bending of the main members will probably require replacement of the entire extension.
- When required, rust spots should be scraped and given a coat of black paint.
- Extensions are best shipped between projects by being left in the leaders, as this protects them from possible damage.
- When extensions are separated from the leaders, they should be handled in a horizontal position with sling around two yokes. They should always be shipped or restored resting on both slide members, never on the yokes or on their sides. When stored alone, place on three timbers, one opposite each yoke.

3.4.5.7.2. Pile Gate

A pile gate is used to help guide the pile into the leaders and to keep it in alignment during driving. Pile gates can either be manually or hydraulically operated. They open to admit the pile and close before driving.

3.4.5.7.3. Stabbing Point

A stabbing point is used with swinging leaders to fix the lower end of the leaders as shown in Figure 3-82. These are very important to assist in the stabilization of the leaders.

Figure 3-82 Stabbing Points

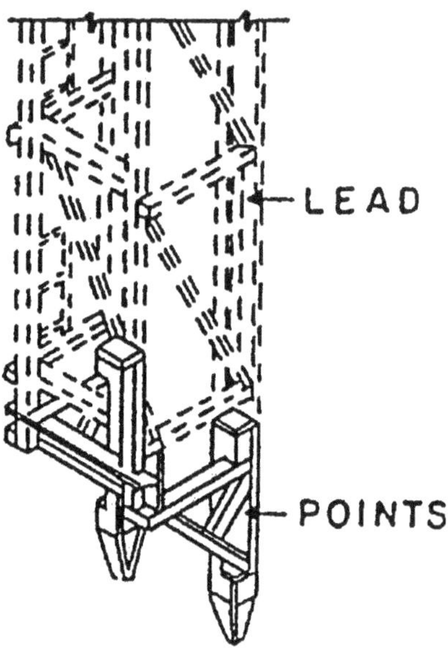

3.5. Specialized Operations and Equipment

Because of the diverse situations under which pile foundations are driven, sometimes specialized operations and equipment are necessary to supplement the regular equipment.

3.5.1. Jetting

Jetting is the use of pressurized fluid to temporarily loosen the bond between pile and soil, thus reducing the resistance of the pile to penetrating the ground. Piles may be installed by jetting that may or may not be accompanied by impact driving or by repeatedly raising and dropping the pile. Jetting is also used to reduce driving stresses, to save time, to obtain increased pile penetration, and to decrease vibrations. Piles should always be driven to their final penetration depth after jetting has been completed. An example of jetting is shown in Figure 3-83.

Figure 3-83 Jetting

Jetting is most effective in fine sands and least effective in clay and coarse gravel. Jetting applications are limited in clay soils where the jets may become plugged, in cohesive soils generally where jetting is not useful or practical, in fine grained and poorly grained soils where jetting may loosen the soil around the pile already driven, and in locations where there is a considerable groundwater and the material disturbed by the jets cannot escape.

Securing reasonable progress requires adequate pressure at the nozzle to excavate or dislodge the material and an adequate volume of water to displace the excavated material to the surface. Jetting may be desirable to penetrate compacted sand and gravel layers above soft materials. A fairly large pipe, perhaps 4 to 6 in. diameter, connected to a nozzle, is used, usually ahead of the pile installation. The pile installation must begin immediately before the soil refills the opening after jetting.

Jetting is typically used when displacement-type piles are required to penetrate strata of dense, cohesionless soils. Exceptions are very coarse or loose gravel where experience shows jetting to be ineffective. Piles, in some cases, have been successfully jetted in cohesive soils but clay particles tend to plug the jets. Jetting aids in preventing structural damage to the pile from overdriving. Water is pumped under high pressure through pipes internally or externally attached to the pile, although air may be used in combination with the water to increase the effectiveness in certain cases. Typically a minimum of 5 to 10 feet of pile penetration should be accomplished with no jetting allowed. Piles that cannot be driven the final 5 to 10 feet without the aid of jetting should be immediately brought to the attention of the design engineer, since the project design criteria may not have been satisfied. When jetting concrete piles, driving should be restricted to a static weight while the water is being injected to prevent damage due to excessive tensile stresses that may be induced by hammer impact. Jetting adjacent to existing structures or piles should be avoided. Although driving vibrations are reduced, caution must be exercised, since jetting causes disturbance of soil material. The design engineer must exercise caution when determining the design capacity for a jetted pile. Adequate provisions must be made for the control, treatment (if necessary), and disposal of run-off water. If jetting is anticipated, test piles should be installed using jetting, with the test piles being installed after the placement of any reaction piles.

The extent (depth) and appropriateness of jetting should be determined by the designer based on site conditions and design requirements. Backfilling should be required if the jetted hole remains open after the pile installation.

Jetting pipes may be encased or cast into the pile, attached to the exterior of the pile or attached to the driving leads and thereby become movable. Moveable jets are preferable, if circumstances do not exclude their use, due to the relative high costs of permanently attached jets. The use of two jets, one on each side the pile, provides the most rapid penetration and best alignment control. When using multiple jets, each should be equipped with its own water source and both should be similarly operated at the same depths and pressures. A single jet placed on one side of the pile may result in excessive pile drift. Experienced personnel should be relied upon when selecting and sizing jetting equipment.

3.5.1.1. Types of Jets

Typical equipment consists of jet pipes, a nozzle, pump, engine and hoses. The equipment must be capable of providing the desired volume of water at the required pressure. Water volume and pressure must be sufficient to allow discharged water to surface along the sides of the pile.

There are two types of jets, fixed and movable.

1. Fixed Jets. Fixed jets are jets that are a permanent part of the pile. Precast jets in concrete piles and concrete sheet piling may be used to avoid off-centre and/or unsymmetrical jetting and the problem of keeping proper alignment. This type of pile is costly but may be desirable where conditions do not permit use of a movable jet.
2. Movable Jets. These are attached to the pile to allow their removal after pile installation. Two jets symmetrically located give the most rapid penetration and best control of the pile path.

3.5.1.2. Pipes and Hose

The diameter of the pipe is essential to allow the required water flow. The diameter of the pipe should be no less than 2" (50.8 mm) and can vary up to 4" (101.6 mm). Nozzle diameters should be from 3/4" (19.1 mm) to 1-1/2" (38.1 mm).

The hose should be approximately 1 inch (25 mm) larger in diameter than the jet pipe but no less than 3" (76.2 mm) in diameter. It should have a protective jacket of canvas, cotton, or steel wire mesh. Hose length should be as short as possible to minimize friction losses.

3.5.1.3. Pumps

A large flow of water is required to successfully jet a pile. This can typically be accomplished by using a jet pump, whose flow should be no less than 250 gallons/minute (946 l/min) and can range up to approximately 1,000 gallons/minute (3785 l/min). Water pressures should generally vary from 100 to 200 psi (6.9 - 13.8 bar) for most soils. However, in gravels the pressures should be set at 100 to 150 psi (6.9-10.4 bar), and 40 to 60 psi (2.8 - 4.1 bar) in loose sands. They should be equipped with only bronze fittings. Jet nozzles may be chrome plated to resist

Figure 3-86 Discharge through Nozzles

DISCHARGE THROUGH NOZZLES

(GALLONS PER MINUTE)

PRESSURE AT NOZZLE P.S.I.	NOZZLE OPENING, INCHES									
	1"	1 1/8"	1 1/4"	1 3/8"	1 1/2"	1 3/4"	2"	2 1/4"	2 1/2"	3"
10	90	115	140	175	200	275	360	450	560	800
20	130	160	200	245	285	390	500	640	800	1150
30	155	200	245	300	350	475	620	780	970	1400
40	175	225	280	340	410	550	720	900	1125	1600
50	200	255	315	380	450	620	800	1020	1250	1800
60	220	275	345	425	500	675	880	1120	1360	1975
80	250	325	400	480	570	780	1020	1280	1560	2250
100	280	360	440	540	650	875	1140	1450	1775	2500
125	320	400	490	600	720	970	1275	1600	1950	2850
150	345	440	540	660	780	1050	1380	1750	2150	3100
175	370	475	580	710	850	1140	1500	1900	2325	3350
200	400	510	630	760	910	1220	1600	2025	2500	3600
250	440	570	700	850	1020	1360	1700	2280	2800	4000
300	480	620	760	930	1110	1480	1950	2500	3000	4400

Figure 3-87 Flow in Pipes[11]

Pressure Loss In Pipe Or Hose Due To Friction Pounds Per Square Inch Per 100 Feet[12]

GPM	2"	2-1/2"	3"	4"	5"	6"	8"	10"	12"	16"
30	1.7	0.6	#	#	#	#	#	#	#	#
50	4.3	1.4	0.6	#	#	#	#	#	#	#
100	16	5.2	2.2	0.5	#	#	#	#	#	#
200	56	19	7.7	1.9	0.6	#	#	#	#	#
300	120	40	17	4.0	1.4	0.5	#	#	#	#
500	300	140	43	10	3.5	1.4	#	#	#	#
1000		370	160	38	13	5.2	1.3	#	#	#
2000				130	47	20	4.8	1.5	0.6	#
4000					150	67	17	5.6	2.4	0.6

Trade Size	O.D. Inches	Inside Diameter, Inches		
		Standard	X-Heavy	XX-Heavy
1"	1.32	1.05	0.96	0.60
1-1/2"	1.90	1.61	1.50	1.10
2"	2.38	2.07	1.94	1.50
2-1/2"	2.88	2.47	2.32	1.77
3"	3.50	3.07	2.90	2.30
4"	4.50	4.03	3.83	3.15
5"	5.56	5.05	4.81	4.06
6"	6.63	6.07	5.76	4.90
8"	8.63	7.98	7.63	6.87

wear. The power source for the pump should have adequate torque and horsepower to pressurize the water, and all fittings, hoses, and orifices should be properly sized to accommodate the flow the water jet at the desired pressure. Figure 3-84 shows a jet pump mounted on a skid and powered by a diesel engine.

Figure 3-84 Jet Pump with Diesel Engine Power

Pressures at the nozzle of 50 psi may be enough for fine sands whereas 200 psi (13.8 bar) or more may be required to successfully jet compacted sand-clay-gravel. Volume must create sufficient velocity to carry the excavated material to the surface; 10 feet per second being adequate for sand sized particles and up to 25 feet per second for gravel. The larger the volume, the faster the jet will advance.

These pumps are suitable for pumping clean water only. They will wear very rapidly if used to pump muddy water, will lose their capacity and require an expensive overhaul. Before beginning any major job, the jet pump should be inspected to make sure it is in good condition.

These pumps have adequate capacity for jetting most piles up to about 16" (406.4 mm). For larger piles, the jetting system should be engineered for each project. For example, one project used four 1000 GPM (3785 l/min) pumps for 36" (914.4 mm) cylinder piles.

3.5.1.4. Jetting Procedure

Typical procedure is to jet a hole at the pile location before driving. Wood and pre-cast concrete piles are frequently jetted during driving to permit reaching the bearing stratum without damaging the pile. The pile is lubricated by water flowing along its sides. Since the pile will tend to walk toward the jet, it is necessary to jet first on one side and then on the other or to use multiple jets. Concrete piles have been cast with a hole in the centre for jetting during driving but this has not always been satisfactory. Jetting batter piles is particularly difficult.

A ring jet is shown in Figure 3-85. Discharge from various size nozzles is shown on Figure 3-86. Under normal circumstances, a 2-1/4" (57.2 mm) nozzle will develop sufficient pressure and provide maximum flow. However, when additional cutting power is needed, a 1-3/4" (44.5 mm) nozzle may be used. In rare circumstances, it may be necessary use a small 1-1/8" (28.6 mm) nozzle when jetting clay hardpan or when using a small jet pump.

Figure 3-85 Ring Jet

One also can have jet nozzles with side openings that tend to erode a larger hole, and this detail may be more effective in sandy clays. Jet nozzles with holes pointing upward may facilitate removal of cuttings. For very hard materials, one occasionally uses nozzles with a chisel point. These are called "chopping bits." In an emergency, a jet can be fabricated in the field by flattening the end of a 4" (101.6 mm) pipe with a sledge.

Jets are usually mounted on the left side of the leaders and handled with a tugger hoist or third drum. A 4" (101.6 mm) hose with "Boss" couplings connects the jet-pipe with a gooseneck in the boom at the midpoint of jet travel. Rigid pipe should be used in the boom. It is not only less expensive, but also less likely to be damaged. Safety chains should be provided to secure the hose at each joint. Caution should be exercised when using old hose unless it has been tested or known to be in serviceable condition. Always check hose and pipe for obstructions, before assembly. Flush the system with water before screwing on the nozzle.

The most common reason for poor performance of a jet is inadequate hose and pipe size. Figure 3-87 shows pressure loss information for various sizes of hose or pipe. As an example, at a flow of 1000 GPM (3785 l/min), there is 38 psi (2.6 bar) pressure drop in each 100' (30.5 m) of 4" (101.6 mm) pipe so with 600' (183 m) from pump to jet tip, the loss will be 228 psi (15.7 bar). If one begins with 250 psi (17.3 bar), such a pressure loss would leave little net pressure for jetting.

The standard set-up for the 4" (101.6 mm) jet with the 1000 GPM (3785 l/min) pump should be not more than 200' of 4" (101.6 mm) hose, 100' (30.5 m) of 4" (101.6 mm) pipe and up to 1000' (305 m) of 6" (152.4 mm) pipe. In calculating the length of pipe, add 10' for each fitting. This pipe/hose combination will provide a residual pressure at the nozzle of 100 psi (6.9 bar) that is adequate for average conditions.

When the pump must be located over 1000' (305 m) from the rig, 8" (203.2 mm) pipe should be used on the ground, rather than 6" (152.4 mm). When this does not give satisfactory performance, use 5" (127 mm) hose and 6" (152.4 mm) pipe in the boom. Increasing one size hose or pipe reduces pressure loss by 60% to 75%. As an alternative, when the water source is far from the rig, a storage tank to feed the jet pump may be placed near the rig and filled at a slower rate between uses.

For larger jets, one may use two, three or four pumps and supply lines must be properly engineered. One 5" (127 mm) jet hose (not over 150' (45.7 bar)) can be used with two pumps and two 5" (127 mm) hoses with three or four pumps. Each 5" (127 mm) hose should be supplied through 6" (152.4 mm) pipe in the boom and 10" (254 mm) pipe on the ground. For very long runs, it may be necessary to go to 12" (304.8 mm) or 16" (406.4 mm) pipe.

The most common reasons for poor jet performance:

1. Inadequate size piping, for the length of run.
2. Poor condition of pump.
3. Obstructions in pipe or hose lines.
4. Too small or clogged nozzle.
5. Loss of water in porous stratum.

3.5.1.4.1. Setting Up

Pumps should be located on a firm and level foundation and as near to water source as possible. It is preferable that the suction be kept flooded by using an elevated tank or other appropriate means. Caution must be exercised in attaching pumps directly to a fire hydrant in a municipal water system. They are of such great capacity that, if attached to a small main, they could cause a dangerous reduction in main pressure and possible infiltration of contaminated ground water.

When pumping from a river, or other open water, provide a suction hose with a foot valve and strainer. Locate a jet pump as far from the jetting operation as possible to minimize the possibility of dirty water entering the suction line. High-pressure jet pumps will wear out more quickly if muddy water is allowed to enter the pump.

The suction hose or pipe should be a minimum of 2" (50.8 mm) larger than the pump inlet, particularly if it is operating under suction lift. The suction line must be free of air leaks; otherwise, the pump will not maintain its prime. The foot valve should be located at least 18" (457.2 mm) below the water level but secured so that it stays several feet above the bottom. The suction hose should slope upward toward the pump; do not allow any loops to trap air. Never place a valve in the suction line when the pump must lift the water. For best results, the pump should be located as near the water as possible but no higher than 10' (3.05 m) above the water.

The discharge pipe for a pump should be 6" (152.4 mm) diameter for runs of up to 1000' (305 m) with a maximum of 300' of 4" (101.6 mm) hose and pipe at the jet. If the water source is over 1000' (305 m) from the driver, the discharge pipe should be 8" (203.2 mm) diameter. An inadequate size discharge pipe is one of the most frequent causes of poor jet pump performance.

Install a 5" (127 mm) gate valve on the pump discharge. Support pipe independently so that weight is not carried on pump casing. Insure that piping is arranged to accommodate of expansion and contraction due to temperature changes. Provide a

1-1/2" (38.1 mm) or 2" (50.8 mm) connection on the discharge line, between the pump and the 5" (127 mm) gate valve, to by-pass water when pump is idling. When using more than one pump, provide a check valve on the discharge side of each pump.

The pump should be located so that there is adequate passage for cooling air to the radiator and ample space around engine, and pump for servicing.

Most pumps are not self-priming and, unless there is positive suction pressure, it must be primed. The pump will be damaged if run dry.

When severe freezing weather is expected, consideration should be given to providing an enclosure for the pump.

3.5.1.4.2. Operation

Before starting up, check radiator, fuel tank, crankcase oil level, air cleaner and perform other routine maintenance as recommended in engine operating manual. Check suction strainer to make sure that it is clean and submerged at least 18" (457.2 mm) below the water. During this step the discharge valve should be closed.

Disengage clutch, start engine and allow warming up at 1000 RPM. If pump has primer, open valve in vacuum line and start vacuum pump; it will take about 1" (25.4 mm) of vacuum for every foot of water lift. Make sure all petcocks and drain plugs in the pump casing are closed. When pump is primed, clutch can be engaged and vacuum pump shut down. When ready to jet, open discharge valve and bring engine up to speed.

When jetting, the pump should not be run any faster than necessary to develop the required pressure; otherwise you are wasting fuel and increase pump and engine maintenance costs. When jetting is completed, throttle engine back, close the 5" (127 mm) discharge valve and open the by-pass valve.

If there is no vacuum pump, or the pump is inoperative, the pump can be primed by attaching a water hose to the connection on the top of the loop above the pump. Open all the petcocks on the top of the casing to allow air to escape. When casing is full of water, engage clutch and observe pressure gauge to check that pressure is rising.

If at any time the pump should lose its prime, disengage the clutch and reprime as explained above. Look for air leaks in suction line, open petcocks in pump casing on air leaks at packing glands.

When operating in freezing weather, it is necessary to thoroughly drain the pump when shutting down for the night. Remove drain plugs from casing and rotate impeller a turn or two to empty it. Raise suction hose and open foot valve; be sure strainer is clean for next days operation. Be sure vacuum system, by-pass line and water-cooled bearings are drained also.

3.5.1.4.3. Maintenance

These pumps will usually operate with very few problems. The engine should receive daily maintenance as indicated in operating manual. The engine will require servicing (oil change, new oil filter, etc.) once every three to six weeks depending on operating conditions. This can often be accomplished when rig is moving, if planned.

The greatest problem with jet pumps, when pumping from open water, is keeping suction strainer clean of trash and out of the mud. If the pump discharge surges, it usually is the result of a clogged suction line. Pumping muddy or dirty water will cause excessive wear inside the pump and loss of pumping capacity.

3.5.1.4.4. Storage

When taken out of service the pump should be flushed with fresh water, all drain plugs removed and pump turned over to drain impellers. Spray the interior of the pump with oil; replace drain plugs and close the suction and discharge openings. Wash exterior of pump, particularly if used around salt water. Prepare engine for storage.

Store pump off the ground so that it is accessible for inventory and maintenance. Lubricate; remove packing and fill packing space with grease. Make minor repairs; report condition of pump and any parts or major repairs needed.

3.5.2. Underwater Driving

Most pile driving for coastal and river structures can be driven from the surface; however, in some cases it is advantageous to drive piling underwater. Underwater driving eliminates the use of pile followers that add weight to the system, longer than needed piles and underwater pile cut-off. Hammers for underwater driving have been available for a long time; Vulcan, for example, manufactured its closed series of differential acting hammers starting in the 1930's; such a hammer driving piles underwater is shown in Figure 3-88.

Figure 3-88 Closed Type Differential Hammer Driving Underwater

Figure 3-89 Underwater Hydraulic Impact Hammer

Today, hydraulic hammers such as the IHC, Menck and Dawson hammers are the best impact hammers for this application. Hydraulic hammers have the advantage of being able to be used in very deep water, as with offshore oil platforms. Such a hydraulic hammer is shown in Figure 3-89. The hydraulic power pack is generally on the barge deck and the hoses extend into the water. In some cases, larger hose reels are used to store the hoses. On larger hammers, the power pack is mounted on the hammer and powered with an electric motor, the cables extending to the generator on the deck.

Hydraulic vibratory hammers can also be used underwater. To use a vibratory hammer underwater or for that matter under the power pack, it is important to both insure that water does not seep into the case and to limit the motor case drain pressure relative to the main vibratory case. The pressure of the motor case must be no higher than 2.75 bar (40 psi) above the vibratory case, lest the motor shaft seal blow oil into the vibratory case.

To determine the effective driving depth of the hammer, measure the vertical distance from the motor case drain outlet up to the power pack reservoir. If this distance is less than 9-12 m (30'-40') and the hammer is not submerged underwater, the hammer should operate satisfactorily as it is. Should the depth be in this range and the hammer be submerged, the case vent should be removed and a hose be attached that runs to the surface, and all leaks of the vibratory hammer must be completely stopped and sealed.

For depths greater than 12 m (40'), the motor shaft seals must be protected. To do this, remove the case vent and attach an air hose to the exciter case. An air compressor on the surface must pressurize this hose, and the case must be pressurized to the pressure of the water surrounding the vibrator at the depth the vibrator is used. The vibratory case pressure must not exceed the motor case pressure by more than 1.4-2 bar (20-30 psi) or the shaft seal(s) will rupture.

3.5.3. Pre-Excavation

Various pre-excavation methods are used in connection with pile driving to penetrate hard upper layers of soil, to prevent soil heave, to reduce driving resistance, to assist in removing or displacing obstructions, to relieve back pressure, to minimize the effect of driving on adjacent structures and to minimize stresses in piles and cores.

Pre-excavation will frequently present problems in disposal of water and spoil. Responsibility for spoil removal should be clearly understood before starting work.

3.5.3.1. Overview of Preboring and Drills

Preboring consists of drilling, augering, or coring a hole in the ground and filling the hole with concrete or driving a pile into the hole. This is generally done with a continuous flight auger. Soil augers or drills may be used where jetting is impractical. Sometimes, predrilling is necessary in order to drive a pile through obstructions such as timbers, boulders, and riprap. Predrilling is also frequently used for pile placement through embankments containing boulders. In such a case it is preferable to minimize predrilling by controlled embankment construction using soil particle sizes limited to 4" or less.

The hole size depends upon the size and shape of the piling and soil conditions. It should be large enough to permit driving but small enough so that the pile will be firmly and solidly supported against lateral movement. Under most conditions, the predrilled hole diameter should be 4 inches less than the diagonal of square or steel-H piling, and 1 inch less than the diameter of round piling. Where piling must penetrate into or through very hard material, it is usually necessary to use a diameter equal to the diagonal width or diameter of the piling. For driven piles, preboring is advantageous when the ground resistance is extremely high. For square concrete piles, the diameter of the bored shaft should be approximately 125% of the nominal pile size. Although preboring will generally reduce the driving resistance, it does so at the expense of shaft resistance, which decreases during the preboring. This influence should be considered during design.

A pilot or prebore hole may be required to penetrate hard nonbearing strata; to maintain accurate location and alignment when passing through materials which tend to deflect the pile; to avoid possible damage to adjacent structures by reducing vibrations; to prevent heave of adjacent buildings, or to remove a specified amount of soil when installing displacement-type piles, thereby reducing foundation heave. It should be noted that on past projects, concrete piles have been successfully driven through man-made fills such as levee embankments without preboring. Preboring through cohesionless soils below the water table is not recommended, since the prebored hole may not stay open and could require a casing. The most widely used method of preboring is by utilizing an auger attached to the side of the crane leads. Oversizing the hole will result in a loss of skin friction and a reduction in the axial and lateral capacity, thereby necessitating re-evaluation of the pile foundation. When extensive preboring is needed, consideration should be given to using a drilled-shaft system rather than a driven-pile system.

A separate pay item for predrilling is usually included in the contract documents.

3.5.3.2. Augering

3.5.3.2.1. Overview

Augering is effective in a wide variety of clayey soils. It consists of a continuous flight auger of suitable diameter and length, with a drilling bit, driven by a hydraulic or air powered drill that is mounted in a carriage. The lower portion of an auger is shown in Figure 3-90.

Figure 3-90 Auger Ready to Drill

The most efficient drilling is when the material is cut up by the bit and carried to the surface by the flights at a steady rate. The auger should be prevented from screwing itself into the ground and stalling. If this happens, the drill should be reversed and the auger backed out.

To select an auger, the "rule of thumb" is to multiply the diameter of the hole in inches by the depth of the hole in inches. Once you have this quantity, multiply it by one of the following factors to compute the amount of torque required in ft-lbs:

- For Loam, Sand, Fine Gravel and Soft Clay: x 1
- For Hardpan, Soft Shale and Gravel: x 1.4
- For Hard Shale and Lime rock: x 2

Moderate speeds, 40 to 80 RPM, are used in augering. In hard formations, the peripheral speed of the auger should not exceed 100 ft/min. The weights of the auger, bit, drill and carriage are usually more than adequate to advance the drill into most materials that must be pre-excavated for piles. Approximately one pound of down crowd for each foot-pound of torque should be sufficient in hard formations.

When augering coarse sands and gravel without a clay binder, there may be a problem keeping the hole open. Difficulty may also be experienced in augering heavy clays due to their tendency to clog the flights. Auger cleaners have been used with some success.

Augers can be expected to stall occasionally when they reach a new stratum or encounter obstructions. When this occurs, the drill should be lifted slightly to allow rotation to continue. Constant stalling overheats the equipment, places an unnecessary strain on all components and slows progress.

Augering can frequently be facilitated by adding water, steam or air through the drill stem to "lubricate" the soil and aid in breaking it up and carrying it to the surface. Before assembling, make sure the hoses and stem are clear. The auger should be worked up and down as necessary to keep the hole open. It may be necessary to make two or three passes to obtain a clean hole.

3.5.3.2.2. Auger Drills

A typical auger drill assembly consists of a continuous flight auger of suitable diameter and length, with a bit, driven by air or hydraulic power. There are two commonly used setups for augers:

1. The drill is generally mounted in a carriage that rides on one side of the leaders on a drill guide frame clamped to the leader columns as shown in Figure 3-91. To reduce load on the drill fairleader, the drill should be reeved a minimum of two parts. Even so, the modern cranes exert sufficient line pull to bend the fairleader if an attempt is made to pull a stuck auger with a main hoist line.
2. The drill can also ride on leaders dedicated for auger operation, as shown in Figure 3-92. The drill is located approximately at the same radius as the hammer. The frames also carry one or more auger guides to keep the auger in line. A fairleader may be mounted on the top leader section for the drill line used in raising and lowering the drill. In order to reduce stress in the leaders when a significant pull is required on the auger, three parts should be used, reeved as shown, and the line run through the boom tip fairleader. When a hydraulic hammer is used with a hydraulic drill, the same power unit can be used for both when properly configured.

Figure 3-91 Auger Drill with Leaders and Hydraulic Impact Hammer

Figure 3-92 Auger Drill Assembly

Note: The modern cranes exert sufficient line pull to damage the rig if an attempt is made to pull a stuck auger with a main hoist line. No attempt should be made to reinforce the drill fairleader or its connection to the leaders.

The augers have a hollow stem and the drill is provided with a water swivel so that water or air can be added in small amounts to facilitate drilling. Augers can be made in a variety of sizes and pitches. Typical auger flighting is right hand, 12" (305 mm) diameter, 3'-8" (1.12 m) thick with 12" (305 mm) pitch and is furnished in 16' (5 m) and 24' (7.3 m) long sections. A 2' (610 mm) bottom auger section has heavy double flighting and is equipped with a cutter head with a pilot bit, shank plates and wisdom teeth, as shown in Figure 3-93.

Figure 3-93 Bottom Auger Section

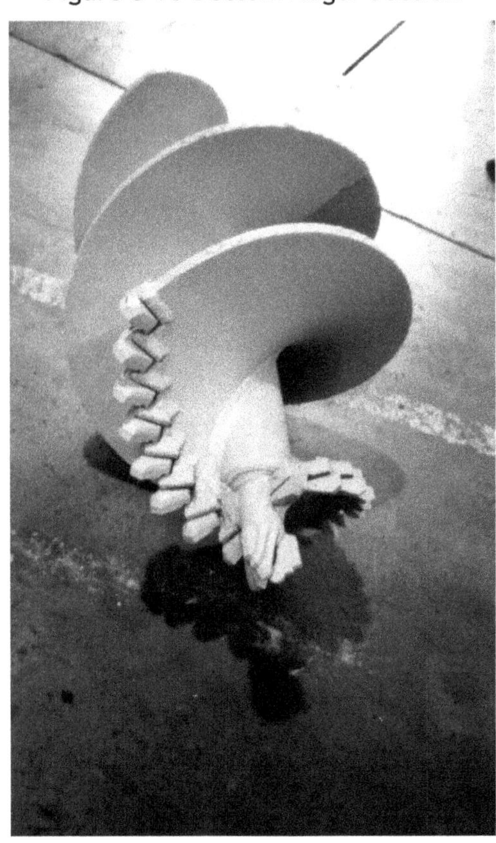

API or similar taper threaded joints join auger sections. Joints should be greased before assembly and a piece of jute or nylon rope wrapped around the pin to form a gasket. For disassembly, heating the box with a torch will expand it and char the jute. If there is fear of the joint loosening, a 1" (25.4 mm) x 1/4" (6.4 mm) strap may be welded across the joint. The top auger section should have a 6-5/8" API Regular box connection to mate with drill. Augers may also be extended using drill pipe.

When time does not permit building a special auger, the standard 12" (304.8 mm) auger can be built up to 14" or 16" (406.4 mm) by welding a 3/8" (9.5 mm) x 1" (25.4 mm) or 3/8" (9.5 mm) x 2" (50.8 mm) strip to the 12" (304.8 mm) flighting. Similarly, 1" (25.4 mm) can be burned off the flighting to reduce its diameter to 10" (254 mm). In either case, the 12" (304.8 mm) bit should be replaced with the proper size. Stepped augers may be used to match shell sizes; a reaming tooth should be located at the bottom of each step and special auger guides may have to be used.

Power for rotating the auger may be supplied by an air, steam or hydraulic motor, the most common drive being hydraulic.

The hydraulic supply piping is valved so that the motors can be run in parallel for normal speed and maximum torque or in series for conditions requiring higher speed and lower torque.

3.5.3.2.3. Maintenance and Lubrication

Drills like all other mechanical equipment require periodic lubrication and routine maintenance to insure efficient operation and free from breakdown.

- Fairleader sheaves, water swivels as well as carriage and guide rollers and centre supports should be lubricated once a day with a grease gun. Water swivels should be greased more frequently when using muddy water. The filler plug should be removed and the oil level visually checked daily in drill gear cases and additional SAE EP-140 gear Oil added, as necessary. Water should be drained from gear case once a week, or more often if indicated. The gear case should be drained and refilled with new SAE 140 gear Oil at the start of each project.
- Auger flights and wet-rotary reamers will wear, particularly in abrasive soils or when bits are not maintained. Diameters should be checked at the start of the job, and periodically thereafter, and built up with weld metal or pieces of hard grade reinforcing rod. Hard surfacing applied to the edge of flighting near the bottom will reduce wear and extend the time between repairs.
- Auger bit teeth and pilot bits should be checked and replaced when worn or dull. When drilling hard materials a spare Bottom Auger Section, with bit, and extra teeth should be kept on hand. Fishtail bits must also be kept sharp and hard surfaced for best operations.
- Drill guide frames and augers should be checked by eye for straightness before using. Bent sections should be straightened or replaced. The guide frames and fairleaders are furnished with elastic stop nuts but all bolts should be checked after the first day's operation and weekly thereafter, and tightened if necessary.
- Hydraulic drills require the usual precautions associated with all hydraulic systems.

Particular attention must be given to keeping the system clean, particularly when coupling and uncoupling hoses, and to replacing filters frequently. Hydraulic drills should be watched for oil leaks both internal and external. Hydraulic system oil can leak past the motor seals and into the gear case and, if allowed to continue, the gear oil will be diluted and gears damaged. Make sure a drain hose is provided and that it is unobstructed. When oil overflows from the gear case, it usually means that hydraulic oil is leaking past the motor seals. The oil level in the hydraulic reservoir must be checked daily.

- Care must be taken with hoses used to power drills. They should be of adequate length so that they are not stretched, sharply bent, kinked or twisted. They should be protected from blows by sharp and heavy objects. All hoses hanging in the air should be secured at both ends so they will not fall should they become uncoupled.
- Before storing, gear cases should be drained and filled with new oil, and all openings plugged or capped to prevent entrance of moisture and dirt. Drills should be stored and preferably shipped in upright position to prevent gear oil from leaking into motors. "Saver-subs" should be kept with drills. If stored outdoors, cover drills with a tarpaulin carefully wired down. Drills, drill stems, augers and related equipment should be blocked up off the ground and accessible for inspection and minor repair.
- One major problem auger users face—especially with auger cast piling—is cleaning the auger. One method widely used is to reverse the auger while it is out of the ground, thus allowing the spoil to fly and impact personnel and property in the work area. This is potentially hazardous. Manual cleaning of the flighting can be time consuming. Auger cleaning equipment is available to facilitate this operation.

3.5.3.3. Wet Rotary Drilling

"Wet Rotary" drilling is used to excavate very deep holes where the power to auger would be excessive. It is also particularly suited for plastic soils that would stick to auger flights and for soils that will collapse unless the hole is left filled with fluid.

For wet rotary drilling, the auger is replaced with a 6" (152.4 mm) or larger x-heavy pipe drill stem. The stem diameter is not so large as to restrict the upward movement of pieces of soil but large enough so that a minimum of 10 feet per second velocity is maintained. When used with an open-centre rotary table, the stem is provided with four 3/4" (19.1 mm) x 1-1/2" (38.1 mm) splines. A suitable fishtail bit is provided at the bottom and a large water swivel is flange-connected at the top.

The drill stem is provided with one or more sets of reamers. These may be of uniform diameter or gradually increase in size to correspond with the diameters of the pile.

Water is supplied through the stem under pressure to carry the excavated material to the surface and to keep the hole open. It is occasionally necessary to recirculate the drilling fluid. If this is the case, a suitable slurry pump will have to be used.

Power to rotate the drill can be provided by a hydraulic power-sub operating in drill guide frames on the left side of the leaders. An open centre rotary table mounted at the bottom of the leaders may also power the drill.

When used with the open-centre rotary table, the drill stem is restrained by one or more centre supports riding on a WF-beam, bolted to the right front leader column.

The large quantity of slurry produced becomes a serious problem and its disposal must be planned. The site should be excavated to a single common elevation (with no pier holes) and adequate draining ditches provided.

On most projects, clear water can be used, but when substantial amounts of granular materials are present, it may be necessary to recirculate the clayey water to aid in carrying the granular particles to the surface. Auger flight may be effective in lifting sand to the surface. Step paddles may be used on the drill stem to shape hole and agitate cuttings.

Adequate quantities of water under pressure may be available from a hydrant. When a pressure supply is not available, the 350 psi (24.2 bar), 1000 GPM (3785 l/min) jet pumps are ideal for furnishing clear water for rotary drilling. One pump will normally be adequate for holes up to 12" (304.8 mm) but as much as 4000 GPM (15,140 l/min) has been required for 30" (762 mm) holes. Adequate size pipe and hose from pump to drill stem is essential.

High pressure jet pumps (such as the 6" (152.4 mm) Peerless, TUT-16) are not designed for pumping mud or water containing sand or solids and will be quickly damaged if so used. A suitable slurry or dredge pump must be employed. A 3' x 3' x 3' (1 m x 1 m x 1 m) screen box should be provided for the slurry pump suction. A vacuum pump is also necessary, as foot-valves do not work well in a slurry pit.

Figure 3-94 Wet Rotary Drill

In wet rotary drilling, speeds of 60 to 100 RPM are used to cut the material into small pieces to make it easier to bring to the surface. Progress will be governed largely by the quantity of water or mud circulated, by proper balance of hole and drill stem size and by maintaining an even drilling pace. At the end of drilling, the bit should be rotated for ½ minute at the bottom of the hole before it is withdrawn, to clear the hole

Fishtail bits should be kept sharp and built up to the proper diameter. A spare bit with coupling to match the drill stem should be kept on hand. Observe the wash water to make sure you continue to bring up material. If material is not coming up, raise the drill and lower it at a slower speed.

Wet rotary drilling can be done either with a carriage mounted hydraulic drill working at the tope of the drill stem or with open centre rotary table mounted near the bottom of the leaders. The open centre rotary table permits extending the drill stem above the top of the leaders, by using a mast, and the drill is more easily serviced at a lower level. It is less practical for drilling batter piles because of possible difficulties with the engine operating on a slope.

When entering a pile in a wet rotary hole, it must be done carefully to allow the displaced slurry to escape. Otherwise, it may collapse the shell or cause the pile to be ejected in projectile fashion, due to flotation.

3.5.3.4. Rotary Drilling

Either a carriage mounted hydraulic drill or an open centre rotary table can be used for drilling soft rock, when equipped with a suitable roller or other type bit. There are many types of bits and a sample of the rock to be drilled should be examined by an experienced drilled before the drill bit is selected. The proper bit is essentially for satisfactory production.

Effective drilling speeds will vary from 20 RPM for hard rock to as high as 150 RPM for very soft materials. An adequate supply of water, air or drilling mud is needed to remove the cuttings. When there is any difficulty keeping the hole open, it may be necessary to increase water supply or to use a larger drill stem to increase velocity of wash water.

The most important factor affecting production in having proper load on the bit. A load of 1000 to 3000 pounds per inch of bit diameter is needed to drill soft rock. The harder the rock, the more load required. Since the weight of the drill, drill stem and bit will total only 4000 to 6000 pounds, it is necessary to add weight by the use of drill collars, a heavier drill stem or pull down cable crowds.

The drilling of hard rock requires pressures of 5000 to 6000 pounds per inch of bit diameter, or even more, and calls for special equipment.

3.5.3.5. Dry Pre-Excavation

A plug of earth may be removed with a dry tube pre-excavator. The open-end tube is driven into the ground with the valve on the top open to the atmosphere. The valve is closed and the tube extracted; the vacuum created at the top helps hold the plug of earth in the tube as it is withdrawn.

The tube is swung over the spoil area and steam or air pressure is applied to push the earth plug out. This simple method is fast and effective in plastic soils to a limited depth. Deeper pre-excavation may require two passes. Try excavating a 10' (3.05 m) plug on the first pass and then to desired depth on the second pass. Care must be exercised not to drive the tube so deep that it is stuck or that the earth plug is too long to be expelled.

3.5.3.6. Spudding

Spudding is similar to preboring, and is ordinarily used to penetrate obstructions that are located more than 10 to 15' (3.0 - 4.6 m) below pile cut-off elevation and cannot be removed more surely and economically by excavation. These layers or obstructions may damage the pile or present unusual driving difficulty. Spudding is accomplished by driving a spud, such as mandrel, heavy steel pipe or H-pile section, to provide a pilot hole. The spud is withdrawn and the pile inserted into the hole and driven to the required depth. Problems may result if the spud is driven too deep, since extraction becomes more difficult as penetration is increased. Spudding may sometimes entail alternately lifting a partially driven pile a short distance and redriving it when very difficult driving is encountered (e.g. for heavy piles). Because this procedure adversely affects the soil's lateral and axial capacity, it should be avoided for friction piles and should never be permitted without the specific authorization of the design engineer.

One can have a number of spuds designed to suit different types of piles. Frequently superintendents make spuds on the job by using an H-beam or a piece of heavy pipe. Occasionally, a short spud is used to make a hole so that a wood pile, H-beam or pipe pile can be lowered far enough to get under the hammer. Spuds may also be effective in displacing small obstructions that cause the piles to run off location or damage shells.

3.5.3.7. Airlifts

Although not strictly methods of pre-excavation, airlifts and air jets are sometimes used in place of water jetting, to clean open-end piles for example. The air jet acts as the water jet in breaking up compacted material and the upward velocity of the escaping air carries the excavated material to the surface. Large volumes of air are required and receivers are used to supplement compressor capacity. An airlift consists of a conductor pipe lowered to the bottom of the pile with means for delivering air at the bottom. The mixture of air and water, being lighter, is forced upward by the water outside carrying soil with it. Airlifts are often combined with water jetting.

3.5.3.8. Screwing

Screw piles consist of a pile casing fitted with one or more turns of helical screw having a larger diameter than the pile. Screwing the casing into the ground to a predetermined level makes the installation. Torque is provided by a capstan or similar device.

3.5.3.9. Pull Down

This is a type of pile jacking where the pile casing is jacked into place and filled with concrete. Where a closed-end casing is used, special equipment is limited to conventional screw or hydraulic jacks. Where an open-end casing is used, a jet or miniature orange peel bucket is used for removing the core. Figure 3-95 shows such a bucket; it can also be used for casings installed by vibration.

Figure 3-95 Bucket for Plug Removal

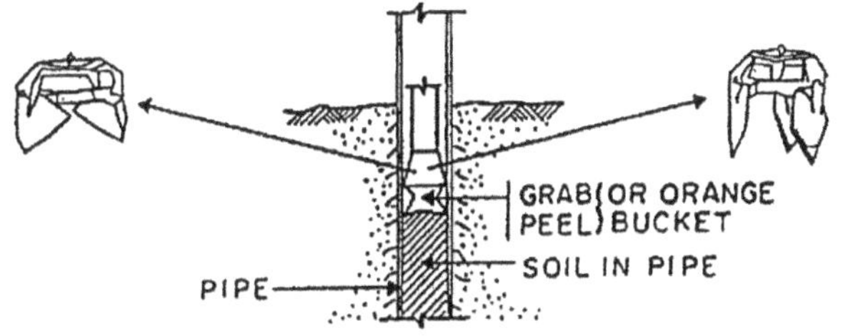

3.5.3.10. Concrete Pile Cutting

When the driving of a concrete pile is complete, the next step is to connect the head of the pile with the structure. It is frequently necessary to have the reinforcing bar protrude above the head of the pile to do this. Two basic ways to accomplish this are (1) to fabricate the pile with protruding reinforcing bar or cable, or (2) to cut off the head of the pile in such a way as to leave the reinforcing bar or cable exposed for connection. If the reinforcing bars are not damaged, piles made by method (1) are ready for connection upon driving. For method (2), it is necessary to cut the concrete pile, and if the reinforcing bars are needed, it is necessary to leave them protruding and undamaged.

1. Manual Methods. Manual methods of concrete pile cutting involve the use of tools adapted for the task. In this case, concrete crushing is performed by jackhammers. Reinforcement bars are then cut by welding, or flame cutting. Concrete pile saws are also used manually; these are rotary saws that simply cut the pile. Manual pile cutting is shown in Figure 3-96.
2. Automatic Methods. These involve the use of hydraulic devices called concrete pile cutters, which crush the pile by effort applied in a transverse direction perpendicular to the pile axis. Because of this effort, the reinforcement bars were exposed and then cut by different ways at a predetermined level. Others combine the bond separation and the pile cracking in the same area of the pile. In the latter method, the cracking takes place both in the same plane as the bond failure and also above and below it. Concrete pile cutters are shown in Figure 3-97 and Figure 3-98.

Figure 3-96 Manual Pile Cutting

Figure 3-97 Concrete pile cutter

Figure 3-98 Concrete Pile Cutter in a Marine Application

Photo Credits

- Berminghammer Corporation (Mr. Mike Justason):
- Dawson Construction Plant: Figure 3-78.
- Giken America (Mr. John Santos): Figure 3-19, Figure 3-20.
- GRL and Associates/Pile Dynamics, Inc.: Figure 3-6, Figure 3-29, Figure 3-57, Figure 3-83, Figure 3-96.
- International Construction Equipment (Mr. Richard Morris): Figure 3-21, Figure 3-24, Figure 3-50, Figure 3-51, Figure 3-55, Figure 3-72, and Figure 3-92.
- Tramac Corp. (Ms. Michele Carline): Figure 3-52, Figure 3-53, Figure 3-54 and Figure 3-56.
- Vulcan Foundation Equipment: Figure 3-7, Figure 3-10, Figure 3-13, Figure 3-17, Figure 3-64, Figure 3-69, Figure 3-79, Figure 3-80, Figure 3-88.
- Don C. Warrington, P.E.: Figure 3-5, Figure 3-8, Figure 3-9, Figure 3-11, Figure 3-13, Figure 3-15, Figure 3-18, Figure 3-22, Figure 3-23, Figure 3-25, Figure 3-31, Figure 3-32, Figure 3-34, Figure 3-36, Figure 3-37, Figure 3-40, Figure 3-38, Figure 3-48, Figure 3-49, Figure 3-65, Figure 3-66, Figure 3-67, Figure 3-68, Figure 3-71, Figure 3-74, Figure 3-76, Figure 3-80, Figure 3-81, Figure 3-90, Figure 3-97.

Footnotes

[1] Note: Some MKT and Menck models use a hollow ram as the cylinder and a stationary piston; this variation does not materially affect the following discussion.

[2] Using the steam or air expansively means that the steam or air is trapped in the cylinder and expanded to power the ram. In diesel hammers, the expansion of the air-fuel mixture after combustion powers the hammer; this is an example of expansive use of a gas in pile driving equipment.

[3] Pile impedance is discussed in Chapter 4.

[4] However, the French company, PTC, built several models that had three eccentrics that worked well. In this type of construction, the centre eccentric has twice the eccentric moment of the other two.

[5] Much of the information in this section has been enhanced by Viking, K. (2002) *Vibro-drivability—a field study of vibratory driven sheet piles in non-cohesive soils.* PhD Thesis, Division of Soil and Rock Mechanics, Royal Institute of Technology, Stockholm, Sweden.

[6] Gumenskii, B.M., and Komarov, N.S. (1959) *Soil Drilling by Vibration.* Moscow: Ministry of Municipal Services of the Russian Soviet Federated Socialist Republic.

[7] Many of the traditionally measured quantities for vibratory hammers such as amplitude, acceleration ratio, etc., are computed using this case. For most conventional vibratory hammers, one can consider the entire system a rigid mass. This is because the relatively low frequency vibrations of most vibratory hammers do not bring the distributed mass and elasticity of the system into play. For sonic pile drivers or very long piles the resonant properties of the system become significant, and the analysis becomes more complicated.

[8] In U.S. units, if the eccentric moment is in in-lbs and the vibrating weight in pounds, the resulting amplitude will be in inches.

[9] Warrington, D.C. (1994) "Survey of Methods for Computing the Power Transmission of Vibratory Hammers." *Pile Buck*, Second August Issue 1994.

[10] It was the basis for the design of many of the Russian vibratory hammers. A modification of this can be found in Erofeev, L.V., Smorodinov, M.I., Fedorov, B.S., Vyazovikii, V.N., And Villumsen, V.V. (1985) *Machines and Equipment for the Installation of Shallow and Deep Foundations.* (In Russian) Second Edition, Mashinostrenie, Moscow, pp. 95-111.

[11] Above table based on Hazen-Williams formula for pipe in average condition c=100. Sizes are i.d. of hose or standard weight pipe.

[12] (#) indicates that loss is less than 0.5 p.s.i. per 100 feet.

Chapter 4. Pile Capacity and Driveability

The ultimate objective of any foundation is to transfer load from a structure to the soil below it. The foundation must do this while neither overstressing the soil to failure nor compromising its own structural integrity. In the case of driven piles, the impact method of installation adds the requirement that the pile not be overstressed during driving as well. The goal of the designer is to achieve all of these objectives in an economical fashion.

There are many books and publications that describe the design of driven piles. This book does not have design as its primary focus but emphasizes issues relating to installation, materials and equipment related to pile driving. Therefore, this chapter is intended to give only an overview of the design process, especially as it relates to the installation of piles.

4.1. Overview

4.1.1. General Considerations for Foundation Selection

A foundation is the interfacing element between the superstructure and the underlying soil or rock. The loads transmitted from the superstructure to the underlying soil must not cause soil bearing failure or damaging settlement. It is essential to systematically consider various foundation types and to select the optimum alternative based on the superstructure and the subsurface conditions.

The following design approach should be used to determine the optimum foundation alternative:

1. Determine the foundation loads to be supported and special requirements such as limits on total and differential settlements, lateral loads, scour, and time constraints on construction.
2. Evaluate the subsurface investigation and the laboratory testing data of the soils or rocks. Ideally, the subsurface investigation and laboratory testing were performed based on the designer's knowledge of the loading and performance requirements.
3. Prepare a "design" soil profile and critical cross sections. Determine soil layers suitable or unsuitable for spread footings

Table 4-1 Foundation Types and Uses

Foundation Type	Use	Applicable Soil Conditions	Unsuitable or Difficult Soil Conditions
Spread footing, Wall footings.	Individual columns, walls, bridge piers.	Any conditions where bearing capacity is adequate for applied load. May use on single stratum; firm layer over soft layer or soft layer over firm layer. Check immediate, differential, and consolidation settlements.	Any conditions where foundations are supported on soils subject to scour or liquefaction. Bearing layer located below ground water table.
Mat Foundation	Same as spread and wall footings. Very heavy column loads. Usually reduces differential settlements and total settlements.	Generally soil-bearing value is less than for spread footings. Over one-half area of structure is covered by individual footings. Check settlements.	Same as footings.
Driven Pile Foundations (Friction, end bearing, or combination.)	In groups to carry heavy column and bridge pier loads. To resist large uplift and/or lateral loads.	Poor surface and near surface soils. Soils of high bearing capacity 25' - 150' below the ground surface. Check settlement of pile groups in clay.	Shallow depth to hard stratum. Sites where pile driving vibrations and heave may adversely impact adjacent facilities (pile jacking may relieve vibration problems.) Boulder fields.
Drilled Pile Foundations (generally end bearing or combination of end bearing and shaft resistance.)	Larger column loads than for piles but eliminates pile cap by using drilled shafts as column extension. Used for bridge piers.	Similar applications to driven piles, subject to limitations in unsuitable soils. Well suited for end bearing on rock.	Deep deposits of soft clays and loose water bearing granular soils. Caving formation difficult to stabilize. Artesian conditions. Boulder fields.

or pile foundations.

4. Consider foundation alternatives and prepare preliminary technically feasible designs. Some of these are described in Table 4-1.
5. Prepare cost estimates for feasible alternative designs including all associated substructure costs.
6. Selecttheoptimumalternative. Generally, the most economical alternative should be selected and recommended. The ability of the local construction force and availability of materials and equipment should be considered. Other factor such as speed of construction, adjacent properties and environmental variables may influence the final selection.

For major projects, if the estimated costs of feasible foundation alternatives (during the design stage) are within 15% of each other, then alternate foundation designs should be consideration for inclusion in the contract documents. Innovative contracting practices such as value engineering, contract substitution, and design-build can alter the original design concept.

Because this book deals only with driven pile foundations, other types of foundations will not be discussed further.

4.1.2. Establishment Of A Need For A Deep Foundation

The first difficult problem facing the foundation designer is to establish whether or not the site conditions dictate that a deep foundation must be used. Typical situations for the use of driven piles as well as additional uses of deep foundations are shown in Figure 4-1.

Figure 4-1 Situations in which Deep Foundations may be Needed[1]

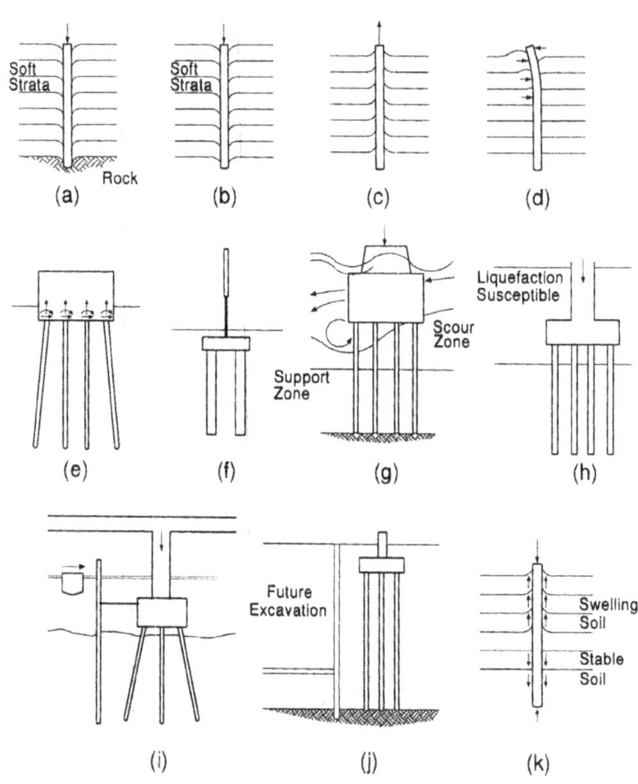

Figure 4-1 (a) shows the most common case in which the upper soil strata are too compressible or too weak to support heavy vertical loads. In this case, deep foundations transfer loads to a deeper dense stratum and act as toe bearing foundations. In the absence of a dense stratum within a reasonable depth, the loads must be gradually transferred, mainly through soil resistance along shaft, Figure 4-1 (b). An important point to remember is that deep foundations transfer load through unsuitable layers to suitable layers. The foundation designer must define at what depth suitable soil layers begin in the soil profile.

Deep foundations are frequently needed because of the relative inability of shallow footings to resist inclined, lateral, or uplift loads and overturning moments. Deep foundations resist uplift loads by shaft resistance, Figure 4-1 (c). Lateral loads are resisted either by vertical deep foundations in bending, Figure 4-1 (d), or by groups of vertical and battered foundations, which combine the axial and lateral resistances of all deep foundations in the group, Figure 4-1 (e). Lateral loads from overhead highway signs and noise walls may also be resisted by groups of deep foundations, Figure 4-1 (f).

Deep foundations are often required when scour around footings could cause loss of bearing capacity at shallow depths, Figure 4-1 (g). In this case the deep foundations must extend below the depth of scour and develop the full capacity in the support zone below the level of expected scour. The geotechnical analysis of bridge foundations should performed on the basis that all stream bed materials in the scour prism have been removed and are not available for bearing or lateral support. Costly damage and the need for future underpinning can be avoided by properly designing for scour conditions.

Soils subject to liquefaction in a seismic event may also dictate that a deep, foundation be used, Figure 4-1 (h). Seismic events can induce significant lateral loads to deep foundations. During a seismic event, liquefaction susceptible soils offer less lateral resistance as well as reduced shaft resistance to a deep foundation. Liquefaction effects on deep foundation performance must be considered for deep foundations in seismic areas.

Deep foundations are often used as fender systems to protect bridge piers from vessel impact, Figure 4-1 (i). Fender system sizes and group configurations vary depending upon the magnitude of vessel impact forces to be resisted. In some cases, vessel impact loads must be resisted by t he bridge pier foundation elements. Single deep foundations may also be used to support navigation aids.

In urban areas, deep foundations may occasionally be needed to support structures adjacent to locations where future excavations are planned or could occur, Figure 4-1 (j). Use of shallow foundations in these situations could require future underpinning in conjunction with adjacent construction.

Deep foundations are used in areas of expansive or collapsible soils to resist undesirable seasonal movements of the foundations. Deep foundations under such conditions are designed to transfer foundation loads, including uplift or downdrag, to a level unaffected by seasonal moisture movements, Figure 4-1 (k).

In many instances either a shallow or deep foundation alternative is technically feasible Under these circumstances, an evaluation of the shallow Foundation should include; (1) the dimensions and depth of shallow footings based on allowable bearing capacity, (2) the magnitude and time-rate of settlement under anticipated loads, and (3) detailed cost analysis including such factors as need for cofferdams, overall substructure cost, dewatering and foundation seals, construction time, construction risk and claims potential. A comparative analysis of feasible deep foundation alternatives should also be made. The cost analyses of feasible alternatives should have a significant role in final selection of the foundation type.

4.2. Basic Load Transfer Concepts for Pile Foundations

Pile groups of a foundation should be proportioned to permit only an acceptable amount of settlement in support of the maximum combination of loads including but not limited to:

- All dead loads, including the weight of the pile cap and any load thereon. Where consolidating soils overlie the bearing strata consideration must be given to possible downdrag, or negative shaft friction;
- The live load, factored as may be permitted by the applicable building code;
- Lateral force and moment reactions, including the effect of eccentricity, if any, between the column load and the load-bearing centre of the pile group;
- Vessel impact loads;
- Uplift and overturning due to wind and hydrostatic loads;
- Seismic loads;
- Ice loads; and
- Loads from moving machinery and traffic vibration.

The allowable load from pile to soil should be determined by a recognized method of analysis. This will include an evaluation of three basic limit states or potential failure modes including:

- Failure of the pile itself (structural capacity);
- Failure of the soil surrounding the pile shaft and under the pile toe (geotechnical capacity); and
- Failure of the pile during driving (installation stresses.)

Piles must penetrate or bear on acceptable material

a sufficient distance to develop the required geotechnical capacity. They must be driven either plumb or to the batter specified. The toe of the pile should be protected from damage during installation. Damaged or broken piles should neither be used nor accepted for payment.

4.2.1. General Considerations

The static load capacity of a pile can be defined as the capacity of the ground to support the loads on the pile imposed by the structure. A static analysis is performed to determine the static load capacity (ultimate and allowable capacities) of individual piles and of pile groups. Static pile capacity computations are necessary to estimate the number of piles and the required pile lengths both for the design of substructure elements and for ordering piles of the correct length from the supplier.

The ultimate static load capacity of an individual pile and of a pile group is the smaller of:

- The structural capacity of the pile(s), or
- The capacity of surrounding medium to support the loads being transferred to it from the pile(s).

Static load tests can also be used to determine pile load capacity. Because load tests are expensive and time consuming, they cannot always be economically justified. Alternatives to static tests include dynamic tests, Osterberg cells or Statnamic™ tests. The designer should carefully evaluate the appropriateness and correct interpretation procedures of the latter two tests.

Sometimes, two static analyses are required for a design. First, it is necessary to determine the number and length of piles necessary to support the design loads. The second analysis is performed to determine the total driving resistance the pile will encounter which helps in determining the necessary capability of the driving equipment. Figure 4-2 and Figure 4-3 illustrate situations that require two static analyses. Figure 4-2 shows a situation in which the piles are to be driven for a bridge pier footing. In this case, the first static analysis is performed to determine the number of piles and pile lengths by neglecting the soil resistance to the estimated scour depth. The second analysis is performed to estimate the total driving resistance encountered by the pile including the soil resistance to the scour depth.

Figure 4-2 Situation Where Two Static Analyses are Necessary Due to Scour

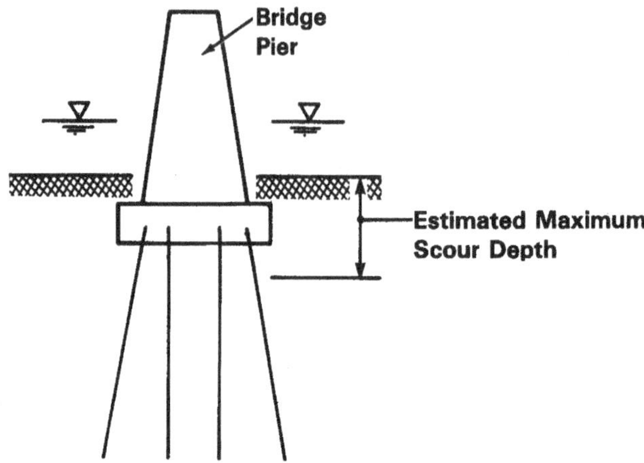

Figure 4-3 shows another frequently encountered situation in which piles are driven through loose uncompacted fill material into the natural ground. The designer has determined that the loose fill material offers unreliable resistance and he or she usually neglects in determining the number of piles and lengths required. The second analysis is performed to determine total resistance encountered by the pile including the resistance in the fill material.

Figure 4-3 Situation Where Two Static Analyses are Necessary due to Fill Materials

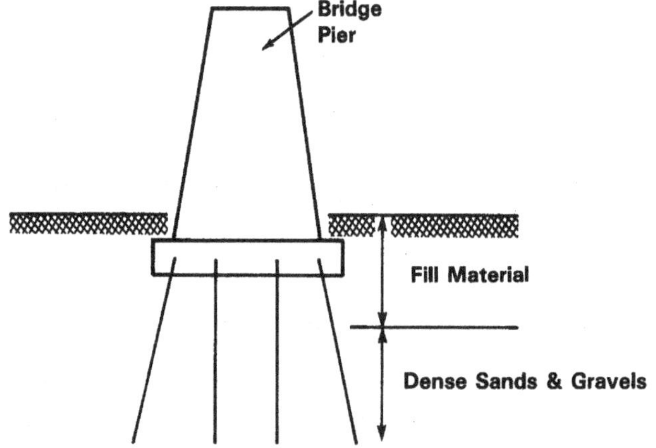

Because of the considerable influence of construction procedures on the behavior of pile foundations, the methods described in this chapter lead to successful designs of deep foundations only if adequate construction monitoring techniques are used. Inspection considerations should be integral part of the design and construction of any foundation. Load tests, wave equation analysis or dynamic monitoring for construction control and monitoring should be used to check the validity of the static design.

4.2.2. Load Transfer

The ultimate bearing capacity Q_u, of a pile in soil or rock may be expressed by the sum of toe resistance Q_p and shaft resistance Q_s, or

Equation 4-1:

$$Q_u = Q_p + Q_s$$

Or more completely

Equation 4-2:

$$Q_u = Q_p + Q_s = q_p A_p + f_s A_s$$

In which q_p equals the unit bearing capacity of the pile toe of area A_p; and f_s equals the average unit shaft friction on the pile shaft of area A_s, Figure 4-4 (A).

Figure 4-4 Load Transfer of a Single Pile

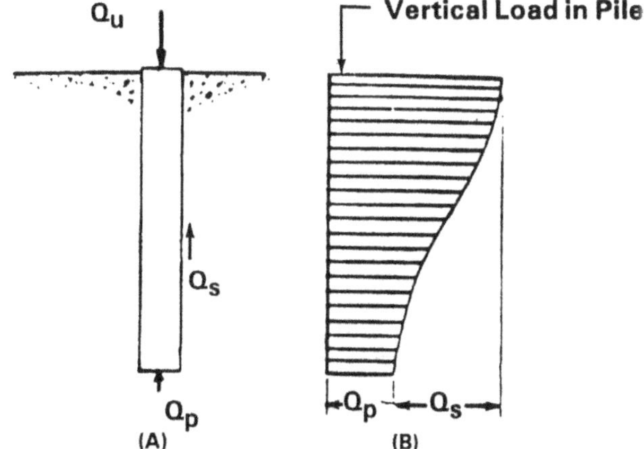

The above equation contains an assumption that both the pile toe and the pile shaft have moved sufficiently with respect to the adjacent soil to simultaneously develop the ultimate toe and shaft resistance. Generally, the displacement needed to mobilize shaft friction is smaller than that required to mobilize toe resistance. Since this is the simplest rational approach for all piles except piles greater than 610 mm (24") nominal dimension, it has been commonly used in static analytical methods. Figure 4-4 (B) illustrates a typical load transfer profile for a single pile. The load transfer distribution can be obtained by installing strain gages or telltale rods at different depths along the pile shaft. Figure 4-4 (B) shows measured axial force in the pile plotted against depth. Q_p represents axial pile load and Q_s represents the pile shaft friction load transferred to the soil medium.

Figure 4-5 (A) shows axial load and shaft friction distributions in cohesive soils. Figure 4-5 (B) shows typical axial load and shaft friction distributions in cohesionless soils. Figure 4-5 (C) illustrates the case of negative shaft friction on a pile shaft.

Figure 4-5 Typical Distributions of Shaft Resistance

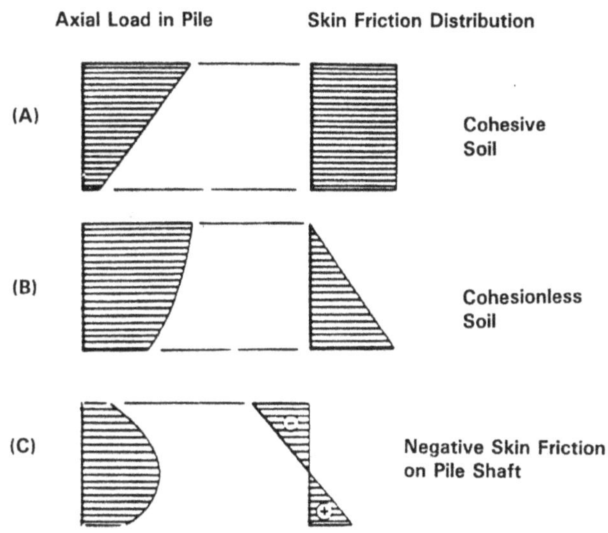

4.2.3. Overburden Pressure

A complete description of soil and rock mechanics, along with descriptions of the various methods of laboratory and field soil testing, is beyond the scope of this book[2]. However, one concept that needs to be discussed is that of effective stress and overburden pressure.

Effective overburden pressure at a given depth below ground is the stress at that depth due to the soil weight above. A plot of effective overburden pressure versus depth is called a "P_o Diagram" such as is shown in Figure 4-6, and is used for pile foundation analysis. The effective overburden pressure (P_o) at any depth is determined by accumulating the weights of all layers above that depth as follows:

Equation 4-3:

$$P_o = \gamma L$$

(For homogeneous soil deposits above the static water table)

Equation 4-4:

$$P_o = \gamma L - \gamma_w L_w$$

(For homogeneous soil deposits below the static water table)

where

- P_o = Overburden Pressure, psf or kPa
- γ = total soils unit weight, lbs/ft^3 or kN/m^3
- L = depth below soil surface, feet or m
- γ_w = unit weight of water, lbs/ft^3 or kN/m^3
- L_w = depth below top of water table, feet or m

Figure 4-6 Effective Overburden Pressure (P_o) Diagram

a) Water Table Below Soil Surface

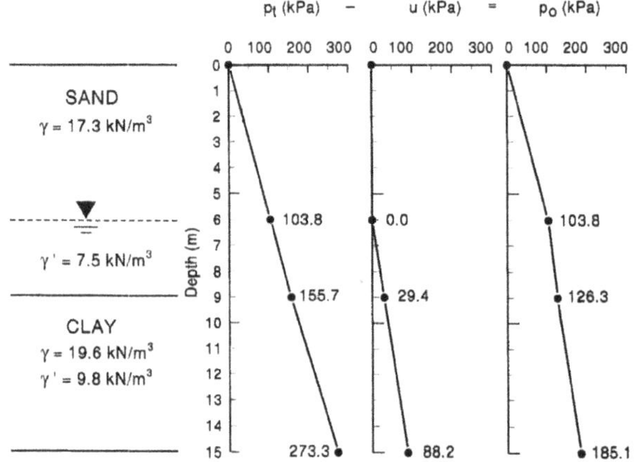

b) Water Table Above Soil Surface

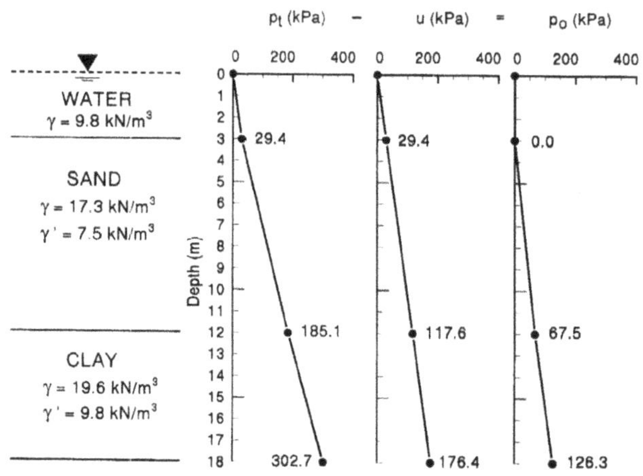

4.2.4. Bearing Capacity of Piles on Rock

Pile foundations on rock are normally designed to carry large loads. For pile foundations which are driven to rock, which include steel H-piles, pipe piles or precast concrete piles, the exact area of contact with rock, the depth of penetration into rock as well as the quality of rock are largely unknown. Therefore, the determination of load capacity of driven piles on rock should be made based on driving observations, local experience and load tests.

Rock Quality Designation (RQD) Values can provide an assessment of rock mass quality as shown in Table 4-2. The RQD is only for NX size core samples (double tube core barrel) and is computed by summing the length of all pieces of core equal to or longer than 4" (101.6 mm) and dividing by the total length of the coring run. The result is multiplied by 100 to get RQD in percent. Fresh, irregular breaks due to drilling and handling should be ignored and the pieces counted as intact lengths. Examples of rock core samples are shown in Figure 4-7.

Figure 4-7 Rock Core Samples

For end bearing piles on rock of fair to excellent quality (Table 4-2), the structural load capacity of the piles (based on allowable (design) stress for pile material) will generally be lower than the capacity of rock to support loads. Therefore, small diameter piles supported on fair to excellent quality rock may be loaded to their allowable structural capacity. For example, if steel H piles are expected to penetrate to rock through soil deposits without obstructions, then an allowable design stress of 83.4 MPa (12 ksi) can be used even without a load test.

Table 4-2 Engineering Classification for In Situ Rock Quality

RQD #	Rock Mass Quality
0-25	Very Poor
25-50	Poor
50-75	Fair
75-90	Good
90-100	Excellent

Piles supported on soft weathered rock, such as shale or other types of very poor or poor quality, should be designed based on the results of pile load tests.

4.3. Subsurface Explorations

The design of a structure's foundation requires adequate knowledge of the subsurface conditions at the construction site. If the designer has the appropriate information, then an economical foundation system can be designed. The absence of a thorough foundation study or adequate geotechnical data often leads to (1) a foundation system with a large factor of safety which is generally a more expensive foundation and in some cases one that may be difficult to construct, or to (2) an unsafe foundation, or to (3) construction disputes and claims.

A thorough foundation study consists of:

- A subsurface exploration program (which includes borings, sampling, groundwater measurements, and in-situ testing);
- Laboratory testing;
- Geotechnical analysis of all data; and
- A determination of design properties; and design recommendations.

This section is an overview of subsurface explorations and especially the more common types that are of specific interest to deep, pile foundations.

4.3.1. Subsurface Exploration Phases

There are three major phases in a subsurface exploration program. These phases are:

(1) Planning the exploration program (office work). The purpose of this phase is to obtain information about the proposed structure and general information on the subsurface conditions. The structural information can be obtained from studying the preliminary structure plan prepared by the bridge design office and by meeting with the structural designer. Approach embankment preliminary design and performance requirements can be obtained from the roadway office. The planning phase prepares the engineer for the field reconnaissance survey, and identifies possible problems and areas to scrutinize.

(2) Completing a field reconnaissance survey. The purpose of this phase is to substantiate the information gained from the office phase and to plan the detailed site exploration program. The field reconnaissance for a structure foundation exploration should include:

a. Inspection of nearby structures to determine their performance with the particular foundation type used.

b. Inspection of existing structure footings and stream banks for evidence of scour (for stream crossings) and movement. Large boulders in a stream are often an indication of obstructions that may be encountered in pile installations.

c. Visual examination of terrain for evidence of landslides.

d. Recording of the location, type and depth of existing structures that may be affected by the new structure construction.

e. Relating site conditions to proposed boring operations. This includes recording the locations of overhead and belowground utilities, site access, private property restrictions, and other obstructions.

f. Recording of any feature or constraint that may impact the constructability of potential foundation systems.

(1) Performing a detailed site exploration program (boring, sampling, and in-situ testing). The purpose of any boring program is not just to drill a hole, but to obtain representative information on the subsurface conditions, to recover disturbed and undisturbed soil samples, and to permit in-situ testing. This information provides factual basis upon which all subsequent steps in the pile design and construction process are based. Its quality and completeness are of paramount importance. Each step in the process directly or indirectly relies on this data. The first step in this phase is to prepare a preliminary boring, sampling, and in-situ testing plan. For major structures, pilot borings are usually performed at a few select locations during the preliminary planning stage. These pilot borings establish a preliminary subsurface profile and thus identify key soil strata for testing and analysis in subsequent design stage borings. During the design stage of major structures, a two phase boring program is recommended. First, control borings are performed at key locations identified in the preliminary subsurface profile to determine what, if any, adjustments are appropriate in the design stage exploration program. Following analysis of the control boring data, verification borings are then performed to fill in the gaps in the design stage exploration program.

Each phase should be planned so that a maximum amount of information can be obtained at a minimum cost. Each phase also adds to, or supplements, the information from the previous phase.

4.3.2. Guidelines For Minimum Structure Exploration Programs

The cost of a boring program is comparatively small in relation to the foundation cost. For example, the cost of one 60 mm diameter boring is less than the cost of one 305 mm diameter pile. However, in the absence of adequate boring data, the design engineer must rely on extremely conservative designs with high safety factors. At the same time, the designer assumes enormous risk and uncertainty during the project's construction. The number of borings required, their spacing, and sampling intervals depend on the uniformity of soil strata and loading conditions. Erratic subsurface conditions require closely spaced borings. Structures sensitive to settlements or subjected to heavy loads require detailed subsurface knowledge. In these cases borings should be closely spaced. Rigid rules for number, spacing, and depth of borings cannot be established. However, the following are general "guidelines" useful in preparing a boring plan.

1. A minimum of one boring with sampling should be performed at each pier or abutment. The boring pattern should be staggered at opposite ends of adjacent footings. Pier and abutment footings over 30 m in length require borings at the extremities of the substructure units.

2. Estimate required boring depths from

data gathered in the planning and field reconnaissance phases. The geotechnical engineer should make confirmation of boring depth suitability for design purposes as soon as possible after field crews initiate a boring program. Although less preferred, it may be possible for field crews to adjust boring depths using a resistance criteria such as: "Structure foundation borings shall be terminated where a minimum SPT resistance of 50 blows per 300 mm has been maintained for 7.5 m." (This rule is intended for preliminary guidance to drillers. For heavy structures with high capacity piles, the borings must go deeper. A resistance criterion may also be inappropriate in some geologic conditions such as sites with boulder fields.)

3. All borings should extend through unsuitable strata, such as unconsolidated fill, peat, highly organic materials, soft fine-grained soils and loose coarse-grained soils to reach hard or dense materials. Where stiff or dense soils are encountered at shallow depths, one or more borings should be extended through this material to a depth where the presence of underlying weaker strata cannot affect stability or settlement of the structure.
4. Standard Penetration Test (SPT) samples should be obtained at 1.5 m intervals or at changes in material with the test data recorded in accordance with ASTM D1586. Undisturbed tube samples should be obtained in accordance with ASTM D1587 at sites where cohesive soils are encountered. The location and frequency of undisturbed soil sampling should be based on project requirements.
5. When rock is encountered at shallow depths, additional borings or other investigation methods such as probes, test pits, or geophysical tests may be needed to define the rock profile. When feasible, borings should extend a minimum of 3 m into rock having an average core recovery of 50% or greater with an NX-core barrel (54 mm diameter core).
6. Drill crews should maintain a field-drilling log of boring operations. The field log should include a summary of drilling procedures including SPT hammer type, sample depth and recovery, strata changes, and visual classification of soil samples. The field log should also include pertinent driller's observations such as location of ground water table, boulders, loss of drilling fluids, artesian pressures, etc. Disturbed and undisturbed soil samples as well as rock cores should be properly labeled, placed in appropriate storage containers (undisturbed tube samples should be sealed in the field), and properly transported to the soils laboratory.
7. The water level reading in a borehole should be made during drilling, at completion of the borehole, and a minimum of 24 hours after completion of the borehole. Long term readings may require installation of an observation well or piezometer in the borehole. More than one week may be required to obtain representative water level readings in low permeability cohesive soils or in bore holes stabilized with some drilling muds.
8. All boreholes should be properly backfilled and sealed following completion of the subsurface exploration program, data collection, and analysis. Borehole sealing is particularly important where groundwater migration may adversely affect the existing groundwater conditions (aquifer contamination) or planned construction (integrity of tremie seals in future cofferdams).

These guidelines should result in subsurface exploration data that clearly identify subsurface stratigraphy and any unusual conditions, allow laboratory assessments of soil strength and compressibility, and document the groundwater table conditions. This information permits a technical evaluation of foundation options and probable costs.

4.3.3. Descriptions of More Common Subsurface Tests for Pile Foundations

The capacity of the soil medium surrounding the pile can be estimated from geotechnical engineering analysis using:

- Shear strength parameters of the soil surrounding the pile;

- Pressuremeter or dilatometer test data;
- Dynamic penetration test data (i.e., Standard Penetration Test); and
- Static penetration test data (i.e., static cone penetrometer).

Because of their importance, we will only outline the last two procedures.

4.3.3.1. Standard Penetration Test (SPT)

The split barrel sampler (Figure 4-8) used in the Standard Penetration Test (SPT) is the primary disturbed sail sampler. The SPT test consists of driving a 51 mm O.D. (35 mm I.D.) split-spoon sampler into the soil with a 64 kg mass dropped 760 mm. The sampler is generally driven 450 mm, and the blow count for each 150 mm increment is recorded. The number of blows required to advance the sampler from a penetration depth of 150 mm to a penetration depth of 450 mm is the SPT resistance value, N. This sequence is generally repeated at vertical intervals of 1.5-3 meters.

Figure 4-8 Split Barrel Sampler

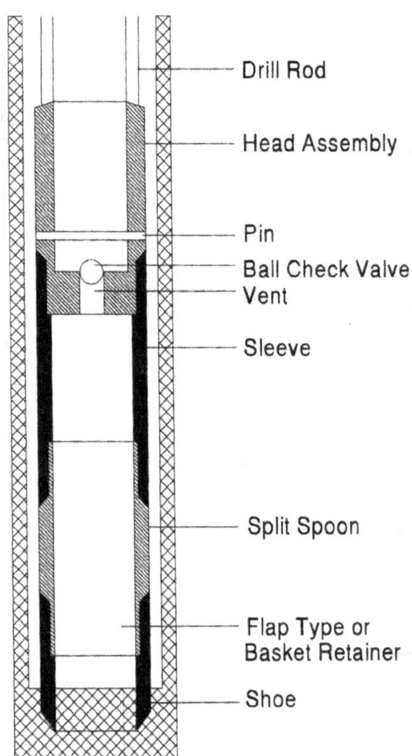

The SPT hammer type and operational characteristics can have a significant influence on the resulting SPT N values. There are two main hammer types currently in use in the US, the safety hammer and the automatic hammer. A third hammer type, the donut hammer, was used almost exclusively prior to about 1970. However, it is seldom used now due to safety considerations. Figure 4-9 provides illustrations of the three SPT hammer types.

SPT tests are subject to common errors, some of which are:

1. Effect of overburden pressure. Soils of the same density will give smaller SPT N values near the ground surface.
2. Variations in the 760 mm free fall of the drive weight, since this is often done by eye on older equipment using a rope wrapped around a power takeoff (cathead) from the drill motor. Newer automatic hammer equipment does this automatically
3. Interference with the free fall of the drive weight by the guides or the hoist rope. New equipment eliminates rope interference.
4. Use of a drive shoe that is badly damaged or worn from too many drivings to "refusal" (SPT N values exceeding 100).
5. Failure to properly seat the sampler on undisturbed material in the bottom of the boring.
6. Inadequate cleaning of loosened material from the bottom of the boring.
7. Failure to maintain sufficient hydrostatic pressure in the borehole during drilling or during drill rod extraction. Unbalanced hydrostatic pressures between the borehole, drill water and the ground water table can cause the test zone to become "quick". This can happen when using the continuous-flight auger with the end plugged and maintaining a water level in the hollow stem below that in the hole.
8. SPT results may not be dependable in gravel. Since the split spoon inside diameter is 35 mm, gravel sizes larger than 35 mm will not enter the spoon. Therefore, soil descriptions may not reflect actual gravel content of the deposit. Also, gravel pieces may jam the end of the spoon that may get plugged and cause the SPT blow count to be erroneously high.
9. Samples retrieved from dilatant soils (fine sands, sandy silts) which exhibit unusually

Figure 4-9 SPT Hammer Types

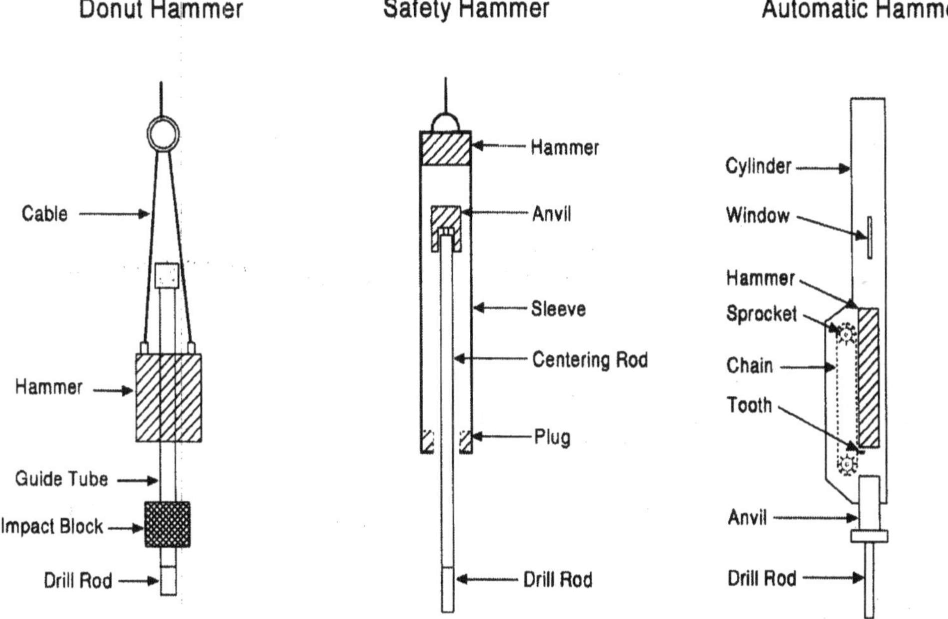

high blow count should be examined in the field to determine if the sampler drive shoe is plugged. Poor sample recovery is an indication of plugging.

10. Careless work on the part of the drill crew.

The use of reliable qualified drillers and adherence to recommended sampling practice cannot be overemphasized. State agencies that maintain their own drilling personnel and equipment achieve much more reliable, consistent results than those who routinely let boring contracts to the low bidder.

With all of these variations, for use in design SPT values should be corrected for at least two factors: the overburden pressure and the efficiency of the hammer.

- Overburden. For SPT values, correct N for overburden using:

Equation 4-5:

$$(N_1)_{60} = C_N N$$

Where

Equation 4-6:

$$C_N = \sqrt{\frac{2\text{ ksf}}{\sigma'_z}} \leq 2 \text{ (U.S. Units)}$$

$$C_N = \sqrt{\frac{100\text{ kPa}}{\sigma'_z}} \leq 2 \text{ (SI Units)}$$

C_N is a correction factor based on the effective overburden stress. Using this correction results in a value of N that would have been measured if the effective overburden stress had been 100 kPa (2 ksf). Correlations of cohesive soil physical properties with N values are crude and, therefore, correction of N values in cohesive soils is not necessary.

- Hammer efficiency. Prior to 1980, the efficiency of the hammer was not well recognized as influencing the blow count and was usually not considered in analysis. SPT tests in the U.S. have been usually been performed with safety hammers with a mechanical efficiency of around 60%. Other types of testing equipment (especially the newer automatic hammers) have different efficiencies. Since N is also sensitive to the energy supplied by the equipment, N_1 should be further corrected to the value at 60% of the input energy, $(N_1)_{60}$. The combined correction is:

Equation 4-7:

$$(N_1)_{60} = C_N \frac{e_{hammer}}{60} N_1$$

Where e_{hammer} = efficiency of the SPT hammer used, percent. Table 4-3 shows representative efficiencies for SPT procedures.

Table 4-3 Energy Ratios for SPT Procedures Country Hammer Type Hammer Release Estimated Rod Energy (%)

Country	Type of Ram	Type of Ram Lifting and Release Mechanism	Efficiency, Percent
Japan	Donut	Free-fall	78
	Donut	Rope & Pulley with special throw release	67
U.S.A.	Safety	Rope & Pulley	60
	Donut	Rope & Pulley	45
	Automatic	Chain or Hydraulic Mechanism	90
Argentina	Donut	Rope & Pulley	45
China	Donut	Free-fall	60
	Donut	Rope and Pulley	50

Generally speaking, most SPT correlations use the $(N_1)_{60}$ values.

Table 4-4 and Table 4-5 show common (but not absolute) relationships between various soil properties and SPT test results (described below.) These tables can be useful for design is results of more accurate laboratory or in situ results are unavailable.

4.3.3.2. Cone Penetration Test (CPT) And (CPTU)

The cone penetration test (CPT) was first introduced in the U.S. in 1965. By the mid 1970's, the electronic cone began to replace the mechanical cone. In the early 1980's, the piezo-cone or cone penetration test with pore pressure measurements (CPTU) became readily available. Since that time, the CPT/ CPTU has developed into one of the most popular in-situ testing devices. Part of this popularity is due to the CPT's ability to provide large quantities of useful data quickly and at an economical cost. Depending upon equipment capability as well as soil conditions, 100 to 350 m of penetration testing may be completed in one day.

Cone penetration testing can be separated into two main categories:

a. Electronic cones.

b. Mechanical cones.

Electronic cones are now the dominant cone type used in cone penetration testing. Hence, mechanical cones will not be discussed further. Electronic cones may be further divided into two primary types, the standard friction cone (CPT), and the piezo-cone (CPTU).

In the CPT test, a cone with a 1000 mm^2 base and a 60° tip attached to a series of rods is continuously pushed into the ground. Typically, a hydraulic ram with 45 to 180 kN of thrust capability is used to continuously advance the cone into the ground at a rate of 20 mm/sec. A friction sleeve with c n surface area of 15000 mm^2 is located behind the conical tip. Built in load cells are used to continuously measure the cone tip resistance, q_c, and the sleeve friction resistance, f_s. The friction ratio, R_f, is the ratio of f_s/q_c and is commonly used in the interpretation of test results.

The piezo-cone (CPTU), is essentially the same as the standard electronic friction cone and continuously measures the cone tip resistance, q_c, and the sleeve friction resistance, f_s, during penetration. In addition to these values, the piezo-cone includes porous filter piezo-elements that may be located at the cone tip, on the cone face, behind the cone tip, or behind the friction sleeve. These porous filter elements are used to measure pore pressure, u, during penetration.

A general schematic of a cone penetrometer is presented in Figure 4-10.

Figure 4-10 Terminology Regarding the Cone Penetrometer

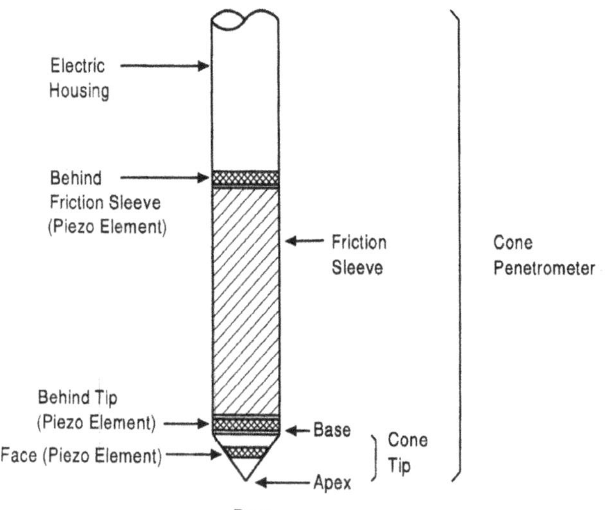

Table 4-4 Relative Density of Cohesionless Soils as a Function of SPT N Values[3]

Nomenclature	Relative Density, Percent	Internal Friction Angle ϕ	Moist Unit Weight	Safety Hammer SPT N_{60} Value (Blow/Foot {300 mm})
Very Loose	0-15%	< 28°	< 15.7 kN/m³ (100 pcf)	Less than 4
Loose	15-35%	28°-30°	14.9-19.7 kN/m³ (95-125 pcf)	4 - 10
Medium Dense	35-65%	30°-36°	17.3-20.4 kN/m³ (110-130 pcf)	10 - 30
Dense	65-85%	36°-41°	17.3-22.0 kN/m³ (110-140 pcf)	30 - 50
Very Dense	85-100%	> 41°	> 20.4 kN/m³ (130 pcf)	> 50

Table 4-5 Relative Consistency of Cohesive Soils as a Function of SPT N Values[4]

Nomenclature	Hand Manipulation Characteristics	Estimated Unconfined Compression Strength q_u	Moist Unit Weight	Safety Hammer SPT N_{60} Value (Blow/Foot {300 mm})
Very Soft	Extruded between fingers when squeezed	< 23.9 kPa (0.50 ksf)	15.7-18.9 kN/m³ (100-120 pcf)	< 2
Soft	Moulded by light finger pressure	23.9-47.9 kPa (0.5 - 1 ksf)	15.7-18.9 kN/m³ (100-120 pcf)	2 - 4
Firm or Medium	Moulded by strong finger pressure	47.9-95.8 kPa (1 - 2 ksf)	17.3-20.4 kN/m³ (110-130 pcf)	4 - 8
Stiff	Readily indented by thumb but penetrated with great effort	95.8-191.5 kPa (2 - 4 ksf)	18.9-22 kN/m³ (120-140 pcf)	8 - 15
Very Stiff	Readily indented by thumbnail	191.5-383 kPa (4 - 8 ksf)	18.9-22 kN/m³ (120-140 pcf)	15 - 30
Hard	Indented with difficulty by thumbnail	> 383 kPa (8 ksf)	> 20.4 kN/m³ (130 pcf)	> 30

Sample cone penetrometer data is shown in Figure 4-52. CPT/CPTU test results are interpreted as follows:

a. CPT/CPTU data can provide a continuous profile of the subsurface stratigraphy. A simplified soil classification chart for a standard electronic friction cone is presented in Figure 4-11.

b. From correlations with CPT/CPTU data, evaluations of in-situ relative density, D_r, and friction angle, ϕ, of cohesionless soils as well as the undrained shear strength, c_u, of cohesive soils can be made. Correlations for determination of other soil properties, liquefaction susceptibility, and estimates of SPT values may also be determined. The accuracy of these correlations may vary depending upon geologic conditions. Correlation confirmation with local conditions is therefore important.

Figure 4-11 Simplified Soil Classification Chart for Standard Electronic Friction Cone

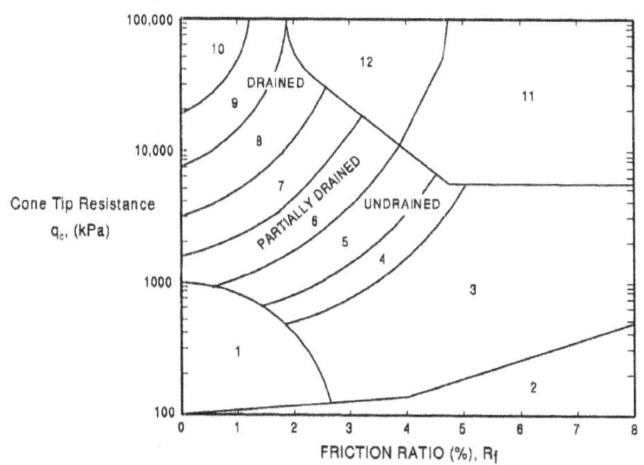

Zone	q_c/N	Soil Behavior Type
1)	2	Sensitive fine grained
2)	1	Organic material
3)	1	Clay
4)	1.5	Silty clay to clay
5)	2	Clayey silt to silty clay
6)	2.5	Sandy silt to clayey silt
7)	3	Silty sand to sandy silt
8)	4	Sand to silty sand
9)	5	Sand
10)	6	Gravelly sand to sand
11)	1	Very stiff fine grained
12)	2	Sand to clayey sand

Advantages and disadvantages of CPT/CPTU test are as follows:

- The primary advantage of CPT/CF 3TU testing is the ability to rapidly develop a continuous profile of subsurface conditions more economically than any other subsurface exploration or in-situ testing tools Determination of in-situ soil strength parameters from correlations with CPT/ CPTU data is another advantage. The CPT/ CPTU test can also reduce the number of conventional borings needed on a project, or focus attention on discrete zones for detailed soil sampling and testing. Lastly, CPT/CPTU results are relatively operator independent.
- Limitations of CPT/CPTU testing include the inability to push the cone in dense or coarse soil deposits. To penetrate dense layers, cones are sometimes pushed in boreholes advanced through the dense strata. Another limitation is that soil samples are not recovered for confirmation of core stratigraphy. Local correlations are also important in data interpretation.

4.4. Design Factors and Criteria

4.4.1. Allowable Stress Design (ASD)

Traditional design of deep foundations - both geotechnical and structural - involved the use of the "allowable stress design" (ASD) technique. This involved first determining the loads and stresses on a given structure or soil, multiplying it by a factor of safety, and then comparing it with an allowable stress or load[5]. All of the methods used in the book are based on ASD.

The factor of safety used in determining the allowable pile load capacity should be selected based on the following considerations:

- o The level of confidence in the input parameters. (This confidence level is a function of the type and extent of subsurface investigations and laboratory testing of soil and rock materials);
- o Variability of the soil and rock;
- o Method of design analysis;
- o Effects of the proposed pile installation method; and
- o Level and type of construction monitoring (dynamic formula, wave equation analysis, dynamic analysis).

The plans and specifications usually require either an allowable or an ultimate pile capacity. The design pile capacity is determined during the design stage by performing a static analysis and by determining allowable stresses in the pile material. The safety factor chosen should be based on the reliability of the method specified for pile capacity during installation. Safety factors increase with the increase in unreliability of the method use for determining pile capacity.

4.4.1.1. Typical factors of safety

Table 4-6a provides typical factors of safety for vertical load behavior. Typical or usual loads refer to conditions that are a primary function of a structure and can be reasonably expected to occur during the service life. Such loads may be long-term, constant, intermittent, or repetitive nature. Deviations from these minimum values may be justified by extensive foundation investigations and testing to reduce uncertainties related to the variability of the foundation soils and strength parameters. Static load tests allow the factor of safety to equal 2 for usual design and may lead to substantial savings in foundation costs for economically significant projects.

4.4.1.2. Other factors of safety

Lower FS are possible for unusual or extreme loads, Table 4-6b, provided soil investigation is thorough and settlement will be within a tolerable range.

- Unusual loads refer to construction, operation, or maintenance conditions that are of relatively short duration or infrequent occurrence. Risks associated with injuries or property losses can be controlled by specifying the sequence or duration of activities and/ or by monitoring performance.
- Extreme loads refer to events that are highly improbable and occur only during an emergency. Such events may be associated with major accidents involving impacts or explosions and natural disasters due to hurricanes. Extreme loads may also occur from a combination of unusual loads.

The basic design for typical loads should be efficiently adapted to accommodate extreme loads without experiencing a catastrophic failure; however, structural damage which partially impairs the operational functions and requires major rehabilitation or replacement of the structure is possible. Caution is required to achieve an efficient design that will avoid unacceptable injuries or property losses.

Table 4-6 Factors of Safety for Bearing Capacity

a) Usual Loads

Construction Monitoring Technique	Loading Condition	Factor of Safety
Static Load Test	Any	2.0
Wave Equation Analysis Calibrated to Results of Dynamic Pile Tests	Compression	2.5
	Tension	3.0
Static Design Analysis Only	Resistance to Uplift	2.5
	Resistance to Downdrag	3.0
	Pile Groups	3.0
	No Static Load Test	3.0
	Soil Profile Containing More than One Type of Soil or Stratum	4.0

b) Influence of Loading Condition

Method of Capacity Calculation	Loading Condition	Minimum Factor of Safety, Compression	Minimum Factor of Safety, Tension
Verified by pile load test	Usual	2.0	2.0
	Unusual	1.5	1.5
	Extreme	1.15	1.15
Verified by pile driving analyzer	Usual	2.5	3.0
	Unusual	1.9	2.25
	Extreme	1.4	1.7
Not verified by load test	Usual	3.0	3.0
	Unusual	2.25	2.25
	Extreme	1.7	1.7

For example, consider a pile with a specified allowable load of 140 kips (623 kN.) The ultimate load of the pile will depend upon the construction monitoring technique:

- For a pile with only a static design analysis, the ultimate load will be (140)(3) = 420 kips (1868 kN).
- If a wave equation analysis is being used to estimate the pile capacity during installation, then the ultimate resistance to which each pile should be driven will be ((140)(2.5) = 350 kips (1558 kN).
- If a static load test is performed to estimate the pile capacity, then each pile should provide an ultimate resistance of (140)(2) = 280 kips (1245 kN).

4.4.2. Load and Resistance Factor Design (LRFD)

Load and Resistance Factor Design (LRFD) methods involve the use of statistical methods to determine the actual combined effect of various types of loads on a structure. The various loads (dead, live, earthquake, etc.) are combined using factors, then compared with a load capacity which itself is factored. Fortunately all of the factors have been determined for a given code or structure type in advance so the designer does not have to deal with the statistical calculations directly.

To begin, the factored load is given by the equation

Equation 4-8:

$$P_u = \sum_{m=1}^{n} \gamma_m P_m$$

Where

- P_u = factored normal load or moment
- γ_m = load factor for a particular type of load m
- P_m = load or moment for a particular type of load or moment

The factored load is then compared with the load capacity by the equation

Equation 4-9:

$$P_u \leq \Phi P_n$$

Where

- Φ = resistance factor
- P_n = nominal normal load capacity, either from force or moment

It should be emphasized that these loads can either be force or moment loads.

Load and resistance factor design will not be

discussed further in this book.

4.5. Structural Design of Driven Piles

4.5.1. Allowable Pile Stresses (ASD)

Any driven pile has to remain structurally intact and not be stressed to its structural limits during both its service life (static capacity) and during driving (dynamic capacity). This requires that limits be placed on:

(1) Maximum allowable design stresses during the service life. With the allowable stress design (ASD) method, dividing the ultimate stress for the pile material by a factor of safety determines an allowable stress. The factor of safety is based on experience and includes the load and structural resistance variations. The limitations on maximum allowable static design stresses for driven piles in various codes generally represent the static load capacity, which is based on pile material, cross section, strength and to some extent site conditions (corrosion potential.)[6] The maximum design stresses apply to long-term ultimate design capacity and ignore the influence of setup and relaxation.

(2) Maximum allowable driving stresses (temporary). In almost all cases, the highest stress levels occur in a pile during driving. High driving stresses are necessary to cause pile penetration in order to develop static pile loads that equate to reasonable static stresses under service loads. The pile must be stressed to overcome ultimate soil resistance in order to be driven to the required length for allowable pile design capacity. The high strain rate and temporary nature of the loading allow a substantially higher driving stress limitation than for the static design case. Pile driving stresses can be estimated by wave equation analysis. These stresses can be measured during driving by dynamic measurements.

The allowable driving stress values shown in Table 4-7 assume that construction control is based on wave equation analysis.

4.5.2. Handling Stresses for Concrete Piles

All piles are subject to stresses during their manufacture, transportation, storage, and lifting into the leaders. Most of these stresses are flexural in nature. All of these conditions should be considered during the design of pile foundations.

For concrete piles, however, pile pick up and placement into leaders is especially critical. Generally speaking, this is done either with one- or two-point pickup. In the case of one point pickup, the handing eye is cast at a distance of about 30% of the pile length from one end. The flexural moment can be estimated for this case by the formula

Equation 4-10:

$$M_{pickup} = \frac{2WL}{49}$$

(Single-Point Pickup)

Where

- M_{pickup} = flexural moment due to pickup, ft-lbs or kN-m
- W = total weight of the pile, lbs or kN, multiplied by a factor 1.5 for impact
- L = length of the pile, ft or m

With two-point pickup, the two lift points are cast about 20% of the pile length from each end of the pile. In this case the maximum flexural moment is

Equation 4-11:

$$M_{pickup} = \frac{WL}{40}$$

(Two-Point Pickup)

The flexural stress can be then computed based on the section modulus of the pile. For allowable handling stresses, see Table 4-7.

Table 4-7 Maximum Allowable Stresses for Various Pile Types

Pile Type	Maximum Allowable Design Stress	Maximum Allowable Driving Stress	Comments
Steel H-Piles	AASHTO: • 0.25 f_y and the engineer confirms the satisfactory results of the load test. • 0.33 f_y if the engineer performs the load test. EI02C097[7]: • 5 f_y/18 with concentric axial tension in the lower part of the pile • f_y/3 using driving shoes • 2 f_y/5 using driving shoes, and using a pile analyzer to verify pile capacity and integrity	0.9 f_y • 223 MPa (32.4 ksi) for ASTM A36 f_y = 248 MPa (36 ksi)) • 310 MPa (45.0) ksi for ASTM A572 or A690, GR50 (f_y = 345 MPa (50 ksi))	Design stress limits assume damaged is unlikely and load tests are performed.
Unfilled Steel Pipe Piles	0.25 f_y (on steel area)	0.9 f_y • 186 MPa (27 ksi) for ASTM A252, GR1 (f_y = 207 MPa (30 ksi)) • 217 MPa (31.5 ksi) for ASTM A252, GR2, (f_y = 241 MPa (35 ksi)) • 310 MPa (40.5 ksi) for ASTM A252 GR3, (f_y = 310 MPa (45 ksi))	Experience with allowable driving stresses for high strength steel is limited. Designer should adequately detail splice requirements in these cases. Welding quality control is more critical with high strength steels as well.
Top Driven, Concrete Filled Steel Pipe Piles (Pipe Piles when filled, Raymond Step-Taper, Monotube, etc.)	0.25 f_y (on steel area) and 0.40 f'_c (on concrete area)	0.9 f_y • 186 MPa (27 ksi) for ASTM A252, GR1 (f_y = 207 MPa (30 ksi)) • 217 MPa (31.5 ksi) for ASTM A252, GR2, (f_y = 241 MPa (35 ksi)) • 310 MPa (40.5 ksi) for ASTM A252 GR3, (f_y = 310 MPa (45 ksi))	Experience with allowable driving stresses for high strength steel is limited. Designer should adequately detail splice requirements in these cases. Welding quality control is more critical with high strength steels as well.
Conventionally Reinforced Concrete Piles	$\frac{f'_c}{3}$ (compression) • Minimum f'_c = 34.5 MPa (5000 psi) Use gross cross-sectional area to determine allowable loads	Compressive driving stress limitation: $0.85f'_c$ psi Tensile driving stress limitation: $0.70f_y$ (of steel reinforcement)	Control of driving stresses is particularly important when driving reinforced concrete piles at high driving stress levels while penetrating through dense soil layers into underlying weaker layers.

Table 4-7 Maximum Allowable Stresses for Various Pile Types (continued)

Pile Type	Maximum Allowable Design Stress	Maximum Allowable Driving Stress	Comments
Prestressed Concrete Piles	$\frac{f'_c}{3} - 0.27f$ (compression) • Minimum effective prestress f_{pe} = 5 MPa (700 psi[8]) • Minimum f'_c = 34.5 MPa (5000 psi) • Use gross sectional area of concrete; pile is assumed fully embedded in soil	Compressive driving stress limitation: $0.85f'_c - f_{pe}$ psi Tensile driving stress limitation (AASHTO): • Normal Environments: $\frac{\sqrt{f'_c}}{4} + f_{pe}$ (SI, MPa) $3\sqrt{f'_c} + f_{pe}$ (U.S., psi) • Corrosive Environments: f_{pe} Handling Stresses: $\frac{\sqrt{f'_c}}{2}$ (SI, Mpa) $6\sqrt{f'_c}$ (U.S., psi) Stresses on gross concrete area	Control of driving stresses is particularly important when driving prestressed concrete piles at high driving stress levels while penetrating through dense soil layers into underlying weaker layers. Allowable tensile driving stresses are AASHTO recommendations. PCI recommendations for these are $\frac{\sqrt{f'_c}}{2} + f_{pe}$ (SI, MPa) and $6\sqrt{f'_c} + f_{pe}$ (US, psi)
Timber Piles	5.5-8.3 MPa (0.8 to 1.2 ksi) depending upon species: • Eastern Hemlock - 5.5 MPa (0.8 ksi) • Red Oak - 7.6 MPa (1.1 ksi) • Douglas Fir - 8.3 MPa (1.2 ksi) • Southern Pine - 8.3 MPa (1.2 ksi)	3 times the allowable design stress (compression and tension) The work of S.M. Grove and G.G. Goble, Driving Stresses in Timber Piles: Investigation of Allowable Levels, National Timber Piling Council, 1988, concludes that 20.7 MPa (3.0 ksi) is a very acceptable level of driving stress for both Southern Pine and Douglas fir piles.	The use of ASTM D-2899-74 Method for Establishing Design Stresses for Round Timber Piles sometimes results in allowable stresses that are recognized by the experts to be too high. It is important to recognize that the engineer can specify species of timber piles but can seldom specify subspecies, which have a range of strengths that must be accounted for. There is a large natural variability of clear wood strength and natural growth imperfections, which significantly affect wood strength.

4.6. Static Analysis of Geotechnical Capacity

In this section the static analysis of the geotechnical capacity of a driven pile is described[9]. In 4.9, we will consider a sample problem based on the geotechnical capacity methods described here and the structural and dynamic analysis methods described in the previous and following sections.

4.6.1. Axial Compressive Capacity of Piles in Cohesionless Soils

The load capacity of piles driven into cohesionless soil depends primarily on the relative density of the soil. During driving, the relative density is increased close to the pile due to vibrations and lateral displacement of soil. The effect is most pronounced in the immediate vicinity of the pile shaft and extends in gradually diminishing intensity over a zone 3 to 5.5 pile diameters around the pile shaft.

The increase in relative density increases the load capacity of single piles and pile groups. The pile type also affects the amount of change in relative density. Piles with large displacement characteristics such as closed-end pipe and precast concrete increase the relative density of cohesionless material more than small displacement steel H-pipes or open-end pipe piles.

The induced increase in horizontal ground stress, which occurs adjacent to the pile, during the driving process, can be lost by relaxation in dense sand and gravels. The relaxation phenomenon occurs as the negative pore pressure generated during driving is dissipated. The negative pore pressures occur because of volume change and dilation of dense sand. The phenomena can be explained by considering Equation 4-12.

Equation 4-12:

$$S = C + (\sigma - u)\tan\phi$$

Where:

- S = shear strength of soil
- C = cohesion
- σ = vertical (normal) pressure
- u = pore water pressure
- ϕ = Angle of internal friction

It can be seen that negative pore pressures temporarily increase the (σ + u) tan ϕ component of shear strength and, therefore, pile capacity. As negative pore pressures dissipate both the shear strength and pile capacity decrease.

The driving process generates high pore water pressures (positive pore water pressures) in saturated cohesionless silts, which temporarily reduce the soil shear strength and the pile capacity. This phenomena is identical to the one described below for cohesive soils. The gain in capacity with time (set up) is generally quicker for silts than for cohesive soils because the pore pressures dissipate more rapidly (silts are more permeable than clays).

The ultimate bearing capacity of a pile in soils is the sum of toe and shaft resistances (Equation 4-1). For calculation purposes, it is generally assumed that the shaft friction resistance and toe bearing resistance can be determined separately and that these two factors do not affect each other. Many analytical and empirical methods have been developed for estimating pile-bearing capacities in cohesionless materials. Table 4-8 describes some of the available methods

4.6.1.1. Method Based on Standard Penetration Test (SPT) Data

Existing empirical correlations between standard penetration test results and static pile load tests can be used for preliminary estimates of static pile capacity for cohesionless soils[11]. These correlations are based on the analyses of numerous pile load tests in a variety of cohesionless soil deposits. For piles driven into a cohesionless stratum, the ultimate unit toe resistance may be approximated by:

Table 4-8 Methods of Static Analysis for Piles in Cohesionless Soils[10]

Methods	Approach	Method of Obtaining Design Parameters	Advantages	Disadvantages	Remarks
Method based on Standard Penetration Test Data (SPT)	Empirical	Results of SPT Tests.	Widespread use of SPT test and input data availability. Simple method to use.	Non-reproducibility of N values. Not as reliable as the other methods presented in this chapter.	Use a factor of safety of 4. Use only for preliminary estimating purposes.
Method based on Static Cone Penetrometer Test (CPT) Data	Empirical	Results of CPT tests.	Testing analogy between CPT and pile. Reliable correlations. Reproducible test data.	Lack of equipment. Lack of experience in USA. Limitations on pushing cone in to dense strata.	Good approach to design short of driving test. Use a factor of safety of 3.
Nordlund Method	Semi-empirical	Charts provided by Nordlund. Estimate of soil friction is needed.	Allows for increased shaft friction resistance of tapered piles and includes effects of pile-soil friction coefficients for different pile materials.	Nordlund recommends no limiting value on unit shaft friction resistance.	Use a factor of safety of 3. Widely used method based on field observations.

Equation 4-13:

$$q_p = 6N$$

(For non-plastic silts)

Equation 4-14:

$$q_p = 8N'$$

(For sands and gravels)

Where

- q_p = ultimate unit toe resistance, ksf
- N' = the average corrected standard penetration resistance per foot or 300 mm, near the pile toe.

Unit toe resistance limiting values are usually very high for dense granular materials. The relationship provided above for determining unit toe resistance usually provides values lesser than the limiting values shown in Figure 4-12.

Figure 4-12 Relationship Between Maximum Unit Pile Toe Resistance and Friction Angle for Cohesionless Soils[12]

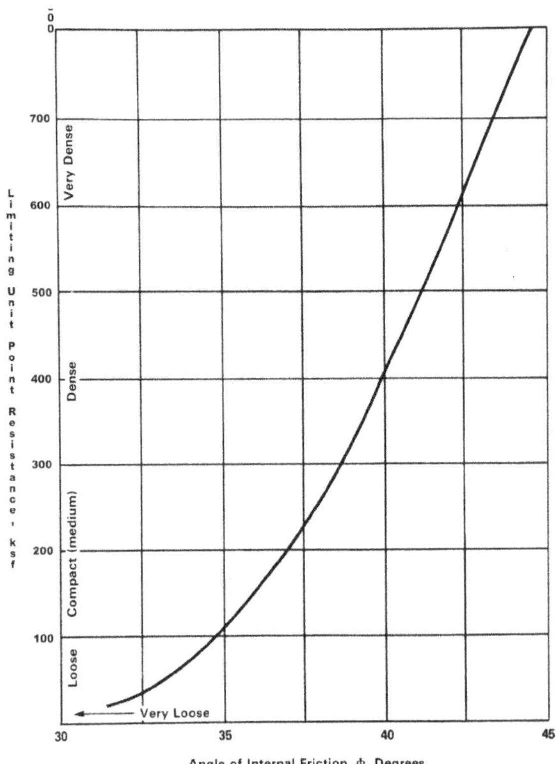

The average unit shaft friction of driven displacement piles, such as closed-end pipe piles and precast concrete piles, is:

Equation 4-15:

$$f_s = \frac{\overline{N}'}{25} \leq 2.0 \text{ ksf}$$

And the average unit shaft friction of driven nondisplacement piles, such as H-piles, is:

Equation 4-16:

$$f_s = \frac{\overline{N}'}{50} \leq 2.0 \text{ ksf}$$

Where

- f_s = average unit shaft friction of driven displacement piles, ksf
- $\overline{N}'$ = the average corrected standard penetration resistance, in blows per foot or 300 mm, within the embedded length of pile.

The Standard Penetration Test is subject to many errors; thus, judgment must be exercised when using the test results. Furthermore, it is recommended to use a factor of safety of 4 if this method is not verified by full-scale pile load tests.

Step By Step Procedure for Using Method Based On SPT Data

1) Correct SPT field "N" values.
2) Compute average SPT corrected "N'" value near pile toe.
3) Compute ultimate unit pile toe resistance, q_p. Use Equation 4-13 or Equation 4-14.
4) Compute ultimate toe bearing capacity:

Equation 4-17:

$$Q_p = q_p A_p$$

Where

- Q_p = ultimate toe bearing capacity, kips;
- A_p = Pile toe area[13] (square feet.)

5) Compute average corrected SPT $\overline{N}'$ within the embedded length of pile. Break into 10

foot to 20 foot layers based on soil density indicated by "N".

6) Compute ultimate unit shaft friction resistance, f_s. Use Equation 4-15 or Equation 4-16.

7) Compute ultimate shaft friction capacity:

Equation 4-18:

$$Q_s = f_s A_s$$

Where

- Q_s = ultimate shaft friction capacity, kips
- A_s = Pile surface area (square feet)

8) Compute total ultimate load capacity, Q_u in kips using Equation 4-1.

9) Compute allowable pile load capacity:

Equation 4-19:

$$Q_{allowable} = \frac{Q_u}{FS}$$

Where

- $Q_{Allowable}$ = allowable pile load capacity, kips, and;
- FS = factor of safety (use a factor of safety of 4 in this case.)

4.6.1.2. Method Based on Static Cone Penetration Test (CPT) Data

The static cone penetrometer, when pushed into cohesionless soil, can be analyzed as a model pile. The cone penetration resistance can then be experimentally correlated with that of a driven full-sized pile. The ultimate unit toe resistance of piles driven into cohesionless soils can be computed by:

Equation 4-20:

$$q_p = q_c$$

Where q_c = cone toe resistance, ksf

The average unit shaft friction of driven displacement piles based on cone penetrometer data can be computed as:

Equation 4-21:

$$f_s = f_c \leq 2.0 \text{ ksf}$$

Where f_c = Average shaft friction measured by cone test, ksf.

The average unit shaft friction of driven nondisplacement piles, such as H-piles, is:

Equation 4-22:

$$f_s = \frac{f_c}{2} \leq 2.0 \text{ ksf}$$

The reasons for limiting unit shaft friction resistance were provided previously. Without shaft friction data (f_c), an estimate of f_s may be made from the following relationships:

Equation 4-23:

$$f_s = \frac{q_c}{100}$$

(for driven displacement piles)

Equation 4-24:

$$f_s = \frac{q_c}{200}$$

(for driven non-displacement piles)

It should be noted that the relationships given above are for a mechanical cone with a friction sleeve referred to as a Dutch-Cone or Begemann Cone. There are several somewhat different cone designs that may have unique pile design capacity relationships. The use of local correlations is strongly encouraged when using cone penetrometer tools for pile predictions. A major disadvantage of a cone test is the difficulty encountered in pushing the cone in very dense sands, gravel and isolated cobbles.

Step By Step Procedure for Using the Method Based on CPT Data[14]

This expanded procedure is necessary in lieu of the simple equations provided above because usually an interpretation of continuous records of CPT is required. This procedure and interpretation are for piles in sands.

1) Determine unit ultimate pile toe resistance q_p

Equation 4-25:

$$q_p = \frac{q_{c1} + q_{c2}}{2}$$

Where q_{c1} and q_{c2} = unit cone toe resistance. Use procedure shown in Figure 4-13 to determine q_p

2) Determine unit ultimate pile toe resistance q_p using Equation 4-17[15].
3) Compute ultimate shaft friction resistance, Q_s, from the following relationship:

Equation 4-26:

$$Q_s = K\left[\frac{\bar{f}_s A_s}{2}_{\text{0 to 8B}} + \bar{f}_s A_{s\,\text{8B to L}}\right]$$

Where:

- K = Ratio of unit pile friction to unit sleeve friction from Figure 4-14.
- $\bar{f}_s$ = Average unit sleeve friction over the depth interval indicated by subscript ($\bar{f}_s \leq 2.0$ ksf)
- A_s = Pile-soil contact area over $\bar{f}_s$ depth interval, ft^2
- B = Pile width or diameter, feet
- L = Pile length, feet
- 0 to 8B = Pile shaft friction resistance for segment from pile head to a depth equal to 8B
- 8B to L = Pile shaft friction resistance for segment from a depth to 8B to the pile toe.

Alternately, if shaft friction data is not available, Q_s can be determined from the following relationship

4) **Equation 4-27:**

$$Q_s = C\sum q_c A_s$$

Where:

- C = 0.01 for driven displacement piles
- C = 0.005 for driven non-displacement piles
- q_c = Average cone toe resistance along the pile length, ksf

5) Determine total ultimate load capacity using Equation 4-1.
6) Determine allowable pile load capacity using Equation 4-19. Use a factor of safety of 3.

Figure 4-13 Procedure for Estimating Pile Toe Capacity (After FHWA Implementation Package, FHWA-TS-78-209)

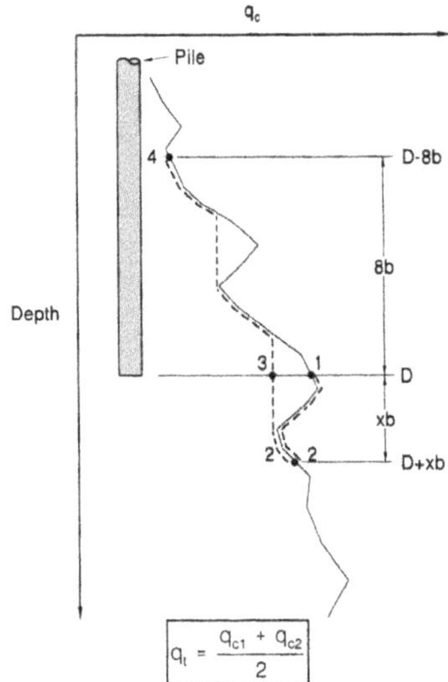

q_{c1} = Average q_c over a distance of xb below the pile toe (path 1-2-3). Sum q_c values in both the downward (path 1-2) and upward (path 2-3) direction. Use actual q_c values along path 1-2 and the minimum path rule along path 2-3. Compute q_{c1} for x-values from 0.7 to 3.75 below the pile toe and use the minimum q_{c1} value obtained.

q_{c2} = Average q_c over a distance of 8b above the pile toe (path 3-4). Use the minimum path rule as for path 2-3 in the q_{c1} computations.

b = Pile width or diameter.

D = Embedded pile length.

Figure 4-14 Penetrometer Design Curves for Pile Side Friction in Sand (After FHWA Implementation Package, FHWA-TS-78-209)

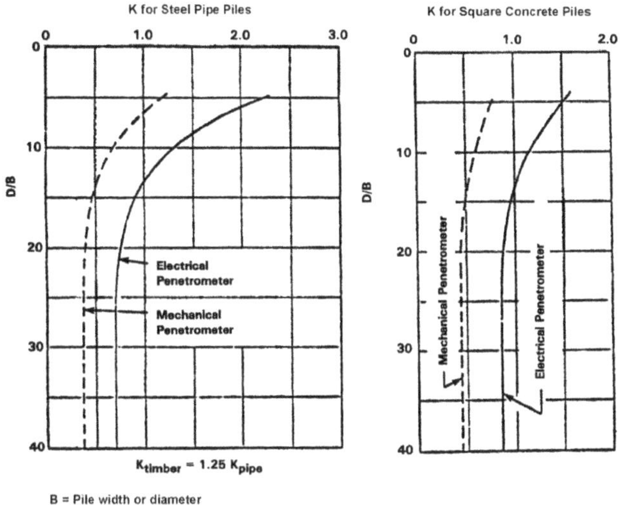

4.6.1.3. Nordlund Method

The Nordlund method[16] is based on field observations. It considers the shape of pile taper and its soil displacement in calculating the shaft resistance. The method accounts for the differences in soil-pile coefficient of friction for different pile materials. The method is based on the results of several load test programs in cohesionless soils. Several pile types were used in these test programs including timber, H, pipe, Monotubes and Raymond Step-Taper® piles.

The ultimate capacity of a pile in cohesionless soil is given in Equation 4-1.

Equation 4-28:

$$Q_u = \alpha N'_q A_p P_D + \sum_{d=0}^{d=D} k_\delta C_f P_d \frac{\sin(\delta + w)}{\cos w} C_d \Delta d$$

Where:

- α = Dimensionless factor dependent on pile depth-width relationship)
- N'_q = Bearing capacity factor
- A_p = Pile toe area
- P_D = Effective overburden pressure at pile toe
- K_δ = Coefficient of lateral earth pressure at depth d
- C_F = Correction factor for K_δ when $\delta \neq \phi$.
- ϕ = Soil friction angle
- δ= Friction angle between pile and soil
- w = Angle of pile taper from vertical
- P_d = Effective overburden pressure at the centre of depth increment d
- C_d = Pile perimeter at depth d
- Δd = Length of pile segment

For a pile of uniform cross section (w=0) and length D, driven in soil layers of the same effective unit weight and friction angle, the above equation becomes:

Equation 4-29:

$$Q_u = \alpha N'_q A_p P_D + K_\delta C_f P_d \sin(\delta) C_d D$$

In developing this method, Nordlund a) suggested using the correlation between corrected standard penetration test "N'" values and ϕ, and b) did not placed a limiting value on the shaft friction resistance. To account for the fact that the pile toe resistance reaches a limiting value at some distance below the ground, it is recommended that if the computed unit toe resistance value exceeds the limiting value obtained from Figure 4-12, then the lower value should be used. The toe resistance reaches a limiting value because of arching in the soil.

Figure 4-15 Nordlund's General Equation for Ultimate Pile Capacity

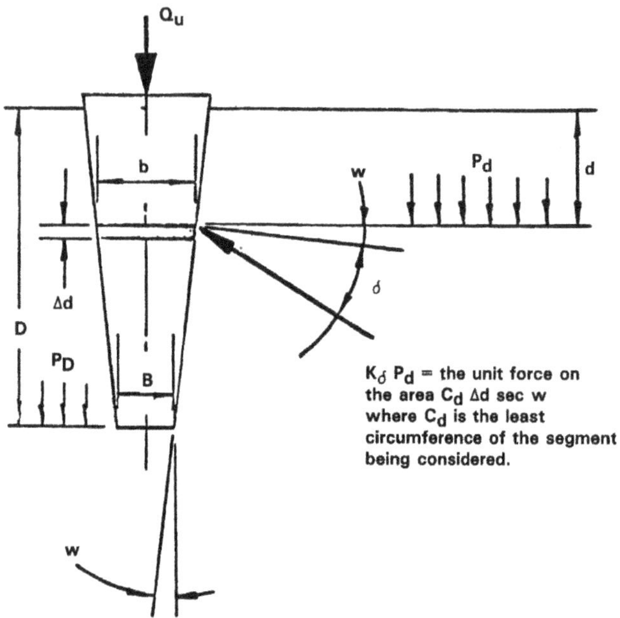

$$Qu = \alpha N'q \times Ap \times P_D + \sum_{d=0}^{d=D} K_\delta \times CF \times P_d \frac{\sin(\delta + W)}{\cos W} \times Cd \times \Delta d$$

Step By Step Procedure for Using Nordlund Method

Steps 1 through 6 are for computing ultimate shaft friction capacity. Steps 7 through 10 are for computing ultimate toe capacity.

1) Determine ϕ angle from SPT data (see Table 4-4.)
   a. Correct SPT field "N" values by using Equation 4-6 and obtaining corrected "N'" values.
   b. Determine average corrected SPT " $\overline{N'}$ " value within embedded length of pile. (Divide pile into sections in each soil layer.)
   c. Use Table 4-4 for estimating ϕ value.
2) Determine K_δ.
   a. Compute volume of soil displaced per unit length of pile (V).
   b. Use Figure 4-16 to determine K_δ for computed "ϕ" and "V" values and pile taper "w". Linear interpolation should be used in determining K_δ values.
3) Determine δ angle.
   a. Use Figure 4-17 to determine δ/ϕ ratio for pile type and displaced volume (V).
   b. Determine angle δ.
4) Determine correction factor C_f if $\delta \neq \phi$.
   a. Use Figure 4-18 to determine correction factor by entering the figure with computed "ϕ" and "δ/ϕ" values.
5) Compute average effective overburden stress (P_d). Compute P_d at midpoint of each soil layer.
6) Compute ultimate shaft friction capacity in each soil layer and total ultimate shaft friction capacity.

Equation 4-30:

$$Q_S = K_\delta C_F P_d \sin \delta C_d D$$

(for uniform pile cross section)[18]

7) Use Figure 4-19 to determine α for computed ϕ angle
8) Compute effective overburden stress, P_d at pile toe.
9) Compute ultimate pile toe resistance by Equation 4-31:

Equation 4-31:

$$Q_p = \alpha N'_q A_p P_D \leq q_{limiting} A_p$$ [19]

   a. The quantity $q_{limiting}$ is determined as follows:
      i. Determine corrected "N'" value near pile toe;
      ii. Estimate ϕ angle by using Table 4-4, and
      iii. Use ϕ angle and Figure 4-12.
10) Compute total ultimate load capacity using Equation 4-1.

Figure 4-16 Design Curves for Evaluation K_δ for Piles

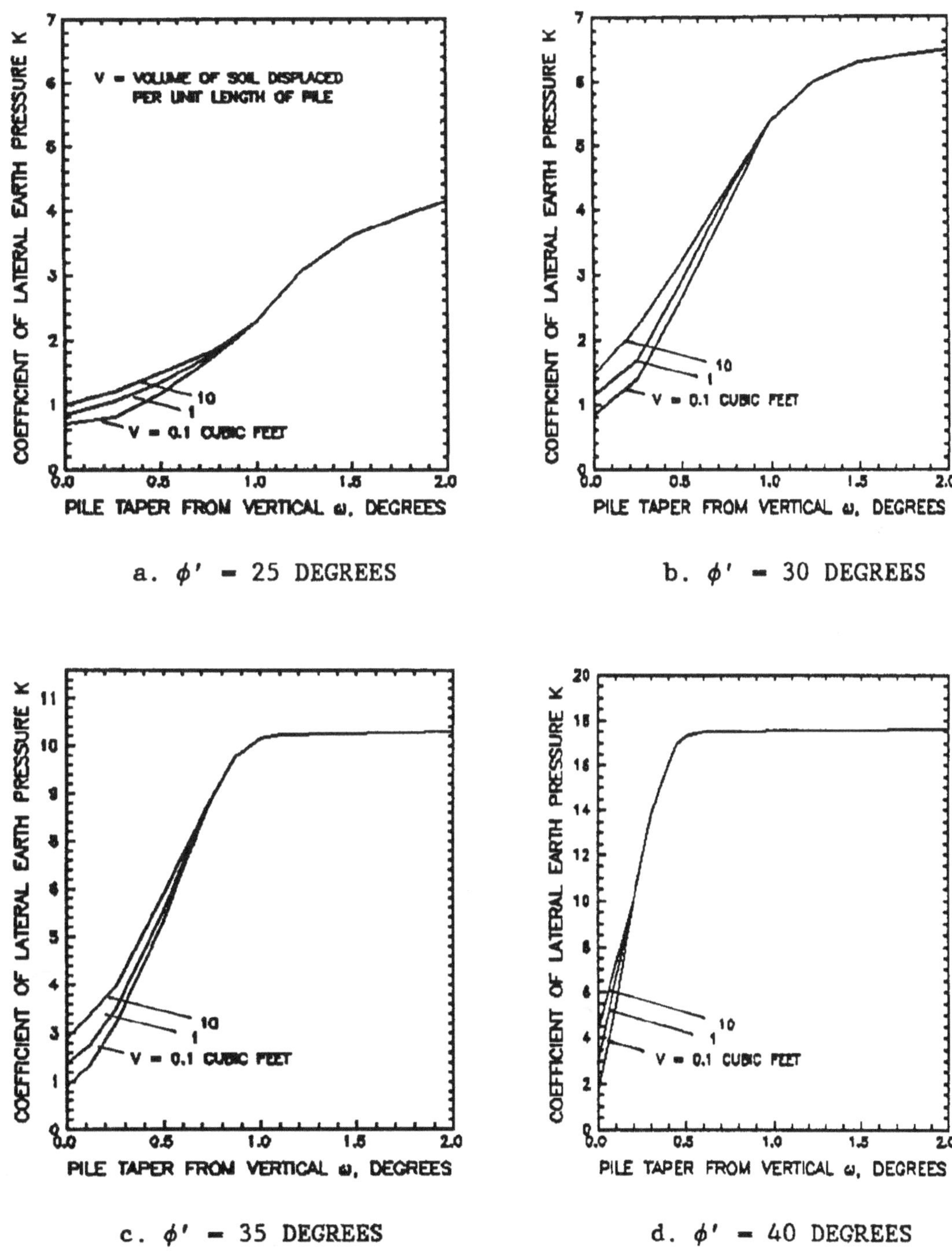

a. φ' = 25 DEGREES

b. φ' = 30 DEGREES

c. φ' = 35 DEGREES

d. φ' = 40 DEGREES

Figure 4-17 Relation of δ/ϕ and Pile Displacement V for Various Types of Piles

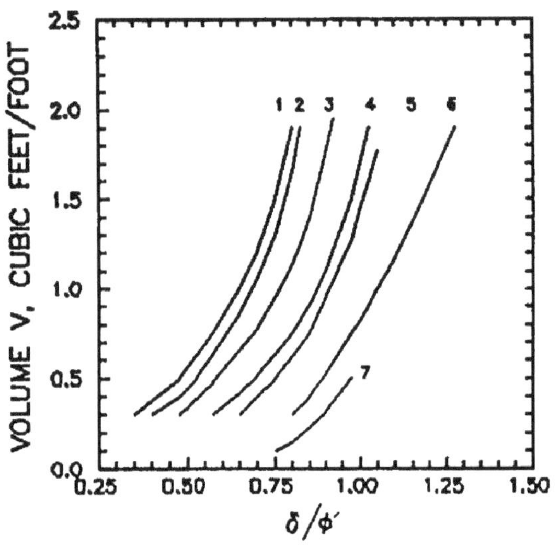

1. Closed end pipe and nontapered portions of Monotube piles
2. Timber Piles
3. Precast Concrete Piles
4. Raymond Step-Taper Piles
5. Tapered portion of Monotube piles
6. Raymond uniform taper piles
7. H-piles

Figure 4-18 Correction Factor for K_δ when $\delta \neq \phi$

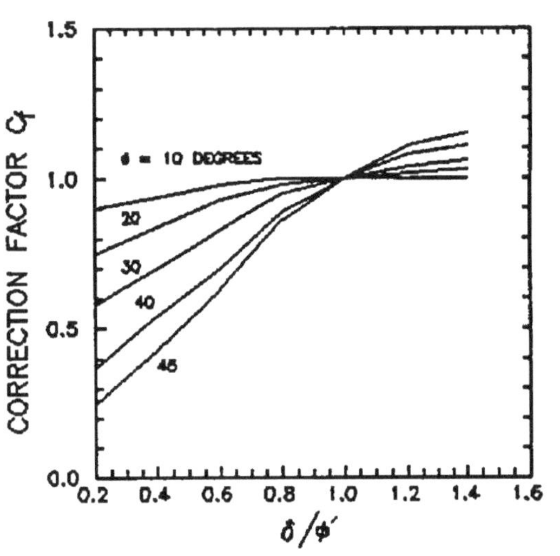

Figure 4-19 Chart[17] for Estimating α Coefficient and Bearing Capacity Factors N'_q

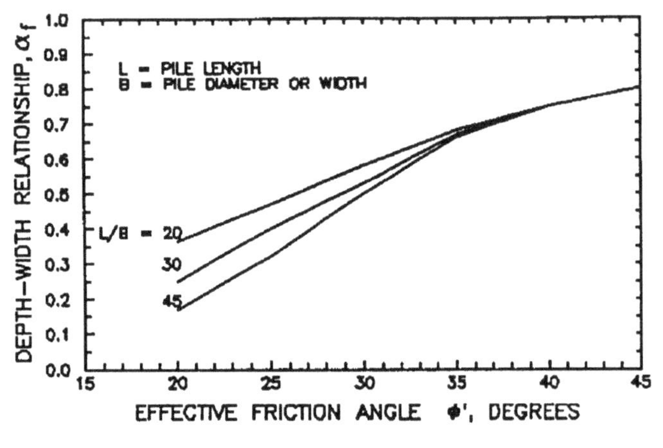

a. α_f

b. N_q

11) Compute allowable pile load capacity using Equation 4-19. Use a factor of safety of 3.

4.6.2. Axial Compressive Capacity of Piles in Cohesive Soils

When piles are driven into saturated cohesive materials, the soil near the piles is disturbed and radially compressed. For soft or normally consolidated clays, the zone of disturbance is within one pile diameter around the pile. For piles driven into saturated stiff clays, there are significant changes in secondary soil structure with remolding and complete loss of previous stress history effects in the immediate vicinity of pile.

The disturbance and radial compression generate high pore water pressures (positive pore water pressures) that temporarily reduce soil shear strength and, therefore, the driving resistance and load capacity of piles. As reconsolidation of clays around the pile occurs, the high pore water pressures are diminished which leads to an increase in shear strength and pile load capacity (set-up). This phenomenon is opposite to "relaxation" described for cohesionless soils. The zone and magnitude of disturbance is dependent on soil properties and soil sensitivity, driving method and the pile foundation geometry. Limited data available for partially saturated cohesive soils indicates that pile driving does not generate high pore water pressures and hence set-up does not occur.

The ultimate bearing capacity of a pile in cohesive soils is expressed by the sum of toe resistance and shaft resistance (Equation 4-1.) A method for determining the ultimate unit toe resistance incohesive soil is outlined in this section.

The shaft friction resistance for piles that are driven into cohesive soils is frequently larger than 80 or 90% of the total bearing capacity. Therefore, for such piles, it is extremely important that the shaft friction resistance be estimated accurately. Design methods for piles in cohesive soils are in some cases of doubtful reliability. This is particularly true for the load capacity of friction piles in clays of medium to high shear strength (c_u > 2 ksf).

Table 4-9 describes two widely used methods for determining the ultimate unit shaft resistance in cohesive soils. Until recently, it was a general practice to evaluate the load capacity of piles in clay using the undrained shear strength (c_u) of the soil. This is known as the total Stress Method (α Method). Modern research has resulted in the development of design methods based on effective stress. The β Method based on effective stress analysis is also presented in this section. Either the α or the β Method can be used for analysis in soft or medium clays. The empirical correlations for "α Method" have not proven entirely reliable for clays with undrained shear strength (c_u) values greater than 500 psf. A large variation in β values occurs for clays with undrained shear strength values greater than 2 ksf. Neither method provides reliable results for piles in stiff clays. Load tests should be required to confirm the bearing capacity of piles in these materials.

4.6.2.1. Toe Bearing Capacity Analysis

The ultimate unit toe resistance in cohesive soil is expressed as:

Equation 4-32:

$$q_p = c_u N_c$$

Where

- c_u is the undrained shear strength at the pile toe.
- N_c is a dimensionless bearing capacity factor that depends on the pile toe diameter.

Table 4-9 Methods of Static Analysis (Shaft Friction Component) for Piles in Cohesive Soils

Method	Approach	Method of Obtaining Design Parameters	Dis-advantages	Remarks
α (Tomlinson) Method	Total stress analysis	Undrained shear strength estimate of soil is needed. α values are provided by charts (Figure 4-20.)	Empirical relationships have not proven entirely reliable for c_u > 500 psf.	Recommended for use in soft and medium clays. Easy to use method. For stiff clays, use α method for preliminary design only. Use a factor of safety of 3.
β Method	Effective stress analysis based on effective stress failure	Undrained shear strength estimate of soil is needed. β values are provided by charts.	Large variations of β for c_u > 2000 psf.	Recommended for use in soft and medium clays. For stiff clays, use β method for preliminary design only. Use a factor of safety of 3.

Values of N_c lie between 7 and 16. A value of 9 is commonly used for N_c for analysis of single piles[20].

Step By Step Procedure for Determining Toe Resistance of Piles in Cohesive Soils

3) Compute ultimate unit toe resistance, q_p using Equation 4-32.
4) Compute ultimate toe bearing capacity, Q_p:

Equation 4-33:

$$Q_p = q_p A_p$$

Where A_p = Area of pile toe[21].

4.6.2.2. Methods for Estimating Shaft Friction Resistance

4.6.2.2.1. "α Method" Based On Total Stress Analysis

This method is also known as the "Tomlinson Method[22]" which assumes that the total shaft friction resistance is directly proportional to the total surface area of the pile.

The shearing stress between the pile and the soil at failure is usually termed the "adhesion", c_a. The average ultimate unit shaft friction, f_s in homogeneous saturated clay, is expressed by:

Equation 4-34:

$$f_s = c_a = \alpha c_u$$

Where α = the empirical adhesion coefficient for reduction of the average undrained shear strength c_u of undisturbed clay within the embedded length of the pile.

The coefficient α depends on the nature and strength of the clay, pile dimension, method of pile installation and time effects. The values of α vary within wide limits and decrease rapidly with increasing shear strength. The values of α can be obtained from Figure 4-20.

Figure 4-20 Adhesion Factors for Driven Piles in Clay (α Method)

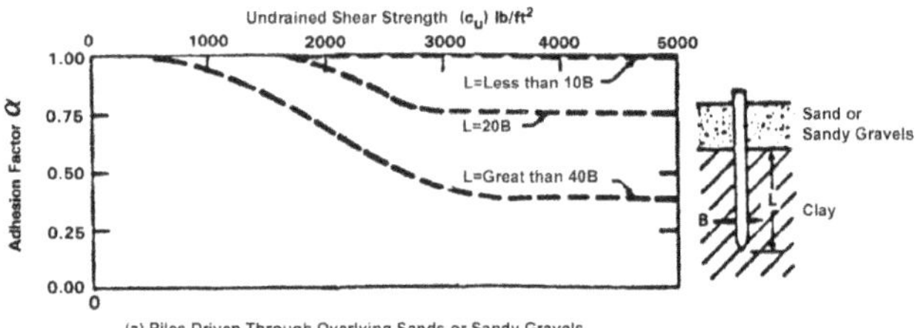

(a) Piles Driven Through Overlying Sands or Sandy Gravels

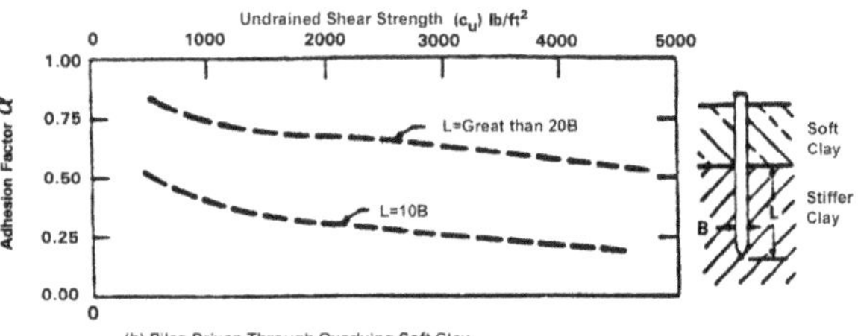

(b) Piles Driven Through Overlying Soft Clay

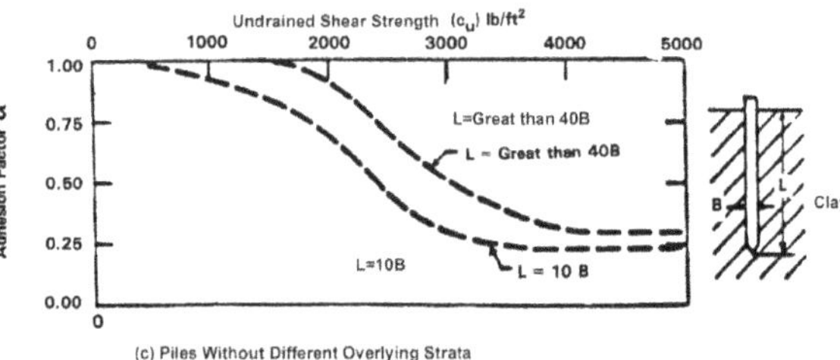

(c) Piles Without Different Overlying Strata

In the case of H piles, there is an uncertainty as to the development of shaft friction along the web and the possibility that the intermixing effects between a soft clay upper layer and stiffer clay may cause a significant reduction in shaft friction. For these reasons, the shaft friction for H piles should be calculated conservatively on a perimeter equal to twice the flange width (the "box" method, described earlier with cohesionless soils.)

Step By Step Procedure for Determining Shaft Friction Resistance By "α Method"

1) Determine adhesion factor α from Figure 4-20. Enter the figure with pile length in clay and undrained shear strength of soil (c_u) in psf. Use appropriate curves for situations (a), (b) or (c) shown in the figure.
2) Compute ultimate unit shaft friction resistance, f_s, using Equation 4-34.
3) Compute total ultimate shaft friction resistance, Q_s:

Equation 4-35:

$$f_s = c_a = \alpha c_u$$

Where: A_s = Pile Surface area[23].

4) Compute ultimate toe bearing capacity, Q_p, using Equation 4-33.
5) Compute total ultimate pile load capacity, Qu, using Equation 4-1.
6) Compute allowable pile load capacity, Q allowable, using Equation 4-19. Use a factor of safety of 3.

4.6.2.2.2. "β Method" Based On Effective Stress Analysis

Static capacity calculations in cohesionless, cohesive and layered soils can also be performed using an effective stress based method. Effective stress based methods were developed to model the long-term drained shear strength conditions. Therefore, the effective soil friction angle φ' should be used in parameter selection.

Let us begin by defining the Bjerrum-Burland beta coefficient as

Equation 4-36:

$$\beta = K_o \tan\phi'$$

where

- β = Bjerrum-Burland shaft friction factor
- φ' = soil friction angle (drained)
- K_o = lateral earth pressure coefficient

The unit shaft friction resistance can then be calculated from the following expression:

Equation 4-37:

$$f_s = K_o \tan(\phi')\bar{p}_o = \beta\bar{p}_o$$

Where:

- $\bar{p}_o$ = average effective overburden pressure along the pile shaft, psf

Available data indicates that for clays with undrained shear strengths less than 2 ksf, β varies from only 0.25 to 0.40 for piles less than 50' (15.2 m) in length. Therefore, for design purposes, a typical value of 0.3 may be used for lengths of 0 to 50' (15.2 m). The β value decreases with increasing pile length. β values for piles longer than 50' (15.2 m) should be obtained from Figure 4-21. Figure 4-22 provides β values for piles in stiff clay with an undrained shear strength in excess of 2 ksf. It can be seen that the β values vary widely from 0.5 to 2.5. Thus, for stiff clays, load tests are recommended for determining pile capacity and the β method should be used for preliminary design only.

Figure 4-21 Values of β For Driven Piles in Soft and Medium Clays (c_u < 2 ksf)[24]

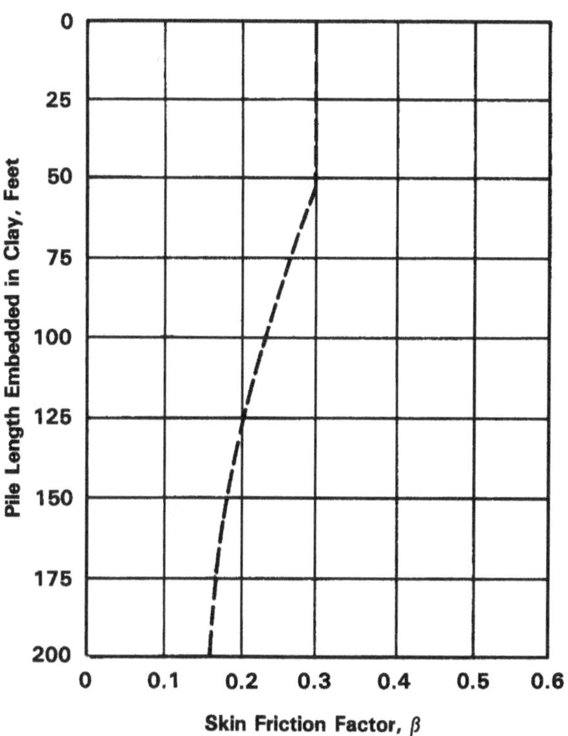

Figure 4-22 Values of β For Driven Piles in Stiff Clay (c_u > 2 ksf)

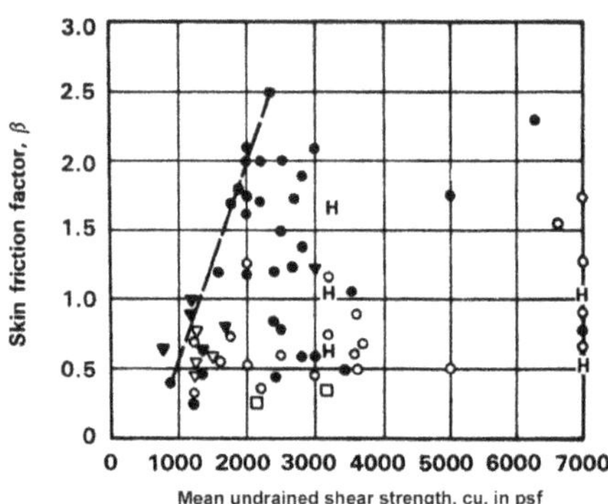

Step-By-Step Procedure For Determining Shaft Friction Resistance By "β Method"

1) Compute average effective overburden pressure, p_o for the pile embedment depth. Use previously described procedure to construct a p_o diagram and determine p_o at the midpoint of the embedment depth.
2) Determine "β" Value. Use Figure 4-21 or Figure 4-22 to determine β for a given pile length and soil undrained shear strength "c_u". (Use $\beta = 0.3$ if c_u < 2 ksf and pile length < 50' (15.2 m))
3) Compute ultimate unit shaft friction resistance, f_s, using Equation 4-37.
4) Compute total ultimate shaft friction resistance, Q_s, using Equation 4-35.
5) Compute ultimate toe bearing capacity, Q_p, using Equation 4-33.
6) Compute total ultimate pile load capacity, Qu, using Equation 4-1.
7) Compute allowable pile load capacity, $Q_{allowable}$, using Equation 4-19. Use a factor of safety of 3.

4.6.3. Pile Group Capacity and Settlement

If piles are driven into cohesive soil (compressible soil) or in dense cohesionless material underlain by compressible soil, then the load capacity of a pile group may be less than that of the sum of the individual piles. In addition, settlement of the pile group is likely to be many times greater than that of an individual pile under and equivalent load. Figure 4-23 (a) for a single pile shows that only a small zone of soil around and below the pile is subjected to vertical stress. Figure 4-23 (b) for a pile group shows that a considerable depth of soil around and below the group is stressed and settlement of the whole group may be large depending on the soil profile. The larger zone of heavily stressed soil for a pile group is the result of overlapping stress zones of individual piles in the group. The overlapping effected is illustrated in Figure 4-24 for an idealized group of friction piles. The group efficiency is defined as the ratio of the ultimate load capacity of a group to the sum of the individual ultimate pile load capacities.

Figure 4-23 Comparison of Stressed Zones beneath End Bearing Single Pile and Pile Group

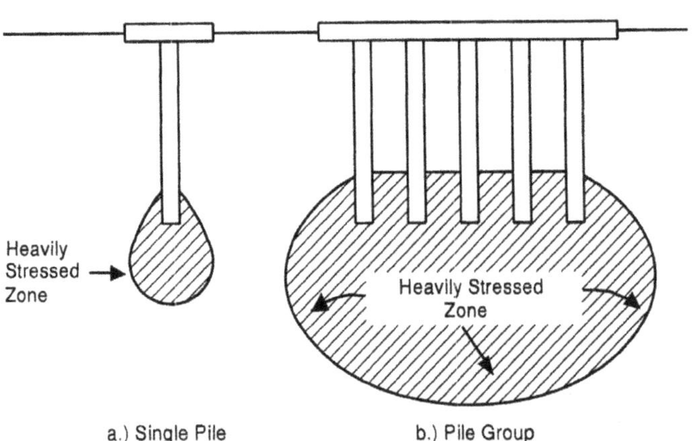

Figure 4-24 Overlapping Stressed Soil Areas for Pile Group

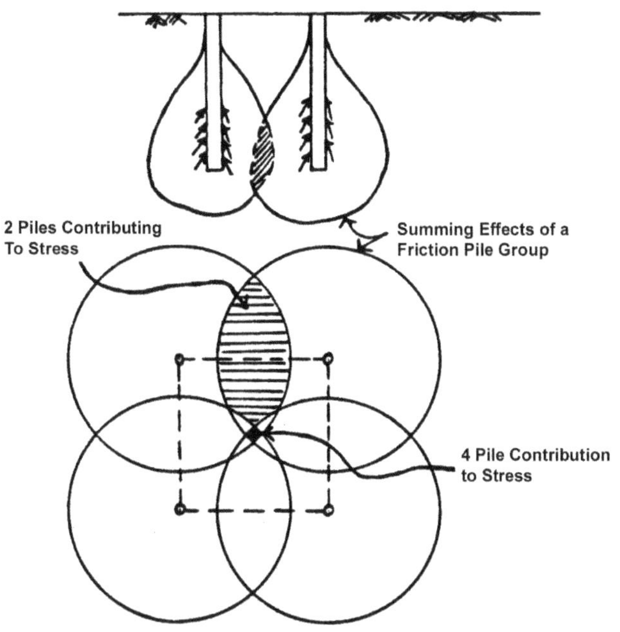

4.6.3.1. Pile Group Capacity in Cohesionless Soils

In cohesionless soils, the ultimate group load capacity of driven piles with a centre-to-centre spacing of less than 3 pile diameters is greater than the sum of the ultimate load of the single piles. The greater group capacity is due to the overlap of individual soil compaction zones near the pile that increases shaft resistance. Piles in groups spaced apart more than three times the average pile diameter act as individual piles.

The following are design recommendations for estimating group capacity in cohesionless soil:

- The ultimate group load in soil not underlain by a weak deposit should be taken as the sum of the single pile capacities.
- If a pile group founded in a firm bearing stratum of limited thickness is underlain by a weak deposit, the ultimate group load is given by the smaller value of either:
    - The sum of the single pile capacities or,
    - By a block failure of an equivalent pier consisting of the pile group and enclosed soil mass punching through the firm stratum into the underlying weak soil.
- From a practical standpoint, block failure can only occur when the pile spacing is less than 2 pile diameters, which is rarely the case. The method shown for cohesive soils (in the next section) may be used to investigate the possibility of a block failure.
- Piles in groups should not be installed at spacings less than 3 times the average pile diameter.

4.6.3.2. Pile Group Capacity in Cohesive Soil

In the absence of negative shaft friction, the group capacity in cohesive soil is usually governed by the sum of the single pile capacities with some reduction due to overlapping zones of shear deformation in the surrounding soil.

The following are design recommendations for estimating group capacity in cohesive soils.

- For pile groups driven in clays with undrained shear strengths of less than 2 ksf (96 kPa) and for spacings of 3 times the average pile diameter, the group efficiency can be taken to be equal to 70%. (If the spacing is greater than 4 times the average pile diameter, than a group efficiency equal to 100% can be used.)
- For pile groups in clays with undrained shear strength in excess of 2 ksf (96 kPa), use a group efficiency equal to 100%.
- Investigate the possibility of a block failure.
- Piles should not be installed at spacings less than 3 times the average pile diameter.

4.6.3.3. Method to Estimate Ultimate Resistance against Block Failure Of Pile Group In Cohesive Soil

A pile group in cohesive soil is shown in Figure 4-25. The ultimate resistance of the pile group against a block failure is provided by the following expression:

Equation 4-38:

$$Q_u = 9c_{u1}BL + 2Dc_{u2}(B + L)$$

Where:

- Q_u = Ultimate resistance against block failure, kips
- c_{u1} = Undrained shear strength of clay below pile toes, ksf
- c_{u2} = Average undrained shear strength of clay around the group, ksf
- B = Width of group, feet
- L = Length of group, feet
- D = Length of piles, feet

Figure 4-25 Pile Group in Cohesive Soil

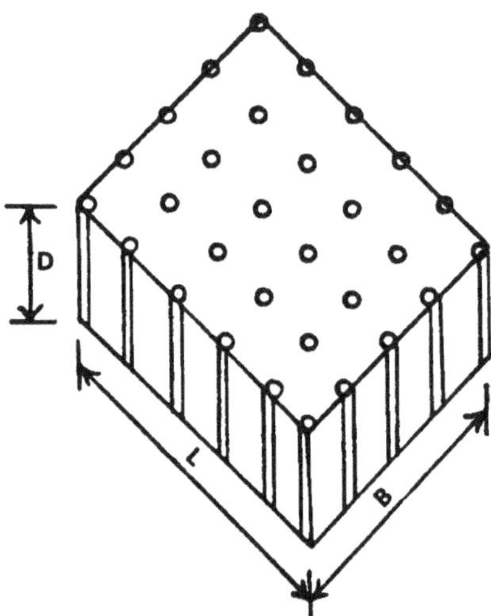

4.6.3.4. Elastic Settlement of Pile Groups

One source of settlement for group piles is elastic compression. The elastic compression can be estimated by the equation

Equation 4-39:

$$\delta = \frac{PL}{AE}$$

- δ = Elastic compression of pile material, inches
- P = Axial load in pile, kips
- L = Length of pile, inches
- A = Pile cross sectional area, in^2
- E = Modulus of Elasticity of pile material
  - E = 207 GPa (30,000 ksi) for steel piles
  - E = 27.8 GPa (4,000 ksi) for concrete piles.

The elastic compression can be included and the result added to the group settlement estimate, especially for longer piles. Equation 4-39 is strictly speaking only valid for purely toe-bearing piles. The topic of elastic compression for single piles is of great interest in static load testing and is covered in conjunction with this in Chapter 6.

4.6.3.5. Immediate Settlements of Pile Groups in Cohesionless Soils

Pile groups supported by cohesionless solid will produce only elastic (immediate) settlements. This means the settlements in cohesionless soils will occur immediately as the pile group is loaded.

4.6.3.5.1. Method Based on SPT Data

The settlement of a pile group in a homogeneous sand deposit not underlain by a more compressible soil at a greater depth may be conservatively estimated by the following expression[25]:

Equation 4-40:

$$S = \frac{pI\sqrt{B}}{\overline{N'}}$$

Where:

- S = estimated total settlement in inches
- B = the width of pile group in feet
- p = the foundation pressure in ksf[26].
- $\overline{N'}$ = the average corrected SPT resistance in blows per foot within a depth equal to B below the pile toes
- I = influence factor for group embedment = 1-D/(8B) ≥ 0.5
- D = pile embedment depth, feet

For silty sands, use the expression

Equation 4-41:

$$S = \frac{2pI\sqrt{B}}{\overline{N'}}$$

4.6.3.5.2. Method Based on CPT Test Data

The following relationship can be used to estimate maximum settlements using the results of static cone penetration tests (for saturated cohesionless soils):

Equation 4-42:

$$S = \frac{2pI\sqrt{B}}{\overline{N'}}$$

Where S, p, B, and I are as defined in the previous method and

- $\overline{q}_c$ = Average static cone toe resistance (ksf) within a depth equal to B below the pile toe.

Figure 4-26 Stress Distribution beneath Pile Group in Clay Using Theoretical Footing Concept

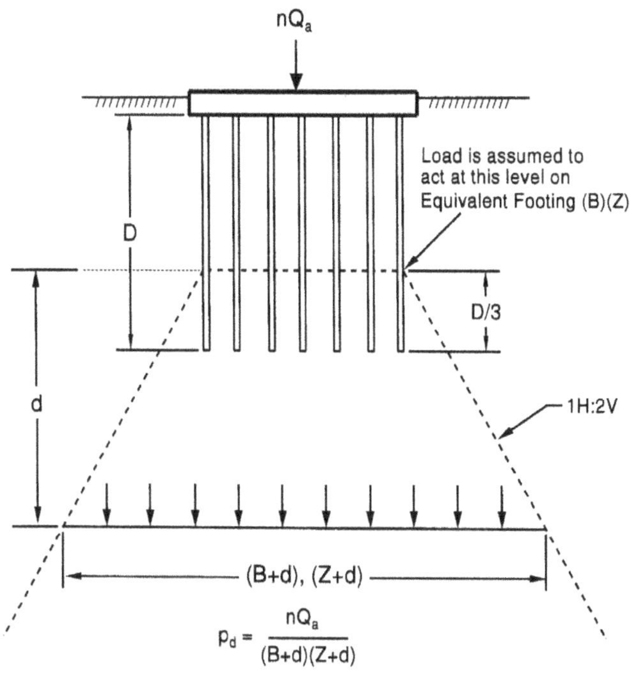

4.6.3.6. Settlement of Pile Groups In Cohesive Soils

Pile groups supported by cohesive solid may produce both elastic (immediate) and consolidation (occurs over a time period) settlements. The elastic settlements will generally be the major amount for over-consolidated clays and consolidation settlements will generally be the major amount for normally consolidated clays.

AmethodproposedbyTerzaghiandPeck,andconfirmed by limited field observations, is recommended for the evaluation of the consolidation settlement of pile groups in cohesive soil. The load carried by the pile group is assumed to be transferred to the soil through a theoretical footing located at 1/3 the pile length up from the pile toe (Figure 4-26). The load is assumed to spread within the frustum of a pyramid of side slopes at 30° and to cause uniform additional vertical pressure at lower levels, the pressure at any level being equal to the load carried by the group divided by the cross-section area of the base of the frustum at that level. This method can be used for vertical or batter pile groups.

To implement the concept shown in Figure 4-26 in nonhomogeneous soils, it is necessary to take in to account the effect of layering. Figure 4-27 shows the effects of various layering situations on the theoretical footing under the pile group.

The consolidation settlement of cohesive soil is usually computed based on laboratory tests. The relationships of the compression indices to void ratio e and pressure are shown in Figure 4-28, which is plotted from consolidated test results. For loadings less than the pre-consolidation pressure $p_{c,}$ settlement is computed using a value of the compression index representing recompression, C_{cr}. For loadings greater than the pre-consolidation pressure, settlement is computed using the compression index, C_c.

The following equation is used for computing consolidation settlement:

Equation 4-43:

$$S = H\left[\frac{C_{cr}}{1+e_o}\log\frac{p_c}{p_o} + \frac{C_c}{1+e_o}\log\frac{p_o+\Delta p}{p_c}\right]$$

Where:

Figure 4-27 Pressure Distribution Below Equivalent Footing for Pile Group, Layered Soils

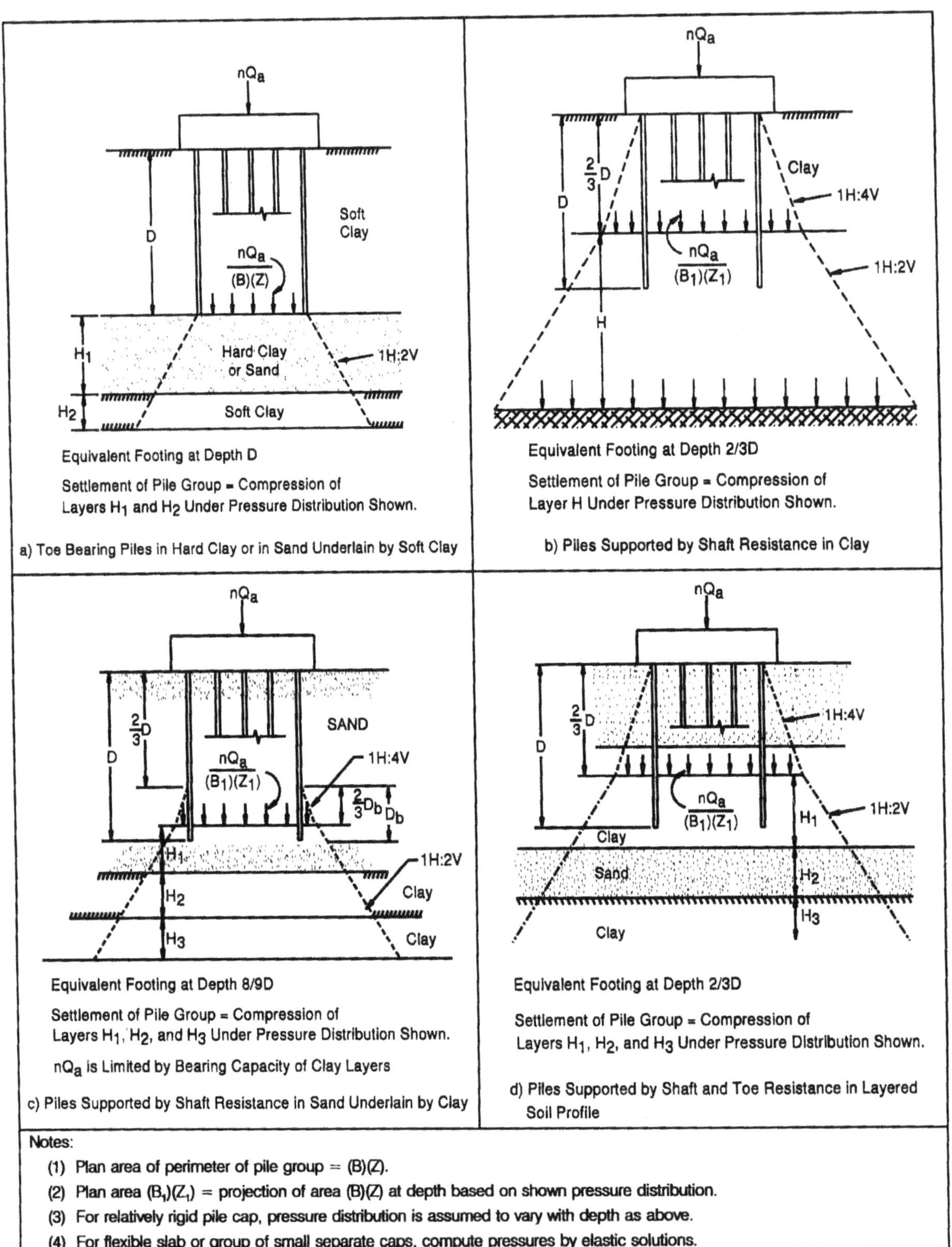

- S = total settlement
- H = original thickness of stratum
- C_{cr} = recompression index
- e_o = initial void ratio
- p_o = average initial effective pressure
- p_c = estimated pre-consolidation pressure
- C_c = compression index
- Δp = the average change in pressure in the compressible stratum considered.

For normally consolidated soils, $p_c = p_o$ and this reduces to

Equation 4-44:

$$S = H\left[\frac{C_c}{1+e_o}\log\frac{p_o+\Delta p}{p_o}\right]$$

Figure -28 "e - Log p" Relationship

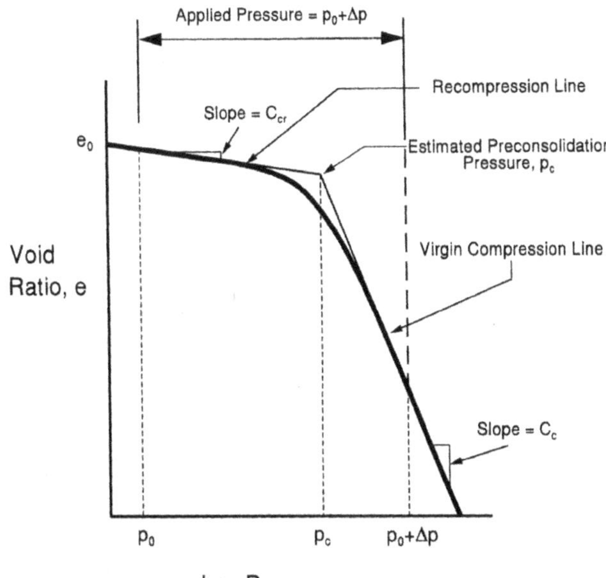

A step-by-step procedure showing the method follows.

Step-By-Step Procedure for Estimating Pile Group Settlement in Cohesive Soil

1) Determine Load Imposed On Soil By Pile Group
   a. Use the method shown in Figure 4-26 or Figure 4-27 to determine the depth at which the additional imposed load by the pile group is less than 10% of existing effective overburden pressure at that depth. This will provide the total thickness of cohesive soil layer to be used in performing settlement computations. Use design load to be applied to the pile group. Do not use ultimate pile group capacity for settlement computations.
   b. Divide the cohesive soil layer determined in (a) into several thinner layers (5 to 10' thick). The layer thickness H is the thickness of each layer.
   c. Determine the existing effective overburden pressure (p_o) at midpoint of each layer.
   d. Determine the imposed pressure (Δp) at midpoint of each layer by using the method shown in Figure 4-28.
2) Determine Consolidation Test Parameters.
   a. Plot results of consolidation test as shown in Figure 4-28.
   b. Determine p_c, e_o, C_{cr} and C_c values from the plotted data.
3) Compute Settlements. By using the settlement equation, compute settlement of each layer. Summation of settlements of all layers will provide the total estimated settlement for the pile group.

4.6.3.7. Uplift Capacity of Pile Groups

The uplift capacity of a pile group is often a significant factor in determining the minimum pile penetration requirements and in some cases can control the foundation design. A few common conditions where group uplift capacity may significantly influence the foundation design include cofferdam seals that create large buoyancy forces, cantilever segmental bridge construction, and seismic, vessel impact, or debris loading. When piles with uplift loads are driven to a relatively shallow bearing stratum, uplift capacity may control the foundation design. AASHTO specifications[27] for the determination of group uplift capacity are presented in 4.6.3.7.1. The AASHTO specifications for group uplift capacity are considered relatively conservative, particularly in cohesionless soils.

In cohesionless soils, Tomlinson's method presented in 4.6.3.7.2 will yield higher group uplift capacities than AASHTO specifications and is recommended for design. Both AASHTO specifications and Tomlinson's method limit the group uplift capacity to the uplift capacity of an individual pile times the number of piles in the group. In the event this limit controls the group uplift capacity, an uplift load test may be cost effective and should be considered. With an uplift load test, a reduced safety factor is used to determine the uplift capacity. This should result in higher individual and group uplift capacities.

In cohesive soils, Tomlinson's method will yield similar results to AASHTO specifications. In the event the uplift capacity of an individual pile times the number of piles in the group limits the group uplift capacity, an uplift load test may again be cost effective and should be considered since an increase in the group uplift capacity would likely result.

4.6.3.7.1. Group Uplift Capacity by AASHTO Code

AASHTO specifications for service load design limit the uplift capacity of a pile group to the lesser value determined from any of the following:

1. The design uplift capacity of a single pile times the number of piles in a pile group. The design uplift capacity of a single pile is specified as 1/2 the ultimate shaft resistance calculated in a static analysis method, or 1/2 the failure load determined from an uplift load test.
2. 2/3 the effective weight of the pile group and the soil contained within a block defined by the perimeter of the pile group and the embedded length of the piles.
3. 1/2 the effective weight of the pile group and the soil contained within a block defined by the perimeter of the pile group and the embedded pile length plus 1/2 the total soil shear resistance on the peripheral surface of the pile group.

4.6.3.7.2. Tomlinson Group Uplift Method

Tomlinson[28] states that the ultimate uplift capacity of a pile group in cohesionless soils may be conservatively taken as the effective weight of the block of soil extending upward from the pile toe level at a slope of 1H:4V, as shown in Figure 4-29. For simplicity in performing the calculation, the weight of the piles within the soil block is considered equal to the weight of the soil. Tomlinson states that a factor of safety of 1 is acceptable in this calculation since the shear resistance around the perimeter of the group is ignored in the calculation. Tomlinson also recommended that the ultimate group uplift capacity determined from this calculation not exceed the sum of the ultimate uplift capacities of the individual piles comprising the pile group divided by an appropriate safety factor. It is recommended that a factor of safety of 2 be used if the ultimate uplift capacity of an individual pile is determined from an uplift load test and a factor of safety of 3 be used if based on the shaft resistance from a static calculation.

Figure 4-29 Uplift of Pile Group in Cohesionless Soil

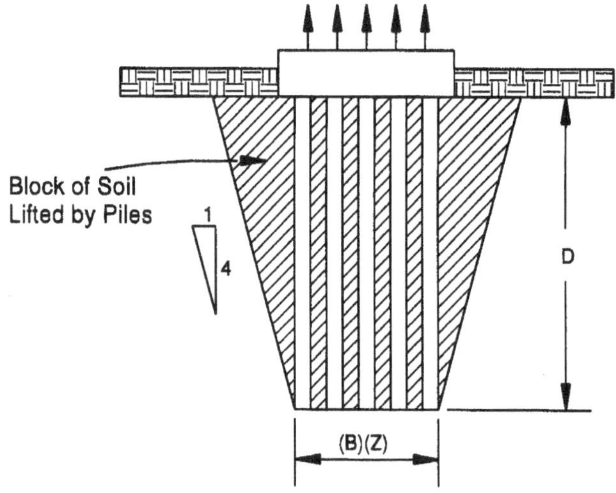

For pile groups in cohesive soils as shown in Figure 4-30, Tomlinson recommends the group uplift capacity be calculated based upon the undrained shear resistance of the block of soil enclosed by the group plus the effective weight of the pile cap and pile-soil block.

Figure 4-30 Uplift of Pile Group in Cohesive Soils

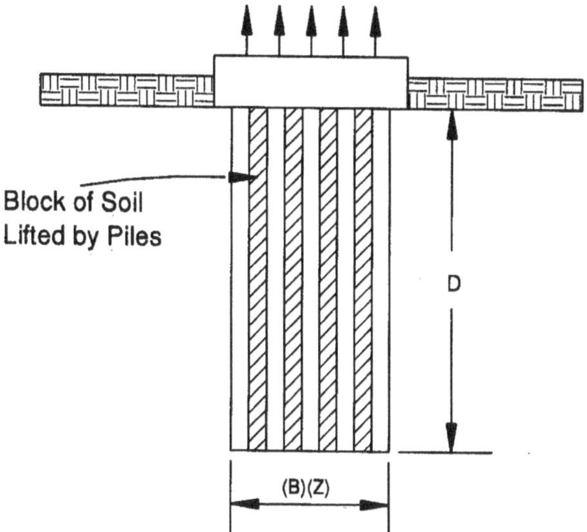

This may be expressed in equation form as:

Equation 4-45:

$$Q_{ug} = 2D(B+Z)c_{ut} + W_g$$

Where:

- Q_{ug} = Ultimate group capacity against block failure in uplift.
- D = Embedded length of piles.
- B = Width of pile group.
- Z = Length of pile group.
- c_{ut} = Weighted average of the undrained shear strength over the depth of pile embedment along the pile group perimeter.
- W_g = Effective weight the pile/soil block including the pile cap weight.

Tomlinson states that a factor of safety of 2 should be used with this calculation to allow for possible weakening of the soil around the pile group as a result of the pile group installation. If long term sustained uplift loading is anticipated, a factor of safety of 2.5 to 3 is recommended. Tomlinson also recommends that the ultimate group uplift capacity determined from this calculation not exceed the sum of the ultimate uplift capacities of the individual piles comprising the pile group divided by an appropriate factor of safety. It is recommended that a factor of safety of 2 be used if the ultimate uplift capacity of an individual pile is determined from an uplift load test, and a factor of safety of 3 be used if based on the shaft resistance from a static calculation.

4.6.4. Negative Shaft Friction

When a soil deposit, through which piles are installed, undergoes consolidation and is forced downward, the resulting downward movement of the soil around piles induces "downdrag" load on the piles. These "downdrag" forces are also called negative shaft friction. Negative shaft friction is the reverse of the usual positive shaft friction developed along the pile surface. This force increases the pile axial load and can be especially significant on long piles driven through compressible soils and must be considered in pile design. Batter piles should be avoided in these situations because of the additional bending forces imposed on the piles, which can result in the pile deformation or damage.

Settlement computations should be performed to determine the amount of settlement the soil surrounding the piles is expected to undergo after the piles are installed. The maximum amount of relative settlement between soil and pile that is necessary to mobilize negative shaft friction is about 10-12 mm (3/8-1/2".) At that movement, the maximum value of negative shaft friction is equal to the soil adhesion or friction resistance. The negative shaft friction cannot exceed these values because slip of the soil along the pile occurs at this value. It is particularly important in the design of friction piles to determine the depth below which the pile will be unaffected by negative shaft friction. Only below that depth can positive shaft friction forces provide support to resist vertical loads.

Figure 4-31 shows two situations where negative shaft friction may occur. Situation (B) is the most common.

The total stress α-method is often used for computing the negative shaft resistance or drag load in cohesive soils. In this approach, the adhesion calculated from the undrained shear strength of the soil times the pile perimeter is equated to the drag load from the consolidating soil layers. Similarly, the drag load from cohesionless layers above a consolidating soil layer is calculated from the shaft resistance in the cohesionless layers.

When selecting the undrained shear strength for calculation of the negative shaft resistance adhesion in the α-method, it is important to

Figure 4-31 Negative Shaft friction Situations

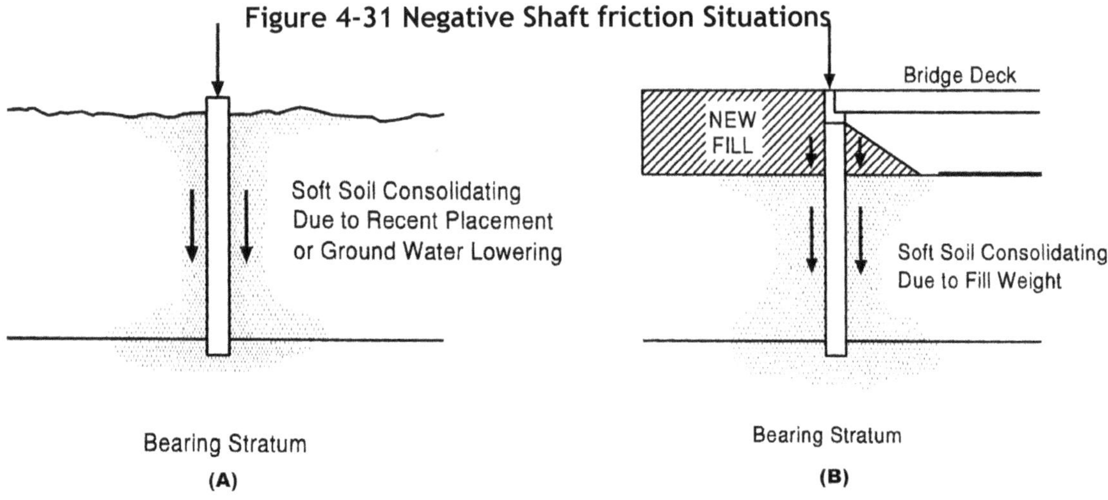

Figure 4-32 Sample Specification for Bitumen Coating on Steel Piles

Description. This work shall consist of furnishing and applying bituminous coating and primer to steel pile surfaces as required in the plans and as specified herein.

Materials.

A. Bituminous Coating. Canal Liner Bitumen (ASTM D-2521) shall be used for the bitumen coating and shall have a softening point of 190 degrees F., to 200 degrees F., a penetration of S6 to 61 at 25 degrees C., and a ductility at 25 degrees C., in excess of 3.5 cm.

B. Primer. Primer shall conform to the requirements of AASHTO M116.

Construction Requirements. All surfaces to be coated with bitumen shall be dry and thoroughly cleaned of dust and loose materials. No primer or bitumen shall be applied in wet weather, nor when the temperature is below 65 degrees F.

Application of the prime coat shall be with a brush or other approved means and in a manner to thoroughly coat the surface of the piling with a continuous film of primer. The purpose of the primer is to provide a suitable bond of the bitumen coating to the pile. The primer shall set thoroughly before the bitumen coating as applied.

The bitumen should be heated to 300 degrees F., and applied at a temperature between 200 degrees F., to 300 degrees F., by one or more mop coats, or other approved means, to apply an average coating depth of 3/8 inch. White-washing of the coating may be required, as deemed necessary by the engineer, to prevent running and sagging of the asphalt coating prior to driving, during hot weather.

Bitumen coated piles shall be stored immediately after the coating is applied for protection from sunlight and heat. Pile coatings shall not be exposed to damage or contamination during storage, hauling, or handling. Once the bitumen coating has been applied, the contractor will not be allowed to drag the piles on the ground or to use cable wraps around the pile during handling. Pad eyes, or other suitable devices, shall be attached to the pile to be used for lifting and handling. If necessary, the contractor shall recoat the piles, at his expense, to comply with these requirements.

A nominal length of pile shall be left uncoated where field splices will be required. After completing the field splice, the splice area shall be brush or mop coated with at least one coat of bitumen.

Method of Measurement. Bitumen coating will be measured by the linear foot of coating in place on the pile surfaces. No separate payment will be made for the primer or coating of the splice areas.

Basis of Payment. The accepted quantities of bitumen coating will be paid for at the contract unit price per linear foot, which price shall be full compensation for furnishing all labor, materials, tools, equipment, and incidentals, and for doing all the work involved in f applying the bituminous coating and primer, as shorn in the plans, and as specified in these specifications, and as directed by the engineer.

Payment will be made under:

Pay Item	Pay Unit
Bitumen Coating	Linear foot

remember that the consolidating cohesive soil will have a higher undrained shear strength with time. The adhesion should be calculated using either the higher adhesion value, determined from the undrained shear strength at the time of the soil borings, or the estimated undrained shear strength of the soil after consolidation. Drag loads equal to 100% of the undrained shear strength of a soft clay, i.e. α =1, have been reported for toe bearing piles driven to a relatively unyielding bearing layer. Engineering judgment should be exercised in determining drag loads so that the drag load is not grossly overestimated, resulting in an expensive foundation design, nor underestimated, resulting in a overloaded foundation.

4.6.4.1. Methods for Reducing Negative Shaft Resistance Forces

In situations where the negative shaft resistance on piles is large and a reduction in the pile design load is impractical, negative shaft resistance forces can be handled or reduced by using one or more of the following techniques:

1. Reduce soil settlement. Preloading and consolidating the soils prior to pile installation can achieve preconsolidation of compressible soils. This approach is often used for bridge foundations in fill sections. Wick drains are often used in conjunction with preloading in order to shorten the time required for consolidation.
2. Use lightweight fill material. Construct structural fills using lightweight fill material to reduce the downdrag loads. Lightweight fill materials often used, depending upon regional availability, include geofoam, foamed concrete, wood chips, blast furnace slag, and expanded shales.
3. Use a friction reducer. Bitumen coating and plastic wrap are two methods commonly used to reduce the friction at the pile-soil interface. Bitumen coatings should only be applied to the portion of the pile that will be embedded in the negative shaft resistance zone. Case histories on bitumen coatings have reported reductions in negative shaft resistance from as little as 47% to as much as 90%. Goudreault and Fellenius[29] suggest that the reduction effect of bitumen may be analyzed by using an upper limit of 10 kPa as the pile-soil shear resistance or adhesion in the bitumen coated zone.

One of the major problems with bitumen coatings is protecting the coating during pile installation, especially when driving through coarse soils. An inexpensive solution to this problem is to weld an over-sized collar around the pile where the bitumen ends. The collar opens an adequate size hole to permit passage of the bitumen for moderate pile lengths in fine-grained soils. Bitumen coatings can present additional construction problems associated with field coating and handling. The use of bitumen coatings can be quite successful provided proper construction control methods are followed. Example specifications for bitumen coatings applied to both concrete and steel piles are provided in Figure 4-32. Bitumen coatings should not be casually specified as the solution to downdrag loading.

The proper bitumen must have relatively low viscosity to permit slippage during soil consolidation, yet high enough viscosity and adherence to insure the coating will stick to the pile surface during storage and driving, and sufficient ductility to prevent cracking and spalling of the bitumen during handling and driving. Therefore, the climate at the time of pile installation should be considered in selection of the proper bitumen coating. Note that these are generic specifications that should be modified to meet the specific needs of each project.

Plastic wrap has proven to be an economically attractive friction reducer, particularly for abutment piles driven behind and before construction of MSE walls. Tawfig[30] performed laboratory tests on 0.15 mm thick polyethylene sheets used as a friction reducer. The laboratory test results indicated plastic wraps reduced the pile-soil shear resistance from between 78% for a one-wrap layer to 98% for a two-layer wrap with mineral oil lubricant of the pile-soil shear resistance. The laboratory test data indicated the pile-soil shear resistance of a one-wrap layer was about 10 kPa and only 1 kPa for the lubricated two-wrap system.

4. Increase allowable-pile stress. In piles where the allowable pile material strength has not been fully utilized, the pile design stress can be increased to offset the negative shaft resistance load. Increased structural capacity can also be obtained by using higher strength pile materials, or in the case of pipe piles, by using an increased wall thickness. Foundation settlement at the increased loading should be computed and checked against the foundation performance criteria.
5. Prevent direct contact between soil and pile. Pile sleeves are sometimes used to eliminate direct contact between pile and soil. Bentonite slurry has been used in the past to achieve the same purpose. These methods are generally more expensive.
6. Use of Spear Piles. Spear piles can be used to reduce downdrag. These are shown in Figure 4-33.

Figure 4-33 Spear Piles for Downdrag Reduction

4.6.5. Lateral Squeeze of Foundation Soil

Bridge abutments supported on piles driven through soft compressible soils (cohesive soils) may tilt forward or backward depending on the geometry of the backfill and the abutment (see Figure 4-34). If the horizontal movement is large, it may cause damage to structures. The unbalanced fill loads shown in Figure 4-34 displace the soil laterally. This lateral displacement may bend the piles, causing the abutment to tilt toward or away from the fill. The following rules of thumb are recommended for either determining whether tilting will occur or estimating the magnitude of horizontal movement.

Figure 4-34 Abutment Tilting Due to Lateral Squeeze

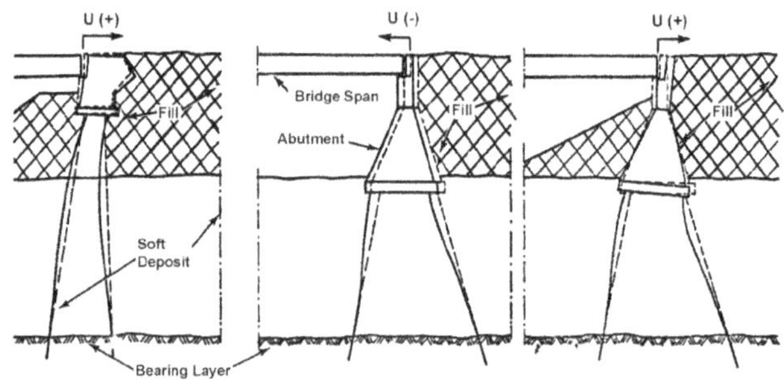

Lateral squeeze (abutment tilting) can occur if

Equation 4-46:

$$\gamma_{Fill} H_{Fill} > 3c_u$$

If abutment tilting can occur, the magnitude of the horizontal abutment movement can be estimated to be 25% of the vertical fill settlement.

4.6.5.1. Solutions to Prevent Tilting

1) Get the fill settlement out before abutment piling is installed (best solution).
2) Provide expansion shoes large enough to accommodate the movement.
3) Use steel H-piles because they provide high tensile strength in flexure.

4.6.6. Lateral Loading of Vertical Piles

4.6.6.1. Overview

In addition to axial compression and uplift loads, piles are routinely subjected to lateral loads. Potential sources of lateral loads on bridge structures include vehicle acceleration and braking forces| wind loads, wave and current forces, debris loading, ice forces, vessel impact loads, earth pressures on the backs of abutment walls, slope movements, and seismic events; These lateral loads can be of the same magnitude as axial compressive loads and therefore warrant careful consideration during design. The foundation deformation under lateral loading must also be within the established performance criterion for the structure.

Historically, designers often used prescription values for the lateral load capacity of vertical piles, or added batter piles to increase a pile group's lateral capacity when it was believed that vertical piles could not provide the needed lateral resistance. However, vertical piles can be designed to withstand significant lateral loads. Modern analysis methods should be employed in the selection of the pile type and pile section. These methods are now readily available that allow the lateral load-deflection behavior of piles to be rationally evaluated. Lateral loads and moments on a vertical pile are resisted by the flexural stiffness of the pile and mobilization of resistance in the surrounding soil as the pile deflects. The flexural stiffness of a pile is defined by the pile's modulus of elasticity, E, and moment of inertia, I.

The design of laterally loaded piles must evaluate both the pile structural response and soil deformation to lateral loads. The factor of safety against both ultimate soil failure and pile structural failure must be determined. In addition, the pile deformation under the design loading conditions must be calculated and compared to foundation performance criteria.

The design of laterally loaded piles requires the combined skills of the geotechnical and structural engineer. It is inappropriate for the geotechnical engineer to analyze a laterally loaded pile without a full understanding of pile-structure interaction. Likewise it is inappropriate for the structural engineer to complete a laterally loaded pile design without a full understanding of how, pile section or spacing changes may alter the soil response. Because of the interaction of pile structural and geotechnical considerations, the economical solution of lateral pile loading problems requires communication between the structural and geotechnical engineer.

Three types of parameters have significant effects on the lateral load capacity of piles. These three types are as follows:

1) Soil Parameters
    a. Soil type and physical properties such as shear strength, friction angle, density, and moisture content.
    b. Coefficient of horizontal subgrade reaction (p_{ci}). This coefficient is defined as the ratio between a horizontal pressure per unit area of vertical surface (psi) and the corresponding horizontal displacement (inches). For a given deformation, the greater the coefficient, the greater is the later load capacity.
2) Pile Parameters
    a. Physical properties such as shape, material, and dimensions.
    b. Pile head conditions.
    c. Method of placement such as jetting or driving.
    d. Group action.
3) Load Parameters
    a. Static (continuous) or dynamic (cyclic).
    b. Eccentricity.

4.6.6.2. Lateral Load Tests

Full-scale lateral load tests can be conducted at a construction site during the design stage. The data obtained is used to complete the design for the particular site. These tests are time-consuming, costly and can only be justified on large projects of a critical nature. They are discussed in more detail in Chapter 6.

4.6.6.3. Analytical Methods

The analytical methods are based on both theory and empirical data. They permit the inclusion of various site parameters. Two available approaches are (1) Broms' method and (2) Reese's methods. Both approaches consider the pile analogous to a beam on an elastic foundation. Broms' method provides a relatively easy, hand calculation procedure to determine lateral loads and pile deflections at the ground surface. Broms' method ignores the axial load on the pile. Reese's methods that are more sophisticated include analysis by computer (COM-624 Program) and a non-dimensional method that does not require compute use. Reese's computer method permits the inclusion of more parameters, provides moment, shear, soil modulus, and soil resistance for the entire length of pile including moments, and shears in the above ground sections.

It is recommended that for the design of major pile foundation projects, Reese's more sophisticated methods be used. For small-scale projects, the use of Broms' method will yield generally acceptable results.

A step-by-step procedure showing the application of Broms' method is provided in the following section.[31] The procedure and the accompanying design tables demonstrate the variable influence of the pile, soil and load parameters.

4.6.6.3.1. Broms' Method

1) Determine the general soil type (i.e., cohesive or cohesionless) within the critical depth below the ground surface (about 4 or 5 pile diameters.)
2) Determine the coefficient of horizontal subgrade reaction k_h within the critical depth from a or b depending upon the soil type.
   a. Cohesive Soils:

Equation 4-47:

$$K_h = \frac{80 n_1 n_2 q_u}{D}$$

Where:

- q_u = unconfined compressive strength in psf
- D = width of pile in feet
- n_1 and n_2 = empirical coefficients taken from Table 4-10.

   b. Cohesionless Soils: Choose K_h from the Table 4-11. (The values of K_h given in Table 4-11 were determined by Terzaghi.)

Table 4-10 Values of Coefficients of n_1 and n_2 for Cohesive Soils

Unconfined Compressive Strength, q_u, ksf	n_1
< 1	0.32
1 to 4	0.36
> 4	0.40

Pile Material	n_2
Steel	1.00
Concrete	1.15
Wood	1.30

Table 4-11 Values of K_h for Cohesionless Soils

Soil Density	K_h, lbs/in³	
	Above ground water	Below ground water
Loose	7	4
Medium	30	20
Dense	65	40

3) Adjust k_h for loading and soil conditions:
   a. Cyclic loading (for earthquake loading) in cohesionless soil:
      - K_h = K_h/2 from Step 2 for medium to dense soil.
      - K_h = K_h/4 from Step 2 for loose soil.
   b. Static loads resulting in soil creep (cohesive soils):
      - Soft and very soft normally consolidated clays: K_h = (1/3 to 1/6) K_h from Step 2.
      - Stiff to very stiff clays: K_h = (1/4 to 1/2) K_h from Step 2.
4) Determine pile parameters:
   a. Modulus of elasticity E (psi).
   b. Moment of inertia I (in.⁴).
   c. Section modulus S about an axis perpendicular to the load plan (in.²).
   d. Yield stress of pile material f_y (psi) for steel or ultimate compression strength f'_c (psi) for concrete.
   e. Embedded pile length L (inches).
   f. Diameter of width D (inches).
   g. Eccentricity of applied load e for free-headed piles - i.e., vertical distance between ground surface and lateral load (inches).
   h. Dimensionless shape factor C_s (for steel piles only):
      - Use 1.3 for piles with circular cross-section
      - Use 1.1 for H-section piles when the applied lateral load is in the direction of the pile's maximum resisting

moment (normal to pile flanges).

- Use 1.5 for H-section piles when the applied lateral load is in the direction of the pile's minimum resisting moment (parallel to pile flanges).

i. M_{yield}, the resisting moment of the pile = $C_s f_y S$ (in.-lb) (for steel piles), or M_{yield} = $f'_c S$ (in.-lb.) (for concrete piles).

5) Determine factor β or n:

Equation 4-48 :

$$\beta = \sqrt[4]{\frac{K_h D}{4EI}}$$

(for cohesive soil)

Equation 4-49:

$$n = \sqrt[5]{\frac{K_h}{EI}}$$

(for cohesionless soil)

6) Determine the dimensionless length factor:

   a. β_L for cohesive soil, or

   b. n_L for cohesionless soil.

7) Determine if the pile is long or short:

   a. Cohesive soil[32]

      - $\beta_L > 2.25$ (long pile)
      - $\beta_L < 2.25$ (short pile)

   b. Cohesionless soil

      - $n_L > 4.0$ (long pile)
      - $n_L < 2.0$ (short pile)
      - $2.0 < n_L < 4.0$ (intermediate pile)

8) Determine other soil parameters:

   a. Rankine passive pressure coefficient for cohesionless soil, $K_p = \tan^2(45+\phi/2)$ where ϕ= angle of internal friction.

   b. Average effective soil unit weight over embedded length of pile, γ (pci).

   c. Cohesion, c_u = one half the unconfined compressive strength, $q_u/2$ (psi).

9) Determine the ultimate (failure) load P_u for a single pile:

   a. Short Free or Fixed-Headed Pile in Cohesive Soil: Using L/D (and e/D for the free-headed case), enter Figure 4-37, select the corresponding value of P_u/C_uD^2, and solve for P_u (lb.).

   b. Long Free or Fixed-Headed Pile in Cohesive Soil: Using M_{yield}/C_uD^3 (and e/D for the free headed case), enter Figure 4-38, select the corresponding value of P_u/C_uD^2, and solve for P_u (lb.).

   c. Short Free or Fixed-Headed Pile in Cohesionless Soil: Using L/D (and e/L for the free-headed case), enter Figure 4-39, select the corresponding value of $P_u/_{Kp} D^3\gamma$ and solve for P_u (lb.).

   d. Long Free or Fixed-headed Pile in Cohesionless Soil: Using $M_{yield}/D^4\gamma K_p$, (and e/D for the free-headed case), enter Figure 4-40, select the corresponding value of $P_u/K_pD^3\gamma$ and solve for P_u (lb).

   e. Intermediate Free or Fixed-Headed Pile in Cohesionless Soil: Calculate P_u for both a short pile (Step 9c) and a long pile (Step 9d) and use the smaller value.

10) Calculate the maximum allowable working load for a single pile P_m from the ultimate load P_u determined in Step 9 (this is shown in Figure 4-41):

Equation 4-50:

$$P_m = \frac{2P_u}{5}$$

11) Calculate the working load for a single pile P_a corresponding to a given design deflection at the ground surface y, or the deflection corresponding to a given design load. If P_a

and y are not given, substitute the value of P_m (lb.) from Step 10 for P_a in the following cases and solve for y_m (inches):

a. Free or Fixed-Headed Pile in Cohesive Soil: Using β L (and e/L for the free-headed case), enter Figure 4-35, select the corresponding value of yK_hDL/P_a, and solve for P_a(lb) or y (inches).

b. Free or Fixed-Headed Pile in Cohesionless Soil: Using nL (and e/L for the free-headed case), enter Figure 4-36, select the corresponding value of $y(EI)^{3/5} K_h^{2/5}/P_aL$, and solve for P_a(lb) or y (inches).

12) Compute or use y or y_m

a. If $P_a > P_m$, use P_m and calculate y_m (Step 11).

b. If $P_a < P_m$, use P_a and y.

c. If P_a and y are not given, use P_m and y_m from Figure 4-36.

13) Reduce the allowable load selected in Step 12 to account for:

a. Group effects as determined by pile spacing Z in the direction of load:

b. Method of installation - for driven piles, use no reduction, and for jetted piles use 0.75 of the value from Step 13a.

14) The total lateral load capacity of the pile group equals the adjusted allowable load per pile from Step 13b times the number of piles. The deflection of the pile group is the value selected in Step 12. It should be noted that no provision has been made to include the lateral resistance offered by the soil surrounding an embedded pile cap.[33]

Figure 4-35 Lateral Deflections, At Ground Surface, of Piles in Cohesive Soils

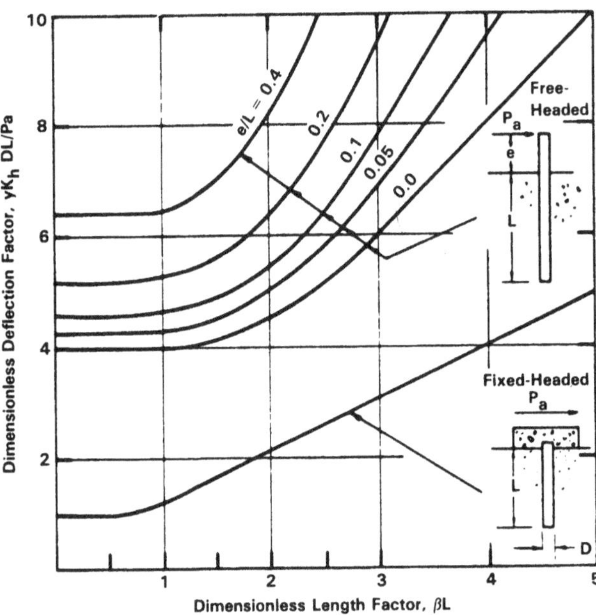

Figure 4-36 Lateral Deflections, At Ground Surface, of Piles in Cohesionless Soils

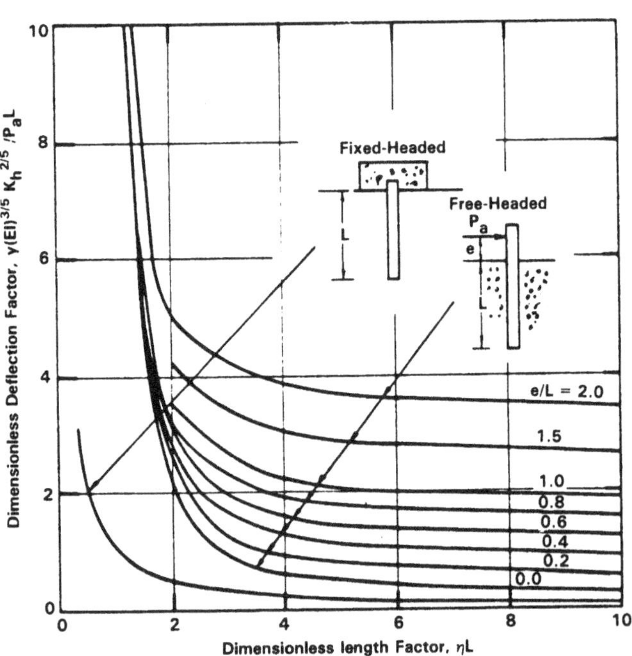

Figure 4-37 Ultimate Lateral Capacity of Short Piles in Cohesive Soils

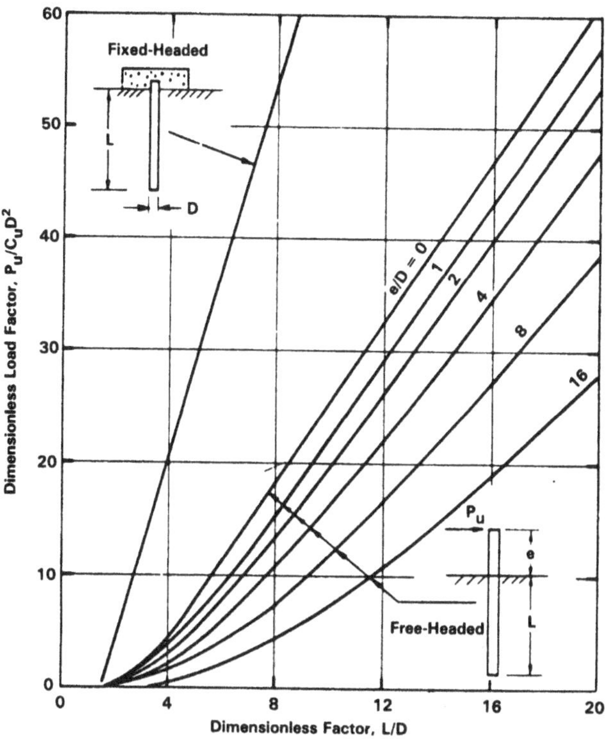

Figure 4-38 Ultimate Lateral Capacity of Long Piles in Cohesive Soils

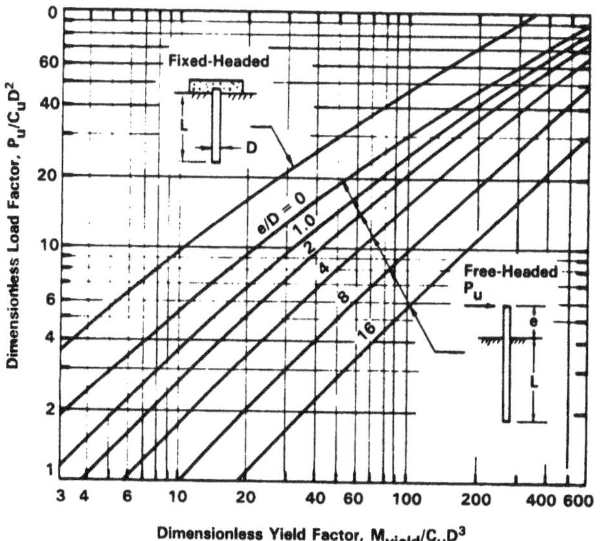

Figure 4-39 Ultimate Lateral Load Capacity of Short Piles in Cohesive Soils

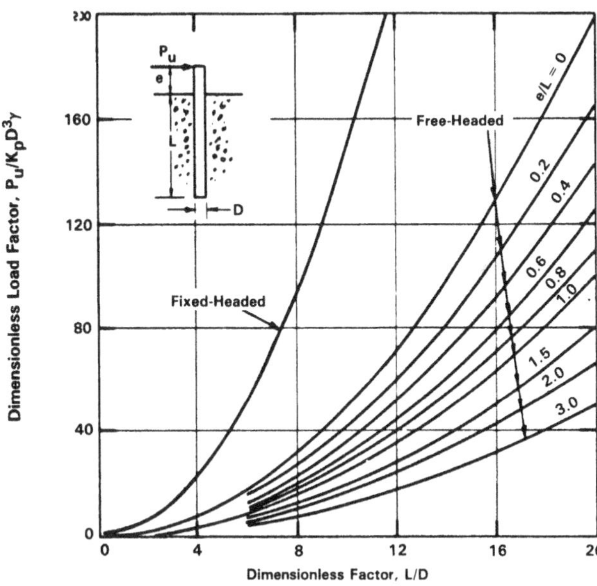

Figure 4-40 Ultimate Lateral Load Capacity of Long Piles In Cohesionless Soils

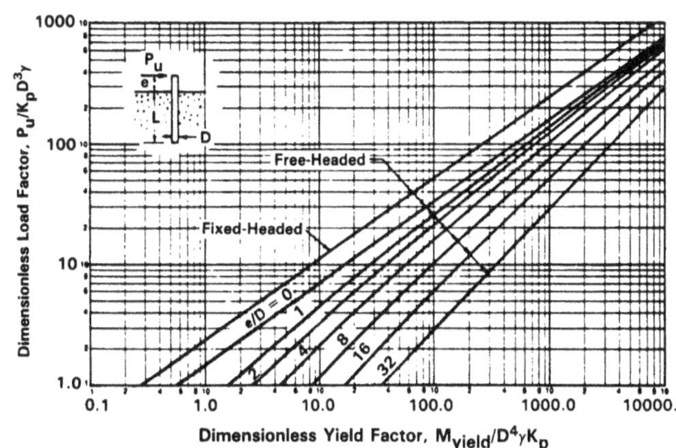

Figure 4-41 Relationship of Load and Deflection

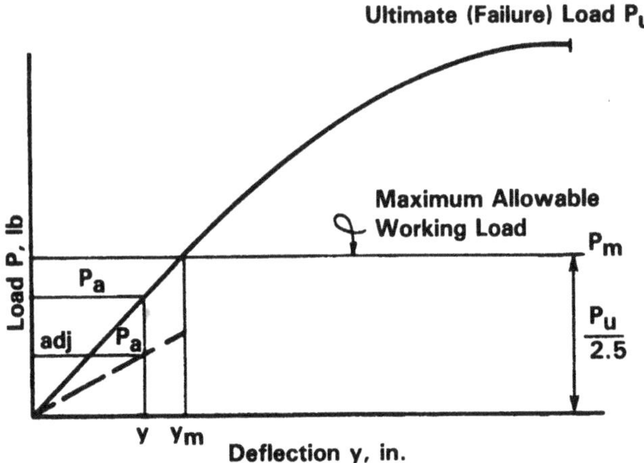

4.6.6.3.2. Reese's Method

Reese's method employs difference equations developed from the basic differential equations of conventional beam theory and non-linear curves relating soil resistance to pile deflections at various depths. To account for the non-linear relationship between pile deflections and soil resistance, an iterative (trial and error) procedure is used by a computer program[34] until satisfactory compatibility is obtained between predicted behavior of the soil and the load-deflection relationship of the elastic pile. Non-computerized solutions for the case with no axial load and constant pile stiffness, using non-dimensional coefficients are also available[35].

The computer program solves for pile deflection, bending moment, soil resistance, and soil modulus values along the pile's length due to lateral loading. It allows the following input options:

1) Inclusion of non-linear relationships between soil resistance p and pile deflection y (p-y curves).
2) Boundary conditions for varying degrees of restrain at the pile head.
3) Lateral load application above ground.
4) Inclusion of a vertical (axial) pile load.
5) Variations in soil modulus as a function of depth.
6) Variations in the flexural rigidity EI of the pile with depth.

Reese has presented procedures for describing the soil response surrounding a laterally loaded pile for various soil conditions by using a family of p-y curves. The procedures for constructing these curves are based on experiments using full-sized, instrumented piles and theories for the behavior of soil under stress.

The soil modulus E_s is defined as follows:

Equation 4-51:

$$E_s = \frac{-p}{y}$$

Where:

- y = lateral soil (or pile) deflection, and
- P = soil resistance per unit pile length

The negative sign indicates that soil resistance opposes pile deflection. P-y curve values can be input into the program or alternately, built in p-y curves in the program can be used. A typical p-y curve is shown in Figure 4-42. The Es value depends primarily on soil properties. It should be noted that the p-y curves are assumed linear between the origin and an ordinate value equal to half the ultimate soil resistance P_u.

Figure 4-42 Typical p-y Curves

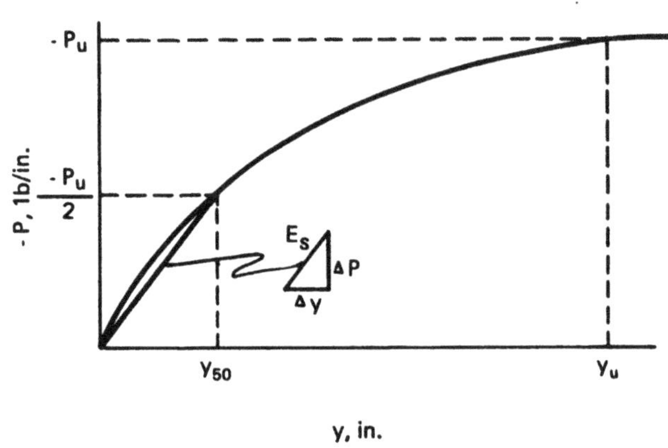

The p-y curves depend upon several parameters including depth, soil or rock stress-strain relationships, pile width, water table location, and loading conditions (static, dynamic or cyclic).

The computer output consists of input information and results of analysis. Input information consists of pile geometry and properties, and soil strength data. Output information includes (1) generated p-y curves at various depths below the pile head and, (2) computed pile deflections, bending moments, stresses and soil moduli as functions of depth below the pile head. This information allows an analysis of the pile's structural capacity and enables the engineer to design against excessive deflections that may affect the integrity of the superstructure.

4.6.6.4. Lateral Capacity of Pile Groups

The ability of a pile group to resist lateral loads from vessel impact, debris, wind, or wave loading, seismic events, and other sources is a significant design issue. The deflection of a pile group under a lateral load is typically 2 to 3 times larger than the deflection of a single pile loaded to the same intensity. Holloway et al.[36] and Brown et al.[37] reported that piles in trailing rows of pile groups have significantly less resistance to a lateral load than piles in the lead row, and therefore exhibit greater deflections. This is due to the pile-soil-pile interaction that takes place in a pile group. The pile-soil-pile interaction results in the lateral capacity of a pile group being less than the sum of the lateral capacities of the individual piles comprising the group. Hence, laterally loaded pile groups have a group efficiency of less than 1.

The lateral capacity of an individual pile in a pile group is a function of its position in the group and the center-to-center pile spacing. Brown et al. proposed a p-multiplier, P_m, be used to modify the p-y curve of an individual pile based upon the piles row position. An illustration of the p-multiplier concept is presented in Figure 4-43. For piles in a given row, the same P_m value is applied to all p-y curves along the length of the pile. In a lateral load test of a 3 by 3 pile group in very dense sand with a center-to-center pile spacing of 3b, Brown found the leading row of piles had a P_m of 0.8 times that of an individual pile. The P_m values for the middle and back row of the group were 0.4 and 0.3, respectively.

Figure 4-43 Illustration of p-multiplier Concept for Lateral Group Analysis

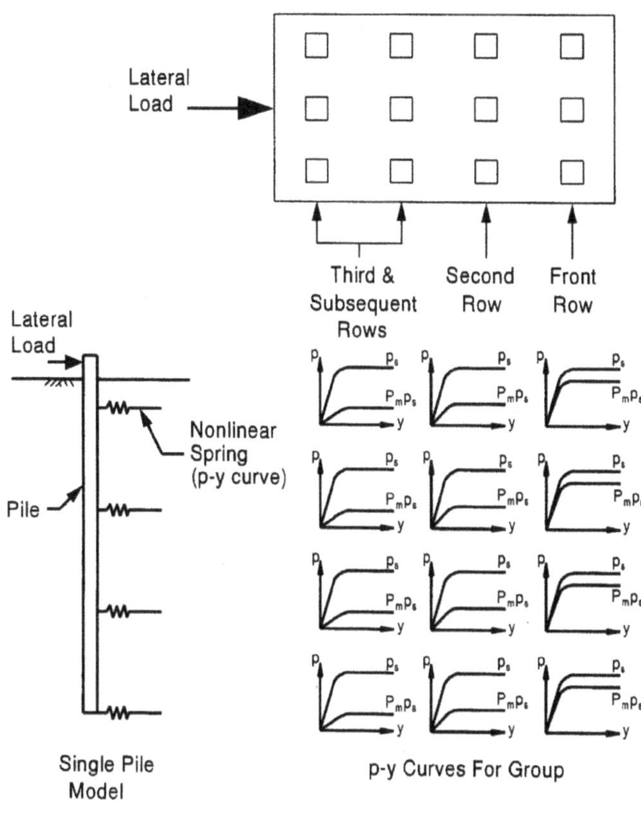

McVay et al.[38] performed centrifuge model tests on a 3 by 3 pile group having center-to-center pile spacings of 3b and 5b. Dense and loose sand conditions were simulated in the centrifuge model tests. For the dense sand case at a center-to-center spacing of the centrifuge model test results were similar to Brown's field results. However, McVay also found that soil density and the center-to-center spacing influenced the P_m values. The P_m results from McVay's centrifuge tests as well as other recent results for vertical piles in 3 x 3 pile groups are summarized in Table 4-12. McVay's centrifuge tests indicated lateral load group efficiencies in sands on the order of 0.74 for a center-to-center pile of 3b and 0.93 for a center-to-center spacing of 5b. Field studies in cohesive soils have also shown that pile-soil-pile interaction occurs. Brown et al.[39] reported P_m values of 0.7, 0.5, and 0.4 for the lead, second, and third row of a laterally loaded pile group in stiff clays.

More recent work on this topic has included full-scale lateral load testing of a 16-pile group in loose sand by Ruesta and Townsend[40] and a 9-pile group in clayey silt by Rollins et al.[41] A scaled model study of a cyclically laterally loaded pile group in medium clay has also been reported by Moss[42]. The center-to-center pile spacing, P_m results, and pile head deflections reported in these studies are included in Table 4-12.

Brown and Bollman[43] proposed a p-multiplier procedure for the design of laterally loaded pile groups. It is recommended that this approach, outlined in the step-by-step procedure that follows, be used for the design of laterally loaded pile groups.

Step-By-Step Design Procedure For Laterally Loaded Pile Groups

STEP 1 Develop p-y curves for single pile.

a. Obtain site-specific single pile p-y curves from instrumented lateral pile load test at site.

b. Use p-y curves based on published correlations with soil properties.

c. Develop site-specific p-y curves based on in-situ test data such as pressuremeter.

STEP 2 Perform COM624P analyses.

a. Perform COM624P analyses using the P_m value for each row position to develop load-deflection and load-moment data. b. Based on current data, it is suggested that P_m values of 0.8 be used for the lead row, 0.4 for the second row, and 0.3 for the third and subsequent rows. These recommendations are considered reasonable for center-to-center pile spacing of 3b and pile deflections at the ground surface of .10 to .15b. For larger center-to-center spacings or smaller deflections, these P_m values should be conservative.

b. Determine shear load versus deflection behavior for piles in each row. Plot load versus pile head deflection results similar to as shown in Figure 4-44(a).

STEP 3 Estimate group deflection under lateral load.

a. Average the load for a given deflection from all piles in the group to determine the average group response to a lateral load as shown in Figure 4-44 (a).

b. Divide the lateral load to be resisted by the pile group by the number of piles in the group to determine the average lateral load resisted per pile. Enter load-deflection graph similar to Figure 4-44 (a) with the average load per pile to estimate group deflection using the group average load deflection curve.

STEP 4 Evaluate pile structural acceptability.

a. Plot the maximum bending moment determined from COM624P analyses versus deflection for each row of piles as illustrated in Figure 4-44 (b).

b. Check the pile structural adequacy for each row of piles. Use the estimated group deflection under the lateral load per pile to determine the maximum bending moment for an individual pile in each row.

c. Determine maximum pile stress from COM624P output associated with the maximum bending moment.

d. Compare maximum pile stress with pile yield stress.

Table 4-12 Laterally Loaded Pile Groups Studies

Soil Type	Test Type	Center to Center Pile Spacing	Calculated p-Multipliers Pm For Rows 1, 2, & 3+	Reported Group Efficiency	Deflection	Reference
Stiff Clay	Field Study	3b	.70, .50, .40	--	51	Brown et al, (1987)
Stiff Clay	Field Study	3b	.70, .60, .50	--	30	Brown et al. (1987)
Medium Clay	Scale Model- Cyclic Load	3b	.60, .45, .40	--	600 at 50 cycles	Moss (1997)
Clayey Silt	Field Study	3b	.60, .40, .40	--	25-60	Rollins et al, (1998)
V. Dense Sand	Field Study	3b	.80, .40, .30	75%	25	Brown et al. (1988)
M. Dense Sand	Centrifuge Model	3b	.80, .40, .30	74%	76	McVay et al. (1995)
M. Dense Sand	Centrifuge Model	5b	1.0, .85, .70	95%	76	McVay et al, (1995)
Loose M. Sand	Centrifuge Model	3b	.65, .45, .35	73%	76	McVay et al. (1995)
Loose M. Sand	Centrifuge Model	5b	1.0, .85, .70	92%	76	McVay et al. (1995)
Loose F. Sand	Field Study	3b	.80, .70, .30	80%	25-75	Ruesta et al. (1997)

STEP 5 Perform refined pile group evaluation that considers superstructure-substructure interaction.

Figure 4-44 Typical Plots of Load vs. Deflection and Bending Moment vs. Deflection for Pile Group Analysis

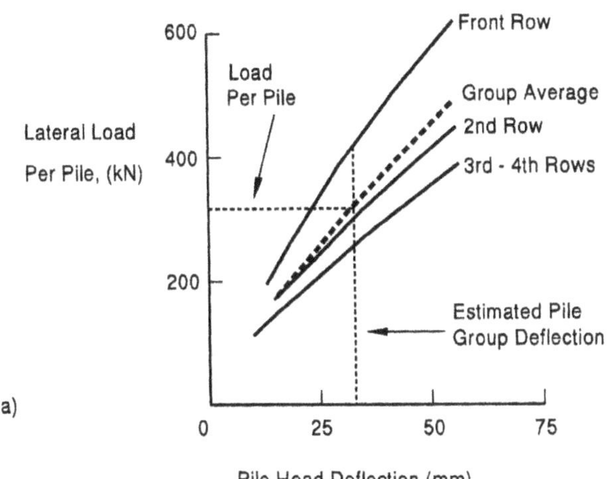

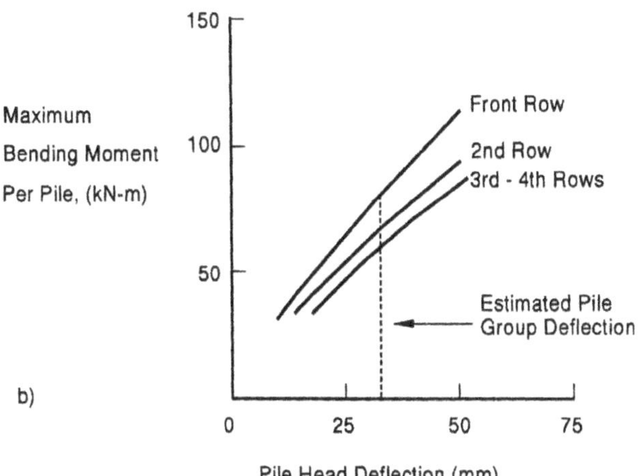

4.7. Dynamic Resistance, Drivability, and Hammer Selection: Impact Hammers

4.7.1. Overview

Piles penetrate the ground by dynamic means such as impact or vibration. Obtaining a successful pile foundation, which meets the design objectives, depends largely on relating the static analysis results presented on the plans to the dynamic methods of field installation. Dynamic analysis can provide answers to the following site-specific questions:

1) Can a given pile be driven to the estimated depth and capacity with a specific hammer?
2) If so, what will be the set (deflection) in the final blows and what will be the maximum stress experienced by the pile?

If driving cannot be accomplished as specified in (1), what hammer characteristics are needed to successfully complete the project with the pre-selected pile? Alternatively, what other pile of the same length can, perhaps, be driven with the hammer first considered?

To answer these and other questions that may appear in connection with a particular piling project, rational analysis based on the hammer-cushion-pile-soil system (dynamic analysis) is useful. Experience alone, however important, is not usually sufficient to answer the above questions. Dynamic analysis should be performed during both the design and construction stages of a project.

The subject of pile dynamics also covers the mechanics of pile driving, including equipment. The dynamic response of pile-supported structures due to imposed dynamic forces generated by earthquakes and vibrating machinery is beyond the scope of this book. The reader is encouraged to seek the assistance of a specialist in the area of foundation vibrations.

The depth, type of pile and soil type control the ultimate soil resistance and pile load capacity. However, the pile stiffness, structural strength and driving system control the depth to which a particular pile can be driven. To achieve economical pile design, the engineer must match soil resistance, pile stiffness, pile strength and driving equipment. Failure to correctly understand the important physical concepts often results in project delays, claims, and additional costs.

4.7.2. Pile Driveability

The limiting pile drivability for a specific pile-soil situation is the maximum soil resistance to which a pile can be driven without damage. The soil resistance developed is a function of the pile dimensions and the subsurface profile. An exception to this maximum is a pile penetrating very soft soil to bear evenly on sound unweathered rock with no transition zone of weathered rock. In this special case, the full structural strength of a pile can be developed without significant driving effort.

4.7.2.1. Factors Affecting Driveability

To perform successfully, a pile must satisfy two aspects of drivability: (1) the pile must have sufficient stiffness to transmit driving forces large enough to overcome soil resistance, and (2) the pile must have sufficient strength to withstand the driving forces without damage. For a given soil condition and pile length, the strength and stiffness of a pile determine its drivability.

It is important to recognize that even if the static structural and static soil capacities allow an increase in pile stresses, it may not be possible to develop the increased pile loads because the resulting driving stresses would exceed allowable pile driving stress limits. It is not possible to drive against an ultimate static soil resistance that is as high as the structural resistance of the pile cross section because of the soil damping forces which are generated during pile driving but no longer present after driving is completed.

The limitations on maximum allowable static design stresses in pile materials by various codes generally represent the static stress levels (static load capacity), which can be consistently developed with common driving equipment and methods.

4.7.2.2. Methods for Determining Driveability

There are three available methods for evaluating drivability:

1) Static Load Tests - Static Load tests are useful for checking drivability prior to production pile driving. Test piles are normally driven to pre-determined lengths and load tested. Load tests can be performed during design and/or construction stages. These are discussed in detail in Chapter 6.

2) Wave Equation Analysis - This method accounts for pile stiffness and predicts driving stresses as well as the relationship of hammer blow count versus ultimate pile capacity. It can be used to check drivability in advance of driving, to design the most economical pile wall thickness or pile section and to select driving equipment.

3) Dynamic Measurements - These measurements and their analysis can be used to measure the driving stresses and static bearing capacity of piles during driving. It can also provide static soil resistance distribution and damping parameters for a wave equation analysis. Hammer and driving systems performance can also be evaluated from the measurements.

Used properly, methods 2 and 3 can yield significant savings in material costs or a reduction of construction delays. These methods make it possible to reduce and possibly eliminate the number of load tests and allow an increase in the maximum allowable design stresses. Dynamic measurements also allow a determination of increase (soil set-up) or decrease (relaxation) in pile capacity by measurement during redriving (retapping). The wave equation computer program allows the user to assess the effects of these time related soil resistance changes based on user input.

Many successful pile contractors use the wave equation analysis to check and improve upon an engineer's pile design, for preparing bids, to avoid construction problems and to select the most appropriate pile hammer for the given project conditions. It is highly recommended that methods 2 and 3 be used for the construction control.

4.7.2.3. Driveability and Pile Type

Drivability should be a consideration in the design of all driven piles. It is particularly critical in the case of cast-in-place concrete filled pipe piles, wherein only the stiffness of the steel casing is available at the time of driving. Although the designer may be temped to specify a thin-wall pipe in order to save material cost, such a pile may lack the stiffness and strength to overcome the encountered soil resistance for reaching the necessary penetration. Wave equation analysis should be used to design the pile section and wall thickness for the required pile capacity.

Single-material solid-section piles such as steel H, precast concrete, and timber are subject to drivability limitations, particularly as allowable design stresses increase. In the case of long prestressed concrete piles, drivability is normally limited by the pile's compressive strength, but tensile strength will be critical in the early driving condition where large hammer energies may be imbalanced to the small soil resistance.

4.7.2.4. Soil Set-Up and Relaxation

Frequently a pile driven in silt, clay or granular soils with high fines gains capacity after driving has been completed. This phenomenon is called soil set-up. Occasionally a pile driven in dense saturated fine sand may exhibit a decrease in ultimate load carrying capacity after the driving has been completed. This phenomenon is called relaxation. Pile capacity gained or lost after driving, due to soil set-up or relaxation, is a function of soil and pile material properties and is not directly related to drivability. Therefore, gains or losses of soil resistance after driving cannot be directly accounted for by dynamic means. These phenomena however, can be estimated or measured based on pile redriving (retapping), load testing, and local experience.

Some rock deposits (including shales, clay stones and clay shales) are compressed and disturbed due to pile driving. Large excess hydrostatic pressures (pore pressures) develop in saturated clays due partly to the disturbance and partly to the high radial compression. The excess hydrostatic pressures reduce the soil shear strength and static pile load capacity during and immediately after driving. The gradients resulting from excess hydrostatic pressures start a consolidation process that provides a regain of shear strength (increase in shaft friction or adhesion) over an extended period. This increase in shaft friction results in a regain of the static soil load capacity and is called soil "set-up" or "freeze." The magnitude of this gain in soil load capacity depends on soil characteristics, pile material and pile dimensions. Load testing of a pile in material that exhibits setup behavior should not be conducted without an awareness of these processes. It is generally advisable to delay load testing or retapping for at least two weeks after driving and preferably for a longer period.

Because the resistance to pile penetration may increase (set-up) or decrease (relaxation) after final set, it is essential that retapping or load testing be performed after equilibrium conditions in the soil or rock have occurred. The time for the return of equilibrium conditions is highly variable and depends on soil type. Piezometers can be employed to determine time required for pore pressure dissipation and strength gain.

4.7.2.5. Driving Resistance Criteria

Most pile driving is controlled by specifying that the pile be driven to a certain number of blows/inch or blows/cm, generally irrespective of the depth of penetration. Sometimes minimum depths of penetration are required in addition to driving resistance. A rational basis should be used for determining a reasonable limit to the number of blows/inch that a pile should be driven. Traditionally, driving criteria were determined by the use of pile driving formulas. As discussed elsewhere in this book, these formulas are fundamentally incorrect and the results are erratic, either in a conservative or an unconservative way.

A wave equation analysis should be used to establish a relationship between the static bearing capacity of driven piles and the number of blows per inch required by a particular hammer and driving system to drive that pile in a given situation. The wave equation should also be used in combination with static bearing capacity theories to select the most suitable pile for the specified structural loadings. The driving criteria established should be verified by static or dynamic load tests whenever possible.

Dynamic monitoring of pile driving and analysis (by the pile driving analyzer developed at Case-Western University) allows the determination of static pile capacity during driving and establishment of a driving criterion. Again, the driving criteria established should be verified by load tests whenever possible. This method of establishing driving criteria provides greater accuracy than the wave equation analysis, but may not be cost effective on all projects.

4.7.2.6. Efficiency in Dynamic Analysis

One factor that is frequently misunderstood by owners, engineers and contractors is the efficiency of the impact hammer. Understanding pile hammer efficiency is essential for correct analysis of any hammer/pile/soil system, irrespective of the dynamic method being employed.

Hammer efficiency can be defined as the ratio of the actual output energy of a hammer to a theoretical potential or rated energy, or

Equation 4-52:

$$e = \frac{E_{act}}{E_{ideal}}$$

Where

- e = efficiency, usually expressed as a percent

(in which case the ratio is multiplied by 100)

- E_{act} = actual energy output
- E_{ideal} = ideal energy output

The ideal output is frequently referred to as the "rated striking energy." The actual output is frequently referred to as the "net striking energy." The difference between the two is the result of a number of factors, two of which are discussed here: mechanical losses and the "cosine effect" for batter driving.

4.7.2.6.1. Mechanical Efficiency

Mechanical losses are inevitable in any type of machinery. In pile driving equipment, they can be due to frictional losses between the moving (ram) and stationary (frame) parts, losses due to backpressure in the valving of air/steam and hydraulic hammers or the exhaust ports of diesel hammers, and other sources. These are a function of both the design of the equipment and the way in which it is maintained. They are best determined from experimental data, although analytical solutions exist and ram impact velocity data can be difficult to obtain for some types of hammers.

The net kinetic striking energy of a plumb hammer at impact (without consideration of batter effects) is given by the equation

Equation 4-53:

$$E_h = E_{act} = \frac{W_h v_h^2}{2g} = eW_h H$$

Where

- E_h = Net Striking Energy of the hammer, in-lbs
- W_h = Hammer ram weight, lbs
- v_h = striking velocity of the ram at impact, in/sec
- g = acceleration due to gravity = 32.2 in/sec^2
- e = mechanical efficiency of the hammer
- H = stroke (single-acting hammer) or effective stroke (double or differential acting) hammer, inches

If this equation is combined with Equation 4-52, the rated striking energy is thus

Equation 4-54:

$$E_{ideal} = W_h H$$

In the case of single acting hammers, the value of "H" is the stroke of the hammer. With double, differential, compound, or any hammer that provides "downward assist" to the ram, the process is more complicated. If the energy developed by the motive fluid or a gas spring (such as is used with double acting diesel hammers and some hydraulic ones) is included, then the basis of the efficiency is the same as with single acting hammers. If some or all of the downward assist is ignored, then the rated energy will be lower and the apparent efficiency of the machine higher. This is the case with some hydraulic hammers, and is part of the reason why these hammers report such high efficiencies.

A recommended way to evaluate hammers based on their impact energies is to compare their net striking energies rather than their efficiency. Of course there are other factors to consider, such as the relationship of the (effective) stroke to the ram weight, type and size of cushion material and the effects of the necessary compression in diesel hammers.

Wave equation analyses are usually based on rated striking energy times an efficiency, which yields a net striking energy. Dynamic formulae vary in this regard; some (such as the Engineering News Formula) have the efficiency included in the formula, others (such as the Hiley Formula) do not. This is discussed in more detail below.

4.7.2.6.2. Batter Piles and the "Cosine Effect"

Another factor in the reduction of output energy occurs when driving batter piles. This is generally considered an "efficiency" factor but strictly speaking is not. It comes from the geometry of the hammer relative to the gravity field, and is most pronounced with single-acting hammers.

If we consider Equation 4-54, if a single acting hammer is operating in plumb, at full stroke operation the ram falls through the gravity field through the full length of the stroke "H". If, however, the hammer is operating on an angle, the actual stroke H is reduced by the equation

Equation 4-55:

$$H' = H\cos\theta$$

Where

- θ = angle from plumb of the hammer/pile system

Such a situation results in a "revised" rated striking energy of

Equation 4-56:

$$E'_{ideal} = W_h H\cos\theta = E_{ideal}\cos\theta$$

This formula gives the name of "cosine effect" to the reduction in rated energy of hammers on a batter.

Batter angles, however, are not generally stated as angles but in the form "X:Y" where X and Y are horizontal and vertical components respectively of the right triangle that defines the slope of the pile from the vertical (as opposed to the slope from the horizontal used in grade and slope analysis.) The angle can be computed from this ratio as

Equation 4-57:

$$\theta = \arctan\left(\frac{X}{Y}\right)$$

Substituting this equation into Equation 4-56 yields

Equation 4-58:

$$E'_{ideal} = \frac{E_{ideal}}{\sqrt{1+\left(\frac{X}{Y}\right)^2}}$$

This equation can only be applied in this form to single-acting hammers. With any hammer with downward assist, the amount of cosine effect depends upon the amount of energy developed with the downward assist. Many hammers with downward assist can operate in a horizontal direction because of this.

For either the dynamic formulae or the wave equation, the cosine effect should be considered as a reduction in rated striking energy rather than a mechanical efficiency loss. Mechanical losses should be considered after cosine effects. Mechanical losses are greater when driving on a batter than in plumb because of the additional friction of the ram-frame interface, although these additional losses are not well quantified in pile hammers.

4.7.2.7. Hammer Size Selection

It is important that the contractor and the engineer choose the proper hammer for efficient use on a given project. A hammer that is too small may not be able to drive the pile to the required capacity, or may require an excessive number of blows. On the other hand, a hammer that is too large may damage the pile. The use of empirical dynamic pile formulas to select a hammer energy should be discontinued because this approach incorrectly assumes these formulas result in the desired pile capacities. Results from these formulas become progressively worse as the complexity of the hammers increase.

A wave equation analysis, which considers the hammer cushion-pile-soil system, is the recommended method to determine the optimum hammer size. For preliminary equipment evaluation, Table 4-13 provides approximate minimum hammer energy sizes for ranges of ultimate pile capacities. This is a generalization of equipment size requirements that should be modified based on pile type, pile loads, pile lengths, and local soil conditions. In some cases, such as short piles to rock, a smaller hammer than indicated may be more suitable to control driving stresses. This generalized table should not be used in a specification.

Table 4-13 Preliminary Hammer Energy Requirements

Ultimate Pile Capacity, kN	Minimum Manufacturers Rated Hammer Energy, kJ	Ultimate Pile Capacity, kips	Minimum Manufacturers Rated energy, ft-kips
<800	16.5	< 180	12
800-1350	28.5	181-300	21
1351-1850	39.0	301-415	28.8
1851-2400	51.0	416-540	37.6
2401-2650	57.0	540-600	42

4.7.3. Dynamic Formulas for Static Capacity Determination

Ever since engineers began using piles to support structures, they have attempted to find rational methods for determining the pile's load carrying capacity. Methods for predicting capacities were proposed using dynamic data obtained during driving. The only realistic measurement that could be obtained during driving was pile set (blow count); thus, energy concepts equating energy delivered by the hammer to the work done by the pile as it penetrates the soil were used to obtain estimates of pile resistance known as pile formulas. Since Wellington proposed the well-known Engineering News Formula in the 1890's, hundreds of dynamic formulas have been proposed.

Subsequent studies have shown that the results of dynamic formulae do not correlate well with actual load tests, except when supplemented with experience. As a result, and because of the development of the wave equation methods, dynamic formulae have been superseded by wave equation analysis for all but the simplest piling applications. In spite of this, they need to be understood both since they are still in many pile driving codes and for historical reasons.

4.7.3.1. Derivation of Dynamic Formulae

Consider the pile driving system as the impact of two rigid masses.[44] The principle of conservation of momentum states that

Equation 4-59:

$$W_h v_h + W_p v_p = W_h v'_h + W_p v'_p$$

Where

- W_h = Weight of ram, lbs.
- W_p = Weight of pile, lbs.
- v_p = velocity of pile before impact, in/sec
- v'_h = velocity of ram after impact, in/sec
- v'_p = velocity of pile after impact, in/sec

The coefficient of restitution[45] for this kind of impact is

Equation 4-60:

$$n = \frac{v'_p - v'_h}{v_h - v_p}$$

Where n = coefficient of restitution

We should also define

Equation 4-61:

$$r = \frac{W_h}{W_p}$$

Where r = ram weight to pile weight ratio

If we solve both Equation 4-59 and Equation 4-60 for v'_h, set the results equal to each other and solve for v'_p, assume $v_p = 0$, solve Equation 4-61 for W_p and substitute this into the result, we have

Equation 4-62:

$$v'_p = v_h \frac{r(n+1)}{r+1}$$

Since from Equation 4-53 we know that

Equation 4-63:

$$v_h = \sqrt{2egH}$$

The kinetic energy of the pile after impact is

Equation 4-64:

$$E_p = \frac{W_p v'^2_p}{2g} = eHW_r r\left(\frac{n+1}{r+1}\right)^2$$

Where E_p = kinetic energy of the pile after impact, in-lbs

This energy is absorbed in one of two ways: 1) elastic compression of the cushion, cap, soil or pile itself, and 2) plastic movement of the soil, i.e., the pile set. This is expressed as

Equation 4-65:

$$E_p = \frac{R_d C}{2} + R_d s$$

Where

- R_d = pile resistance (without consideration of a factor of safety), lbs.

- C = aggregate spring constant of elastic elements of cushion/pile/soil system, lbs/in
- s = plastic set of pile, inches

A factor of safety is necessary. This is defined as

Equation 4-66:

$$FS = \frac{R_d}{R_a}$$

Where

- FS = Factor of safety
- R_a = "Design" load or resistance of piles, lbs. The "design" load may be equal to or less than the allowable load.

Combining Equation 4-64 and Equation 4-65 and substituting the design load of Equation 4-66, and solving for that design load, we have at last

Equation 4-67:

$$R_a = \frac{eW_hH}{FS\left(\frac{C}{2}+s\right)} r\left(\frac{n+1}{r+1}\right)^2$$

Equation 4-67 forms the basis of many of the dynamic formulae. For example, if we assume FS = 3 and change the ram drop to feet, we have

Equation 4-68:

$$R_a = \frac{4eW_hH_{ft}}{\left(\frac{C}{2}+s\right)} r\left(\frac{n+1}{r+1}\right)^2$$

(Hiley Formula)

Where H_{ft} = ram stroke or equivalent stroke, feet

If we assume that FS = 6, n = 1, C = 2c, r = 1 and e = 1, and change the drop height to feet, we have

Equation 4-69:

$$R_a = \frac{2W_hH_{ft}}{s+c}$$

(Engineering News Formula)

Where c = factor for various elastic components of the system, inches = 1 for drop hammers, 0.1 to 0.3 for air/steam hammers

In American practice, the Engineering News Formula is the best known and most widely used of the dynamic formulae, although there are literally hundreds of these in existence.

4.7.3.2. Use of the Engineering News Formula

Since it is so important in engineering practice in the U.S. - especially with timber piles -- the application of this formula is presented. In actual use, is generally written as

Equation 4-70:

$$R_a = \frac{2W_hH_{ft}}{s+c}$$

(Single-Acting Hammers)

Or

Equation 4-71:

$$R_a = \frac{2E_{en}}{s+c}$$

(Hammers with Downward Assist)

If solving for the desired energy, Equation 4-70 is written as

Equation 4-72:

$$E_{en} = W_hH_{ft} = \frac{R_a(0.1+s)}{2}$$

(Single Acting Hammers)

And if solving for the blow count,

Equation 4-73:

$$s = \frac{2W_h H_{ft}}{R_a} - 0.1 = \frac{2E_{en}}{R_a} - 0.1$$

Consider the example of a Vulcan #1 hammer, with a ram weight of 5 kips and a stroke of 3' for a total rated striking energy of 15 ft-kips. For a pile with a design capacity of 60 kips, the blow count will be

$$s = \frac{(2)(3)(5)}{(60)} - 0.1 = 0.4 \text{ inches/blow} = 30 \text{ blows/ft}$$

4.7.3.3. Strengths and Weaknesses of Dynamic Formulae

Dynamic formulae represented an important step forward for driven piles as they:

- Established a correlation for the energy of the hammer, the set of the pile, and the capacity/resistance of a pile.
- Accurately determined that the blows per foot vs. driving resistance of a pile were a "saturation growth" model, which both actual driving and the wave equation analysis confirm.
- Made it possible to use the hammer as a measuring instrument, which increased the credibility of the load carrying capacity of driven piles.
- Were simple enough to be used widely in times when computational power was limited to purely manual calculations or slide rules.

However, dynamic formulae have some serious weaknesses:

- They model the pile as a rigid, solid mass. For shorter piles (such as timber piles) this was reasonable, but as piles became longer, the importance of wave propagation in the piles became more apparent.
- The modeling of the soil response and the various elements of the driving system was rudimentary or nonexistent, based in many cases on local experience (which did not always transfer to other places) or educated guesses. This is best illustration by considering the assumptions of the Engineering News Formula (Equation 4-69.) It is interesting to think about the fact that the most widely used formula in the U.S. assumes that:
    - Impact between ram and pile is perfectly elastic.
    - The weight of the ram and the pile is the same.
    - The hammer has an efficiency of 100%.
    - The various hammer/pile configurations' elastic properties can be reduced to a simple variable.

Given all of these, it is little wonder A.M. Wellington chose a factor of safety of 6 when developing the formula.

The assumption of a single mass pile does not take into account tension waves in the piles during driving, which may lead to tension cracking of concrete piles. The potential for concrete pile breakage was one of the main motivations to investigate stress wave propagation in piles. Dynamic formulae do not address the issue of driving stresses in piles.

The various ways the variables of Equation 4-67 could be assumed led to the proliferation of formulae, which led to widely varying results for a single case depending upon the formula used and thus undermined the credibility of dynamic formulae.

These difficulties begged for a more comprehensive solution, which was provided for impact driven piles by the wave equation.

4.7.4. Dynamic Analysis by the Wave Equation

4.7.4.1. Introduction

The wave equation analysis[46] is now the standard method of predicting drivability in anticipation of pile driving. The wave equation is used to describe how stress waves are transmitted in a long rod when a force is applied at one end of the rod. The idea of applying the wave equation to pile driving first came from the Australian civil engineer David Victor Isaacs in 1931. Although he and others in the 1930's basically set forth the theory and most of its potential use, the theory behind the wave equation was not implemented then because the equations involved in the calculations were too difficult due to complications from the actions of the ram, the hammer cushion, the pile, and the soil.

The development of high-speed digital computers permitted the wave equation to be applied to practical pile driving problems. E.A.L. Smith, who used a numerical solution to investigate the effects of such factors as ram weight, ram velocity, cushion and pile properties, and the dynamic behavior of the soil during driving, started widespread interest in the method in 1960. Since then, considerable research has been completed, and numerous full-scale pile tests have been correlated which now permit engineering judgment to be coupled with the mathematical accuracy of the wave equation. Both state transportation departments and private contractors have used the wave equation extensively to predict the ability of specific pile driving hammers to successfully install pile foundations.

In general, the computer solution is used to obtain the following information for a single blow of the hammer:

1. To predict the driving stresses induced in the pile.
2. To determine the resulting motion of the pile during the impact.
3. To determine the resistance to penetration afforded by the soil at the time of driving.
4. To estimate the transferred energy delivered to the pile top.

This information then enables the engineer to answer such questions as:

1. Can a given hammer drive the pile to the required depth?
2. What rate of penetration will the hammer provide, i.e., how long will it take to install the pile?
3. To what maximum penetration can the pile be driven?
4. What is the maximum soil resistance to penetration that the hammer can overcome?
5. Will excessive stresses be generated in the pile or hammer during driving?

The wave equation is also often used as an aid in design. For example, it is commonly used:

1. To indicate the blow count required for penetration of the pile afforded by an estimated soil resistance at the time of driving. Note that the wave equation only predicts the resistance to penetration at the time of driving since soil set up, group effect, negative friction, and other time effects may influence the long-term bearing capacity. Judgment tempered by experience from geotechnical engineers is necessary to assess sol resistance at the time of driving compared with long term load carrying capacity. The wave equation can be used in conjunction with restrikes to evaluate long-term pile capacity.
2. To optimize the cushion, i.e., to determine which cushion will effectively limit the driving stresses induced in the hammer and pile, and yet will still produce the maximum possible permanent pile set per blow of the hammer.
3. To determine the correct size of the driving hammer. This reduces the chance of picking selecting a very large and expensive hammer whose both its capacity is not needed and pile damage would result, as well as the more unfortunate situation of selecting a small hammer whose driving capacity is found to be inadequate to drive the pile to the required resistance or depth.
4. Since the driving stresses can be estimated, the pile section and strength can be designed. For example, tensile cracking of prestressed concrete piles, and the buckling of pipe piles are but two examples of driving failures that have been corrected by use of the wave equation. The choice of pile dimensions not only affects the driving stresses, but the drivability of the pile. For example, in some cases, a pile with a small cross-sectional area cannot be driven to the required depth, whereas a pile having a larger cross-sectional area can. Thus, with the use of the wave equation the economic merit of being able to drive the stiffer pile to a greater depth can be evaluated.
5. To determine the influence of the driving accessories. It has been shown that in many cases the driving accessories absorb a major portion of the total energy output of the

hammer. In some cases, these accessories account for a 50% reduction in the energy output of the hammer. The use of the wave equation enables the selection of optimum driving accessories required to minimize these losses.

6. The wave equation is also a powerful engineering aid for the foundation designer, since numerous alternative designs can be quickly studied at very little expense. Such a study greatly increases the probability that the final design will be economical and to installation problems will be minimized.

A longitudinal wave propagates along the pile axis when a hammer applies an impact load to a pile. As the ram impact occurs, a force pulse is developed. The amplitude and duration of the force pulse depend on the properties of the hammer-pile-soil system. The force pulse in the pile travels downward toward the pile toe at a constant velocity, which depends on the wave speed of the pile material. When the force pulse reaches the portion of the pile embedded in soil, its amplitude is reduced due to static and dynamic soil resistance forces along the pile. The force pulse reaching the pile toe will generate a reflected force pulse (tension or compression) governed by the soil resistance at the pile toe. The pile will penetrate into the soil and have a permanent set when the peak force generated by the ram impact exceeds the combined static and dynamic resistance at the pile toe.

4.7.4.2. Wave Equation Analysis: Numerical Solution

The pile and driving system Figure 4-45 (A) are represented by a series of masses and springs. The soil is modeled by a spring (R, Static) and a dashpot (R, Dynamic) attached to each mass. The soil resistances, shown in Figure 4-45 (C), are linear elastic plastic for the spring where the maximum force, R_u, is reached at a displacement q, called the quake, and linearly proportional to the element velocity for the dashpot (commonly known as the damping force).

The analysis proceeds by giving the ram an initial velocity. At each element, the displacement is calculated for a small time increment with element velocities determined from the previous time increment. With these displacements and velocities, the forces acting on each mass can be determined. Forces on each mass arise from the pile spring deformations, and dashpot forces. Using a discretization of Newton's Second Law, the mass-accelerations can be calculated. Integration of computed acceleration yields the velocity. The computation then proceeds to the next time increment.

Figure 4-45 Wave Equation Model

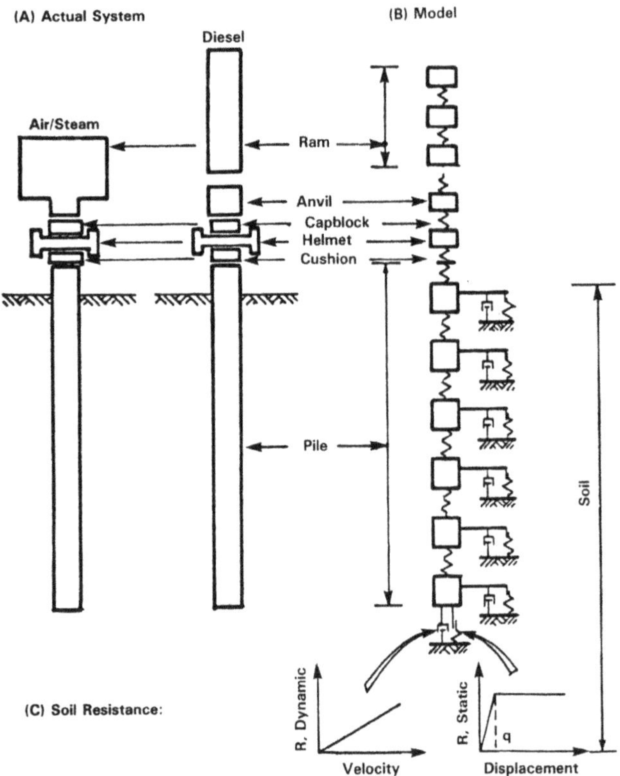

(a) the system to be analyzed;
(b) the wave equation model;
(c) the components of the soil resistance model

In application, sets of soil forces R_u and damping forces are assigned at each element. Then the ram is given its rated impact velocity and the dynamic computations outlined above are continued through successive time increments until all soil forces are less than R_u. The total permanent displacement will have then been calculated.

In practice, the wave equation can be used to develop a relationship between the blow count and the resistance of the pile, as is the case with dynamic formulae. In addition, driving stresses can be rationally limited. While the shape of the two curves is quite similar, the differences are substantial. A particular wave equation bearing graph is associated with a single driving system, hammer stroke, pile type, soil profile and a particular pile length. If any one of the above items is changed, the bearing graph changes.

4.7.4.2.1. Available Computer Programs

Wave equation programs fall into three basic families:

- The Smith/TTI Family. This includes the original Smith program, which still exists and is in limited use. More widely disseminated is the TTI program, developed in the late 1960's at the Texas Transportation Institute at Texas A&M University. The TTI program was the first wave equation program to be widely disseminated outside of the Raymond organization. It is still available as the MICROWAVE program, which runs under MS-DOS. The ZWAVE program developed by Vulcan Iron Works in the 1980's was derived indirectly from the TTI but included other improvements.
- The WEAP Family. The WEAP program was originally developed by GRL engineering organization for the FHWA in the 1970's and 1980's, and has been updated by the organization for commercial sale. They include programs such as WEAP, WEAP86, WEAP87[47], and GRLWEAP, which is available for the Microsoft Windows platform. Although this program is similar in theory to the TTI program, its main - and very important - improvements are a) the realistic modeling of diesel pile hammers and b) the inclusion of a database of hammers along with many other user conveniences. GRLWEAP has also been enhanced with many improvements that relate to changes in both pile driving equipment and piles.
- The TNO Family. This was developed by the TNO organization in the Netherlands and is available as the TNOWAVE program. It has many of the convenience features of the WEAP family. A version of this program, IHCWAVE, is freely distributed by the IHC Hydrohammer organization for use with their equipment. It includes simplified hammer, pile and soil input as well.

4.7.4.2.2. WEAP Program Input Parameters

Before a wave equation analysis is performed, all input data should be assembled. The wave equation analysis results are strongly affected by the quality and accuracy of the input data. This is typically organized on forms such as is shown in Figure 4-46.

Input for a wave equation analysis generally requires the following kinds of information:[48]

1. Title. The name of the job.
2. Analysis Options
    a. Output Option. The type of output desired.
    b. Residual Stress Analysis Option. Commonly wave equation analysis assumes that there are no stresses in the pile or soil at the time of impact. In some cases, this is not so; this allows such an analysis to be performed.
    c. Maximum Analysis Time (usually by default).
    d. Number of Iterations (usually by default).
    e. Critical Time Increment Ratio (usually by default).
3. Helmet/Hammer Cushion Information
    a. Type (diesel, air/steam, hydraulic) and model of hammer.
    b. Weight of the helmet. Although this item is frequently included with hammer manufacturer's data, contractors frequently use caps that are not made by the hammer manufacturer. When possible this should be determined and, if not the manufacturer's cap, the weight should be determined. In some cases, a helmet is not used (for example, if a diesel hammer anvil impacts the pile directly.)
    c. Area of the Hammer Cushion, usually taken from manufacturer's data. Some hammers do not have a hammer cushion.
    d. Elastic Modulus of the Hammer

Figure 4-46 Pile and Driving Equipment Data Form

Contract No.: ____________________ Structure Name and/or No.: ____________________

Project: ____________________

____________________ Pile Driving Contractor or Subcontractor: __________

County: ____________________ ____________________

(Piles driven by)

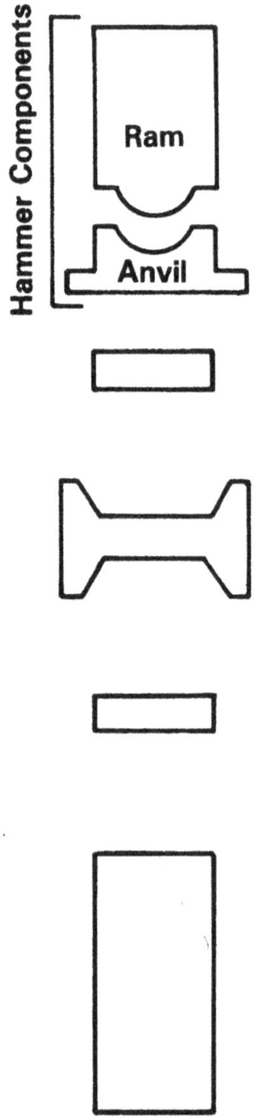

Hammer Components

Ram

Anvil

Hammer

Manufacturer: __________ Model: __________

Type: __________ Serial No.: __________

Rated Energy: __________ at __________ Length of Stroke

Modifications: ____________________

Capblock (Hammer Cushion)

Material: ____________________

Thickness __________ Area: __________

Modulus of Elasticity — E __________ (P.S.I.)

Coefficient of Restitution-e __________

Pile Cap — Helmet / Bonnet / Anvil Block / Drivehead — Weight: __________

Pile Cushion

Cushion Material: ____________________

Thickness: __________ Area: __________

Modulus of Elasticity — E__________ (P.S.I.)

Coefficient of Restitution __________

Pile

Pile Type: ____________________

Length (in Leads) — ____________________

Weight/ft. ____________________

Wall Thickness: __________ Taper: __________

Cross Sectional Area __________ in²

Design Pile Capacity: __________ (Tons)

Description of Splice: ____________________

Tip Treatment Description: ____________________

Distribution
One Copy Each To:

- ☐ State Bridge Engineer
- ☐ State Soils Engineer
- ☐ District Engineer
- ☐ Resident Engineer

Note: If mandrel is used to drive the pile, attach separate manufacturer's detail sheet(s) including weight and dimensions.

Submitted By: ____________________ Date: __________

Cushion.

   e. Thickness of the Hammer Cushion. The manufacturer usually provides this.
   f. Coefficient of Restitution for the Hammer Cushion. The manufacturer usually provides this.
   g. Round-Out Deformation for the Hammer Cushion.

4. Pile Cushion Information
   a. Area of the Pile Cushion. Generally, this is equal to the cross-sectional area of the pile head.
   b. Elastic Modulus of the Pile Cushion.
   c. Thickness of the Pile Cushion.
   d. Coefficient of Restitution for the Pile Cushion.
   e. Round-Out Deformation for the Pile Cushion.
5. Pile Head Properties
   a. Total Pile Length above and below the ground surface.
   b. Cross-Sectional Area at Pile Head.
   c. Elastic Modulus of Pile Head Material.
   d. Specific Weight of Pile Head Material.
   e. Coefficient of Restitution of Pile Head (usually default)
   f. Round-out deformation of Pile Head (usually default)
   g. Number of Pile Segments.
   h. Uniform Pile Option. For piles that have a constant cross-sectional area, this option can be selected. If not, then the properties (a) through (d) of each different cross section of the pile must be input, along with the number of segments for each cross-section. In the case of tapered piles, the pile is divided into segments of decreasing cross sectional area and the pile is treated as a "Step-Taper®" pile.
   i. Pile Damping (usually default). For numerical stability, some material dampening is introduced into the pile, although engineering materials invariably have some material dampening in any case.
6. Pile Segment Information.
7. Hammer Information. Most common wave equation programs have hammer libraries that obviate the need for manual input of this information. One simply selects the hammer desired and the information is passed to the program. The hammer's estimated mechanical efficiency may be included in this data.
   a. Hammer Override Values. In the case of a hammer not in the database, or with special features or modifications, the following values for hammer properties are of interest:
   b. Stroke Options (primarily for diesel hammers; for WEAP, the default option varies the stroke until it matches the pile response.)
   c. Hammer Stroke.
   d. Hammer Efficiency.
   e. Hammer Fuel Setting (for diesel hammers.)
   f. Reaction Weight.
   g. Combustion Delay or Ignition Start Volume (for diesel hammers.)
   h. Hammer Dampening.
8. Soil Parameters
   a. Selection of dampening type (Smith, Case or other dampening).
   b. Shaft soil quake. Quake is the distance a segment of the pile can move relative to the soil and still experience elastic soil response. Beyond this, the soil becomes plastic and permanent set of the pile becomes possible.
   c. Toe soil quake.
   d. Shaft soil dampening. This is primarily the ability of the soil to dissipate the

pile's stress wave into the soil mass though radiation, and secondarily dampening due to fluid movement.

e. Toe soil dampening.

f. Shaft Friction Distribution. The shaft friction of a pile can be distributed in a number of ways along the length of the pile, and can begin either at the pile head or at any point below. Depending upon the program being used, this data can be input using simplified distributions, from SPT data or from programs that analyze the ultimate static capacity of the pile. In addition, the portion of the pile resistance that is mobilized at the pile toe is input at this point.

9. Splice/Slack Options. For piles that are spliced, during driving the splices can momentarily separate. This option allows the user to place the splices at the required locations. It is necessary to properly set up the pile segments so the splices appear at segment boundaries.

10. Ultimate resistances. This can be one or more pile resistances to be analyzed. Usually, more than one is used to construct a "bearing graph," which shows the relationship between the resistance and the blow count. Newer wave equation programs have the ability to include set-up and relaxation factors in the soil. These also have the ability to analyze drivability by checking the blow count at multiple depths; in some cases these programs can estimate driving time as well.

11. Output Options. Shows the various output options for the user to view, both tabular and graphical.

4.8. Drivability and Hammer Selection: Vibratory Hammers

High installation rates can be achieved with vibratory hammers under certain conditions. This makes them attractive for installing piles. The influence of the type of vibratory hammer was discussed in Chapter 3. The influence of soil type is as follows:

1) Loose, wet, granular soils (including gravels): Vibratory hammers are extremely effective in this type of soil.

2) Soft and low plasticity clays: Vibratory hammers generally work well.

3) High plasticity clays, hardpan, and decomposed rock: Vibratory hammers perform poorly in these types of soil because of their limited "chopping" effect (as compared to an impact hammer). If use of a vibratory hammer is desired in such soils, a heavier model with a large amplitude of oscillation should be specified.

For vibratory drivability studies, the first thought is to employ the wave equation as is done with impact hammers. Both GRLWEAP and TNOWAVE have been adapted to analyze the drivability of piles using a vibratory hammer. However, the use of the wave equation for vibratory pile drivability is restricted by two factors:

1. The shortness of piles relative to the vibration frequency does not induce wave propagation in most piles during vibratory driving.[49]
2. The interaction of the pile and soil is not as well quantified as it is with impact hammers.

This has led to the development of a large variety of methods for the selection of vibratory hammers. These can be broken down in the several categories:

- Rule of thumb:
    - 0'-30' piling - small vibratory hammer (eccentric moment approximately 1200 in-lbs., frequency 1200-1600 RPM)
    - 30'-50' piling - medium vibratory hammer (eccentric moment approximately 2400 in-lbs., frequency 1200-1600 RPM)
    - Over 50' piling - large vibratory hammer (eccentric moment approximately 4800 in-lbs., frequency 1200-1600 RPM)
- Scientific:
    - Add the weight of the piling to the vibratory mass in the formula for amplitude and if the resultant is less

than 1/4 inch, select the next larger vibratory hammer. Such a method follows below.

- Experience:
    - What has worked in previous situations with similar soil conditions and piling lengths.

Whatever the method, soil conditions have a greater affect on the size requirement than pile length. Is it granular? What is the moisture content? What are the "blow counts" or "N" values on a standard penetration test? Examine the boring logs - if the material is loose or medium coarse sand with some moisture, driving should be easy and a smaller vibratory hammer is required. On the other hand, if the material is very stiff or hard clay, driving will usually be difficult. Clay doesn't necessarily mean that a vibratory hammer will be effective. Where there is sufficient moisture content, plasticity is low, and the vibratory hammer is large enough. Here amplitude is important; a very satisfactory result can be achieved with a vibratory driver.

The following is a method that is a composite of others developed over the years. It is based on both experience and theory. Even so, it should be kept in mind that method presented here is at best a very sophisticated "rule of thumb" and should be supplemented by local experience with both actual piles and soils and good engineering judgment. One item not considered here is the static weight of the system this can be increased by mounting bias weights on top of the vibratory hammer. This can increase the speed of pile penetration.

For a vibratory hammer to be suitable for a particular application, the required dynamic force can be computed by the equation

Equation 4-74:

$$F_{dyn} = \frac{\aleph}{\psi}\left(\beta_o R_{so} + \beta_i R_{si} + \beta_t R_t\right)$$

Where

- F_{dyn} = Dynamic Force of Vibrator, kN or kips
- β = Beta Factor for Soil Resistance (general)
- β_i = Beta Factor for Soil Resistance (outside shaft)
- β_o = Beta Factor for Soil Resistance (inside shaft)
- β_t = Beta Factor for Soil Resistance (toe)
- R_{si} = Inside Pile Shaft Soil Resistance, kN or kips
- R_{so} = Outside Pile Shaft Soil Resistance, kN or kips
- R_t = Pile Toe Soil Resistance, kN or kips.
- ψ = Pile Factor (0.8 for concrete piling and 1 for all other piling.)
- $\aleph$ = Soil Resilience Coefficient (should be between 0.6 and 0.8 for vibration frequencies between 5 and 10 Hz and 1 for all other frequencies.)

Suggested values for β are given in Table 4-14.

Table 4-14 Values of β_n [50]

Type of Soil	β
Round Coarse Sand	0.10
Soft Loam/Marl, Soft Loess, Stiff Cliff	0.12
Round Medium Sand, Round Gravel	0.15
Fine Angular Gravel, Angular Loam, Angular Loess	0.18
Round Fine Sand	0.20
Angular Sand, Coarse Gravel	0.25
Angular/Dry Fine Sand	0.35
Marl, Stiff/Very Stiff Clay	0.40

The toe resistance and the outside and inside (where applicable with open ended pipe and cylinder pile) shaft resistance should be computed using methods similar to those employed for impact hammers. For extraction, this formula is altered to read

Equation 4-75:

$$F_{dyn} = \frac{\aleph}{\psi}\left(\beta_o R_{so} + \beta_i R_{si} + \beta_t R_t\right) - F_{ext}$$

Where

- F_{ext} = Extraction Force of Crane, kN or kips.

Once this is known, a possible vibratory hammer for the project can be selected based on minimum permissible dynamic force in kN or kips. The parameter of dynamic mass (the dynamic mass includes any mass of the vibrator not dampened from vibration, the clamp and any mass of the pile) should be noted, along with the frequency and eccentric moment of the machine.

Next the basic parameters of the vibratory hammer-pile system must be checked. The first is the peak acceleration, whose value is computed using the equation

Equation 4-76:

$$n = 102\frac{F_{dyn}}{M_{dyn}}$$

(SI Units)

Equation 4-77:

$$n = \frac{F_{dyn}}{W_{dyn}}$$

(English Units)

Where

- n = peak acceleration, g's.
- M_{dyn} = Dynamic Mass of System, kg
- W_{dyn} = Dynamic Weight of System, kips.

Minimum values for this acceleration are given in Table 4-15.

Table 4-15 Minimum Peak Acceleration Values

Type of Pile	Minimum Peak Acceleration n, g's
Steel Sheet Piling H-piles	9
Open Ended Pipe Piles Caissons Closed Ended Pipe Piles Heavy Wall Pipe Piles	5
Concrete and Wood Piles	3

If the peak acceleration figure is too low, a vibratory hammer with a larger dynamic force must be chosen and the peak acceleration recomputed. This iteration must continue until the minimum peak acceleration value is achieved.

Next, the required amplitude of the vibratory system is computed using the equation

Equation 4-78:

$$A = \frac{2000K\psi}{M_{dyn}}$$

(SI Units)

Equation 4-79:

$$A = \frac{2K\psi}{W_{dyn}}$$

(English Units)

Where

- A = Total Cycle Displacement Amplitude, mm or inches
- K = Eccentric moment, kg-m or in-lbs

The recommended amplitude values are shown in Table 4-16 for SI units and Table 4-17 for U.S. units. Combinations of pile, soil and frequency where no value is provided are cases where the pile should not be vibrated at the given frequency range and soil condition. Larger amplitudes than those shown are generally permissible.

Table 4-16 Amplitude Requirements, SI Units

Type of Pile and Soil	Amplitude, mm, for various frequency ranges		
	5-12 Hz	12-17 Hz	17-27 Hz
Steel Sheet Piling, Open Ended Pipe Piles, H-Piles and Other Piles with A_t < 150 cm²			
Sandy Soils		16-20	8-12
Clayey Soil		20-24	12-16
Closed End Steel Pipe Piles, A_t < 800 cm²			
Sandy Soil		20-24	12-16
Clayey Soil		24-30	16-20
Reinforced Concrete Piles, Square or Rectangular Section, A_t < 2000 cm²			
Sandy Soil	24-30		
Clayey Soil	30-40		
Reinforced Concrete Cylinder Piles of Large Diameter, Driven with Soil Plug Removed			
Sandy Soil	12-20	8-12	
Clayey Soil	16-24	12-20	

Table 4-17 Amplitude Requirements, U.S. Units

Type of Pile and Soil	Amplitude, inches, for various frequency ranges		
	300-720 RPM	720-1020 RPM	1020-1620 RPM
Steel Sheet Piling, Open Ended Pipe Piles, H-Piles, and Other Piles lighter than 80 lb/ft			
Sandy Soils		0.63-0.80	0.32-0.50
Clayey Soil		0.80-0.93	0.50-0.63
Closed End Steel Pipe Piles lighter than 420 lb/ft			
Sandy Soil		0.80-0.95	0.50-0.63
Clayey Soil		0.95-1.18	0.63-0.80
Reinforced Concrete Piles, Square or Rectangular Section, Cross Sectional Area less than 310 in²			
Sandy Soil	0.95-1.18		
Clayey Soil	1.18-1.57		
Reinforced Concrete Cylinder Piles of Large Diameter, Driven with Soil Plug Removed			
Sandy Soil	0.50-0.80	0.32-0.50	
Clayey Soil	0.63-0.95	0.50-0.80	

If the amplitude is insufficient, then a new vibratory hammer with the same or greater dynamic force as the previous one and greater eccentric moment should be chosen and the amplitude checked again.

Finally, the peak velocity should be checked. It is computed by the equation

Equation 4-80:

$$v_{dyn} = \frac{1.561n}{\theta}$$

(SI Units)

Equation 4-81:

$$v_{dyn} = \frac{307.24n}{RPM}$$

(English Units)

where

- v_{dyn} = peak dynamic velocity of vibrating system, m/sec or ft/sec. This should be between 0.5-0.8 m/sec (1.7-2.6 ft/sec), but it can be higher if necessary.
- θ = frequency of vibrations, Hz
- RPM = frequency of vibrations, rev/min

4.9. Sample Problem

The methods used in this sample problem are the same as described earlier in the text. When step-by-step descriptions are used, the steps correspond exactly.

4.9.1. Statement of the Sample Problem

The Bridge Division has estimated that the maximum design load to be imposed on a single pile in both pile groups is 120 kips (534 kN). It is also estimated that the total design load on each pile group is 1700 kips[51]. All of the piles are 12" (304.8 mm) x 12" (304.8 mm) square prestressed concrete piles, having an estimated trial length (based on an interpretation of the subsurface profile) of 30' (9.1 m) at the north abutment and 65' (19.8 m) on the south abutment. For the concrete piles, f'_c = 7000 psi and f_{pe} = 700 psi. The preliminary plan is shown in Figure 4-47, and the preliminary profile is shown in Figure 4-48[52].

Figure 4-47 Preliminary Plan, Sample Problem

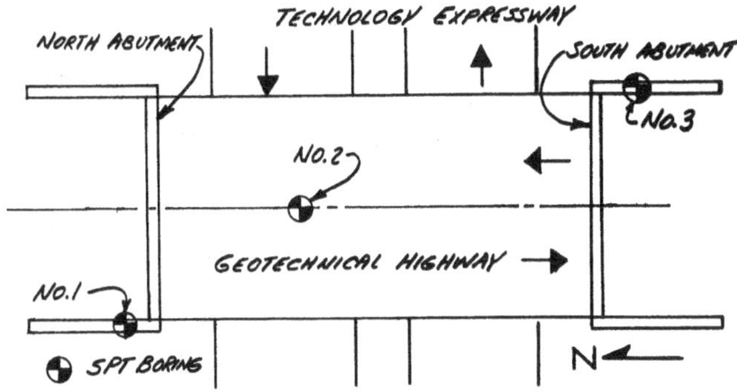

Figure 4-48 Preliminary Profile, Sample Problem

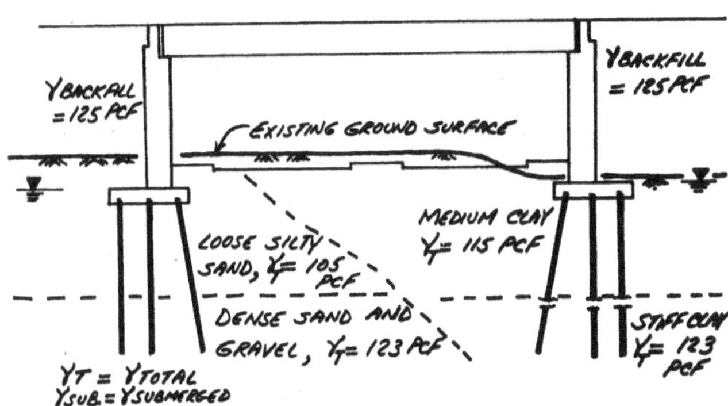

The pile foundation layout is shown in Figure 4-49.

Figure 4-49 Pile Foundation Layout, North and South Abutments

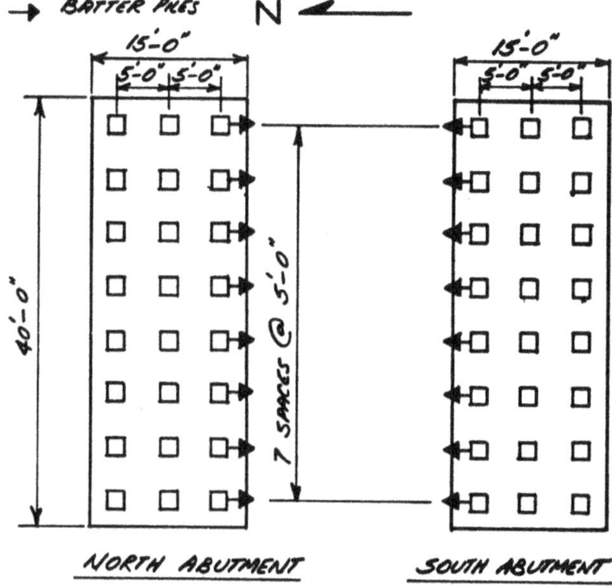

The boring logs for the north and south abutments are

shown in Figure 4-50 and Figure 4-51 respectively.

Figure 4-50 Log of Boring No. 1, North Abutment

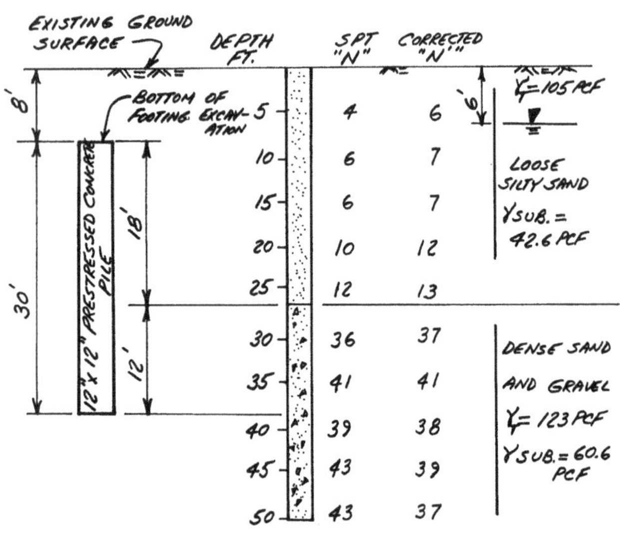

Figure 4-51 Log of Boring No. 3, South Abutment

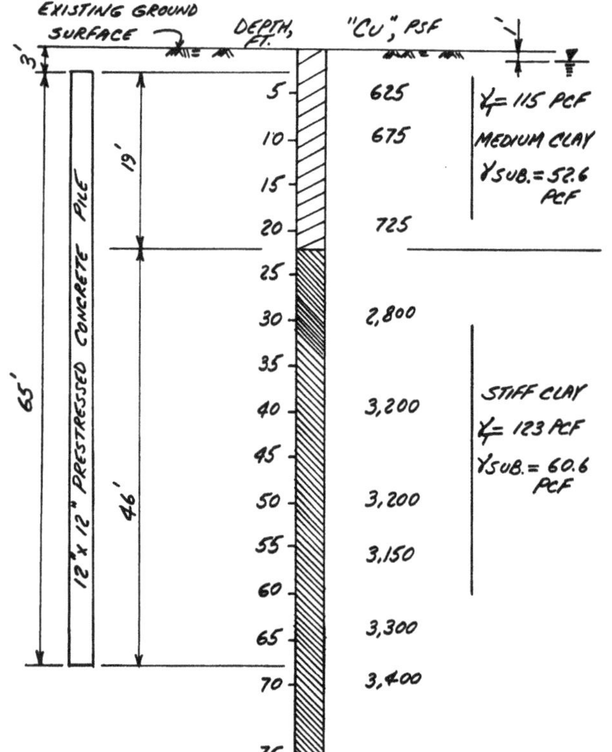

The cone penetrometer data for the north abutment is shown in Figure 4-52.

Figure 4-52 CPT Data for North Abutment

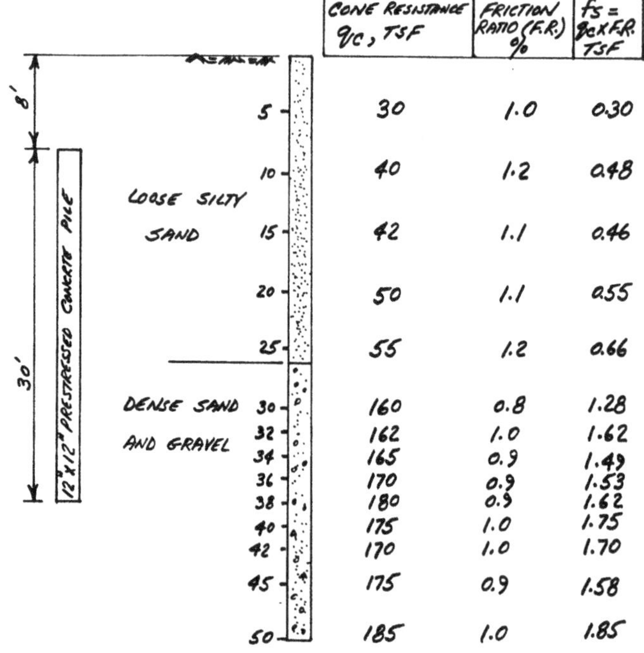

The goal is:

- Determine the ultimate axial, single pile capacities and compare these results to the design loads.
- Determine pile group loads and settlements.
- Analyze lateral squeeze of the abutment.
- Analyze negative shaft friction and propose a solution, if necessary.
- Perform dynamic analysis (both dynamic formulae and wave equation.)

4.9.2. Structural Capacity

Using the formula in Table 4-7, the allowable axial compressive stress in the pile is

$$\sigma_{max} = \frac{7000}{3} - (0.27)(700) = 2144 \text{ psi}$$

The maximum allowable compressive load in the pile is thus

Equation 4-82:

$$P_{allow} = \sigma_{max} A_{pile} = (2144)(12)(12) = 308,784 \text{ pounds} = 309 \text{ kips} > 120 \text{ kips}$$

Where

- P_{allow} = allowable axial load for structural

purposes

- σ_{max} = maximum allowable stress in total cross-sectional area of pile
- A_{pile} = cross-sectional area of pile

We note that we are considering only an axial compressive load here. Combination loads with bending loads are beyond the scope of this book.

4.9.3. Axial Capacity for North Abutment Piles (Cohesionless Soils)

4.9.3.1. SPT Method

1) Correct SPT "N" values. The effective overburden pressure (P_o) is needed to correct these values. Figure 4-53 shows this correction process.

Figure 4-53 Corrections of "N" Values for Overburden Pressure

THICKNESS	γ (UNIT WT.)	PRESSURE	CUMULATIVE PRESSURE (P_o)
6 FT.	105 PCF	630 PSF	630 PSF
20 FT.	105-62.4=42.6	852	1482
24 FT.	123-62.4=60.6	1454	2936

DEPTH (FT.)	P_o (T.S.F.)	FIELD SPT "N"	CORRECTION FACTOR	CORRECTED "N'" (FIELD SPT "N" X CORR. FACTOR)
5	0.26	4	1.40	6
10	0.40	6	1.22	7
15	0.51	6	1.20	7
20	0.61	10	1.15	12
25	0.72	12	1.10	13
30	0.86	36	1.04	37
35	1.01	41	1.00	41
40	1.17	39	0.97	38
45	1.32	43	0.90	39
50	1.47	43	0.85	37

2) Compute average SPT corrected "N'" near pile toe, within 10' below estimated toe:

$$(38+39+37)/3 = 38$$

3) Compute ultimate unit pile toe resistance:

$$q_p = 8N' = (8)(38) = 304 \text{ ksf}$$

4) Compute ultimate pile toe bearing capacity:

$$Q_p = (304)(1.0')(1.0') = 304 \text{ kips}$$

5) Compute average corrected SPT $\overline{N'}$ value within embedded pile length:

   a. Average $\overline{N'}$ for loose silty sand layer (8' to 26'):

   $$(7+7+12+13)/4 = 10$$

   b. Average $\overline{N'}$ for dense sand and gravel layer (26' to 38'):

   $$(37+41)/2 = 39$$

6) Compute ultimate unit shaft friction resistance:

   a. Loose silty sand layer:

   $$f_s = 10/25 = 0.4 \text{ ksf}$$

   b. Dense sand and gravel layer:

   $$f_s = 39/25 = 1.56 \text{ ksf}$$

Neither of these values exceeds the 2 ksf limits of the equations.

7) Compute ultimate shaft friction capacity:

   a. Loose silty sand layer:

   $$Q_{s1} = (0.4)(4.0')(18.0') = 28.8 \text{ kips}$$

   b. Dense sand and gravel layer:

   $$Q_{s2} = (1.56)(4')(12') = 74.88 \text{ kips}$$

   c. Total shaft friction:

   $$Q_s = 28.8 + 74.88 = 103.68 \text{ kips}$$

8) Compute total ultimate pile load capacity:

$$Q_u = 304 + 103.68 = 407.68 \text{ kips} \approx 408 \text{ kips}$$

9) Compute allowable pile load capacity:

$$Q_{allowable} = 407.68/4 = 101.92 \approx 102 \text{ kips}$$

Which is below the design capacity of 120 kips (534 kN). Normally, this would require the lengthening of the pile; however, since the SPT Method is preliminary, and we plan to use two other methods of analyses, we will proceed to analyze the pile using these methods before determining whether the piles need to be lengthened or not. At this point we would normally perform a group capacity analysis, but we will defer this to the other static methods as well.

4.9.3.2. CPT Method

The steps here correspond to the steps above. We first need to plot the CPT data from Figure 4-52. Such a plot is shown in Figure 4-54. Note that the values for q_c in Figure 4-52 and Figure 4-54 are in tons per square foot but our computations are in kips per square foot.

Figure 4-54 Plot of CPT Data

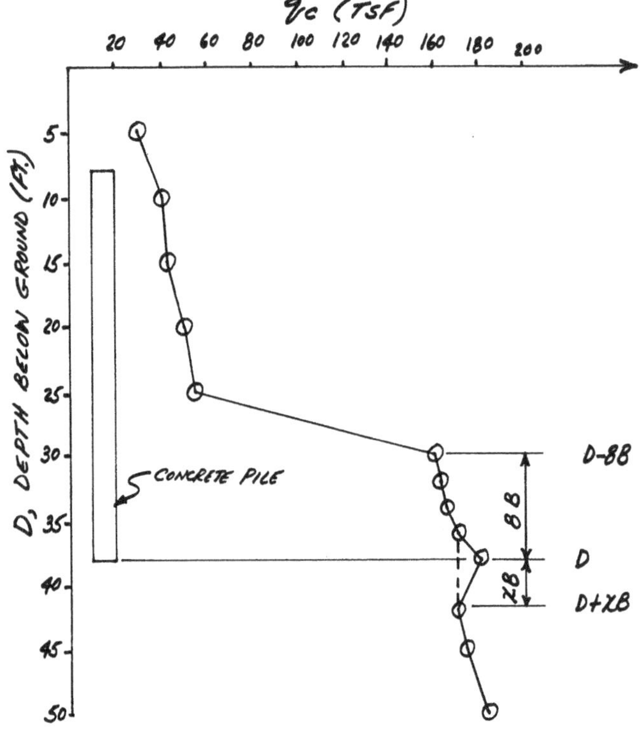

1) Determine unit pile toe resistance.

   a. Compute q_{c1} for x-values from 0.7B to 3.75B below the pile toe and use the lowest value. Computations are shown below.

X	x_B (feet)	D[53]+x_B (feet)	q_{c1} Computation[54]	q_{c1} (ksf)
0.7	0.7	38+0.7=38.7	2(180)/1	360
3.0	3.0	41.0	2(180 + 170 + 170)/3	347
3.75	3.75	41.75	2(180 + 180 + 170)/3	347
			Recommended q_{c1}	347

   b. Compute q_{c2} over a distance 8 times the pile width above the pile toe (8 x 1' = 8' above pile toe.)

$$q_{c2} = (2)(170+170+165+162+160)/5 = 330.8 \text{ ksf}$$

   c. Compute q_p.

$$q_p = (347+330.8)/2 = 339 \text{ ksf}$$

2) Compute total ultimate toe resistance.

$$Q_p = (339)(1.0')(1.0') = 339 \text{ kips}$$

3) Compute ultimate shaft resistance.

   a. Use Figure 4-14 to determine K. The D/B ratio = 30/1 = 30. For square concrete piles with this ratio, K=0.45.

   b. Compute the shaft contact areas for the two regions of the pile. The area for each is the perimeter times the embedded length. The first is from 0 to 8' below the pile head; the area of this is (8')(4') = 32 ft^2. The second is below 8' and to 30' for this pile; the area of this embedment is (30'-8')(4') = 88 ft^2.

   c. Compute $\bar{f}_s$:

$$\bar{f}_s\Big|_{0 to 8'} = \frac{2(0.48+0.46)}{2} = 0.94 \text{ ksf}$$

$$\bar{f}_s\Big|_{8' to 30'} = \frac{2(0.55+0.66+1.28+1.62+1.49+1.53+1.62)}{7} = 2.5 \text{ ksf} > 2 \text{ ksf}_{limit}$$

Since the value below 8' exceeds the limit, the limiting value of 2 ksf will be used in this region.

4) Compute the ultimate shaft resistance.

$$Q_S = 0.45\left[\frac{(0.94 \times 32)}{2} + (2 \times 88)\right] = 86 \text{ kips}$$

5) Determine total ultimate load capacity.

$$Q_u = 339 + 86 = 425 \text{ kips}$$

6) Determine allowable pile load capacity.

$$Q_{allowable} = 425/3 = 142 \text{ kips}$$

This exceeds the design capacity of 120 kips (534 kN).

7) Compute the allowable pile group capacity. In this case, piles are spaced at more than two times the pile diameter. Because the dense sand and gravel strata is not underlain by a weak stratum, there is no possibility of block failure; therefore, the allowable group capacity is taken as the sum of the single pile capacities, or (in this case) the product of the allowable single pile capacity times the number of piles in the group.

$$Q_{allowable\text{-}group} = (24)(142) = 3408 \text{ kips}$$

The maximum design load to be imposed on the pile group is 1700 kips. Therefore, the total number of piles provided in the group is more than sufficient. The number of piles in the middle and rear rows may be reduced.

4.9.3.3. Nordlund Method

Applying the step by step process that was previously discussed:

1) Determine ϕ value from SPT data.
   a. We have already corrected the "N" values for the overburden pressure (see Page 40.)
   b. Determine average corrected values:
      i. For loose silty sand:

$$\overline{N'} = \frac{7+7+12+13}{4} = 10$$

      ii. For dense sand and gravel:

$$\overline{N'} = \frac{37+41}{2} = 39$$

   c. Use Table 4-4 to estimate ϕ values. For loose silty sands, $\overline{N'}$ =10, thus ϕ = 30°. For dense sand and gravel, $\overline{N'}$ =39, thus ϕ = 40°
2) Determine K_δ.
   a. Compute volume of displaced soil per unit length. V = (1.0)(1.0)(1.0) = 1.0 ft^3.
      i. For loose silty sands, use Figure 4-16 with V = 1.0 and w = 0. K_δ=1.15.
      ii. For dense sand and gravel, use Figure 4-16 with V = 1.0 and w = 0. K_δ =3.0.
3) Determine δ angle.
   a. Use Figure 4-17, V = 1.0 and curve C for precast concrete piles:

$$\delta/\phi = 0.77$$

      i. Loose silty sand layer: δ = (30°)(0.77) = 23°
      ii. Dense sand and gravel layer: δ = (40°)(0.77) = 30.8°
4) Determine correction factor for C_f. This is necessary because $\delta \neq \phi$. Use Figure 4-18.
      i. Loose silty sand layer: c_f = 0.92 for ϕ = 30° and δ/ϕ = 0.77.
      ii. Dense sand and gravel layer: c_f = 0.86 for ϕ = 40° and δ/ϕ = 0.77.
5) Compute average effective overburden stress P_d. Compute P_d at the midpoint of each soil layer from the P_o diagram previously developed.
      i. Loose silty sand layer: P_d at midpoint of pile embedment (17' below existing ground) = 1099 psf = 1.099 ksf.
      ii. Dense sand and gravel layer: P_d at midpoint of pile embedment (32' below existing ground) = 1845 psf = 1.845 ksf.
6) Compute ultimate shaft friction capacity. C_d = pile perimeter = 4'
   a. Loose silty sand layer:

$$Q_{s1} = (1.15)(0.92)(1.099((\sin 23°)(4.0)(18.0) = 32.7 \text{ kips}$$

   b. Dense sand and gravel layer:

$$Q_{s2} = (3.0)(0.86)(1.845)(\sin 30.8°)(4.0)(12.0) = 117 \text{ kips}$$

   c. Total:

$$Q_s = Q_{s1} + Q_{s2} = 32.7 + 117 = 149.7 \text{ kips}$$

7) Determine bearing capacity factor using Figure 4-19 and ϕ = 40° (pile toe is in dense sand and gravel): N_q' = 160.
8) Determine α factor, using Figure 4-19 and ϕ = 40°, α = 0.75.
9) Compute effective overburden stress P_d at pile toe. Use previously defined P_o diagram.

P_d at 38' below the existing ground surface
P_d = 2209 psf = 2.209 ksf.

10) Compute ultimate pile toe resistance.

$$Q_p = (0.75)(160)(1.0')(1.0')(2.209) = 265 \text{ kips}$$

a. Average corrected SPT "N" value within 10' below estimated pile toe is (38+39+37)/3 = 38. From Table 4-4, Φ = 40° for this N' value. From Figure 4-12, $q_{limiting}$ = 200 tsf = 400 ksf for this case.

$$Q_{p\text{-}limiting} = (400)(1.0')(1.0') = 400 \text{ kips}$$

Use Q_p = 265 kips (smaller of the two values obtained.)

11) Compute total ultimate pile load capacity:

$$Q_u = 149.7+265=414.7 \approx 415 \text{ kips}$$

12) Compute allowable pile load capacity:

$$Q_{allowable} = 415/3 = 138 \text{ kips}$$

This exceeds the design capacity of the pile of 120 kips (534 kN).

13) Compute allowable pile group load capacity:

$$Q_{allowable\text{-}group} = (24)(138) = 3312 \text{ kips}$$

The comments from the CPT method apply here also.

4.9.3.4. Summary for Ultimate Geotechnical Capacity in Cohesionless Soils

Table 4-18 summarizes the capacity results of the three methods. Even though the computed ultimate pile capacity by the method based on SPT test data is close to those computed by other methods, it is emphasized that this method is not as reliable as the other two methods. It is not appropriate to draw any definitive conclusions from these results. Relative results between design methods will vary depending upon pile type, pile dimensions and soil density.

Table 4-18 Summary of Sample Problem Results for Ultimate Capacity in Cohesionless Soils

Method Used for Estimation of Pile Capacity	Estimated Ultimate Static Pile Capacity, kips	Factor of Safety Used for Method	Estimated Allowable Static Pile Capacity, kips
Method based on SPT Test Data	408	4	102
Method based on CPT Test Data	425	3	142
Nordlund Method	415	3	138

4.9.4. Axial Capacity for the South Abutment Piles (Cohesive Soils)

4.9.4.1. "α" Method

The steps are the same as the step-by-step outline.

1. Determine adhesion factor α from Figure 4-20.

   a. For pile length in the medium clay (19' embedment), L/B = ratio of pile length to pile width = 19.0/1.0 = 19. The average undrained shear strength of medium clay:

$$c_u = (625 + 675 + 725)/3 = 675 \text{ psf} = 0.675 \text{ ksf}$$

   From Figure 4-20 (c), α = 1.0.

   b. For pile length in the stiff clay (46' embedment), L/B = 46/1 = 46. The average undrained shear strength of the stiff clay:

$$c_u = (2800 + 3200 + 3200 + 3150 + 3300 + 3400)/6 = 3175 \text{ psf} = 3.175 \text{ ksf}$$

   From Figure 4-20 (b), α = 0.70.

2. Compute ultimate unit shaft friction resistance:

   a. Medium clay: f_{s1} = (1.0)(0.675) = 0.675 ksf

   b. Stiff clay: f_{s2} = (0.70)(3.175) = 2.223 ksf

3. Compute ultimate shaft friction resistance:

   a. Medium clay: Q_{s1} = (0.675)(4.0 x 19.0) = 51.3 kips

   b. Stiff clay: Q_{s2} = (2.223)(4.0 x 46.0) = 409.032 kips

c. Total:

$Q_{s\text{-}ultimate}$ = **51.3 + 409.032 = 460.332 ≈ 460 kips**

4. Compute ultimate toe bearing capacity:

Q_p = **(9)(3400)(1.0) = 30,600 lbs. = 30.6 kips**

5. Compute total ultimate pile load capacity:

Q_u = **460 + 30.6 = 490.6 ≈ 491 kips**

6. Compute allowable pile load capacity:

$Q_{allowable}$ = **491/3 = 163.3 ≈ 163 kips**

7. Compute allowable pile group load capacity (see 4.6.3.6.) Since pile spacing is greater than four times the pile width and the major portion of the embedded length is in stiff clay (c_u > 2 ksf) use a group efficiency equal to 100%.

$Q_{allowable\text{-}group}$ = **(24)(163) = 3912 kips**

8. Investigate the possibility of a block failure. Using Equation 4-38:

Q_u = **(9)(34)(15)(40) + (2)(19)(0.675)(15+40) - (2)(46)(3.175)(15 + 40) = 35,836 kips**

Keep in mind that the pile group is 15' (4.6 m) wide and 40' (12 m) long. Since Q_u against block failure is larger than the computed ultimate pile group load capacity (3912 x 3 = 11,736 kips), block failure cannot occur.

4.9.4.2. "β" Method

1. Compute average effective overburden pressure, p_o for the pile embedment depth. Neglect the effect of the footing excavation. The plot of this is shown in Figure 4-55.

Figure 4-55 Plot of Overburden Pressure

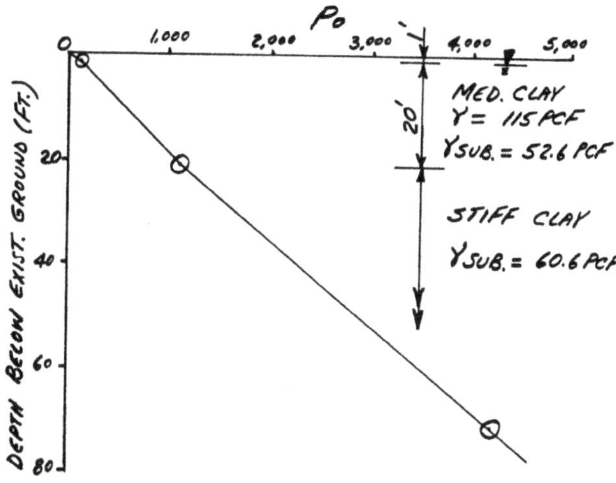

a. For Pile Length in Medium Clay (19' embedment): Midpoint of embedded pile length is 12.5' below existing ground. The average effective overburden pressure at this depth P_o = 720 psf = 0.72 ksf.

b. For Pile Length in Stiff Clay (46' embedment): Midpoint of embedded pile length is 45' below existing ground. The average effective overburden pressure at this depth P_o = 2621 psf = 2.621 ksf.

2. Determine β value.

a. Medium Clay: Use Figure 4-21. Since c_u < 2 ksf and pile embedded length < 50' (15.2 m), β = 0.3.

b. Stiff Clay. Use Figure 4-22 since c_u > 2 ksf. Since the average c_u = 3.175 ksf (determined using α method) and using plotted values of the open circles for cylindrical piles, a conservative β value of 0.75 can be obtained[55].

3. Compute ultimate unit shaft friction resistance.

a. Medium Clay: f_s = (0.3)(0.720) = 0.216 ksf.

b. Stiff Clay: f_s = (0.75)(2.621) = 1.966 ksf.

4. Compute total ultimate shaft friction resistance.

a. Medium Clay: Q_{s1} = (0.216)(4.0)(19.0) = 16.4 kips.

b. Stiff Clay: Q_{s2} = (1.966)(4.0)(46.0) = 361.7

kips.

c. Total: $Q_s = 16.4 + 361.7 = 378.1 \approx 378$ kips.

5. Compute ultimate toe bearing capacity. This is the same as before:

$$Q_p = (9)(3400)(1.0) = 30{,}600 \text{ lbs.} = 30.6 \text{ kips}$$

6. Compute ultimate pile load capacity: 378 + 30.6 = 408.6 ≈ 409 kips.
7. Compute allowable pile load capacity:

$$Q_{allowable} = 409/3 = 136.3 \approx 136 \text{ kips}$$[56]

8. Compute allowable pile group capacity. As previously discussed, the group efficiency here is 100%.

$$Q_{allowable\text{-}group} = (24)(136) = 3{,}264 \text{ kips}$$

9. Investigate the possibility of a block failure. The result is the same as with the α method.

4.9.4.3. Summary for Geotechnical Capacity in Cohesive Soils

A summary of the results of the two methods is given in Table 4-19.

Table 4-19 Summary of Sample Problem Results for Geotechnical Capacity in Cohesive Soils

Method Used for Estimation of Pile Capacity	Estimated Ultimate Static Pile Capacity, kips	Estimate Allowable Static Pile Capacity, kips (FS = 3)
α (Tomlinson) Method	491	164
β Method	409	136

4.9.5. Pile Group Settlement

4.9.5.1. SPT Method (North Abutment)

1. Compute unit foundation pressure (design pressure) imposed by the piles at their toes[57]. Keep in mind that the design load of the pile group is 1700 kips[58]. Pile group area = (36')(11') = 396 ft^2.[59] Net foundation pressure 1700/396 = 4.3 ksf.

2. Determine average corrected SPT $\overline{N'}$ value below pile toes.

$$\overline{N'} = (38 + 39)/2 = 38$$

3. Determine influence factor I for pile embedment depth D = 30'.

$$I = 1 - (30)/((8)(11)) = 0.66$$

4. Determine estimated settlement for the pile group.

$$S = (4.3)(0.66)(\sqrt{(11)})/38 = 0.25''$$

5. Determine elastic compression of pile material under design load on each pile (120 kips (534 kN).) Because the piles are friction piles, the pile axial load P varies along the length. For this reason, compute the average axial load in each pile segment as shown in Figure 4-56. The pile cross sectional area = 12" x 12" = 144 in^2. The pile modulus of elasticity is 3,000 ksi.

Figure 4-56 Pile Axial Load Diagram, Cohesionless Soils

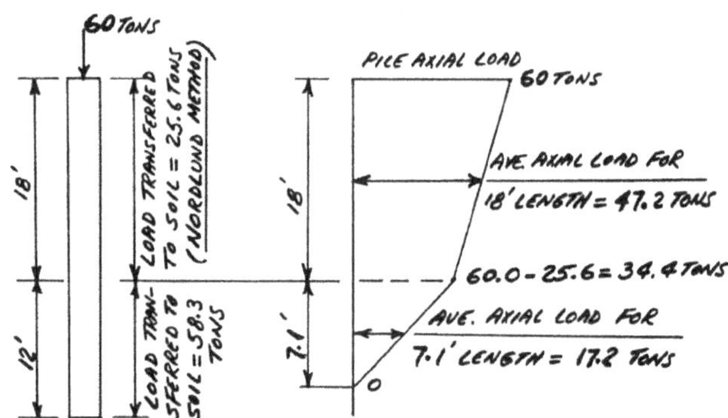

$$\delta = (47.2)(2)(18)(12)/((144)(3000)) + (17.2)(2)(7.1)(12)/((144)(3000)) = 0.0540''$$[60]

6. Compute total group settlement, which is the sum of the elastic compression of the pile and the compression of the soil

$$\delta_{total} = 0.25 + 0.0540 = 0.304''$$

4.9.5.2. CPT Method (North Abutment)

1. Compute unit foundation pressure imposed by the piles at the pile toes = 4.3 ksf (as before.)
2. Determine average static cone toe resistance within a depth equal to 11' (pile group width) below pile toes.

$$\overline{q_c} = 2((180 + 175 + 170 + 175 + 185)/5) = 354 \text{ ksf}$$

3. Determine influence factor, I = 0.66 (as

before.)

4. Determine estimated settlement for the pile group.

S = (4.3)(11.0)(0.66)/((2)(354)) = 0.044"

5. Determine elastic compression of the pile material under the design load for each pile = 0.054" (as before.)

6. Compute total group settlement.

δ_{total} = 0.044 + 0.054 = 0.098"[61]

4.9.5.3. Cohesive Soil (South Abutment)

1. Determine load imposed on soil by pile group.
    a. Use the method shown in Figure 4-26 to determine loads imposed at various depths. See Table 4-20.

Figure 4-57 Pile Settlement Diagram, Cohesive Soils

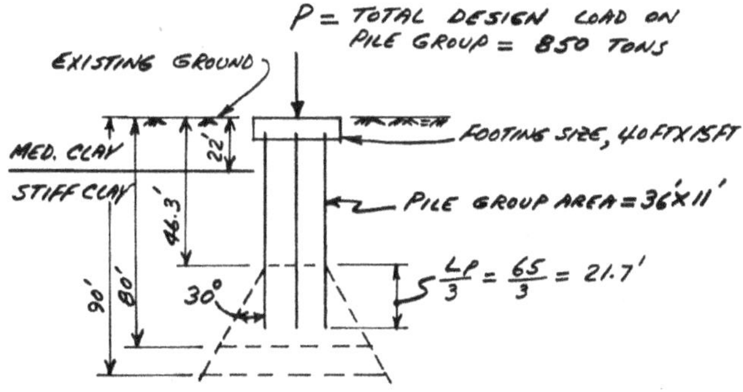

Table 4-20 shows that at 80' depth below the existing ground, the additional load imposed by the pile group is less than 10% of the existing overburden pressure at that depth. Therefore, the thickness of the clay layer to be considered for settlement computations is from 46.3' to 80' below the existing ground (layer thickness = 33.7')

    b. Divide the clay layer into 5' (1.52 m) thick layers as shown in Table 4-21.
    c. Determine existing overburden pressure P_o at the midpoint of each layer. Use Figure 4-55. See Table 4-21.
    d. Determine imposed pressure ΔP at the midpoint of each layer by using method described above. See Table 4-21.

2. Determine consolidation test parameters. The laboratory consolidation test data on the undisturbed stiff clay sample was plotted as "Log Pressure P vs. Void Ratio e" in a manner similar to Figure 4-28. The following data was obtained from the graph:
    a. p_c = preconsolidation pressure = p_o (normally consolidated soil)
    b. e_o = initial void ratio = 0.95
    c. c_c = compression index = 0.15
    d. c_{cr} = no recompression because soil is normally consolidated

3. Since c_{cr} =0, compute settlements using Equation 4-44. See Table 4-21 for results. Total pile group settlement from the table = 4.4".[62]

4. Determine elastic compression of the pile material under the design load. Because the piles are friction piles, the pile axial load P varies along its length. For this reason, compute pile average axial load in each pile segment as shown in Figure 4-58[63]. Pile cross-sectional area and pile material modulus of elasticity are the same as with cohesionless soils.

Figure 4-58 Pile Axial Load Diagram, Cohesive Soils

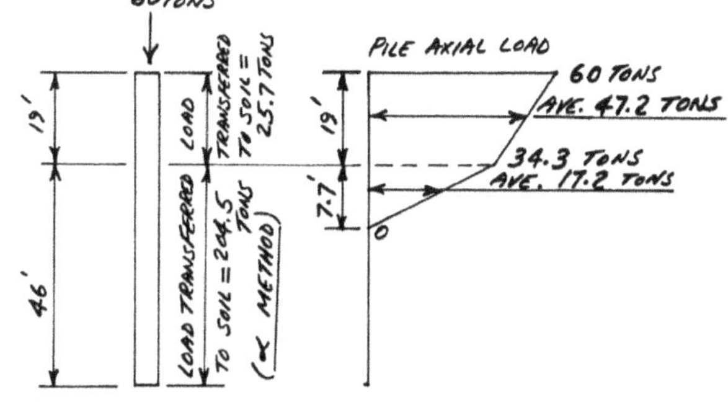

NOTE: FOR ELASTIC PILE COMPRESSION COMPUTIONS, IT IS ASSUMED THAT ALL PILES IN THE GROUP ARE LOADED WITH 60 TON DESIGN LOAD. THIS ASSUMPTION IS CONSERVATIVE BECAUSE PILES IN THE MIDDLE AND REAR ROWS WILL BE LOADED WITH SMALLER LOADS.

δ = (47.2)(2)(19)(12)/((144)(3000)) + (17.2)(2)(7.

Table 4-20 Table for Computing Imposed Loads at Various Depths

DEPTH BELOW EXIST. GROUND (FT.) z	WIDTH OF LOAD DISTRIBUTION AREA (FT.) B_z	LENGTH OF LOAD DISTRIBUTION AREA (FT.) L_z	LOAD DISTRIBUTION AREA (SQ.FT.) $B_z \times L_z$	IMPOSED DISTRIBUTED LOAD FROM PILE GROUP $P_z = \frac{850 \times 2000}{B_z \times L_z}$ (PSF)	EFFECTIVE OVER-BURDEN PRESSURE P_o* (PSF)	$\frac{P_z}{P_o} \times 100$	
48.8	12.9	37.9	489	3476	2844	122 %	** CLAY LAYER THICKNESS TO BE USED FOR SETTLEMENT COMPUTATIONS IS FROM 46.3' DEPTH TO 80' DEPTH (↕ 46.3' to 80')
53.8	18.7	43.7	817	2081	3147	66 %	
58.8	24.4	49.4	1205	1411	3450	41 %	
63.8	30.2	55.2	1667	1020	3753	27 %	
68.8	36.0	61.0	2196	774	4056	19 %	
73.8	41.8	66.7	2788	610	4359	14 %	
78.2	46.8	71.8	3360	506	4625	10.9 %	
78.8	47.5	72.5	3444	494	4662	10.6 %	
80.0	48.9	73.9	3614	470	4735	9.9 %	
83.8	53.3	78.3	4173	407	4965	8.2 %	

* P_o IS OBTAINED FROM THE P_o DIAGRAM DEVELOPED PREVIOUSLY FOR USING "β METHOD" FOR ESTIMATING PILE LOAD CAPACITY.

** BECAUSE THE LOAD IMPOSED BY THE PILE GROUP ON SOIL BELOW 80' DEPTH IS LESS THAN 10 % OF THE EXISTING P_o, SETTLEMENT COMPUTATIONS ARE PERFORMED TO 80' DEPTH.

Table 4-21 Settlement Computations Table

SOIL LAYER	LAYER THICKNESS (H)	DEPTH @ LAYER MIDPOINT #	P_o @ LAYER MIDPOINT ##	ΔP (P_z FROM PREVIOUS TABLE) ###	$\frac{P_o + \Delta P}{P_o}$	$\log \frac{P_o + \Delta P}{P_o}$	$\frac{C_c}{1+e_o}$	LAYER SETTLEMENT $H\left[\frac{C_c}{1+e_o} \log\left(\frac{P_o+\Delta P}{P_o}\right)\right]$
46.3' TO 51.3'	5 FT	48.8 FT	2844 PSF	3476 PSF	2.22	0.35	$\frac{0.15}{1+0.95} = 0.077$	5 × 0.077 × 0.35 = 0.13 FT
51.3' TO 56.3'	5 FT	53.8 FT	3147 PSF	2081 PSF	1.66	0.22	0.077	0.08 FT
56.3' TO 61.3'	5 FT	58.8 FT	3450 PSF	1411 PSF	1.41	0.15	"	0.06 FT
61.3' TO 66.3'	5 FT	63.8 FT	3753 PSF	1020 PSF	1.27	0.10	"	0.04 FT
66.3' TO 71.3'	5 FT	68.8 FT	4056 PSF	774 PSF	1.19	0.08	"	0.03 FT
71.3' TO 76.3'	5 FT	73.8 FT	4359 PSF	610 PSF	1.14	0.06	"	0.02 FT
76.3' TO 80.0'	3.7 FT	78.2 FT	4625 PSF	506 PSF	1.11	0.05	"	0.01 FT

TOTAL SETTLEMENT, $\sum S = 0.37' = 4.4''$

\# THESE VALUES ARE SHOWN AS DEPTHS z IN THE "TABLE FOR COMPUTING IMPOSED LOADS AT VARIOUS DEPTHS."

\## THESE VALUES ARE TAKEN FROM THE PREVIOUS "TABLE FOR COMPUTING IMPOSED LOADS AT VARIOUS DEPTHS"

\### THESE VALUES ARE SHOWN AS P_z VALUES IN THE PREVIOUS "TABLE FOR COMPUTING IMPOSED LOADS"

7)(12)/((144)(3000)) = 0.057"[64]

5. Compute total group settlement, which is the sum of the soil compression and the elastic compression of the pile:

$$d_{total} = 4.4 + 0.057 = 4.457"$$

4.9.6. Negative Shaft Friction

4.9.6.1. "α" Method (Modified)

1) Establish a simplified soil profile and estimate values for total unit weight and other soil properties needed for computing settlements. Figure 4-59 shows this graphically.

Figure 4-59 Soil Profile for Negative Shaft Friction Problem

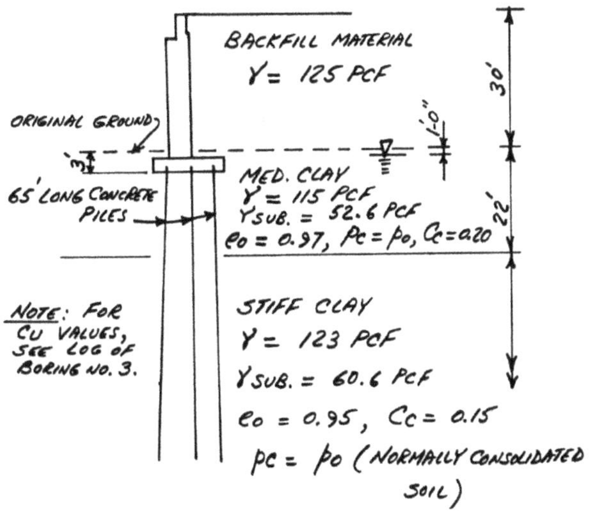

2) Estimate the settlements that will occur and determine the pile settlement length which will experience negative shaft friction.

   a. Determine additional overburden pressure Δp due to the backfill placed behind the abutment. The pressure diagram shows the effective overburden pressure p_o line before the backfill is placed and the effective pressure line $p_o + \Delta p$ after the backfill placement. Δp values shown in Table 4-22 were computed by using the pressure coefficients from the pressure distribution chart. The following input values were used in using the pressure distribution chart:

      i. Embankment Top Width = 40'.

      ii. Side Slopes: 1 V on 2 H.

      iii. Embankment Height = 30'

      iv. Value of "b" to use chart = 40/2 + 60/2 = 50'.

The point where settlements are needed is shown in Figure 4-60 (mid point of end slope.)

Table 4-22 Overburden Pressure Increments for Negative Shaft Friction Problem

DEPTH BELOW EXISTING GROUND FT	PRESSURE COEFFICIENT K	$\Delta p = K \times \gamma_{FILL} \times H_{FILL} = K \times 125 \times 30$ PSF
ORIGINAL GROUND SURFACE	0.5	1875
0.2×b = 0.2×50 = 10	0.5	1875
0.4×b = 20	0.5	1875
0.6×b = 30	0.5	1875
0.8×b = 40	0.4	1500
1.0×b = 50	0.4	1500
1.2×b = 60	0.35	1313
1.6×b = 80	0.30	1125

Table 4-22 (continued)

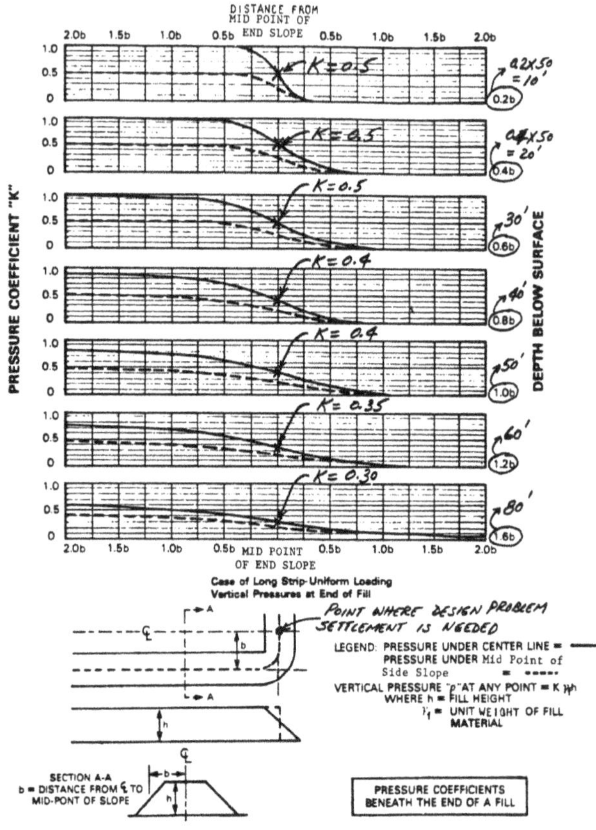

Figure 4-60 Overburden Pressure Diagram

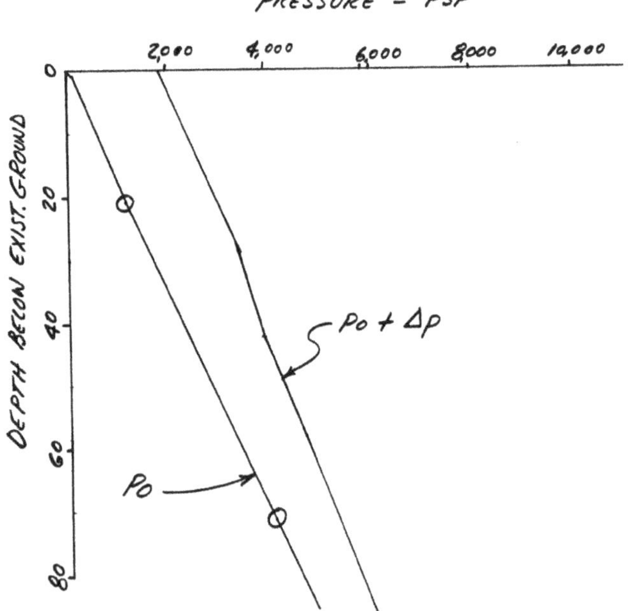

b. Perform settlement computations and determine the pile segment that will experience negative shaft friction. Table 4-23 shows detailed settlement computations. e_o and c_c for medium and stiff clays were obtained from consolidation tests. Since both clay layers are normally consolidated, Equation 4-44 is used for estimating settlements. Because the total long-term settlement of the clay is very high (26.8"), it is assumed that placing a temporary surcharge for the necessary time prior to pile installation will preload the soil. It is also assumed that 90% consolidation of the clay will be achieved prior to pile installation. The table shows that the relative soil movement (after 90% consolidation has occurred) below the 42' depth is less than 0.5". Therefore, the pile length from 35' below the pile head (pile head will be 3' below the existing ground line) will experience negative shaft friction. The remaining length will provide positive shaft friction (capacity to sustain loads from the structure.)

3) Determine the magnitude of negative shaft friction.

a. For pile length in medium clay (19' embedment), use Figure 4-20 (c).

i. L/B = 19/1 = 19

ii. Average shear strength of medium clay c_u = 675 psf

iii. For L/B = 19 and c_u = 675 psf, α = 1.0.

iv. f_s = (1.0)(675) = 675 psf = 0.675 ksf

v. $Q_{s1\text{-}negative}$ = (0.675) (4.0) (19) = 51.3 kips

b. For pile length in stiff clay: This segment will experience negative shaft friction from 19' to 39', or 20' in length. Use Figure 4-20 (b).

i. L/B = 46/1 = 46

ii. Average shear strength

of stiff clay c_u = 3,175 psf

iii. For L/B = 46 and c_u = 3,175 psf, α = 0.70

iv. f_s = (0.7)(3175) = 2223 psf

v. Q_{s2} = (2223)(4.0)(20) = 177,800 lbs. = 177.8 kips

c. Compute total ultimate negative shaft friction:

$Q_{s\text{-}negative}$ **= 51.3 + 177.8 = 229.1 kips**

4) Compute pile capacity provided by positive shaft friction from 39' below pile head to pile toe and the toe bearing capacity.

a. The pile length contributing to positive shaft friction is the total pile length less the pile segment length that may experience negative shaft friction = 65' - 39' = 26'.

b. Using α method described previously, the ultimate positive unit shaft friction is 2,223 psf.

c. The total ultimate positive shaft friction is

$Q_{s\text{-}positive}$ **= (2223)(4.0)(26) = 231,192 kips = 231.2 kips**

d. The ultimate pile toe bearing from this method was previously computed to be 30.6 kips. The total ultimate positive pile capacity is

$Q_{u\text{-}positive}$ **= 231.2 + 30.6 = 261.8 kips**

5) Determine the net pile capacity to resist the imposed loads.

$Q_{u\text{-}net} = Q_{u\text{-}positive} - Q_{u\text{-}negative}$ **= 261.8 - 229.1 = 32.7 kips**

The net ultimate pile capacity is very small. Therefore, alternatives to obtain higher pile capacities must be considered.

4.9.6.2. Reducing Negative Shaft Friction

For this problem, we need to consider alternatives to obtain higher pile capacity.

1) Use bitumen coating on piles to reduce negative shaft friction[65]. Assume 75% reduction of negative shaft friction by applying a bitumen coating (this is a conservative value.)

Table 4-23 Settlement Computation Table, Negative Shaft Friction Problem

SOIL LAYER	LAYER THICKNESS FT	DEPTH @ LAYER MID POINT FT	P_o @ LAYER MIDPOINT (FROM PRES. DIA.) PSF	IMPOSED PRESSURE Δp (FROM PRES. DIA.) PSF	$\frac{P_o+\Delta p}{P_o}$	$\log\frac{P_o+\Delta p}{P_o}$	$\frac{C_c}{1+e_o}$	LAYER SETTLEMENT $H\left[\frac{C_c}{1+e_o}\log\left(\frac{P_o+\Delta p}{P_o}\right)\right]$ FT	10% OF LAYER SETTLEMENT IN	DEPTH BELOW EXIST. GROUND FT	REC. SOIL MOVEMENT DUE TO 10% SETTLEMENT IN
0 TO 10 FT	10	5	350	1875	6.36	0.80	$\frac{0.20}{1.97}$ = 0.10	10x0.10x0.80 = 0.80 (9.6")	0.96	EXIST. GROUND LINE 10	2.67 2.67-0.96=1.71
10 FT TO 20 FT	10	15	850	1875	3.21	0.51	0.10	10x0.10x0.51 = 0.51 (6.1")	0.61	20	1.71-0.61=1.10
20 TO 22 FT	2	21	1167	1875	2.61	0.42	0.10	2x0.10x0.42 = 0.08 (0.96")	0.096	22	1.10-0.096=1.00
22 FT TO 32 FT	10	27	1550	1875	2.21	0.34	$\frac{0.15}{1.95}$ = 0.08	10x0.08x0.34 = 0.27 (3.24")	0.324	30	1.00-0.324=0.68
32 TO 42 FT	10	37	2150	1650	1.77	0.25	0.08	10x0.08x0.25 = 0.20 (2.40")	0.24	42	0.68-0.24=0.44
42 TO 52 FT	10	47	2750	1500	1.55	0.19	0.08	10x0.08x0.19 = 0.15 (1.80")	0.18	52	0.44-0.18=0.26
52 TO 62 FT	10	57	3350	1400	1.42	0.15	0.08	10x0.08x0.15 = 0.12 (1.44")	0.14	62	0.26-0.14=0.12
62 TO 72 FT	10	67	3350	1275	1.32	0.12	0.08	10x0.08x0.12 = 0.10 (1.20")	0.12	72	0.12-0.12= 0
								Σ = 2.23' = 26.76"	Σ = 2.67"		

*THIS PILE SEGMENT WILL EXPERIENCE NEGATIVE SKIN FRICTION (REC. MVMT. > 0.5")

POSITIVE SKIN FRICTION

* PILE SEGMENT LENGTH WHICH WILL EXPERIENCE NEGATIVE SKIN FRICTION = (42') - (3') = 39.'

NOTE THAT THE PILE TOP WILL BE 3' BELOW THE EXISTING GROUND LINE.

a. $Q_{s\text{-}downdrag}$ = (0.25)(229.1) = 57.3 kips

b. $Q_{u\text{-}net}$ = $Q_{u\text{-}positive}$ - $Q_{u\text{-}negative}$ = 261.8 - 57.3 = 205.4 kips

c. $Q_{allowable\text{-}single}$ = 205.4/3 = 68.2 kips

d. $Q_{allowable\text{-}group}$ = (24)(68.2) = 1636 kips

e. The maximum design load to be imposed on each pile is 120 kips (534 kN) and the total design load is 1700 kips. Therefore, a larger number of piles will be needed for this alternative.

2) Use longer piles driven to a stiffer or dense layer. This alternative will provide greater toe bearing and shaft friction capacities but may cost more.

3) A stub abutment instead of a full height abutment may be a better choice for the south abutment. The stub abutment could be supported on a spread footing with specified embankment material and density control in the foundation area. A stub abutment with a pile foundation is another alternative available for consideration.

A cost analysis of all alternatives should be performed before making the selection.

4.9.7. Lateral Squeeze

1) Determine if abutment tilting can occur.

a. γ_{fill} = 125 pcf.

b. H_{fill} = 30'.

c. $c_{u\text{-}average}$ = 675 psf

d. $\gamma_{fill}H_{fill}$ = (125)(30) = 3,750 psf

e. $3c_u$ = (3)(675) = 2,025 psf

f. Since Equation 4-46: $\gamma_{fill}H_{fill} > 3c_u$, is fulfilled, abutment tilting can occur.

2) Determine the magnitude of the horizontal movement.

a. If piles are placed before any soil compression occurs. Vertical fill settlement from computations performed previously for negative shaft friction = 27". The estimated horizontal movement is estimated to be 25% of the vertical movement = 0.25 x 27" = 6.75". This lateral movement is not tolerable.

b. If piles are driven after 90% of estimated, vertical settlement has occurred. Estimated vertical fill settlement after 90% settlement has occurred = 2.67". The estimated horizontal movement is 25% of this, or 0.67". This movement is tolerable but provisions should be made in the bridge shoe and expansion joint design to accommodate the displacement.

4.9.8. Wave Equation Analysis

The last part of our sample problem will be the wave equation analysis of our pile. Let us assume that the successful bidder of the project had started the hammer selection process by using the Engineering News formula. The design capacity is 120 kips (534 kN). Using a minimum set of 0.1" (120 BPF) and Equation 4-72, substituting yields

$$E_{en} = (120)(0.1 + 0.1)/2 = 12 \text{ ft-kips}$$

In this example, we input the design (or allowable) load into the Engineering News formula. Some more modern texts (such as FHWA HI 97-103) recommend using the ultimate capacity in dynamic formulae, but this is not the way in which most of these formulae (and this includes the EN formula) were originally intended to be used, as our earlier derivation of these formulae shows. We will assume that the contractor uses the allowable load and makes a preliminary selection of a Vulcan #1, which has a rated striking energy of 15 ft-kips.

If we use an allowable load of 425 kips based on our static methods, the required hammer energy would be

$$E_{en} = (425)(0.1 + 0.1)/2 = 42.5 \text{ ft-kips}$$

This would imply the use of a Vulcan 016 (48.75 ft-kips) for example.

For this example, only the North Abutment will be considered.

Before we begin the wave equation analysis, we need to evaluate the maximum allowable driving stress of the 12" concrete piles we are driving. Using the criteria in Table 4-7, the maximum compressive stress is

$$\sigma_{max\text{-}compressive} = (0.85)(7000) - 700 = 5250 \text{ psi (compressive)}$$

and the maximum tensile stress (AASHTO criterion) as

$$\sigma_{max\text{-}tensile} = 3\sqrt{(7000)} + 700 = 951 \text{ psi (tensile)}$$

The data is entered into the program (in this case GRLWEAP[66]) and the input data is processed and summarized by the program as shown in Figure 4-61.

Figure 4-61 Input Screen for North Abutment Piles, Vulcan #1 Hammer

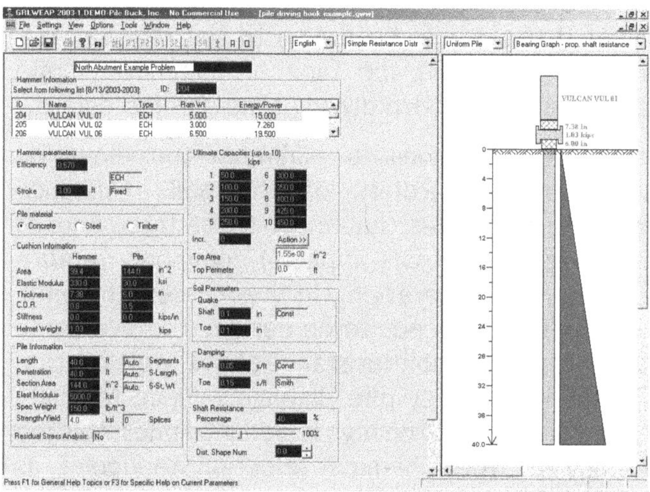

A few comments on this data:

1) The piles are assumed 40' (12 m) long. The concrete is assumed to have a modulus of elasticity of 5,000 ksi and a unit weight of 150 lbs/ft^3.
2) The shaft friction is assumed to act along the entire shaft length of the pile in a triangular distribution.
3) The toe is assumed to have approximately 60% of the pile resistance/capacity, based on the results of the Nordlund Method.
4) The quake and damping properties are standard for cohesionless soils.
5) All hammer data is taken from the program's instruction manual and hammer database. The hammer cushion is 1" (25 mm) thick Micarta plates interleaved with 1/2" (13 mm) thick aluminum plates. The pile cushion is 6" (152 mm) thick.
6) A successful wave equation analysis would indicate that the hammer selected is able to drive the pile to the ultimate capacity without damage to the hammer and pile. Damage is limited by the following:
   a. Limiting the blow count of the hammer. A very reasonable blow count is 120 blows per foot, corresponding (but not exactly) to a set of 0.1".
   b. Limiting the pile stress, as previously discussed.

Based on the static analysis methods used, the ultimate capacity is in the range of 400-425 kips. We could simply analyze this resistance; however, we need to insure the hammer has some reserve capacity in the event the actual hammer in the field is less efficient than modeled or a larger soil resistance is encountered than estimated. We will thus input a series of resistances that vary from 50 kips (222 kN) to 450 kips (1998 kN) as shown.

GRLWEAP can generate a bearing graph, which plots the blow count against a range of resistances. Figure 4-62 shows the bearing graph along with the maximum tensile and compressive stresses using the Vulcan #1 hammer.

Figure 4-62 Results of Wave Equation Analysis for Vulcan #1 Hammer

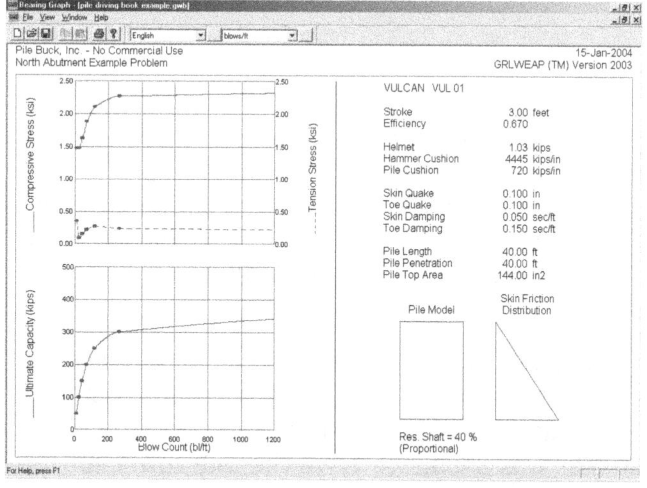

The results clearly show that neither the tensile nor the compressive stresses exceed their respective limits. Unfortunately the results also indicate that the Vulcan #1 cannot drive these piles to the ultimate resistance, with blow counts exceeding 300 blows/ft at an ultimate resistance of less than 300 kips.

The most straightforward solution to this problem is to use a larger hammer. The larger hammer considered next is the Vulcan 016, with more than triple the ram weight and rated striking energy. This was also the hammer suggested by the EN formula using the ultimate capacity of the pile. In addition to changing the hammer selection in the program, the cushion configuration and the cap weight were also changed, as the Vulcan 016 uses different driving accessories and has a different cushion configuration. The pile and soil configurations were unchanged.

The results for the Vulcan 016 analysis are shown in Figure 4-63.

Figure 4-63 Results of Wave Equation Analysis for Vulcan 016 Hammer

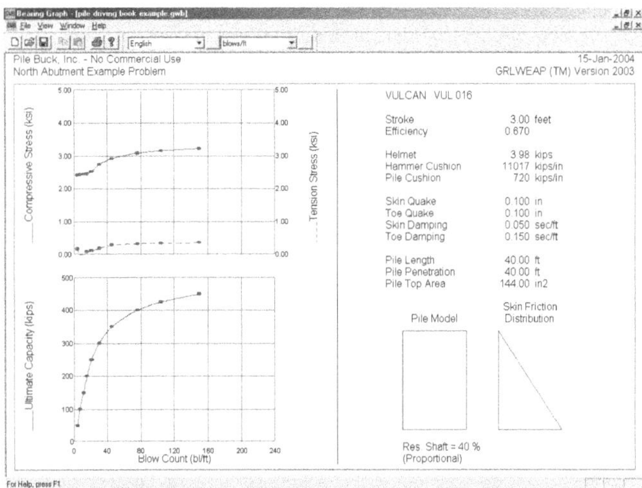

As was the case with the #1, the pile stresses are within acceptable limits, although the compressive stresses are higher. The capacity situation improves considerably, however, as the 425 kip ultimate resistance goal is reached at 104 BPF, an acceptable result.

At this point we have reached our goal; however, variations in actual soil properties and hammer performance could lead to a situation where the hammer was unable to drive the pile to its 40' toe elevation. This may or may not be detrimental to the performance of the foundations; the actual factor of safety we have here is 425/120 = 3.5, which is above the factor of safety for most static analyses. This leads to a common situation with driven piles, i.e., the situation where the required axial capacity of the piles may be obtained before the design toe elevation is achieved. To understand this issue thoroughly is essential in proper analysis and oversight of pile driving.

One of the advantages of driven piles is that the pile hammer is used as a measuring device to verify the capacity of the pile at the time of installation. Before the development of soil mechanics theories and static pile capacity estimation, hammer blows were the only way to estimate (other than a static load test) the capacity of a pile to carry axial load. Even with the methods described here, driving the pile may actually reveal the need either to lengthen the pile or to shorten it to achieve the desired capacity. Lengthening the pile is a favorite object of contractor claims and disputes, but shortening can also create problems, especially if the engineer insists that the pile be driven to the required toe elevation, even in the face of hammer or pile damage and excessive driving time and job delays.

In certain situations, a minimum toe elevation is required regardless of the soil resistance encountered. Piles that must resist lateral loads (wind, seismic, vessel impacts) commonly require a minimum toe elevation. In these cases, the pile/hammer system needs to be designed to overcome all the estimated soil resistance above toe elevation without hammer or pile damage. Since laterally loaded piles tend to be high impedance piles, this job is simplified from a hammer selection standpoint, as high impedance piles transmit the stress wave from the hammer more efficiently.

In situations where only axial load capacity is required, provisions need to be made for a proper test pile program that includes in situ verification of the resistance of the soil to the pile. Such verification can include dynamic monitoring (which can also be done during production driving,) static load tests, and Statnamic™ testing. Additionally one could specify an estimated toe elevation as opposed to a minimum one. In this way, the owner's need for piles capable of carrying the load can be achieved without an undue burden on the contractor.

Our sample analysis did not examine the South Abutment. In that case achieving toe elevation is important because the toe resistance offsets the loss of effective capacity from negative shaft friction.

One other result of the wave equation analysis is the force-time and velocity-time histories during impact, as shown in Figure 4-64.

Figure 4-64 Force-Time and Velocity-Time Curves for Vulcan 016 Hammer

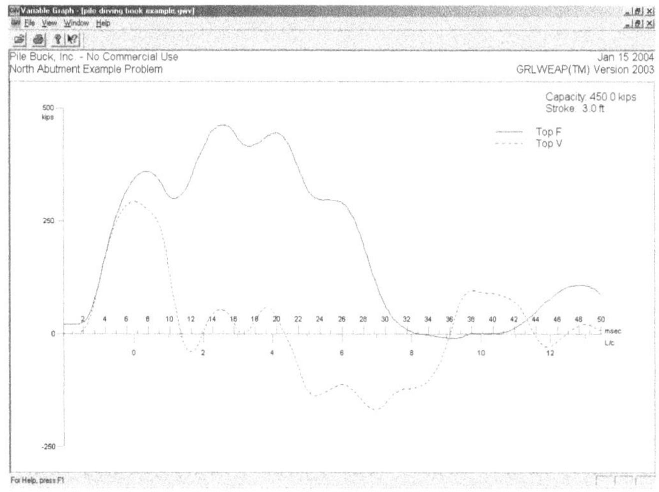

4.10. Design of Sheet Pile Walls

Straight walls built of Z-type or Arch-web type piles are designed as cantilevered or supported beams. The supports include the ground into which the pile wall is embedded, and either internal structural bracing or external anchorages. The pressure of the soil retained, the pressure of any water present, or the combination of the two generates the load on the "beam". Vertical pressures are converted to horizontal loads against the wall, which in turn, generate bending moments that must be resisted by the beam. Design is therefore a combination of geotechnical science and structural engineering.

4.10.1. Soil Pressure

A detailed discussion of soil mechanics leading to design of a working sheet pile structure is well beyond the intent of this chapter, however some basic considerations and methods can be provided. The reader is referred to publications such as the Pile Buck Sheet Piling Design Manual and the technical references cited therein for an in-depth discussion of sheet pile design.

Soils are broadly classified as coarse grained or fine grained, the former typically gravels and sands and the later clays and silts. In between there are many gradations and combinations of those general classifications.

Any proposal to design and construct a sheet pile structure should include a comprehensive subsurface investigation of the site. Standard penetration tests should be made and samples obtained for laboratory analysis. The results of this investigation will provide the input for much of the design, or may indicate that the structure originally planned is unfeasible.

The most important soil properties coming from the tests and for sizing the wall are the unit weight of the soil, the angle of internal friction and the cohesion. The weight of the soil affects the vertical pressure at any given depth of the soil mass. The angle of internal friction is a characteristic of the soil regarding angularity and roughness of the grains and the state of compaction. It is a decisive factor in the effective pressures exerted against vertical walls. Most sands and gravels have angles of internal friction of 30° to 40°. Cohesion as a property is generally associated with the fine-grained soils and particularly the clays, however even fine sand may exhibit some cohesion due to inclusion of clay particles in its makeup. Cohesion is the capacity within the soil to resist shearing stresses. Calculations of earth pressure forces include the effect of cohesion. Cohesion is generally ignored when dealing with the coarse-grained soils group. Furthermore, the numerical value of cohesion, even in truly cohesive soils, is generally ignored for permanent retaining structures. Design input that may have been valid at the time of construction of the wall, may result in an unsafe situation at some future time. So-called "drained strength" conditions must be evaluated for fine-grained soils and an "effective" friction angle determined by testing. This strength is available to compute earth forces for permanent earth retaining structures.

Fresh water weighs 62.4 pounds per cubic foot; for each foot of water retained by a wall, the pressure increases by another 62.4 pounds per square foot. The pressure at a point ten feet below the surface is thus 10 x 62.4 of 624 pounds per square feet. The pressure distribution is triangular starting at zero and increasing to 624 pounds. Earth pressure is also the force per unit area, exerted in this case by the soil on the sheet pile structure. However, unlike water, the magnitude depends on the physical properties of the soil and its interrelationship with the wall stiffness and method of restraint (cantilevered or braced.)

Three states of stress in the soil are of interest, (1) the in-situ state of stress "at rest," (2) the active state and (3) the passive state. If a wall is driven into soil the pressure exerted by the soil against the wall at any point below the surface will be a product of the effective weight of the soil and the depth, similar to the water case previously described. The earth is in "at rest" condition. Now, excavate the earth from one side of the wall so that the pressures are no longer balanced. If the wall is able to move even a very small distance away from the soil behind, that soil will enter an "active" state of stress which is the lowest state of lateral stress when the full strength of the soil is mobilized against shear failure. Steel sheet pile walls are flexible and bend somewhat, under load, enough to develop the active condition. Note that active pressures are only a fraction of "at rest" pressures.

If a wall moves toward the soil rather than away, as is the case with the embedded portion of the sheet pile wall, the lateral pressure will increase. When the full strength of the soil is mobilized a "passive state" of stress exists. Classical theories developed by Rankine and Coulomb for practical application of these phenomena are still in use today.

Rankine assumed (somewhat incorrectly) that the soil failure surface is a plane rather than curved and that the friction between soil and wall is of no consequence. When the Rankine state of failure has been reached, active and passive failure zones develop as shown.

The active and passive earth pressures for these states are expressed as follows:

Equation 4-83:

$$P_a = \gamma Z K_a - 2C\sqrt{K_a}$$

$$P_p = \gamma Z K_p + 2C\sqrt{K_p}$$

Where

- P_a and P_p = unit active and passive earth pressure at depth z
- γZ = vertical pressure at depth z due to weight of soil above
- c = unit cohesive strength
- K_a and K_p = Coefficients of pressure

The coefficients K_a and K_p are functions of the angle of internal friction and the slope of the backfill. For level backfill the values reduce to the equations

Equation 4-84:

$$K_a = \tan^2\left(45^\circ - \frac{\phi}{2}\right)$$

$$K_p = \tan^2\left(45^\circ + \frac{\phi}{2}\right)$$

The Coulomb Theory recognizes the effects of friction between the soil and the wall on vertical shear stresses in the soil. Lateral pressures, which include wall friction, are different from those assuming no friction.

The Coulomb Theory, like the Rankine, assumes that the surface of sliding is a plane whereas the actual failure surfaces are curved. This produces an error that is insignificant in the active case but is unsafe in the passive case. The Log Spiral Theory more closely evaluates the failure surface profile and is an alternate approach to analysis.

In summary, for the determination of lateral earth pressures on sheet pile walls:

1. Compute active pressures based on the Coulomb or the log-spiral theories. For simple cases, Rankine earth pressure coefficients are adequate.
2. Passive pressure should be computed using the Coulomb Theory (with wall friction) or the Log-Spiral method.

Computer software, such as Pile Buck's SPW 911, are especially helpful for complex situations.

4.10.2. Soil Properties

The properties of soils at the site should be determined from laboratory tests of representative samples. For preliminary designs and cost estimates, Table 4-4 relates Standard Penetration Resistance to the various properties of soil, which are of interest to the designer.

Earth pressure is time dependent in nature. This is particularly true in clay soils where the values of cohesion and internal friction change with time. Sheet pile structures in clayey soils should be designed for both the period immediately after construction and long-term conditions. The long-term value of c approaches zero and ϕ somewhere between 20° and 30°. The long-term case thus approaches that for sheet piling in granular soil. Earth pressures (both active and passive) at various depths are then converted to total active loads and passive restraining pressure for final analysis of the sheet pile "beam."

4.10.3. Surcharge Loads

Design of sheet pile walls should include the loads other than soil or water pressure to which the wall might be subjected. Equipment operating adjacent to the wall, railroad or highway loads, piles of ore or stone, buildings and crane rails, are examples of surcharge loads. The most common way of accounting for these additional loads is to allow for a "uniform surcharge" applied at the surface. Whereas the pressure distribution from granular soil is triangular in intensity from the surface to depth Z, the lateral pressure due to uniform surcharge is uniform laterally to depth Z, and is equal to the assumed surcharge in pounds per square foot times K_a or K_p depending on whether the active or passive case is being investigated.

Other surcharge loads may be described as point loads, line loads, or strip loads. A point load might be a concentrated wheel load or perhaps a footing near the proposed wall. A line load may be a continuous wall footing of narrow width parallel to the structure. A strip load might be a highway or railroad parallel to the wall. Calculations to convert these loads to lateral pressures against the wall are based on semi-empirical methods, and are conservative when applied to flexible walls. Formulas for these cases can be found in *Sheet Piling Design by Pile Buck*.

4.10.4. Unbalanced Hydrostatic Head

A sometimes forgotten load condition in design is the effect of waves, tidal action, storm surges and heavy rainfall. Since the locks of sheet piling are more or less tight when installed and become more watertight as soil is drawn in, water can be trapped behind the wall causing a head imbalance and greatly increasing the total load, even though temporary. Under certain design conditions, this imbalance can also affect the effective unit weight of soil at the foot of the wall and in turn, reduce the passive resistance that had originally been counted upon to restrain the toe. Where unbalanced head has been neglected in design, the effects have been seen more in the anchorage system or in toe movement than in bending failure of the sheets. The magnitude and potential for an imbalanced water force should be examined on an individual basis.

4.10.5. Other Lateral Loads

In addition to the above, other loads that should be investigated for possible consideration include wave forces, ship impact, mooring pull, earthquake forces and ice thrust.

4.10.6. Structural Aspects of Sheet Pile Design

The objective in specifying a section of sheet piling is to obtain sufficient resistance to bending moment. Allowable bending moment is computed by the formula

Equation 4-85:

$$M_{allow} = Z\sigma_{allow}$$

where

- M_{allow} = Allowable bending moment
- Z = Section Modulus
- σ_{allow} = Allowable Stress of the Material

Both the allowable bending moment and the section modulus are specified as per lineal foot or meter of wall. The section modulus is strictly a function of the physical shape of the material. The sheet pile manufacturer normally furnishes this. Equation 4-85 is only valid for pure bending stresses. Situations where other considerations such as deflection (bending and shear deflection) and transverse bending should be considered separately.

4.10.7. General Design Considerations

The design of sheet pile retaining walls requires several successive operations:

- Evaluation of the forces and lateral pressures that act on the wall;
- Determination of the required depth of penetration of the piling for stability;
- Computation of the maximum bending moments in the piling.

For this purpose, a one-foot wide section of the wall is isolated and analyzed. The piling is modeled as a beam on supports or a cantilevered beam. Computation of stresses in the wall and selection of the appropriate section and steel grade follows. Finally, the design of the waling and anchorage system is completed. The location and elevations of the structure will also be important to the design input as well as the elevation of the top. The dredge line and the proposed anchorage system must be pre-established. Any lowering of the dredge line due to scour or overdredging should be accounted for during design.

4.10.7.1. Cantilever Walls

In cantilevered wall construction, the piling must be driven to a sufficient depth into the ground to become fixed as a cantilevered beam in resisting lateral pressures. This wall type depends entirely on embedment for stability and is generally suited only for walls of moderate height up to about 15' (4.6 m).

The design of cantilever walls in granular soils is treated differently than in cohesive soils. When granular backfill can be used behind a wall, this will aid in simplifying the design. Figure 4-65 is a typical earth pressure diagram for a cantilevered wall.

Figure 4-65 Resultant Earth Pressure Diagram

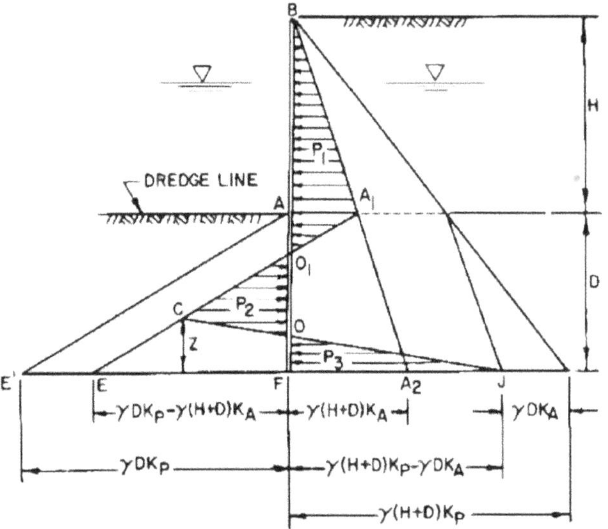

Procedure begins by assuming a trial depth D of from H to 2H depending on the soil density. Determine the active and passive lateral pressures. Because of deficiencies in the theory, Coulomb passive earth pressure coefficients should be used with care, as the results will be unconservative in many cases. Unit pressures are products of soil weights, the active and passive pressure coefficients K_a and K_p and the depths. All requirements for static equilibrium must be satisfied or the trial depth adjusted and new calculations made. The maximum bending moment occurs at the point of zero shear. Penetration should be increased to provide an added safety factor.

Pressure distributions on walls completely in cohesive soils are different than granular soils. This is because the cohesive strength of the soil (as opposed to the lateral earth pressure from the effective stress) governs the design for temporary structures. However, the same general conditions for equilibrium must be met.

A detailed discussion of design methods and sample problems for both coarse and fine-grained soils are provided in *Sheet Pile Design by Pile Buck*.

Figure 4-66 Open Type Dock Concrete Deck on Steel Piles with Sheet Pile Cut-off Wall

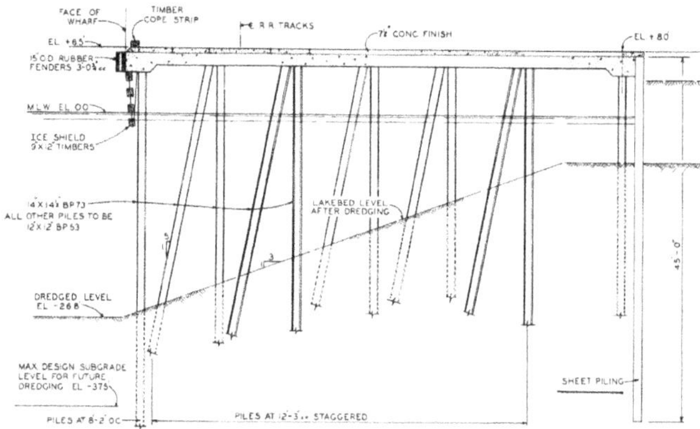

4.10.7.2. Anchored Walls

Anchored sheet piling walls derive their stability from passive pressure on the front of the embedded portion of the wall and an external anchorage system at or near the top. The anchors are generally steel tie rods connected to the front wall by a system of wales and structural hardware. This system is then tied to an anchor wall, as some distance behind the face wall. Most anchored bulkheads are constructed where there is sufficient room behind the face to accommodate this method since the rear wall may have to be placed as much as 60' (18.29 m) from the wall face. Where this cannot be done, other anchor methods such as battered steel tension piles or soil or rock anchors might be feasible. While the determination of bending moments in the wall is important, an equally important consideration is the safe depth of penetration and the anchor forces. Most wall failures are the result of the toe or the anchor system failure rather than bending failure of the piling. With the availability of Z-type sheet piling with section modulus of about 60 cubic inches per foot, bulkheads for about 35' of water can be constructed in good soil. Using high strength steels or HZ walls, these limits can be expanded to 40 or 50' (15.2 m) of water. Backfilling with select fill material can reduce active soil pressures.

Where a solid face wall is not feasible (economics, local soil and site conditions, loading conditions or local preference) steel sheet piling is often used as a "cut-off" wall behind a dike and platform on vertical piles.

In cases where dense soil prevents the piling from behind driven deep enough for stability (for example hard pan or soft rock) the toe can be pinned with steel bars or the bottom excavated and then tremie grouted.

There are several approaches to anchored bulkheads. For toe stability, the pile must be driven to a definable minimum depth. Figure 4-67 shows the general relationship between depth of penetration, lateral pressure distribution and elastic line shape.

Figure 4-67 Effect of Depth of Penetration on Pressure Distribution and Deflected Shape

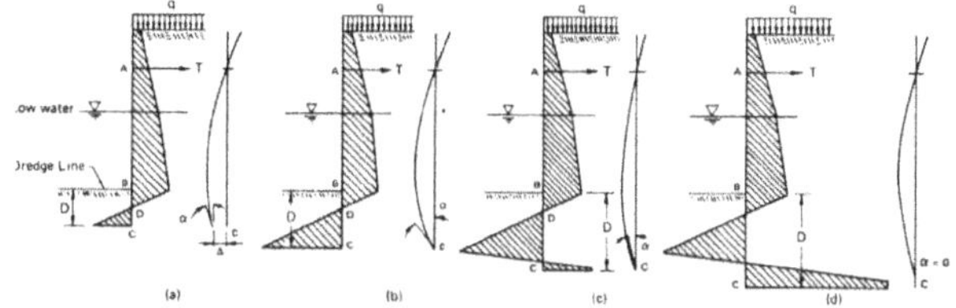

- Case (a) illustrates a wall driven just deep enough to where it is in a state of imminent failure due to insufficient passive resistance at the toe. The moment curve for this wall is open at the bottom since the only support for the wall is the anchorage at the top.
- Case (b) is an extension of (a), but the wall has now been driven deep enough to where it is structurally stable and the sum of moments and horizontal forces are zero. This case illustrates the Free Earth concept of bulkhead design. Free Earth Support is a classical approach that does not recognize the ability of the passive soil to produce restraint capable of generating negative bending moments. The results of this approach produce very high calculated positive moments in the wall. After a minimum depth of penetration for stability is achieved, the toe is generally increased by 20 to 40% for safety, which may be sufficient in many cases to approach the Fixed Earth condition. Free Earth analysis will generally produce ultra-conservative designs that are also uneconomical. There are now procedures for reducing moments derived from Free Earth analysis based on the ability of steel piling to flex under load and decrease the bending moment[67].
- Case (c) indicates a wall that has penetrated further to where passive resistance is mobilized on both sides of the embedded portion.
- In Case (d), the toe of the pile at point C becomes fixed with no further rotation. This is the Fixed Earth Support concept. The typical elastic line curve of a flexible wall with sufficient embedment for fixity reverses its curvature at a point of contraflexure and becomes vertical at the toe. The wall will act as a partially built-in beam subject to bending moments.

German engineer H. Blum first proposed that an "equivalent beam" replace the elastic line, which would be simpler to analyze. This method is illustrated in Figure 4-68. A hinge is assumed at the point of contraflexure; the part of the beam above the hinge is treated as a beam on two supports (the ground at the hinge and the anchor at the top). The beam below the hinge to the toe can also be treated as a beam on two supports. Various assumptions are made in Blum's approach regarding the location of the point of contraflexure and the required depth of penetration that have been found to be sufficiently accurate. Tschebotarioff has simplified the equivalent beam further by assuming the hinge is at the dredge line and setting a depth of penetration at 0.3 H. These methods are limited to sheet pile walls in coarse-grained materials.

Figure 4-68 Blum's Equivalent Beam Method

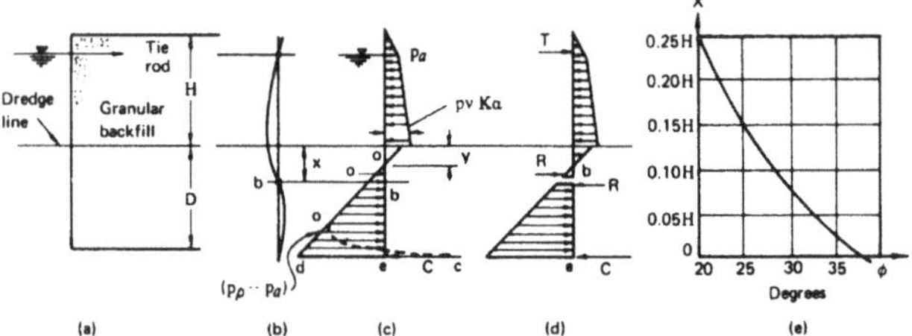

In cases of complex or irregular loading, graphical design methods are advantageous. Lateral pressures are determined by conventional methods, and then converted to active and passive force vectors to a predetermined scale on the graph. A moment diagram can be constructed. Anchor tension and toe reaction diagram can be constructed. Anchor tension and toe reaction is obtained from a vector diagram. Maximum moments are scaled from the moment diagram. This information is also useful for cases where reinforcing plates are to be added to conventional piling to increase the section modulus over part of the length.

Once the maximum bending moment is obtained from any design method, the minimum section modulus required for the wall is determined by dividing the moment by the allowable unit design stress. Since there are three grades of steel to choose from, this provides considerable flexibility in selection. There is obviously considerable competition among the many producers of sheet piling. This has fostered a trend by owners toward specifying a bending moment to be satisfied rather than a pre-selected piling section. In other cases, bidders in lieu of the one specified may propose alternate grades and shapes.

Where the design requirements cannot be satisfied using "as-rolled" sheet piling, stiffer sections can be fabricated by adding reinforcing plates to those portions of the flanges where needed to develop the section modulus required.

4.10.8. Cellular Design

Sheet pile cells are massive, freestanding, gravity structures, capable of withstanding large horizontal loads from water, earth, or external forces such as those imposed by ship mooring operations. A principal application for cofferdams in deep water is the enclosure of large areas that must be kept clear of bracing to facilitate construction of large structures. An example are the navigation locks and dams along rivers such as the Ohio and the Arkansas Rivers which were built inside open cofferdams consisting of sheet pile cells as high as 60' (18.29 m) and up to 65' (19.8 m) in diameter. The cofferdams were over 1000' (305 m) long in order to accommodate the navigation locks built inside.

The basic type of cell is a circle. The number of piles needed to close a circle is given by the equation

Equation 4-86:

$$N_{piles} = \frac{\pi D}{W_{pile}}$$

where

N_{piles} = total number of piles in the circle[68].

D = diameter of circular cofferdam, feet

W_{pile} = width of pile from interlock to interlock, feet

For a given number of piles, the resulting diameter of the cofferdam is solved to

Equation 4-87:

$$D = \frac{N_{piles} W_{pile}}{\pi}$$

As an example, consider a 60' (18.29 m) diameter circular cell utilizing the Domestic Section PS 27.5, with a width of 19.69" (1.641'). Using Equation 4-86, the minimum number of piles to close the structure is

$$N_{piles} = (\pi)(60)/(1.641) = 114.9 \approx 116 \text{ piles}$$

The actual circular structure diameter with this given number of piles:

$$D = (116)(1.641)/\pi = 60.59'$$

After erection, the cells are filled with select materials and become self-supporting. In order to make the wall continuous, "connecting arcs" connect adjacent cells and these sub-units are filled. As the cells expand under the pressure of the fill, the interlocks of the sheeting tighten and become watertight.

Figure 4-69 Side View of Cellular Cofferdam

Cofferdams are designed to resist the overturning moment from water pressure and/or soil pressure on the wet side of the cofferdam. These forces may also cause sliding of the cofferdam on its base. Like straight wall design, a one-foot strip of wall is analyzed. Classical design methods check the location of the resultant of driving and resisting forces on the wall. If the resultant falls within the middle third of the base, the overturning criteria had been met. Beginning in the 1930's, the Tennessee Valley Authority built a large number of masonry flood control and hydroelectric dams in the Southeast, most of which employed large cellular cofferdams in their construction. TVA engineers concluded that cell failure would occur due to shearing failure of the fill within the cell rather than by overturning. The success of historical structures designed by classical methods was due more to good fortune and the quality of fill used. Some failures could be traced to poor fill. Present design practice therefore is based on either vertical or horizontal shear failure in the fill or examination of both possibilities. Other investigations include sliding on base, base stability, and shear failure between sheeting, fill, and interlock tension.

Cellular cofferdams are generally designed to handle normal high water conditions. When overtopping is possible, floodgates are opened and the cofferdam flooded until the water recedes. In high water, seepage into the cell is very probable and the effective unit weight of the fill is reduced. It is important to provide weep holes on the inside sheets and to keep those free draining so that the cell does not fill up with water. Saturation lines for design purposes are assumed to slope from 1 on 1 to 1 on 3 depending on the quality of fill. The pressure of fill in the cell along with any water pressure present results in hoop tension in the periphery and thus across the sheet piles and their interlocks. The tension is a function of the cell radius and the internal pressure and is computed at some point above the base. The maximum interlock tension in the main cell is given by the formula

Equation 4-88:

$$t = \sigma_T R$$

Where

- T = hoop tension, kips/in
- σ_r = internal pressure of the cofferdam, ksi
- R = radius of the cofferdam, inches

For a sheet pile such as PS 27.5 or PS 31 with minimum ultimate interlock strength of 16 kips/inch, design pull should be limited to 8 kips/inch.

Figure 4-70 Two Varieties of Circular Cells

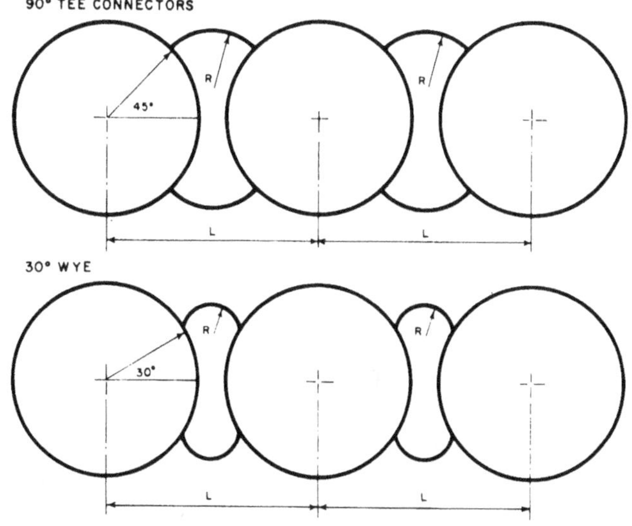

Figure 4-71 Working Inside a Cellular Cofferdam

Figure 4-72 Two Varieties of Diaphragm Cells

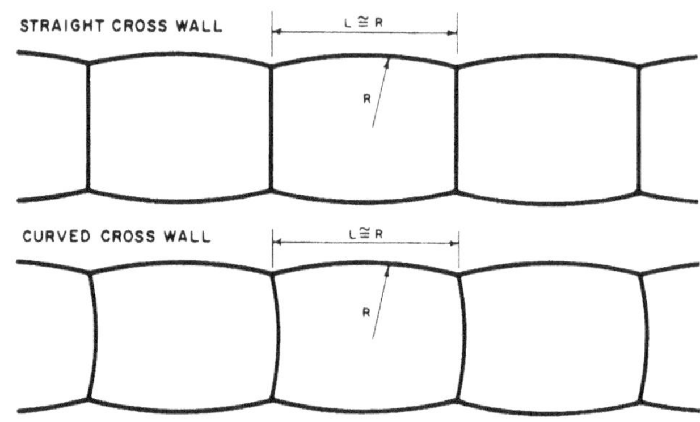

As a rule of thumb, for preliminary design, cell diameters may be assumed to be about 1.25 times the height. For 50' (15.2 m) of water, this translates to cells of about 62' (18.9 m) in diameter. When cells get much beyond 65' (19.8 m) in diameter interlock tensions approach potentially dangerous levels. If sufficient space is available inside the cofferdam, interior berms can be added that will help stabilize the cell not only from the sliding and internal stability standpoint but also reduce interlock tensions. If cells are filled either hydraulically or at low water or both, they are pre-exposed to probably the highest stresses they would experience during use. This provides a pre-test of the cell and also tensions the locks and expands the cells to the maximum that is favorable to stability. While failure in a cell at any time is costly to repair, most cell failures due to overstressing have occurred during filling where the net effect is much less serious.

Another type of cellular cofferdam is the diaphragm type. Rather than complete circles, this type uses arcs of smaller radius for the face wall and rear wall and straight walls to connect them. By reducing the radii, lighter weight sheets such as the PSA 23 can be used with economies based on weight savings. This type is unfortunately less reliable than the circular cell type since failure of one cell could affect adjacent cells. Diaphragm construction has been used more for permanent bulkheads or dock walls than temporary cofferdams. Only the face sheets need be full length and economies can be obtained by shortening the lengths of sheets that are back in the fill.

Cellular cofferdams must be built on or in rock or soils, which can support their great weight. When built on deposits containing large boulders, an effort should be made to remove the boulders by pre-dredging. If this effort is not made, the sheets may be damaged during driving, will not provide a reliable cut-off of seepage under the dam, and be difficult to remove later. It has been generally assumed that granular materials such as crushed rock, sand and gravel mixes were the only proper fill for cells and this is probably true for coffer-cells where free drainage is important. However there are cases of permanent docks being constructed of cells where cohesive soils were left in place and became part of the cell fill after excavation. Presumably, some consolidation was conducted by surcharge loading prior to final use.

Photo Credits

- GRL and Associates/Pile Dynamics, Inc.: Figure 4-33.
- Don C. Warrington, P.E.: Figure 4-7.

Footnotes

[1] Modified from Vesic, A.S. (1977). *Design of Pile Foundations*. National Cooperative Highway Research Program, Transportation Research Board, National Research Council, Washington, D.C., Synthesis of Highway Practice No. 42.

[2] This information can be found in *Soil Mechanics*, available from Pile Buck.

[3] It is emphasized that for soils containing gravel sized particles, this table may yield unreliable results. In those cases, the correlations should be used for rough estimation purposes only.

[4]The undrained shear strength is one half of the unconfined compressive strength. Correlations of N

values to undrained shear strength of clavs are crude and unreliable for design. It should be used only for preliminary estimating purposes. Undisturbed cohesive samples should be obtained for laboratory determination of accurate shear strength and unit weight.

[5] Any load design criterion should be applied in conjunction with a settlement analysis.

[6] The allowable pile stresses recommended in this chapter are generally taken from a number of sources: these include the following:

- AASHTO (1994) *Standard Specifications for Highway Bridges, Fifteenth Edition.* Washington, DC: American Association of State Highway and Transportation Officials.
- Prestressed Concrete Institute (1993) *Recommended Practice for Design, Manufacture and Installation of Prestressed Concrete Piling.* Chicago, IL: Precast/Prestressed Concrete Institute.

The recommendations may not necessarily be applicable to other types of pile foundations, which are governed by local or national building code requirements.

[7] More information on allowable steel stresses in structures can be obtained from the American Institute of Steel Construction (1980), *Manual of Steel Construction.* Eighth Edition. Chicago, IL: AISC.

[8] AASHTO value. PCI recommends prestress values of 4.8-8.3 MPa (700-1200 psi) for piles longer than 12.2 m (40') or 2.7-4.8 MPa (400-700 psi) for piles shorter than 12.2 m (40'.)

[9] As is the case with most geotechnical problems, there are frequently several methods to analyse a particular problem. The suitability of a particular method (or implementation of a method; these too vary) may depend upon local conditions, code requirements, or experience with similar piles. The methods presented here are not meant to be a comprehensive summary of all of the methods used to analyse driven piles; more complete descriptions of these methods and others can be found in texts on foundation design in general and deep foundation design in particular.

[10] All three of these methods will tend to overpredict (be unconservative) for pile sizes greater than 610 mm (24"). In these cases, a method such as Dennis and Olson may be more appropriate. A description of this method is given in Schroeder, W.L, Dickenson, S.E. and Warrington, D.C. (2003) *Soils in Construction*, Fifth Edition. Upper Saddle River, NJ: Prentice-Hall.

[11] This method is primarily derived from Meyerhof, G. G., 1976. "Bearing Capacity and Settlement of Pile Foundations." American Society of Civil Engineers, *Journal of the Geotechnical Engineering Division*, March, 1976.

[12] This chart is taken from G.G. Meyerhof, "Bearing Capacity and Settlement of Pile Foundations". Journal of the Geotechnical Engineering Division. New York: American Society of Civil Engineers, 1976.

[13] For steel H and unfilled open end pipe piles, use only steel cross section area at pile point unless there is reasonable assurance and previous experience that a soil plug will form at the tip. The assumption of a soil plug would allow the use of a box area at H pile tip and total pipe cross section area for pipe pile

[14] This method is largely based on Schmertmann, J.H., *Guidelines for Cone Penetration Test, Performance and Design.* Implementation Package Prepared for the Federal Highway Administration, FHWA-TS-78-209, 1977.

[15] For steel H and unfilled open-end pipe piles, use only the steel cross section area at the pile point unless there is reasonable assurance and previous experience that soil plug would form at the tip. The assumption of a soil plug would allow the use of a box area at H pile toe and total cross section area for pipe pile.

[16] R.L. Nordlund originally presented his method in his paper, "Bearing Capacity of Piles in Cohesionless Soils", *Journal of the Soil Mechanics and Foundations Division* ASCE Vol. 89, May 1963. He updated the method in the paper, "Point Bearing and Shaft Friction of Piles in Sand," Fifth Annual Fundamentals of Deep Foundation Design, University of Missouri-Rolla, 1979. Most of the figures in this section come from the second paper.

[17] This chart is modified from J.E. Bowles, *Foundation Analysis and Design.* New York: McGraw-Hill and Company, 1977.

[18] For H-piles in cohesionless soils, the "box" area should generally be used for shaft resistance calculations.

[19] For H-pile and unfilled open-end pipe piles, use only the steel cross-sectional area at the pile toe unless there is a reasonable assurance and previous experience that a soil plug will form at the pile toe.

[20] For justification of using less than 9 for N_c, see Coduto, D.P. (2001), *Foundation Design: Principles and Practices.* Second Edition. Upper Saddle River, NJ: Prentice-Hall.

[21] For both H piles and open end pipe piles, a soil plug may form at the H pile tip and the point bearing capacity may be calculated using the gross cross section area (i.e., flange width x web depth). This design assumption should be verified in the field.

[22] This method is largely derived from Tomlinson, M. J. (1995) *Foundation Design and Construction*, Sixth Edition. Burnt Mill, England: Longman Scientific and Technical.

[23] For H piles, use perimeter = 2 X flange width ("box" method).

[24] Both this figure and the next are taken from Meyerhof (1976).

[25] Both this method and the next are taken from Meyerhof (1976).

[26] Design load to be applied to the pile group divided by the group area.

[27] American Association of State Highway and Transportation Officials [AASHTO], (1992). *Standard Specifications for Highway Bridges*. Division 1 and 2, Washington, D.C.

[28] Tomlinson, M.J. (1994). *Pile Design and Construction Practice*. Fourth Edition, E & FN Spon, London, 411.

[29] Gouderault, P. and Fellenius, B.H. (1994). *UNIPILE Program Background and Manual*. Unisoft, Ltd., Ottawa.

[30] Tawfig, K.S. (1994). Polyethylene Coating for Downdrag Mitigation on Abutment Piles. *Proceedings of the International Conference on Design and Construction of Deep Foundations,* Vol. 2, 685-698.

[31] This is taken from *Lateral Load Capacity of Vertical Pile Groups*, Report NYSDOT-ERD-77-RR47. Albany, NY: New York State Department Of Transportation, Engineering Research And Development Bureau, 1977. The immediately preceding table is also taken from this source.

[32] It is suggested that for β L values between 2.0 and 2.5, both long and short pile criteria should be considered in Step 9, and then the smaller value should be used.

[33] Inspection of Figure 4-39 and Figure 4-40 for cohesionless soils indicates that the ultimate load P_u is directly proportional to γ, the effective soil unit weight. As a result, the ultimate load for short piles in submerged cohesionless soils will be about 50% of the value for the same soil in a dry state. For long piles, the reduction in P_u in somewhat less than 50% due to the partially offsetting effect that the reduction in γ has on the dimensionless yield factor. In addition to these considerations, it should be noted that the coefficient of horizontal subgrade reaction K_h is less for the submerged case (Table 4-11) and thus the deflection will be greater than for the dry state.

[34] The computer program is called COM-624 and is available from various U.S. government agencies. It has been produced in an MS-DOS version, which can be downloaded with documentation from http://www.vulcanhammer.net. A complete description of both the computer program and the theory of Reese's Method is found in L.C. Reese, L.A. Cooley and N. Radhakrishnan, *Laterally Loaded Piles and Computer Program 624G*. Technical Report K-84-2. Vicksburg, MS: U.S. Army Corps of Engineers, Lower Mississippi Valley Division, April 1984. A commercial program entitled LPILE embodies the same analytical method as COM624G but has a superior user interface, built-in "standard" soil and rock models based on more current load tests and other technical advantages. It is available from Ensoft, Austin, TX.

[35] One of these is the method of Evans and Duncan, described in Coduto, D.P. (2001) *Foundation Design: Principles and Practices*. Upper Saddle River, NJ: Prentice Hall. These are best employed for preliminary analysis or as a check for the computer solution.

[36] Holloway, D.M., Moriwaki, Y., Stevens, J.B. and Perez, J-Y (1981). "Response of a Pile Group to Combined Axial and Lateral Loading." *Proceedings of the 10th International Conference on Soil Mechanics and Foundation Engineering*, Boulimia Publishers, Stockholm, Vol. 2, 731-734.

[37] Brown, D.A., Morrison, C. and Reese, L.C. (1988). "Lateral Load Behavior of Pile Group in Sand." American Society of Civil Engineers, *Journal of Geotechnical Engineering*, Vol. 114, No. 11, 1261-1276.

[38] McVay, M., Casper, R. and Shang, T-l. (1995). "Lateral Response of Three-Row Groups in Loose to Dense Sands at 3D and 5D Pile Spacing." American Society of Civil Engineers, *Journal of Geotechnical Engineering,* Vol. 121, No. 5, 436-441.

[39] Brown, D.A., Reese, L.C. and O'Neill, M.W. (1987). "Cyclic Lateral Loading of a Large-Scale Pile Group in Sand." American Society of Civil Engineers, *Journal of Geotechnical Engineering*, Vol. 113, No. 11, 1326-1343.

[40] Ruesta, P.F. and Townsend, F.C. (1997). "Evaluation of Laterally Loaded Pile Group at Roosevelt Bridge." American Society of Civil Engineers, *Journal of Geotechnical and Geoenvironmental Engineering*, Vol. 123, No. 12, 1153-1161.

[41] Rollins, K.M., Peterson, K.T., and Weaver, T.J. (1998). "Lateral Load Behavior of Full-Scale Pile Group in Clay." American Society of Civil Engineers, *Journal of Geotechnical and Geoenvironmental Engineering*, Vol. 124, No. 6, 468-478.

[42] Moss, R.E.S. (1997). *Cyclic Lateral Loading of Model Pile Groups in Clay Soil, Phase 2B*. Master Thesis Research, Utah State University.

[43] Brown, D.A. and Bollman, H.T. (1993). "Pile-Supported Bridge Foundations Designed for Impact Loading." Appended Document to the *Proceedings of Design of Highway Bridges for Extreme Events*, Crystal City, Virginia, 265-281.

[44] The complete derivation of these formulae is found in Jacoby, H.S. and Davis, R.P., *Foundations of Bridges and Buildings*, Third Edition. New York: McGraw Hill and Company, 1941.

[45] This is not to be confused with the coefficient of restitution of cushion materials that is used in wave equation analysis.

[46] Much of this introduction to wave equation analysis

comes from Lowery, L.L, *Pile Driving Analysis by the Wave Equation*, Bryan, Texas: Wild West Software, 1993.

[47] This program can be downloaded from McTrans (http://mctrans.ce.ufl.edu) or The Wave Equation Page for Piling (http://www.vulcanhammer.net/wave). The MICROWAVE program can be downloaded from this last site or from http://lowery.tamu.edu.

[48] Much of this data—especially the hammer data—is usually furnished with the wave equation program being used. The user, however, should check all data before use. For example, hammer data can be altered by contractor modifications or variations in the manufacturer's product that are not reflected in the wave equation data.

[49] Viking, K. (2000) "Full-scale field-test study of dynamic soil resistance of vibratory driven sheet piles." Presented at the Sixth International Conference of the Application on Stress-Wave Theory to Piles, Sao Paulo, Brazil, 11-13 September 2000.

[50] Where n = i, o, or t depending upon relative position of pile and soil in question.

[51] The design load on front row piles is 120 kips (534 kN) whereas the middle and rear row piles have smaller design loads.

[52] For simplicity, each pile group is assumed to have an equal number of piles in each row. Usually, abutment pile groups require smaller number of piles in the middle and rear rows than those required in the front row.

[53] This is the depth from the existing ground surface to the pile toe.

[54] The computations start at the pile toe and these values are taken at points where q_c was measured.

[55] This figure shows a range of values from 0.5 to 1.20 for cylindrical piles 50' (15.2 m) to 100' (30.5 m) long.

[56] This allowable pile capacity does not include negative shaft friction load.

[57] Assume that all load is transferred at the pile toes. See Figure 4-26 or Figure 4-27.

[58] Settlements should be computed for the design load to be imposed on the pile group. Settlements should not be computed for the ultimate or allowable pile group capacities.

[59] Pile group area is used, not pile cap area.

[60] Elastic pile compression is very small which is why it is normally neglected.

[61] It should be noted that the SPT method for estimating group settlement provided a total estimated settlement of 0.3" and is the more conservative method (results in a higher settlement) than the CPT method.

[62] Preloading the soil with additional, temporary surcharge for the necessary time prior to pile installation can reduce long-term settlement. Another alternative is to design the bridge superstructure to tolerate the estimated settlement. In this case, provisions would have to be made to adjust the bridge shoe heights periodically. An expensive alternative would be to drive the piles down to an incompressible layer. In this case, the incompressible layer is at great depth and therefore not recommended.

[63] For elastic pile compression computations, it is assumed that each of the piles in the group are loaded with the 120 kip design load. This assumption is conservative because piles in the middle and rear rows will be loaded with smaller loads.

[64] Elastic compression is small; therefore it can usually be ignored, especially for short piles.

[65] Bitumen coating should only be applied to the top 39' of pile length. Batter piles should be avoided if possible to avoid excess bending stresses and pile breakage.

[66] Permission to use GRLWEAP, University Version 2003, for this book and the screen shots of the program output are courtesy of GRL Engineers.

[67] Such methods include Rowe's Moment Reduction Method.

[68] Always use an even number of piles. In addition, the minimum number of piles should be such that no pile spans an arc more than ten degrees from interlock to interlock. This means that the minimum number of piles for a cofferdam should be 360°/10° = 36.

Chapter 5. Installation of Driven Piles

Few fields of construction can generate more unpleasant surprises than the installation of deep foundations in general and pile foundations in particular. These "unpleasant surprises" can result in lost time and money; they can and have broken contractors and suppliers and put owners in untenable situations. Fortunately, most of these problems can be avoided with proper planning, contingency provisions and competent, experienced personnel on site.

5.1. Before The Project: Bidding, Specifications and Estimating

5.1.1. Specifications

For any project in which there is an interest, obtain the plans as soon as they are available and review them promptly. This will give an idea of what is to be done and the crews and equipment needed. Consideration can be given to what people and machines may be available when the project must start. It can be determined if a firm may be in a competitive position. Sometimes it may be desirable to fully estimate and price a project it is not practical to bid to keep abreast of the market.

Bid documents can be improved; any contractor's estimator, field engineer or lawyer will gladly confirm this, as will the engineer or owner's field forces. Even the specification writer might be enthusiastic about an opportunity to do a better job.

Encouragement for cooperation among all concerned parties can help. Specification writers can get help from their own field forces who have had to work with their previous documents. Contact should be established early with the field supervisor; he can be asked to mark up a set of project documents for return to the specification writers. Contractors and engineer/estimators will usually be pleased to review proposed materials and discuss on a friendly basis what can be done to minimize problems. Competent contractors are willing to take the time for this as a contribution to reducing cost of construction and obtaining early knowledge of a project.

Many times change of a phrase or suggestion of a different item can avoid problems for the project and save money for the owner. Often on foundation work, a review of the borings and draft specifications by an experienced pile person can result in improved requirements or elimination of expensive and unnecessary stipulations. If later change orders, with their attendant paperwork, can be avoided, all parties have more time for other activities.

A foundation project of any magnitude should have specifications specially written by a person with technical knowledge of the foundation problems of the project. Too often specifications are assembled by cutting sections from previous documents and pasting them into a mixed assembly that just invites misunderstanding and litigation. If a foundations specialist and field person cannot be assigned to help with the writing, they should review the draft documents - with definite instructions to be critical and to make suggestions. Of course, final arrangements should be made by the professional specifications writer.[1]

5.1.1.1. Supply All Information

Place borings in the bid documents. Reference where and how the geotechnical report can be reviewed. Soil and rock samples can be included if available. All known information pertaining to a construction site should be made available to bidders. Courts have held that information either an owner or the engineer has or should have had about subsurface conditions is pertinent to a contract. Not supplying all available information can be a basis for a claim. If a project presents difficult site conditions for pile installation (such as boulders), say so in the bid documents. If all parties are aware of difficulties, the project should be staffed and equipped for it. Instead of an adversary relationship, the field forces will work together to meet problems. Courts may consider such warnings in case of litigation.

5.1.1.2. Changed Conditions

A changed conditions clause in a contract can be helpful to the contractor and save money for the owner. With such a clause, an estimator can minimize the contingency reserve. If there is no problem, the owner saves. If there is a problem without a change of conditions clause, the contractor will file a claim. Litigation is an expensive and lengthy process whose only real victors are usually the attorneys.

A changed conditions clause suggested by the Association of Soil and Foundations Engineers follows:

> Subsurface Conditions Clause
>
> The owner has had a subsurface investigation performed by a foundation consultant, the results of which are contained in the consultant's report. The consultant's report presents his or her conclusions on the subsurface conditions based on his or her interpretation of the data obtained in the investigation. The contractor acknowledges that he has reviewed the consultant's report and any addenda thereto and that his or her bid for earthwork operations is based on the subsurface conditions, as described in that report. It is recognized that a subsurface investigation may not disclose all conditions as they actually exist and further, conditions may change, particularly ground water conditions, between the time of a subsurface investigation and the time of earthwork operations. In recognition of these facts, this clause is entered in the contract to provide a means of equitable additional compensation for the contractor if adverse unanticipated conditions are encountered and to provide a means of rebate to the owner if the conditions are more favorable than anticipated.
>
> At any point in time during earthwork operations that the contractor encounters conditions that are different from those anticipated by the foundation consultant's report, he shall immediately (within 24 hours) bring this fact to the owner's attention. If the owner's representative on the construction site observes subsurface conditions that are different from those anticipated by the foundation consultant's report, he shall immediately (within 24 hours) bring this fact to the contractor's attention. Once a fact of unanticipated conditions has been brought to the attention of either the owner or the contractor, and the consultant has concurred, immediate negotiations should be undertaken between the owner and the contractor to arrive at a change in contract price for additional work or reduction in work because of the unanticipated conditions. The contractor agrees that the (following) unit prices would apply for additional or reduced work under the contract.

5.1.2. Pay Schedules and Assigned Risk

An equitable schedule of payment for piling can result in the best overall price to the owner and minimize adversary developments. Money, of course, is the problem. Specification writers and designers should understand that there is a substantial difference in conditions where piles are known to drive easily and uniformly and where obstructions may be encountered. The cost of splicing also can vary enormously. Provision for equitable pay will bring the best bid price.

Most specifications state that the contractor must take all the risk. However, he does not; he puts in as much contingency as he thinks the project will stand and hopes for the best. If trouble develops, he can ask for relief.

Specifications may state, for example, that if toe protection is needed the contractor will supply and use them at no increase in the contract price. This is unfair to the contractor and actually specifies controversy on the project. The owner selected the land, with its obstructions. The engineer has had weeks or months to study borings and obtain additional data if needed. The contractor has the plans for a relatively short time, usually with little opportunity for additional investigation. Besides, it would be wastefully expensive if each bidder made a separate subsurface investigation.

5.1.3. Mobilization

For foundation work, a separate bid item for mobilization of equipment should be used. Mobilization is put into the bid as one lump sum item. Then, if quantities of excavation and length of piles vary, the usual inequities of changed quantities are minimized. If a Mobilization item is not used, the contractor is likely to "load" the items on which he can get the earliest return, that is, excavation and piling. These are also the items most likely to vary in quantity. If a high unit price is assigned to piling bid on a per foot of length and quantities increase, the contractor makes a windfall profit. If piling is short, the contractor gets much less return to apply to bringing in the equipment required for pile installation. He will immediately ask for renegotiation of a changed condition. Most specifications have an automatic renegotiation clause for a 15% quantity change.

5.1.4. Prompt Payment

Provide in the bid document for prompt partial and final payment. On large projects, partial payment for materials on hand should be helpful. Such provisions indicate an awareness of contractor's needs and financing problems. The owner probably obtains financing at much lower rates than the contractor so the project should be bid at a lower overall cost to the owner.

5.1.5. Negotiating as a Pile Subcontractor

Several items require consideration and discussion in developing a subcontract for pile installation: time to start and complete a project; space for pile storage and an office; labor arrangements should be compatible; excavation for pile caps-if done by others-to be such that pile equipment can move where required; who provides security for the subcontractor's equipment and for driven piles. Most of these items, and many more, are covered in standard forms of subcontracts.

Be sure to have an understanding with the owner's comptroller that what is promised should be done. Arrange in advance for the mechanics of preparing, routing and processing the estimates. Having money for payroll and for taking discounts on invoices makes a big difference in a contractor's financing and his or her attitude toward the project.

5.1.6. Estimating

"Do it yourself" is the best way to estimate, plan and bid a project. There are competent quantity take-off organizations that will provide accurate quantities, but not full knowledge about the project. If a firm gets the work, its personnel then have to go through this carefully. The commercial take-off quantities can be invaluable as a check on materials needed-and sometimes a reminder of an item that may have been missed. However, such firms are fallible.

Read the entire specification; make marginal notes and dog-ear or paper clip pages for restudy and perhaps discussion with associates. Don't take anything for granted.

While taking off quantities, planning the project can be done: How many drivers should be needed? What hammers will probably be most effective? Will fixed or hanging leads be most suitable? Most important, how many working shifts should be required?

For foundation work, examine the borings. For all projects, review all subsurface investigation reports. A prospective bidder is entitled to all subsurface information that is known to exist.

5.1.7. Noise, Neighbors, Workers and OSHA

Excavators and pile drivers are the first on a project. They are likely to feel the full thrust of environmental enthusiasts' activities. It is good business to investigate the existing regulations and potential for increased control of noise and air pollution at any site where construction is planned.

Pile driving is accompanied to varying extent by noise generation. Noise levels at 23' (7m) are typically 90-115 dBA for impact hammers and 70-90 dBA for vibratory ones. Moreover, one can expect at the same distance 100 dBA from the crawler crane and 85 dBA from the air compressor; however, the impulsive noise of an impact hammer is more disturbing that the steady noise of, say, the air compressor. If the pile project is in an area where pile-driving noise may annoy the neighbors, special record keeping and precautions may be in order. Additionally, OSHA (Occupational Safety and Health Administration) has very stringent rules on protecting the hearing of workers on construction and industrial projects. Air pollution from equipment emission and from job dust and debris also is a problem.

Noise from construction is being severely limited to not more than 85 dbA at 50' (15 m) from the machines. Compressors for use with steam/air hammers have achieved this and lower. One method that can be used with air/steam hammers to mitigate the noise is an exhaust muffler. An example of this is shown in Figure 5-1. Steel piles, where exposed for some length above the ground, have a loud metallic ring when struck with a hammer. Enclosing hammers - or at least the impact area and pile, as shown in Figure 5-2 - in acoustically lined shrouds helps.

Figure 5-1 Exhaust Muffler for Air/Steam Hammers

Figure 5-2 Noise Shroud for Pile Driving

5.1.8. Considerations Due to Pile Driving Induced Vibrations

Since impact or vibratory hammers drive piles, ground vibrations of some magnitude are almost always induced into the surrounding soils during pile installation. Damage to nearby structures can result from vibration induced soil settlements or from the effects of vibrations on the structure itself. If a sensitive existing structure is located within approximately 150 meters of the pile driving location, vibrations or vibration induced soil densification may result in settlement damage to the existing structure. In many highway projects, vibrations are of limited concern, as surrounding structures are often greater than 150 meters from the location of pile driving.

For projects in urban areas, and for widening of existing bridges, the proximity of existing structures is often within the zone of potential damage. Careful evaluation of the pile driving procedures and/or monitoring of ground vibrations during pile installations should be performed for these projects. Wiss[2], reported "safe" levels of ground vibration have typically been recommended between 12 and 100 mm per second. Lacy and Gould[3] found that vibration induced soil densification settlements and structural damage can occur at peak particle velocities much less than 50 mm per second and that soil gradation is an important factor in this phenomenon. For a specific project, the ground vibration level where structural damage may occur will be dependent upon the type of soils, pile type(s), pile hammer, pile installation techniques, as well as the condition and type of existing structure.

If the potential for damaging ground vibrations is high, pile installation techniques should be specified to reduce vibration levels. Specifications could require predrilling or jetting as well as use of a different pile type or use of a specific type of pile hammer. Since predrilling and jetting influence compression, uplift, and lateral pile capacities, a determination of probable vibration levels and remediation measures should be evaluated in the design stage. A case history illustrating how a change in pile installation procedures reduced vibration induced densification and off-site settlement damage was reported by Lukas and Gill[4].

Woods[5] provides a synthesis of pile driving induced vibrations and typical mitigation practices. This synthesis noted that vibration problem management is the key to minimizing vibration damage, delays and claims. Two important elements in vibration management are a vibration specification with limits on the maximum peak particle velocity and a predriving survey of surrounding structures. An example vibration specification that details the requirements of a preconstruction survey as well as particle velocity controls is included. The predriving survey needs to document conditions within the potential effected area. Woods reported that vibration damage a distance greater than one pile length away from driving is relatively uncommon but settlement damage in loose clean sands can occur up to 400 meters away. Woods also concluded that piles with low impedances, EA/c, tend to transmit the hammer energy to the soils along the pile shaft and thus increase ground vibrations, whereas piles with higher impedances tend to more effectively transmit the hammer energy to the pile toe resulting in lower ground vibration levels. Hence, selection of a stiffer pile section at sites where vibrations are a concern may reduce vibration problems.

5.2. Planning and Project Organization

A good start is one of the best ways to achieve a successful completion. The contractor must know what he or she is required to do, how he or she is going to do it and be prepared to cope with difficulties that bound to arise.

Pre-project planning should be as detailed as time permits. Unfortunately, the contractor may not have time to plan as thoroughly as his or her judgment indicates. Even the simplest project should have every point covered in this section considered.

With a quantity estimate and a general idea of the project, a visit to the site to study of field conditions is next in order. Experienced field forces can be most helpful in planning the project. If an engineer and superintendent who may get the project assignment can assist with the estimating, it will bring practical input. They can suggest the optimum work force and outline the equipment most suitable for the project. Just as important, the field supervisors will then have an added responsibility to make their ideas work and understand why some were not used. They should be part of the team from the start.

5.2.1. Study the Project

The superintendent should receive both a written and an oral briefing on the project scope and be given copies of the following documents:

1. Contract

2. Specifications
3. Plans and Borings
4. Estimate and estimate work sheets
5. Project Data or Information Sheet
6. Method of Pricing; Results hoped for by Estimator
7. Copies of orders, shipping documents, etc.
8. Set of forms
9. Insurance Certificates and Safety Information

All of these items should be reviewed completely and the parts applicable to contracted work studied in detail, before starting. Items 4 and 6 should be treated as confidential and not left lying around the project office or easily accessible in the computer.

5.2.1.1. Contract

The contract with the client outlines what the contractor must do or not do and how it should be paid for its work. Be sure the superintendent understands:

1. Scope of the work; number of piles and base length or aggregate footage; any variations between estimated quantities and bid quantities.
2. Other items of work, such as test piles, load tests or reinforcing.
3. Method of payment for various items of work.
4. Provisions for payment for longer or shorter piles.
5. Time of starting and completion; specified progress, if any; penalty clauses.
6. Provisions concerning order in which project is to be done, if any exist.
7. Any delays anticipated and provisions in the contract for the reimbursement for them.
8. Who is responsible for grading, excavations, sheetingandbracingremovalofpre-excavated or heaved material (if contemplated), pumping and drainage and protection of adjacent structures, provision for payment for piles driven through overburden and who cuts off piles after excavation.
9. Responsibility for permits, access and ramps, maintenance of access roads for rig and material trucks, provision of storage for shells and pile materials, sites for offices, parking and tool houses, and utilities.
10. Provisions for removal of overhead, lateral or underground obstructions, payment for delays if not done in advance, acceptance of piles hitting obstructions, responsibility for footing redesign and replacement or additional piles.
11. Responsibility for furnishing boiler, compressor, drilling or jetting water.
12. Responsibility for staking out piles, maintaining stakes and giving cut-off grades for each pile, logging piles and making "as built" survey and drawings.
13. Special provisions for furnishing other services or materials, such as reinforcing steel.
14. Provisions covering extra work or countercharges by either party.
15. Inspection and tests; who performs and who pays for these services.

As the superintendent studies the contract, he or she should make notes about anything that is not clear or which he or she thinks will cause a problem. He or she will then be able to discuss it with his or her superior(s) before the work is started.

5.2.1.2. Specifications

Specifications are usually part of the contract by reference. Specifications should be read and understood. Are piles to be driven to resistance only, or is a specified bearing stratum to be reached? Is a minimum penetration (tip elevation or penetration into bearing stratum) required? What alternatives are permitted? What is the intended function of the piles (resistance to downward load; resistance to lateral load, resistance to uplift, compaction of soil, etc.)? Project specifications should present this information. Study the table of contents and check the sections that apply to your work, such as:

1) General Conditions
    a) Authority of architect or engineer to amend specifications (usually only in writing).

b) Notice required to be given on discovery of errors or omissions in plans or specifications, unforeseen soil conditions, etc., to whom, when and how.

c) Provisions regarding delays and extra work and procedures to be followed before payment (such as having orders in writing).

d) Requirements concerning interruptions to nearby operations - railroad tracks, power lines, production lines, etc.

e) Safety restrictions, noise limitations, limitation on open flames, permits required for welding and cutting. Etc.

f) Description of special hazards, underground hazards, buried cables, gas mains, etc.

g) Right of owner or contractor to suspend work.

h) Equal Opportunity requirements.

i) Payroll submittal requirements.

2) Piling Specifications

a) Driving criteria and size of hammer to be used.

b) Minimum pile length, specified toe elevation or minimum penetration into bearing stratum, if any; alternatives if these cannot be reached without excessive driving.

c) Pile type, minimum head, toe, or average diameter.

d) Hammer cushion specification or limitation.

e) Provisions or limitations on jetting, spudding or pre-excavation.

f) Driving tolerances as to plumbness, location, or curvature.

g) Procedures to be followed if pile hits obstruction.

h) Concrete placing, concreting radius, dewatering requirements.

i) Pile capacity testing criteria

j) Test pile specifications; number, location, driving criteria, cutting off, concreting, etc.

k) How load tests, if any, are to be conducted; can driving be continued during load testing?

3) Material Specifications

a) In addition to driving specifications, material requirements pertaining to the specific types of material you are supplying should be reviewed[6]:

b) Concrete materials; mix design and approval; strength requirements.

c) Precast piles, design and fabrication

d) Reinforcing steel and placing reinforcing steel or dowels.

e) Structural steel for H-pile piles; pipe for piles; sleeves and plates; splicing or welding.

f) Structural timber; wood piles; treatment; handling; shoes or caps.

g) Tests of materials and reports to be furnished; concrete test cylinders, mill certificates for H-pile and pipe piles and inspection reports on precast and wood piles.

4) Plans and Borings

a) Review general plan to determine overall scope of project; review notes.

b) Review sheet or sheets covering foundation or piling layout and carefully read all notes.

c) Study foundation detail drawings:

i) Check notes; check dimensions on piling sketches

ii) Study pier layouts and pile spacing

iii) Be alert for isolated piles or piles with unusually high or low cut-off

iv) Note pile cut-off elevations and where you may have to drive through overburden or in deep excavations

v) Study plans, notes and symbols looking for batter piles.

vi) Count number of piles of each type or length and check with contract and estimate; if difference exists count again, being sure you find isolated piles, resolve differences.

vii) Plot location of borings on pile plan to scale and note boring numbers.

viii) Indicate on boring logs the average

cut-off elevation in area of the boring.

ix) Indicate on boring logs probable pile toe locations, based on requirement of specifications and your experience.

x) Note the excavation needed at various locations with relation to existing grade or street level and pile cut-off elevation.

xi) Compare your probable pile lengths with those anticipated by the estimator.

xii) Estimate sheet gauges required and compare with those estimated and on initial order, if any.

xiii) Make note of all items you want to discuss with your superior.

5.2.1.3. Estimate and Data Sheet

Review the schedules of the estimate and project data sheet with particular reference to:

1. Number, type, loading, length of piles.
2. Driving progress and planned sequence of driving.
3. Schedule, including allowance for on and off, tests, weather holidays and delays.
4. Shell gauges figured; allowance for shell waste and initial shell order placed.
5. Labor crew and rates.
6. Equipment estimated, equipment planned and its source; allowance for freight and set up.
7. Concrete, re-steel and other pile material required; anticipated sources and orders placed to date.
8. Special insurance requirements or hazards.
9. Compare all the above with contract, plans and specifications and make note of any differences to discuss with general superintendent.
10. Review allowance made in estimate for other items and note any you think you may improve on or may be inadequate and why you think so.

5.2.1.4. Site Visit

A site visit should be made before commitment of equipment and labor. Preferably, both the general superintendent and the Project superintendent should make it together. Specific items that should be covered include:

1. Progress made to date by others, excavations, access roads, etc. and order in which site should be available to us.
2. Access to site; road conditions, truckload limits on bridges, height or width restrictions, etc.
3. Overhead obstructions at or near site which might interfere with setting up or operation driver; power lines, overhead pipelines, etc.
4. Nature and condition of adjacent structures or facilities that might warrant special precautions or observations of possible movement; need for pre-project survey.
5. Nature of nearby buildings; hospitals, instrument factories, etc., that might be affected by noise or vibration of driving.
6. Adjacent or nearby activities that might present unusual fire hazard, extraordinary public liability exposure or danger to the employees or equipment.
7. Stability of the ground for safe movement of rig.
8. Evidence of possible underground facilities such as pipelines, cables or tunnels.
9. Availability of space for setting up equipment, temporary buildings, storage of material, welding yard, etc.
10. Availability of water and other utilities.
11. Determine ability of concrete plant and other local suppliers to meet specifications and delivery requirements.
12. Labor availability.
13. Plans and schedules of the owner, the general contractor and others for doing work related to the contract.
14. Discuss your requirements and plans with local representative of the client.

15. Street address of site and how to reach from nearest main highway and both light and heavy loads.
16. Location and designation of nearest railroad siding and name of railroad, if any shipments may come by rail.
17. Mail address and name of nearest post office; make application for post office box when needed.
18. Telephone number of client or general contractor's field office; make application for telephone service if so instructed by general superintendent.

Edmund Prentis, a founder of the underpinning and foundation firm of Spencer, White and Prentis, was fond of telling that if a firm he knew had raised all bids ten percent they would have missed the toughest foundation projects and, over the years, made more money. Prentis also tells of being very low on a bid for a foundation project at Columbia University. The estimate had been made several months previously, and then shelved because of postponements. With a new request, the old estimate was dusted off and used. What a then five-cent subway ride would have disclosed was that 30,000 yd^3 (23,000 m^3) of rock from other excavation had meanwhile been dumped in the area. The moral of this incident is clear: always visit the site and check everything.

If the owner-engineer arranges a prebid conference, be sure to attend-with all the questions developed in estimating and planning. Do not hesitate to suggest an addendum to the specifications if it should be helpful. Nevertheless, remember, all bidders will share your ideas.

5.2.2. Overall Plan

Having studied the project, formulate a plan for doing the work that answers:

1. How can you best approach this project to minimize costs and maximize profit?
2. What items of equipment should be required?
3. What labor and supervision should be needed?
4. What pile materials need to be ordered?
5. What are the best methods to follow?
6. What progress can you expect?
7. What is the best sequence or schedule for the work?
8. What problems are you likely to encounter and how should you prepare now to cope with them?
9. How can you reduce the risks inherent in the project?

Pile locations should be accurately established by survey and plainly marked ahead of driving. A chart should be prepared identifying each pile by number; this identification system shall be used in making a record of length, final driving resistance, location of the driven pile and any unusual occurrence with its installation.

The best answer to some of these questions may require improvisation or compromise to fit the people, materials and equipment available. Those parts of the plan that affect others will have to be discussed and any differences eliminated. Finally, the superintendent and general superintendent should develop a workable plan, acceptable to everyone.

5.2.2.1. Equipment and Personnel Planning

Your construction manager should normally have completed a broad overall equipment planning. Depending on the type of pile and driving criteria, the hammer and driving accessories were first selected and determination made on the need for drills or jets, etc. The leader type and height depended principally on type of pile and its length, with provision for mandrel insertion when applicable. An air compressor and/or power pack are selected to meet needs of hammer and drill. When these requirements were determined, the general superintendent consulted the capacity charts of available rigs to select an appropriate crane and boom combination. The spotter, auxiliary and support equipment are selected to be compatible with the basic rig.

Based on these general plans, the superintendent should prepare detailed equipment lists by reference to the following check lists and to other parts of this book. Then, in consultation with the general superintendent, the specific pieces of major equipment should be identified, by number, present location and date available, and designates the source and availability of all other items. This checklist is a guide; different situations and equipment set-ups may require more or fewer items.

1) Equipment List - Driver
    a) Basic
        i) Core, adapter and point
        ii) Core head; core pins; core sling and clips
        iii) Followers or driving accessories.
        iv) Hammer; hammer modifications
        v) Extensions
        vi) Hammer cushion
        vii) Steam, air or hydraulic hose
        viii) Spare cores; spare hammer cushion shields and plates; spare core pins
    b) Drills and Jets
        i) Drill or rotary table
        ii) Water swivel
        iii) Stem or auger
        iv) Bits - auger, roller, fishtail - incl. Spares
        v) Guide frames, guide beam
        vi) Fairleaders and reeving
        vii) Auger or stem guides or supports
        viii) Jets and nozzles
        ix) Water or jet hose; gooseneck
        x) Hydraulic hose (incl. Drain)
        xi) Boom hose or piping
        xii) Spuds
    c) Leaders
        i) Main Section
        ii) Top and Bottom Sections
        iii) Intermediate Sections
        iv) Leader connecting bolts
        v) Boom-leader connecting pins
        vi) Fairleader - top, front, drill
        vii) Pile Lifting Bracket
        viii) Sliding Frame and Guides
        ix) Warning lights or special painting
    d) Boiler, Air Compressor or Power Pack
        i) Boiler and fittings - lubricator, burner, etc.
        ii) Boiler Platform with compressor
        iii) Straps, turnbuckles, cross-beam and adapters for mounting on rig
        iv) Counterweight plates
        v) Catwalks and railing
        vi) Firebrick, oil and water tanks
        vii) Hood and Stack
        viii) Piping and valves, boiler to boom
        ix) Air Compressor
        x) Hydraulic Power Pack
        xi) Mounting and mechanical connection to engine
        xii) Gauges, valves and controls, incl. Spares
        xiii) Safety devices, if required
        xiv) Temperature and pressure reducing of steam by others
    e) Crane and Boom
        i) Basic Crane
        ii) Independent Swing
        iii) Is the third drum needed?
        iv) Check capacity tables
        v) Boom - butt, inserts and top sections
        vi) Fairleader; boom-leader connecting link
        vii) Boom bolts, seat pins, sheaves and top pin
        viii) Topping Lift adapter or straps
        ix) Pendant cables to suit boom length, with pins
        x) Hydraulic requirements and capabilities
        xi) Steam or hydraulic lines, rear of crane to boom butt
        xii) Steam, hydraulic and water line in boom
        xiii) If rental; terms, insurance requirement, time, responsibility for modifications, etc.
    f) Spotter and Moonbeam
        i) Spotter
        ii) Moonbeam, long or short
        iii) Mounting plates for crane turntable
        iv) Spotter brace to suit crane

- v) Pins, front and back
- vi) Wire rope or hydraulic hoses
- vii) Hydraulic controls and control hoses
- viii) Platforms and handrails
- ix) Safety beams and cables

g) Driver Accessories

- i) Tugger hoists
- ii) Mats and slings
- iii) Blocking for cribbing and trestles
- iv) Steam, hydraulic and jet hose
- v) Pile lifting slings or gear
- vi) Cables; hammer, boom, #1 line, tuggers, 3rd drum, safety
- vii) Water supply hose, pipe and pumps
- viii) Fire extinguishers

h) Driver man's Tool Box

- i) Siphons with hose
- ii) Bailers
- iii) Shell Wrenches
- iv) Shell denesting tools
- v) Cutting outfit with spare tips
- vi) Slings and chokers
- vii) Cable clips; shackles
- viii) Manila rope
- ix) Pile Chain
- x) Leader-bolt socket wrench (2-1/4" across corners)
- xi) Pipe wrenches and other hand tools

2) Equipment List - Driver Support

a) Jetting and pre-excavating

- i) Water supply and jet pump
- ii) Pump mounting and suction hose with foot valves and strainer
- iii) Shut off and check valves; priming pump
- iv) Piping, valves, hose-pump to rig
- v) Other pre-excavators
- vi) Equipment for disposal of water and spoil

b) Concreting and Cutting-off

- i) Batching and Mixing equipment
- ii) Concrete pump, buggies and runs or crane and bucket
- iii) Steep sided concrete funnel, chutes, and dump tube
- iv) Wheelbarrow, hoe, shovels and grout buckets
- v) Slump cone, rod and scoop, trowel, cylinder forms and curing container
- vi) Torch, saw, internal cutter or other special cut-off tools
- vii) Dewatering equipment; bailer, siphon, etc.
- viii) Vibrator
- ix) Mirror and light for inspection

c) Water and Fuel Supply Equipment

- i) Transportation
- ii) Storage - tanks or barrels
- iii) Handling
- iv) Transfer Pumps

d) Shell or Pile Rack Equipment

- i) Shell Racks
- ii) Power and hand shell wrenches and denesting tools
- iii) Welding equipment
- iv) Welding and pipe racks
- v) Lifting equipment
- vi) Handling slings
- vii) Trimming or beveling equipment
- viii) Pile Heading equipment
- ix) Hauling equipment

e) Maintenance

- i) Lube equipment
- ii) Welders
- iii) Other shop equipment
- iv) Lubrication and maintenance manuals

f) Tools

- i) Driver man's Tool Box
- ii) Shop and mechanic's tools
- iii) Grease guns and fuel handling
- iv) Carpenters and dock builders - forms falsework, caps, etc.

v) Cement finishing and mason tools
vi) Plumbing and electrical
vii) Welders' tools and accessories

g) Test Equipment
i) Test Load Beam
ii) Test Load Platform
iii) Test Weights
iv) Plates, shims, and jacking columns
v) Jack with calibration chart
vi) Pumps and Gauges
vii) Dial gauges, scales, mirror piano wire, pulleys, surveyor level and rod
viii) Tarpaulins, shelters and protection
ix) Reinforcing for uplift pile
x) Other test equipment

h) Service and Administration
i) Office Trailer or Shanty
ii) Change Trailer or Shanty
iii) Tool Trailers; shop facilities
iv) Fuel and Lube handling and delivery
v) Shelters for inspectors, security guards, visitors, etc.
vi) Heaters or air-conditioners
vii) Office furniture; office machines and computers
viii) Surveying equipment
ix) Automobile and trucks
x) Transportation for people - buses, crew boats
xi) Telephone or other communication equipment
xii) Sanitary facilities
xiii) Safety and first-aid
xiv) Camps and feeding facilities

3) Material Check List

a) Pile Materials - General
i) Quantities required per week or month
ii) Storage space available at site
iii) Uploading facilities
iv) Possibility of change in pile length, size, specifications
v) Orders already placed
vi) Quantities yet to be ordered
vii) Number of piles each length
viii) Inspection, tests and samples
ix) Rail, truck or water shipment
x) Moving from unloading sit to rig
xi) On site processing; welding, splicing, trimming, denesting, etc.

b) Shells
i) Number of sections and type of each section
ii) Length and gauge of each section
iii) Nesting and bundling

c) Pipe
i) Specification for pipe
ii) Outside diameters and wall thicknesses
iii) Delivered lengths and tolerance
iv) Type of end - square cut or beveled
v) Boot plates
vi) Sleeves
vii) Mill test certificates

d) Wood piles
i) Kind of wood
ii) Quality - knots, twists, bark, etc.
iii) Head and toe diameter
iv) Length tolerances - straightness
v) Where and when inspected
vi) Pointing - shoes
vii) Top collars or plates
viii) Preservative treatment
ix) Changes in length
x) Marking and identification
xi) Treating butts after cut-off

e) Precast Piles
i) Concrete specifications
ii) Mixing and placing concrete
iii) Curing; strength requirements for driving
iv) Reinforcing steel

v) Number and type bars

vi) Size and shape

vii) Change in length

viii) Marking

ix) Pick up hardware

x) Cushions

xi) Protection of dowel holes

xii) Storage

xiii) Shop Drawings

xiv) Splicing

xv) Tests and reports

f) H-piles and Monotubes

i) Size and weight per foot

ii) Length and tolerances

iii) Kind of ends

iv) Plates

v) Type and gauge

vi) Mill Test Certificates

g) Concrete

i) Type - conventional or special mix

ii) Check total requirements

iii) Delivery rate - average and maximum

iv) Access to site

v) Delivery time - (nights, weekends?)

vi) Concrete tests and reports

vii) Waiting time

h) Fuel

i) Type - boiler, diesel, gas engine

ii) Estimated daily usage

iii) Storage available

iv) Delivery by vendor so as not to interfere with operations

i) Lubricants

i) Lube oil and hydraulic oil

ii) Gear oil

iii) Grease

iv) Steam Cylinder or Air Compressor Oil

v) Estimated usage of each

vi) Delivery and storage

j) Water

i) Quantities required for boiler

ii) Jetting and driving operations

iii) Mixing Concrete and Grout

iv) Sanitary use

v) Drinking water and ice

k) Lumber and Timber

i) Walkways

ii) Barricades and fences

iii) Scaffolding and staging

iv) Planks for driver support and leveling

v) Planks for concrete trucks and buggies

vi) Foundations for trailers and shanties

vii) Pile and Material Racks

viii) Temporary structures

ix) Timber for cribbing and trestling

x) Forms and falsework

xi) Plywood or planks for protection or surfaces to be crossed by rig

l) Other Consumables

i) Hammer cushion Plates - rate of use and availability

ii) Plastic slate, bentonite, jute or other joint material

iii) Welding rod - piles, maintenance, temporary structures

iv) Oxygen and acetylene - trim piles, cut-off, maintenance

v) Oil Filter cartridges - check engine manuals for type

vi) Air Filter cartridges

vii) Hydraulic filter cartridges

viii) Replacement hose - steam, air, water, jet, hydraulic

ix) Packing for pumps, engines, hammers

x) Hand hole gaskets for boilers

xi) Rags, waste, cleaning supplies

xii) Boots, gloves, raincoats

xiii) Flashlights, batteries and mirrors

- xiv) Replacement rope and cable
- xv) Hard hats, goggles and other safety supplies
- xvi) Concrete test cylinder forms

4) Personnel Check List
   - a) Supervision
     - i) Superintendent
     - ii) Field and Shift Superintendents
     - iii) Project Engineer
     - iv) Master Mechanic
     - v) Office Manager
   - b) Working Conditions
     - i) Union agreements; AGC books
     - ii) Crew size
     - iii) Hours of work and rates of pay
     - iv) Holidays and overtime
     - v) Business agents
     - vi) Shift work
   - c) Driver Crews (one per rig shift)
     - i) Supervisor
     - ii) Driver man
     - iii) Shell rack crew
     - iv) Operator
     - v) Fireman or oiler
     - vi) Dock builders
     - vii) Winch men and deckhands
   - d) Pile or Shell Unloading - Pile Racks
     - i) Supervisor
     - ii) Crane Operator and Oiler, if required
     - iii) Truck Drivers - Lift Drivers
     - iv) Welders
     - v) Laborers and Helpers
   - e) Concrete Gang
     - i) Concrete Supervisor
     - ii) Laborers
     - iii) Pump or Mixer Operator
     - iv) Concrete Truck Drivers
   - f) Bull Gang
     - i) Supervisor
     - ii) Pile drivers
     - iii) Laborers
   - g) Other than Pile Driving
     - i) Carpenters
     - ii) Ironworkers
     - iii) Electricians
     - iv) Boat Operators
   - h) Field Engineering
     - i) Surveyors and assistants
     - ii) Quantities Engineer
     - iii) Inspectors
     - iv) Test lab technicians
     - v) Cost Engineers
     - vi) Safety Engineer and Inspectors
   - i) Equipment Maintenance
     - i) Shop Supervisor and Equipment Supervisor
     - ii) Mechanics and helpers
     - iii) Lube Truck drivers and field mechanics
     - iv) Equipment Pool Drivers, Operators and Helpers
     - v) Equipment and Parts Clerk
   - j) Administration
     - i) Chief Clerk
     - ii) Timekeeper, clerks and stenographers
     - iii) Personnel Man and assistants
     - iv) Purchasing Agent and buyers
     - v) First Aid man
     - vi) Warehouseman, lift truck operators and helpers
     - vii) Watchmen and Guards
     - viii) Camp Manager - Cooks, Janitors, Housekeepers
     - ix) Chauffeurs and Bus Drivers; crew boat pilots
   - k) General
     - i) Number of people of each craft needed
     - ii) Availability of necessary skilled personnel
     - iii) Labor rates and area practices
     - iv) Training
     - v) Transportation

vi) Feeding

vii) Housing

l) Organization

i) Organize so that tasks to be performed can be assigned to one crew to do.

ii) Everyone should understand his or her responsibilities and have only one "boss".

iii) On smaller projects, the superintendent usually performs the engineering and administrative functions; rig supervisor will take care of maintenance.

iv) A full-time supervisor may be needed for as few as four inexperienced people doing complex tasks; the same supervisor may efficiently supervise ten or twelve experienced people doing repetitive work.

v) Organize by craft where skilled people are working independently or where work is limited.

vi) Organize by team where several crafts need to coordinate their work or when crew geographically separated; typically, the driver crew, tugboat crew or pile yard crew.

vii) Keep organization simple. Until project has over 25 or 30 people, there normally needn't be more than one supervisor between the superintendent and any person on the project.

viii) When a supervisor has less than ten people to supervise, he should usually be a "working" supervisor.

ix) Organize to take advantage of the capabilities, and minimize the limitations, of available personnel.

5.2.2.2. Owned, Purchased or Rented Equipment

At one time most pile driving equipment was bought and owned by contractors; however, with changes in the market and the advent of vibratory equipment, this has changed, and now a good proportion of equipment is rented from both manufacturers and equipment dealers. Although rental is an ideal solution in many instances to satisfy the contractor's equipment needs, it is not a panacea for all the problems.

The decision to rent or buy equipment should be based on some basic considerations, such as:

1. The frequency a contractor uses a certain piece of equipment. If a contractor has a long range need for a certain piece of equipment, he or she should purchase it. Rental equipment is ideal for the occasional user of pile driving equipment, or where the type or size being rented does not have a potential for repeated use.
2. The willingness and ability of a contractor to invest in the maintenance infrastructure, both human and capital, to maintain a fleet of hammers. When a piece of equipment is rented, the contractor is also renting the service capabilities of the lessor, especially for problems beyond normal wear and tear. For equipment that is owned, the contractor must provide these himself to insure continuous equipment operation.
3. The tax effects of the transaction, which are invariably complex and change from year to year.

With rented equipment, it is important for the contractor to perform basic maintenance on the equipment and to return it to the lessor in the same condition that it was sent to him, normal wear and tear excepted. Failure to do so will result in serious back charges and, in the case of some dealers, will result in the equipment being removed from the project, with potentially catastrophic consequences for the contractor.

First, of course, consider what equipment the organization owns that can be made available. Some of it may be on projects where need for it may be diminished. However, do not mobilize equipment in poor condition.

Equipment of all kinds and capacities is available on a rental basis, usually with an option to purchase. Renting is, in effect, instant equipment; the contractor gets what he needs, when and where he needs it, quickly. Rental firms are developing their heavy-duty inventories, offering more services than ever before and expanding their product lines. They have a selection of hammers making it practical to rent one that most nearly fits the soil, pile and driving conditions. Of course, owning equipment can cost less if the hammer, for example, can be kept busy but often the next pile project may be sometime in the future and a different hammer might be more suitable. Using a hammer because it is available may be unwise. In addition, rental equipment is promptly returned so others store, repair and finance. A contractor's funds are kept more flexible. Depreciation write-off is not questioned.

The economics of purchase vs. rental should be reviewed for major equipment for any project. If work for a major item can be foreseen for a year, purchase probably is economical. Purchase is also sensible if the equipment can be utilized for subsequent projects.

Sometimes leasing through a contract with your bank or equipment distributor may be desirable. This conserves capital and bonding capability; it may allow payment in tomorrow's inflated dollars. However, leasing may be more expensive than borrowing to purchase. Figure it both ways.

5.2.2.3. Subcontracting

Pile installers are likely to be a subcontractor rather than do much subcontracting. However, frequently the pile contractor must locate and stake the piles to be driven. This may best be subcontracted. Occasionally he may be required to employ (subject to approval) and compensate the pile inspectors.

5.2.2.3.1. Negotiating As a Pile Subcontractor

Several items require consideration and discussion in developing a subcontract for pile installation:

- Time to start and complete a project;
- Space for pile storage and an office;
- Labor arrangements should be compatible;
- Excavation for pile caps - if done by others - to be such that pile equipment can move where required;
- Who provides security for the subcontractor's equipment and for driven piles?

Most of these items, and many more, are addressed in standard forms of subcontracts.

Try to get agreement for payment in full in a specified reasonable time after the pile driving is completed. This can be far better for the sub than waiting until the last door is hung and painted, plus a few months, for the final payment. A prospective sub can offer a substantial price differential for this early payment, which releases money and bonding capability needed for other work.

5.2.2.3.2. Selecting a Surveyor

The pile contractor now usually must locate the piles - with his or her own forces or by contract. One crew can serve several drivers: sometimes the field engineer can do this for one or two rigs. On a subcontract, the general contractor may be able to handle the staking. There are competent firms that do this as a business. Do not just hope that survey work will happen. Arrange and get it in writing that it should be done and in advance of need. Also, get in writing, with either the general contractor or a survey organization how errors and omissions should be handled. Probably this is covered in "fine print" of a contract but read it so the share of responsibility of the pile subcontractor is known.

5.2.2.4. Getting the Picture

Pile installation brings noise and perhaps vibration to a neighborhood. In a sensitive area, particularly where structures and people are nearby, it may be worthwhile to make a site and noise survey before starting work.

5.2.2.4.1. Photos can be Evidence

Cameras are available that even the inexperienced can use to take pictures with sharp detail. Most organizations have a "shutterbug" that can take pictures. Use of a digital camera eliminates the need for film and processing, although the results should be printed. Pictures from a good camera can be enlarged to verify details and support memory of project forces. In addition, with a simple filing system the photos can be available after personnel have gone to other work. However, insist that pictures be captioned, dated and filed so they can be found and positively identified. If you are the project shutterbug, ask the supervisor about taking the pictures.

Take pictures of the site before equipment moves in and the excavation starts. Particularly take detail pictures of structures that are near pile driving. This is especially necessary of masonry structures in poor condition. Whether or not pile driving may hasten deterioration, it is likely to be claimed that it did. If cracks exist, take plenty of pictures and get the owner-engineer forces to verify.

Interior condition of structures should be examined to the greatest extent possible. Office and public buildings that are open to visitors' examination of hallways and open areas can probably be made on a casual basis. If there is any evidence of cracks or settlement, take detailed photos. Taking photos should be done with the knowledge and consent of the building owner. Ideally a representative should accompany the photographer-investigator and verify the conditions found.

The building owner has no obligation to permit such an inspection. If he does not cooperate, document this and transmit a copy to the principal having the new structure built. Keep a record of the reply, including the date. If there is no reply in a reasonable time, send a message in a way that delivery can be verified. Refusal to permit such an inspection can be an important factor in any subsequent litigation.

If cracks or other evidences of movement are found, a piece of tape or a crack gauge can be placed across them, marked and photographed. Further movement can then be verified. It is beyond the scope of this book to detail responsibility for damage to adjacent structures. This varies in different states and jurisdictions. If there may be a problem, it should be discussed with an attorney.

5.2.2.4.2. Photos as Project Records

Arrangements should be made to take photos regularly for the contractor's records as well as any required by contract documents. A video camera can be set up to record up to six hours of site activity. Studying the pictures with the pile supervisor and project superintendent often will show how lost time can be minimized. Showing the tape to the crew gets their cooperation and may increase production.

Pictures of interesting and unusual activities should be taken and processed promptly throughout the life of the project. They should show the good as well as being a record of problems encountered. Send good pictures, with captions, to a favored semi-technical magazine. Editors are pleased to have an opportunity to consider use of pictures and complete construction articles.

When the project approaches completion, place selected photos into an album with critical comments on how to do it better. It should be a good project record and help in bidding and constructing the next project.

5.2.3. Operating Plan

Now the superintendent is ready to get down to the planning of his or her day-to-day work. The superintendent should visualize each operation and anticipate the problems and difficulties that will arise. The better job done at this stage, the better able the superintendent should be to meet the unexpected and, rather than having to solve a "crisis", he or she will avoid any potential crisis.

The successful outcome and profitability of the project will depend on how thoroughly he does his or her planning.

1) Schedule and Progress
    a) Move on
    b) Set up
    c) Drive test piles
    d) Shut down for load tests (if required)
    e) Driving and normal rig delays
    f) Unplanned delays
    g) Dismantle and Move off
2) Move on and off
    a) Loading equipment elsewhere
    b) Move to site in proper order for erection
    c) Unloading, inspection, lubrication and minor repair
    d) Coordination with parts or accessories coming from other locations
    e) Availability of crews, unloading crane fuel and other supplies when needed
    f) Erect equipment; will service crane be needed?
    g) Move to area of work
    h) Move to dismantling area
    i) Dismantle
    j) Loading out or storing
3) Planning rigs moves
    a) Determine maximum safe operating radius and batter limitations
    b) Minimum moving necessary to reach all piles; use scale template of driver and move around pile plan

c) Avoid overhead obstructions, pier holes or other excavations, soft ground, hazardous areas, bracing, previously driven piles

d) Consider how concreting operations should be conducted and make sure that rig operations, particularly pulling piles and shells to rig will not interfere with concreting or cutting off

e) Rig moves should be coordinated with work of others, excavation and staking out ahead, pouring caps and setting steel behind us

f) Consider test pile locations, pile types, pile lengths, reaction piles

g) Begin and end in accessible places; don't box yourself in

h) Plan for easy delivery of piles or shells to rig without interfere with other work

i) Avoid interference between rigs if more than one to be used; keep as far apart as practical

j) Give consideration to water supply for boiler and jetting and disposal of water and spoil from pre-excavation

k) Requirements of contract or prior understandings with client regarding sequence of work

l) Location of access roads and ramps both for rig and for material delivery

m) Does nature or condition of adjacent structure require that nearest piles be driven first?

n) Avoid having to set up and dismantle more than once, if possible

o) With multiple rigs, use capabilities of each rig to best advantage

4) Driving Cycle

a) Total blows affected by pile type and length, hammer energy, soil conditions, driving criteria and any drilling, jetting or pre-excavation

b) Driving time is total blows divided by average blows per minute, plus time for pile to penetrate under weight of the hammer and stops for splicing, adding shell, etc.

c) Time allowance for raising hammer (and core, if used)

d) Swing, pick-up pile or shell, swing and shell up; this should be affected by point of delivery of pile or shell to vicinity or rig

e) Lay mats and move to site of next pile

f) Swing and spot for next pile; batter leads, if required

g) Drilling, jetting or pre-excavation

h) Raise drill, swing and spot pile

i) Retapping, if required

5) Normal rig delays

a) Planned maintenance, fuelling and equipment inspection

b) Hook up, disconnect and move water and jet lines and hoses

c) Start up and shut down, morning, noon and night

d) Change hammer cushions

e) Change equipment, such as cores and hammers, for reasons other than breakdown

f) Unpredictable obstructions or stuck cores

g) Extra moves

h) Load tests

i) Training and safety talks

j) Weather

k) Holidays

6) Unplanned rig delays

a) Breakdown of equipment

b) Run out of shells or piles

c) No stakes

d) Waiting for instructions on what to do next

e) Running out of water, fuel or steam

f) No access to pile locations

g) Lack of permits or licenses

h) Accidents and injuries

i) Strikes or other work stoppages

7) Pile Assembly Operations

a) Output normally based on driver requirements

b) Unloading facilities should be based on peak rates of arrival of material or allowance made for rail car, barge or truck demurrage charges

c) Storage before processing to allow for peak rates of arrival and interruption to transport - weather, poor condition of access roads, high or low water, accidents, etc.; delivery dates should lead dates needed as required

d) After-processing storage should be sufficient to take care of above or below average driving rates and pile assembly delays, such as breakdowns.

- e) Since project requirements, such as pile lengths, will vary, the lead-time on delivery and processing should be the minimum that will allow a reasonably safe margin.

8) Concreting
   - a) Should be kept as close behind driving as specifications and common sense permit, but out of heave range
   - b) Plan for maximum driving rate anticipated; if you get behind it is doubly difficult to catch up
   - c) Discharging directly from truck into pile is always best, if it can be done
   - d) Are access roads for concrete trucks adequate? A stuck truck hurts everyone
   - e) If truck access to piles impractical, alternate delivery requires expert advice; get help
   - f) Grout: receipt and storage of material; getting it to pile locations; mixing, placing
   - g) Dewatering; siphon, steam supply; bailing; blowing; pumping
   - h) Re-steel and cages, receipt and storage; make-up; moving to piles; placing
   - i) How cut-off? Moving air and gas or cut-off equipment around project; how dispose of cut-offs?
   - j) Concrete tests slump tests; making cylinders; temporary project storage, transportation to lab; testing

9) Pile Tests
   - a) Driving criteria
   - b) Concrete and cut-off
   - c) Set up Test
   - d) Test Methods
   - e) Instrumentation
   - f) Records and reports
   - g) Need for further tests
   - h) Need to drive extra piles in case original test fails

10) Special Problems
   - a) Night operations - lighting, access, transportation of people, security, availability of inspectors and other support personnel
   - b) Engineering - establishing base lines, bench marks and building lines, pile layout, maintaining stakes, cut-off elevations; checking pile locations, "as-built" drawings
   - c) Weather; equipment modification or protection; maintain water lines; boiler protection; protection of personnel
   - d) Security - daytime, nights and week-ends; visitors and public
   - e) Getting up steam or other start-up problems
   - f) Furnishing water - source, hauling or pumping, storage, delivery
   - g) Adjacent activities - special scheduling account of noise, rail movements, production operations; unusual hazards
   - h) Protection of adjacent building and public from hammer exhaust or other operations

11) Safety
   - a) Can all areas of project be crossed safely with or without mats or other aids available?
   - b) Must core be carried on brake during some moves or must it be removed and, if so, how will the core be transported?
   - c) Must rig operate near edge of slopes or near power lines; if so, what precautions should be taken?
   - d) What special safety warnings should be given to crews; protective clothing required?
   - e) Will material unloading and storage facilities be safe? Checking stockpiles
   - f) How will one insure that equipment and tools are inspected daily for safety defects?
   - g) Have local authorities licensed boilers and other equipment; if not, what are the procedures?
   - h) Will weather present some unusual safety problems - icing, sunstroke, mud, etc.?
   - i) Are there special public liability hazards present; is there need for survey or special precautions?
   - j) How will the need for working safely be brought home to each employee - talks, posters, etc?
   - k) Are adequate first-aid supplies available and someone available who knows how to use them?
   - l) Are there facilities available for transportation and treatment of the seriously injured?
   - m) Will barricades, fences and other safety devices have to be removed to facilitate operations? Any problems with restoring them?
   - n) Will open piles present a problem?
   - o) Will refueling engines present special hazards,

such as inability to shut down?

p) Any hazardous materials to be handled?

q) Will hazardous areas require no open fires, no smoking, etc.?

12) Maintenance

a) Operations and Parts Manuals available for all commercial equipment?

b) How will fuelling and lubrication be accomplished, are proposed equipment and supplies adequate?

c) How will filters, V-belts and other spares are stored so they are available when needed but not lost?

d) Do the operating schedules allow adequate time for planned maintenance?

e) What maintenance back-up support can be expected from nearest yard or equipment dealers; will it be adequate?

f) Schedule boiler washing, oil and filter changes

13) Unexpected Problems

Are any of the following problems likely to arise and if so, how will one cope with them (outside help, additional equipment, alternate plans)?

a) Piles running substantially longer or shorter than anticipated

b) Obstructions, buried cables, water lines, etc.

c) Unusual weather, high water, flooding

d) Equipment breakdown

e) Collapsing or tearing shells, leaking shells, broken piles; dog-leg piles

f) Excessive driving needed to meet some requirement

g) Excessive number of open piles

h) Piles driven out of proper location due to improper layout, carelessness or other cause

i) Delays by others; awaiting instructions from owner, engineers or inspectors; waiting for pile layout, pumping, excavation, or water

j) Stuck Cores

k) Heaving, freezing or relaxation of piles

l) Failure of load tests

m) Settlement or movement of nearby structures

n) Shut down by owner due to development of existing problem

o) Shut down by authorities due to environmental problems; noise, smoke, discharge of dirty water, etc.

p) Shut down due to interference with traffic, owners' production activities or other contractors

14) Relationships

Understand and plan proper relationships, if any, with each of the following:

a) Owner

b) General Contractor

c) Client

d) Other Sub-Contractors

e) Architect

f) Structural Engineer

g) Geotechnical Engineer

h) Resident Engineer and inspectors

i) Building Inspector or other public officials

j) Local Office of Insurance Company

k) Labor Unions

l) Public in general

5.3. Moving On and Setting Up

5.3.1. Scheduling

The ideal schedule is one that brings the rig, materials and operating personnel all together, ready to drive, at the proper time. All scheduling should be back figured from the date one expects to drive the first pile. Since the superintendent has more control over setting up the rig than over other elements, he will have to make allowance for the uncontrollable factors - weather, failure of others to meet promises, breakdowns, delays in transportation, etc.

A site visit, before ordering out crews or shipping equipment, is essential. It will indicate whether the work of others should be completed on schedule, particularly excavation, sheeting, dewatering, staking out piles, removal of obstructions, provision of access roads, ramps, rig set-up area and water.

Field operations are generally scheduled in the following order:

1. Arrival of Office Trailer and providing utilities for same; securing post-office box or

arranging mail delivery; erecting appropriate signs at entrance, if necessary.

2. Notifying offices of project address and telephone numbers, and address and telephone number of superintendent's living quarters.
3. Setting up computer equipment; making sure LAN is functional and all software, hardware and operating systems are working properly; making provisions for either modem or broadband (the latter for a construction site will probably be wireless) connection to the Internet; testing of Internet connection and of Internet-based project software (if any).
4. Notifying local union of expected starting date and arranging for crews.
5. Contacting local office of insurance company; obtaining telephone number of doctor, hospital, ambulance, police and fire departments and others to be contacted in emergency.
6. Obtaining home telephone numbers of general superintendent and others to be contacted in an emergency.
7. Arrival of tool boxes, tool shanties or trailer and light support equipment, such as welders, pumps, generators and repair facilities.
8. Setting up personnel support facilities - change shanties, sanitary facilities, transportation, parking area, etc.
9. Floodlighting, if night work is anticipated. Moving on and setting up at night should be avoided.
10. Obtaining water, fuel, steam or air cylinder oil and other lubricants, air and gas, hand tools, lumber and similar supplies.
11. Arranging for permits not provided by others including local boiler permit, if required; arranging for concrete mix, design and tests.
12. Setting up material unloading facilities, shell and welding racks; unloading and start processing pile material.
13. Unloading mats, cribbing and timbers, particularly those needed to erect rig.
14. Unloading and assembling the basic crane, including crawlers and boom.
15. Adding counterweight and boiler, if removed for transportation.
16. Assembling leaders, drill or jet guides, pile lifting brackets
17. Erecting leaders.
18. Installing hammer, core, follower, drills and jets.
19. Setting up jet pumps, testing equipment, concreting equipment and racks for assembly of reinforcing cages, if any.

5.3.2. Loading from Storage

If you are using stored equipment, you should:

1. Visit storage site and inspect the equipment that you plan to use; check condition; check components for standard dimensions at points where they are assembled; check that all necessary pins, bolts and links are on hand or on order. Take only what you need and don't cannibalize other equipment, except in emergency. Make note of and report the condition of remaining equipment and any parts or repair required.
2. Lubricate, paint and make minor repairs so that equipment will arrive on the project in creditable condition.
3. Review loading arrangements with the supervisor and truck driver; plan loads to balance weight and capacity of trucks. Drain wet spots and fill potholes, if necessary, to make access road passable.
4. Arrange for minimum fuel and tool requirements; arrange for crane, if required. Schedule arrival of trucks and trailers or arrange for spotting of rail cars or barge.
5. Dismantle or separate equipment as required for transportation; local truck driver should be able to advise you on legal limits, bridge limits, low clearances, etc.
6. Load out equipment in order required at project site; if this is not feasible, arrange for extra trailers so that items can be loaded without rehandling and still arrive on project in order needed for erection.

7. Leave storage site in a clean and acceptable condition.

5.3.3. Transportation

1. During the site visit, select access roads after inspecting for low load limit bridges, low headroom, limited clearances, unsuitable grades, etc. Plan with experienced truck person, if possible. Note exit numbers, route numbers, street names and landmarks to guide drivers to site over proper routes.
2. If rail transportation is planned, locate nearest appropriate siding for unloading. Secure the proper name of the siding and delivering railroad. Check siding on shell ordering ticket; start tracing cars.
3. If barge delivery is planned, contact towing company to determine how and when the barge should be delivered and any problems in bringing it to the desired unloading point.
4. Before making commitments to a trucker, check with your general superintendent to be certain you are dealing with a reputable carrier who will give good service at competitive rates. Not all trucking firms have the proper equipment, give equal service or charge the same rates.
5. Advise the yards and other projects that will supply equipment of proper destination and route to be followed. Maintain contact until equipment arrives to be sure you will have crews and facilities available for unloading.

5.3.4. Unloading

The most important consideration is selecting the proper site for unloading each piece of equipment and class of material. A good unloading site should:

1. Be accessible by all-weather roads, with adequate parking for vehicles waiting to unload, without undue interference with local traffic or other operations.
2. Permit equipment and materials to be moved to point of use with minimum of interference with other project operations and with a minimum of re-handling.
3. Provide sufficient space for temporary storage of maximum quantity of pile material expected to be on hand at any one time. This will require careful planning, particularly on multi-rig and multi-shift projects and where high driving rates are anticipated.
4. Provide for alternate sites as project progresses or in the event the first site is no longer usable.
5. The site for unloading the pile driver should preferably be as near as practicable to where the first pile should be driven.

Schedule crews and support equipment so that unloading can proceed smoothly with no lost time for crews and minimum of delays to trucks and rented cranes.

Whether a crane is used to assist in unloading and assembly will depend on availability of cranes and on economics. While most of the pile driver cranes can assemble themselves, it will usually pay to have a crane available to assist if the crane has been disassembled. A helper crane may be needed to raise the leaders.

5.3.5. Inspection and Repair

Check equipment as it arrives against your equipment list. Each item should be inspected, cleaned, lubricated, minor repairs made and painted, if not previously done. Particular attention should be paid to the following:

1. An item requiring major repair or replacement should be reported immediately.
2. Inspect leaders, spotters and booms for structural damage.
3. Check that smaller parts, such as pins, bolts and links, required to connect equipment, are available. This is a tedious but a most important task if delays in assembly are to be avoided.
4. Inspect hammer for loose or worn parts; is size and stroke appropriate; is hammer cushion complete and proper size; is spare hammer cushion available?
5. Determine if there are hydraulic and steam or air hoses of proper size and length in good condition to connect all components.

6. Check ladders and catwalks for safe condition; clean leader platforms, spotters and walkways, if necessary; secure or tighten all loose fittings and connections.
7. Check and lubricate all sheaves, particularly leader headblock and fairleaders; inspect and clean Raymond signs on rig, leaders and boom.
8. Check turntable pad eye locations to be certain that they are standard for the spotter and boiler platform.
9. Inspect boiler, wash and punch flues, if needed; have fireman fire up as soon as possible.
10. Inspect, clean and replace oil and filters in all hydraulic equipment. Inspect drill; check auger, bits and water swivel as to size, length and usable condition.
11. Inspect cores and check for cracks; check tendons and sockets for wear.
12. Make sure lubrication and instruction books are available for cranes and other major pieces of equipment; if not, order them immediately.
13. Have operator perform all required maintenance on crane, check condition of all frictions, brakes, clutches and make repairs or adjustments, if necessary.
14. Check condition of all wire rope and make sure that all cables, including pendants, are of proper size and length.
15. Check oil and radiator levels in all engines; note if oil and filters require changing; don't forget gear cases, torque converters, chain cases, etc. Check condition of all batteries; make sure generators are charging; replace or repair, if needed.
16. Check that adequate spares are on hand, on order, or obtainable locally, including hammer cushion plates, steam and hydraulic hose, hydraulic system and engine filters, V-belts, gaskets, packing, safety valve, injectors and injector check valves.
17. Check toolbox, particularly for shell wrenches, denesting tools, siphon, bailing bucket, concrete funnel or dump tube, grout mortar box and hoes.

5.3.6. Erection

Rigs are normally erected at the point where they should begin driving. The usual steps in erecting the drivers are as follows:

1. Choose an accessible, level a firm site that will permit assembly of various components, including the leaders. Keep away from the edge of excavations at least two feet per foot of depth.
2. Place sufficient good mats to provide firm support to the rig; usually about six should be required. Particular attention should be given to the mat under the front edge of the crawlers, as it should be heavily loaded when leads are raised. Be careful not to disturb pile stakes with mats.
3. Unload or walk crane on to center of the mats; the side of the crawlers should be kept back at least five feet from the edge of any mat.
4. Raise gantry to operating position if lowered for traveling. Attach boom, counterweight and boiler in that order. Truck with boiler, or the crane, can be raised on planks until heights match.
5. Assemble required sections of leaders alongside rig on blocking, so that boom-leader connecting pin should be under boom tip with the bottom end of the leaders just to the outside of one crawler. Install sliding frame and hammer extensions at bottom of leaders; secure to prevent movement during raising.
6. Install drill or jet guides, pile lifting brackets, fairleaders and other hardware on leaders; make an inventory of leader and boom sections and all hardware for future reference and for ordering parts.
7. Arrange for a service crane to assist in raising leaders, if previous planning indicates that such a crane should be necessary.
8. Lower boom and connect the clevis of the tapered pin in the leaders to the appropriate place on the crane boom. Steam, air, hydraulic and jet hoses should be attached at the boom tip and the free ends tied off temporarily alongside the boom.

9. Reeve hammer, drill and free lines and tie off at bottom of leaders. Connect a tag line to bottom of leaders to control them as they clear the ground.
10. Raise boom and leaders. Watch rear track rollers for lifting from pads. Do not add extra counterweight or tie rear of rig down as this may result in failure of the topping lift or gantry that will normally be loaded to capacity when raising the leaders.
11. Install spotter and moonbeam and connect to leaders. Connect hydraulic hoses or reeve cable spotters and test operation. Boom up and bring spotter in to minimum radius for balance of assembly.
12. Reeve hammer, pick up hammer and lower extensions to hold hammer in leaders; secure the top of the extensions to hammer with a short wire rope sling. Install core head, or follower and hammer cushion; connect steam or air hose.
13. Unless rig must move a considerable distance or down a ramp, core may be installed at this time. Core sections are normally picked up with the free line, one at a time, and pinned off to the core head.
14. Install drill and auger guides, or jet, and hook up hoses; raise drill and connect auger or drill stem; test drill circuit.
15. Have person go aloft and check for anything that may have been loosened during erection. Make certain all lines have proper leads.
16. Hook up jet or drilling pumps or other auxiliary equipment, if any, and test. Set hammer or core on block and test hammer operation.

5.3.7. Project Management

Proper management is very important. It is easy for the crews to be disorganized at the start and firm control by the superintendent is essential. Under favorable conditions, ideal weather or having a crew that has worked together for a long time, the operations should be completed much faster. Under adverse conditions, unusual bad weather, confined space, poor access roads, or lack of power to assist in unloading, more time will have to be allowed for some operations.

The superintendent must make a schedule indicating when the start and completion of each operation. This should form the basis for determining the arrival of trucks and support services. Review the schedule with your supervisor. Listen to his or her suggestions as to how to better the allotted time. As the operation progresses, track your schedule in the remarks section of the daily report and adjust the balance of the schedule accordingly. If you fall behind, analyze what went wrong and decide how it can be avoided in the future.

Sometimes, failure to meet your goals should be the result of factors beyond your control. However, too often delays are the result of inadequate planning and supervision. Be sure this doesn't happen on your project!

Remember that the goal you are trying to achieve is an assembled pile driver ready to go to work in a minimum amount of time. Refer to the notes in this section to avoid overlooking little items.

5.4. Personnel Management and Labor Relations

5.4.1. Crews and Work Regulations

Work conditions and construction crew makeup can be substantially different in different parts of the country. It may be customary or required by union agreement to have persons in the crew that appear to be non-essential. An example is an operator at high hourly pay and overtime stipulations to watch a compressor or welder run. In some areas, reinforcing bars come to the project "Bent, Bundled and Tagged"; in other areas all cutting and bending should be done on the project. Point reinforcing for piling can sometimes be attached offsite at lower cost. However, check before shipping; some jurisdictions require that all such work be done locally.

The narrative that follows assumes the use of union labor, although in some locations it is certainly possible to use non-union personnel. Much of what follows is applicable to both union and non-union labor.

Pile driving operations are under the jurisdiction of three unions, the Operating Engineers, Pile Drivers (Dock Builders) and Laborers. Occasionally, one will deal with other unions, such as when one sets reinforcing cages, hauls materials etc.

It is essential that the superintendent understand local labor practices and working conditions prevailing in the area. From time to time, union business agents or project stewards may show you "rule books" to back up some claim. You should not be guided by such rules, as one may have a different agreement than some other contractors in the area.

The pile driver crew is usually established by local agreement and no additional people should be used, under any circumstances, without consulting your general superintendent. To do otherwise might well set a precedent that could affect similar situations in the future where additional personnel were unnecessary.

The minimum crew for a crane pile driver normally consists of a supervisor, an operator and two driver people (a front-end person and a lofts person). When one makes up shells, two additional people are used on the shell rack.

The other crews required for the project will vary depending on the type of pile driven, the other work in the contract and the methods and equipment employed. The size and make-up of the crews should be planned with your general superintendent.

Your supervisor is the key person in personnel management and will make the crew a smoothly working team. He not only requires technical skill but also must have the ability to lead his or her people. He represents the Company but the good supervisor will also represent the people's needs to you.

The superintendent's skill in handling other phases of the project improves relations with the people. Materials and the proper equipment should be on hand when needed, operations planned and followed up and clear decisions made. The people will not take much interest in a project that they see is poorly supervised.

The superintendent's personal habits will also affect the entire project. He or she should be on the project well before the starting time in order to assure the day gets off to a good start. He should limit his or her own lunch period to that of the people and remain on the project until after quitting time.

In order to meet a requirement of the contract it is occasionally necessary to speed up production. Usually, the least costly methods are to work a second shift, to extend the workday to 10 hours or, work six days per week. This will significantly increase your unit costs and must not be undertaken until discussed with, and approved by, your general superintendent.

Except for short periods in a real emergency, a workday in excess of 10 hours, or a workweek over six days will result in low efficiency. A seven-day week may result in less total production than can be secured in six days.

5.4.2. Employment

For union labor, one normally depends on Referral Halls as ready sources of skilled employees. These Referral Halls normally operate in close cooperation with the Building Trades Unions but in compliance with Federal regulations governing discrimination.

To be an "Equal Opportunity Employer" is a desirable and often a mandated goal. A company policy to this end could read as follows:

> "It has been and will continue to be the policy of ________________ never to discriminate against any applicant for employment, employee, subcontractor, supplier or vendor because of race, color, religion, sex or national origin. Applicants and employees should be compensated, trained, advanced, demoted, terminated, hired and transferred solely on the basis of their skill, devotion, loyalty, honestly, reliability, integrity and the economic needs of the organization."

Special attention should be focused on projects performed under municipal, state or federal funding. These projects have specific Equal Employment Opportunity Clauses in the contracts. These require affirmative actions, and should be thoroughly studied and understood.

The employment of persons less than 18 years of age is restricted in most states. When in doubt, ask for proof of age; if under 18, you should not hire without the approval of your general superintendent. You should not refuse employment to an older person solely because of his or her age, if he is physically and otherwise qualified for the project. However, working around a pile driver requires considerable physical effort and an alert mind. If you have reason to question an applicant's physical or mental condition, you should suggest that he be examined by a doctor who understands the type of work he should be doing and furnish you with a doctor's statement that he is physically and mentally fit. In the U.S., the Americans with Disabilities Act also govern treatment of these factors.

When working in remote areas, you may have to recruit your own people. Other contractors will probably be able to advise you of sources of labor or may "lend" you some of their people for a limited period. Your general superintendent may be able to give you the names of people who have worked for us in the past. The arrival of the pile driver should be known and you should be alert for contacts from former employees.

It is sometimes necessary to assign a supervisor and few key people from another area when qualified people are not available locally. In the interests of good labor relations, it is desirable to discuss this situation with your general superintendent and the local union before importing outsiders.

At the time of employment, each employee must complete and sign a number of federally and state mandated forms. These are essential if the person is to be paid. Do not wait to do it later; the person could quit or be injured and you would not have the required data.

Depending on the new employee's experience, the superintendent or supervisor may have to orient him before beginning work. He should understand his or her rate of pay, working hours, which is his or her boss, the conduct expected of him and proper work habits.

When signing up a new person, be sure he is adequately clothed with work shoes and work gloves. Issue him a hard hat and any other special safety equipment, such as goggles or rubber gloves. Instruct him in the recognition and avoidance of unsafe conditions as required by the Federal Health, and Safety Regulations. Give each new employee a copy of the safety rules.

5.4.3. Payrolls

Part of the your records should be a daily list of each employee and the hours worked.[7] From this information, the payroll can be prepared. The payroll checks (or pay envelopes) should be sent to the project. When the payroll week closes, it may be necessary to telephone in the last day's hours to save time. The closing day and other details will vary somewhat depending on project location and local custom. If the crew is paid in cash, the superintendent must get a signed receipt from each person.

Preparing payrolls requires careful coordination between the project and office. Most union agreements provide that the people should be paid waiting time, if their pay is not ready on time. Even on a large project, where there is a resident clerk, the superintendent still has the overall responsibility to assure the payroll is accurate and properly handled. The payroll disbursements represent the largest financial outlay that is under your direct control.

5.4.4. Training

The superintendent is responsible that all people are properly trained to perform their duties in a correct, efficient and safe manner. While his or her supervisor will do this on a daily basis, the superintendent should continually watch for improper work habits and assure they are corrected. This should always be done in a constructive manner and not to embarrass the person involved, "criticize privately, and praise publicly."

New supervisors will frequently need guidance in planning ahead, record keeping and similar skills. The new supervisor will tend to do too much work himself while the crew watches. When the opportunity is available for private discussion, point out how the activity might have been carried out more efficiently. Only when it is urgently necessary, should the superintendent correct or question the orders of his or her supervisor in front of the crew.

Superintendents and supervisors should observe people in the crew who have leadership ability and give them a chance to use their initiative when the opportunity arises. They should be rotated between various projects so that they can develop the several skills of their craft. Advise your general superintendent about supervisor and crew members that you think are ready for promotion.

When operating a multi-crew project, training should be expedited by proper crew assignments, placing the less experienced people with the most experienced supervisor.

5.4.5. Working Conditions

The superintendent must ascertain what working conditions prevail in the area. The following points should be clarified prior to starting work:

1. Rates to be paid per hour for each classification.
2. Overtime rates - beyond 8 hours, Saturdays and Sundays.
3. Which classifications are paid straight time.
4. Reporting or waiting time.
5. Travel time, if any.
6. Working hours; normally 8:00 A.M. to 4:30 P.M. but varies in some areas.
7. Are operators required on pumps and compressors and, if so, starting with what size?
8. Any peculiar customs, such as larger or smaller than normal crew size?
9. What holidays are observed and are they observed by all crafts? Is it customary to work on any of the minor holidays and, if so what rates are paid to each craft?
10. What are provisions for shift work, if more than one shift is contemplated.
11. Any contractual obligations or local regulations regarding hours of work, smoking, use of alcohol, special safety gear, etc.

5.4.6. Morale

A good attitude on the part of employees is essential to efficient production. No longer are people motivated by just having a job; they must feel that your company is a good company with which to be associated.

Many morale building factors are included in union agreements but even if not, the following should be considered in accord with local customs:

1. A change shanty or trailer, heated in winter, with adequate space to hang work clothes to dry and for people to get shelter from the weather. On large projects, separate shanties may be provided for each craft, particularly those that provide their own tools.
2. Drinking water and toilet facilities, properly maintained and complying with Federal Regulations and local codes.
3. Boots, raincoats and other foul weather gear, particularly for people who will have to keep working in bad weather.
4. Adequate tools in good condition, other than those normally provided the individual.
5. Safe working conditions.
6. In remote areas, arrange for food service.
7. Parking space convenient to work area to the extent this is feasible.
8. Both superintendent and supervisor should be readily accessible to the people, listen to their "gripes" and, when reasonable, do something about them. Small annoyances, if not taken care of, may grow into major grievances, with possible loss in production. Any claimed shortage in wages should be resolved promptly in cooperation with the supervisor involved. If it cannot be immediately settled, it should at least be narrowed to the specific hours or days in dispute.
9. The superintendent will occasionally have to discipline an employee for repeated tardiness or for more serious infractions, such as drinking or violation of safety rules. This will not hurt morale, if you observe the following:
10. Be fair; treat everyone alike.
11. Explain to the individual how he is not only hurting himself but may also be putting a burden on or possibly endangering others.
12. Try to correct the employee's attitude before the situation reaches serious proportions.
13. If possible, give the employee a formal warning and advise his or her supervisor and project steward that you have done so.
14. If your efforts to reform him are unsuccessful and you must discharge him, be certain you have a sound reason. You may have to justify your actions before an arbitrator.

5.4.7. Working with Unions

Dealing with organized labor is not always straightforward and you should be faced with problems that will have to be settled on a day-to-day basis.

The most important thing is for you to know the labor agreements covering each craft thoroughly. If any questions arise as to how such agreements apply to specific areas of the work, consult your general superintendent. You may be surprised how many problems may have already been resolved. If you are new in the area, you should be particularly careful not to change something that has been previously settled.

You should be fair in your dealings with the union representative, not withholding benefits to which the people are entitled but at the same time, being certain that you don't give away rights that properly belong to management.

While you should get to know the local business representatives for the various crafts, most of your day-to-day dealings should be with the project steward. It is the steward's job to represent the people and protect their rights. The capable steward is neither a "tough person" nor a "timid soul" but will know his or her people's rights and protect them without hesitation.

The capable steward will try to understand your viewpoint and win his or her point without embarrassing you. You should be prudent to act in the same fashion. The capable steward doesn't expect you to try to outwit, outmaneuver or out-shout him so, if you do, you will lose his or her respect. If you lose your temper you will probably also lose your case.

Most labor agreements have provisions for handling disputes. While it is desirable to handle these at the project level, you won't be able to solve all differences. In a few cases, the steward will have to make a claim "just for show," and may not push it, if you know your rights. In other cases, he will think he is on strong ground but settling may be a matter of policy, which you should refer to your general superintendent.

Resolve grievances as quickly as possible. Do not stall. If you must make an unpopular decision, delay will only make the action more difficult. Don't make threats or permit them to be made. Bluffing is one of the most shortsighted and dangerous tactics management can employ.

Know your contract well. When a steward raises a grievance, you must immediately recognize whether a union member has been treated unfairly or not. You should try first to resolve the grievance between the steward, the supervisor and yourself. Whether you have the employee present should be decided between you. Personalities have no place in labor relations. Letting your emotions control your decisions can only create animosity between you and the union.

There is no faster way to discredit yourself than to make an issue of events when the facts have not been fully verified. The first step in solving any difference in to agree on the facts before discussing the merits of the case. You may not be able to arrive at a solution but you should be able to agree with the project steward on the facts.

In most instances, you or the steward will recognize that you have no case and you should be well advised not to push such poor cases. In borderline cases, you may both feel that justice is on your side. In such instances, try to agree on as many facts as possible and reduce the grievance to writing. This is particularly important should the case go to arbitration.

One of the most common problems you will encounter is that of jurisdiction between crafts. The regular crews are pretty well agreed on the jurisdiction of the Operating Engineers, Pile Drivers and Laborers. You will occasionally be approached by the steward of another craft claiming jurisdiction over some work being done by the people.

You should immediately contact your general superintendent who can advise you how such questions have been handled in the past. You should also advise the steward of the craft performing the work and the representative of the client, if he should be affected.

There is a National Joint Board for the Settlement of Jurisdictional Disputes, which has responsibility for assignment of work between crafts. Appeals can also be made to the National Labor Relations Board. Your general superintendent can advise you on what to do it your dispute cannot be handled locally.

Almost all agreements have a "no strike," "no lockout" clause. However, to make such a provision effective you will have to be diplomatic without compromising management's rights. If you cannot reach immediate agreement with union representatives, do not lose your temper or push them into losing theirs. If settlement is beyond your authority, do not hesitate to admit it and ask for time for it to be considered by company management.

5.5. Safety and Accident Prevention

The superintendent is responsible for safety and accident prevention. He must instill an attitude of safety consciousness in the minds of supervisors and all other people on the project. He should be alert to causes of accidents and remove such causes when they come to his or her attention.

Persistent efforts will have to be made by the superintendent and Supervisor to insure that each person understands his or her job, how it is done safely, learns to recognize a circumstance that could lead to an accident and learns that it does not pay to be careless or take chances. Everyone should be constantly alert to insure that their orders or methods of operation to not jeopardize the safety of others.

Accident prevention works both for the contractor and for the individual. Most obvious is the direct relationship between accident rates and insurance rates. However, hidden losses -loss of a trained worker, delays on the project, etc. - come directly out of project profits. Drive home Emphasize to the people that an injured person receives less than his or her usual income while he recovers and may have impaired his or her earning capacity for the rest of his or her life.

When a Safety Engineer visits your project, supplement his or her suggestions with your practical experience and detailed knowledge. Post the safety bulletins provided by the insurance company and assure they are changed often enough to be noticed.

5.5.1. Starting the Project

1. Contact local office of insurance company; obtain telephone number of doctor, hospital, ambulance, police and fire departments and others to be contacted in emergency.
2. Check that a fully stocked first-aid kit is available. Unless you have had certified training, first-aid should be limited to minor cuts and bruises, artificial respiration and emergency action.
3. If the project is remote from doctor or hospital, assure there is a trained first-aid person - plant aid station, police, etc. - locally available in an emergency.
4. Review safety aspects of your operating plan.
5. Review safety information relating to all construction equipment.
6. Check that adequate safety equipment is on hand; gloves, hard hats, goggles and welder's helmets, cart for air and gas bottles, fire extinguishers, etc.
7. ReviewtheOSHArecordkeepingrequirements. Make sure all mandated posters are placed where they can be seen. If you have a multilingual workforce, make sure as many languages as needed are represented with the posters.
8. Place safety posters on rig and other appropriate places.
9. The superintendent should personally conduct the first safety meeting, prior to driving the first pile, if possible. Explain the safety rules to all employees and give each employee a copy of the safety rules.

5.5.2. Safety

5.5.2.1. Equipment

Everyone on the project, including operators, firemen and laborers, must wear brimless plastic hard hats, with peak. Spares should be available for visitors and for replacements. They should be purchased from a reliable supplier and conform to American National Standards Institute Z 90.1-1969.

Hard hats should be inspected before being issued and any that are split, dented, drilled or damaged in any way, should be discarded. Check interiors and reject any that have suspensions that are broken, worn or deteriorated. Check that there is 1-1/4" (31.8 mm) clearance between the suspension and the shell.

Welding goggles, either eyecup type or cover spec type, with shade 3 tinted lenses should be worn by burners cut ting shells, light wall pipe and light H-piles; shade 4 should be used for heavier cutting. Chipping goggles with clear lenses should be worn when chipping or grinding or working in blowing dust or sand.

Welders must wear welding helmets with a shade 10 lens for light work and shade 12 when using 3/16" (4.76 mm) or larger electrodes, as shown in Figure 5-3. When extensive welding operations are required, employees who may be accidentally exposed should be protected by portable shields and provided with shade 2 spectacles with side shields.

Figure 5-3 Welder with Welding Helmet

Neoprene coated canvas work gloves should be worn by people handling plastic slate, creosoted piles, cement or other irritating materials.

When working over or near water, employees shall be provided with life jackets and a boat should be readily available for rescue operation.

Fire extinguishers should be provided as follows:

1. One multi-purpose dry chemical extinguisher, 6#, type 2A-10BC (Fry-Fyter model 6MA or equal) in office, tool house, or change shanty area.
2. One dry chemical extinguisher, 2#, type 5BC on each crane or other major piece of equipment, including heavy trucks.
3. One multi-purpose dry chemical extinguisher, 6#, type 2A-10BC for each welding or burning operation; inside buildings or adjacent to formwork, partitions or other combustible material, larger extinguishers should be provided.

Approved safety cans, with spring release spout and vent, and dispensing units shall be used for gasoline and other flammable liquids.

Employees should be observed to insure that they are properly clothed for their job. This requires work shirt with long sleeves, sturdy long trousers or overalls, work shoes (preferably with hard toes) and gloves.

5.5.2.2. Accident Prevention Program

The following is a "model" accident prevention program. Although it contains many helpful suggestions for jobsite safety, you may wish to add items and to make it more specific for your company or operation.

5.5.2.2.1. Objectives

1. To furnish a safe place of employment free of recognized hazards for all employees.
2. To reduce to a minimum the injury and suffering of employees due to work accidents.
3. To promote maximum efficiency and to effect savings by the reduction of business interruption from occupational accidents.

5.5.2.2.2. Methods

1. Planning all work to minimize personal injury, property damage and loss of productive time.

2. Maintaining a system of prompt detection and correction of unsafe practices and conditions.
3. Making available and enforcing the use of personal protective equipment and mechanical guards.
4. Establishing an effective system of tool and equipment inspection and maintenance.
5. Establishing an educational program to maintain interest and cooperation of all levels of employment through:
6. New employees to receive accident prevention instructions at the time of hire.
7. Weekly safety meetings (additional meetings to be held as required).
8. Daily safety inspection of project site, equipment, tools and personnel by superintendent, supervisor or operators.
9. Investigation of all accidents to determine cause and initiating corrective action to eliminate recognized hazards.

5.5.2.2.3. Safety Program

1. Project superintendents should conduct ten minute "Tool Box" talk on project safety each week.
2. Install metal frame on bottom leader section to hold safety posters, or attach safety posters to crane body.
3. General superintendent to direct safety discussion when visiting project site.
4. Construction office to inform project superintendents and supervisors monthly of current accidents in the region with comments on how to prevent similar accidents.
5. Safety posters and accident prevention information shall be included with project documents and distributed periodically throughout project.
6. Superintendents on jobs will investigate all lost-time accidents thoroughly and report to the general superintendent why and how accident happened and what steps taken to correct unsafe condition.
7. Pile supervisor, labor supervisor, and operators receive letter from Construction Office regarding accident under his or her supervision.
8. Accident reports sent to the proper recipients on day of accident.
9. Project personnel complete a "Safety Check List" and return to project superintendent.

5.5.3. Site Conditions

The work area and walkways should be kept free of loose materials, debris, pile cut-offs and scrap lumber. Containers should be provided for oil rags, combustible materials, trash and garbage and emptied at regular intervals. Combustible materials should be stored in a safe manner. Equipment not in use should be stored out of the working area.

For night work, construction areas should be adequately lighted, the general intensity not less than 3 foot-candles. Temporary lights shall be installed in accordance with the National Electrical Code and shall be supplied by heavy-duty electric cords maintained in a safe condition.

Employees should be provided with suitable drinking water containers and paper cups. Toilet facilities should be available, complying with local codes and properly maintained.

Suitable precautions should be taken to protect employees when working in areas where there are harmful gases, vapors, fumes or dusts or inadequate ventilation. Special care should be taken when working in plants where ground is contaminated with flammable liquids or other dangerous substances or when driving in areas where there are underground pipes carrying such materials.

The superintendent should assure work areas and walkways are maintained in a safe condition. Muddy areas should be drained to the extent feasible and plank runways provided if necessary, particularly across trenches or other hazardous areas; the minimum being two cleated 2 x 10's. Slippery work areas and walkways should be sanded; salted in freezing weather. Ladders should properly built and maintained in a safe condition. Side rails should extend 36" (914.4 mm) above a landing. Platforms and walkways more than four feet above ground or in any hazardous location shall be provided with a secure handrail with toe board.

Fences, barricades or other suitable means should

be provided to exclude unauthorized persons from work areas.

5.5.4. Material Storage

All materials should be stacked, blocked or otherwise secured to prevent sliding, falling, or collapse. When storing shells, use stakes or chocks to prevent spreading. Aisles or passageways should be provided between piles and kept clear. When storage area permits, material should not be stacked higher than five feet, particularly adjacent to roadways where they might block the view of opposing traffic.

When materials are stored on platforms or in buildings, care should be taken not to exceed safe floor loading. Avoid stacking shells or other materials close to the edge of an embankment. When assembled shell sets are rolled into an excavation, it should be done with care so they do not roll into an area where people are working or pile up in a manner hazardous to the person hooking up the pile chain.

5.5.5. Material Handling

Strains and hernias from improper lifting methods require education of workers in the proper way to lift with their legs. Proper work clothing, particularly gloves and shoes, can minimize cuts, scrapes, bruises and skin burns. This is especially important when handling creosoted piles.

Shells and other pile material should be unloaded from trucks with caution, making certain that load is safely held while stakes are removed. Load binders should be released from the backside of the load after stakes have been removed from the opposite side; use temporary auxiliary binders or lines, if necessary.

Unloading and handling or large timbers, long woodpiles, H-piles and other structural shapes should be done with tongs when available.

Always inspect rigging equipment before each use to be certain that it is in good condition. Use only first quality factor made slings, hooks, etc. Don't use homemade hooks and fasteners or project-assembled slings. Discard questionable items.

Everyone should stand clear when handling materials; do not swing loads over heads of workers. Tag lines should be used to control loads. Do not permit people to ride on crane hooks, on top of loads, or in the cargo section of a truck.

5.5.6. Wire Rope

The strength of any wire rope is dependent on rope size, type, condition, type of end connections and how they are applied. The 6 x 19 wire rope, which is one commonly used has the characteristics shown in Table 5-1.

Table 5-1 Characteristics of 6x19 Improved Plow Steel Fiber Core Wire Rope

Diameter, Inches	Weight, Lbs. Per Foot	Breaking Strength, Tons	Safe Load, Lbs. (F =5)
1/4	.10	2.74	1100
5/16	.16	4.26	1700
3/8	.24	6.10	2400
7/16	.32	8.27	3200
1/2	.42	10.7	4200
5/8	.66	16.7	6600
3/4	.95	23.8	9500
7/8	1.29	32.2	12,800
1	1.68	41.8	16,700
1-1/4	2.63	60.0	24,000
1-1/2	3.78	85.0	34,000
2	6.72	155.0	62,000

Wire rope should be inspected frequently and should be discarded when there is significant wear of outside wires, kinks, bird cages, corrosion, evidence of crushing or heat damage or reduction in diameter of 1/32" or more. Running ropes should be discarded when you can find six randomly distributed broken wires or three broken wires in one strand. Standing ropes, such as pendants, should be replaced when you can find more than one broken wire at an end connection or more than two broken wires elsewhere.

The strength of wire rope end connections is shown in Figure 5-4.

Figure 5-4 Strength of Wire Rope End Connections

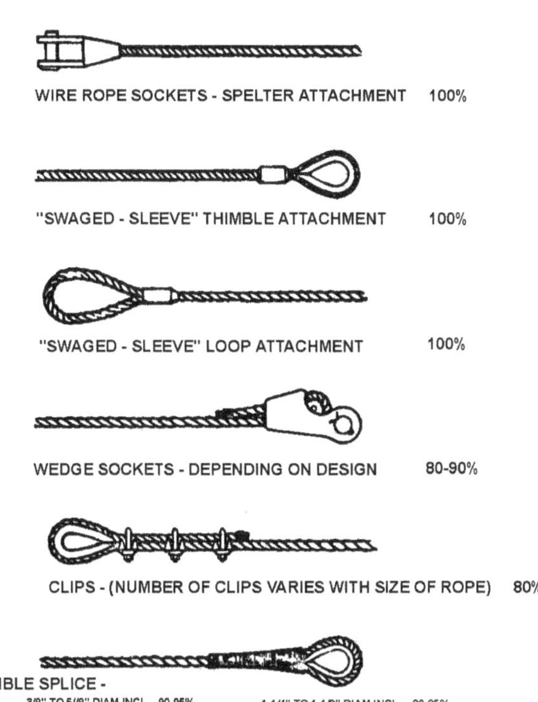

The desirability of using factory applied ends can be seen as only a swaged or zinc poured socket will develop the full strength of the rope.

An eye formed with cable slips will develop about 80% or the rope's strength if a sufficient number of clips are properly applied and they are properly spaced; see Table 5-2. A 12" (304.8 mm) wrench is adequate for tightening clips on 3/4" (19.1 mm) and smaller rope; a larger wrench should be used on large clips.

Table 5-2 Use of Cable Clips

Size of Wire Rope	Number of Clips Needed	Minimum Clip Spacing, inches (Center To Center Of Clips)	Length Of Rope Turned Back, inches (Exclusive Of Eye)
3/8	2	3-1/2	7
1/2	3	4	12
5/8	3	4-1/2	14
3/4	4	5	20
7/8	4	5-1/2	22
1	5	6	30

Cable clips should be placed so that the U-bolt bears on the short end, if the maximum of 80% strength is to be developed. Other methods of applying clips will result in lower capacity, as shown in Figure 5-5. A thimble should protect the rope.

Figure 5-5 Methods of using cable clips

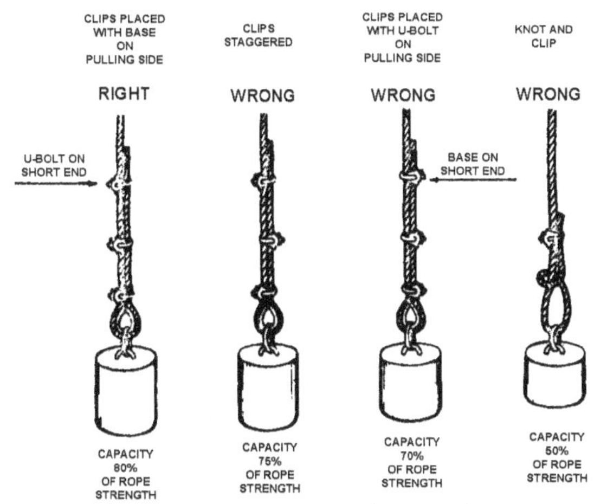

Loose strand ends should be covered or blunted. Wire rope should not be knotted. If a wire rope is broken in service, it should be replaced and not spliced.

5.5.7. Slings

Loads should be picked up with slings; never wrap a hoist line around a load. Slings should be purchased with factory made eyes with hooks and rings of adequate capacity. Capacities of various type wire rope slings are shown in Table 5-3, Table 5-4 and Table 5-5.

Slings should be inspected before each use and

Table 5-3 Capacity of wire rope chokers[8]

Wire Rope Diameter, inches	Wire Rope Construction	Rated Load, lbs.	Minimum diameter of loop "D", inches
1/4	6x19	740	5"
5/16	6x19	1,140	6-1/4"
3/8	6x19	1,600	7-1/2"
7/16	6x19	2,200	8-3/4"
1/2	6x19	2,800	10"
9/16	6x19	3,400	11-1/4"
5/8	6x19	4,200	12-1/2"
3/4	6x19	5,800	15"
7/8	6x19	7,800	17-1/2"
1"	6x19	10,000	20"
1-1/4	6x37	14,800	25"
1-1/2	6x37	20,000	30"

Table 5-4 Capacity of Basket Slings

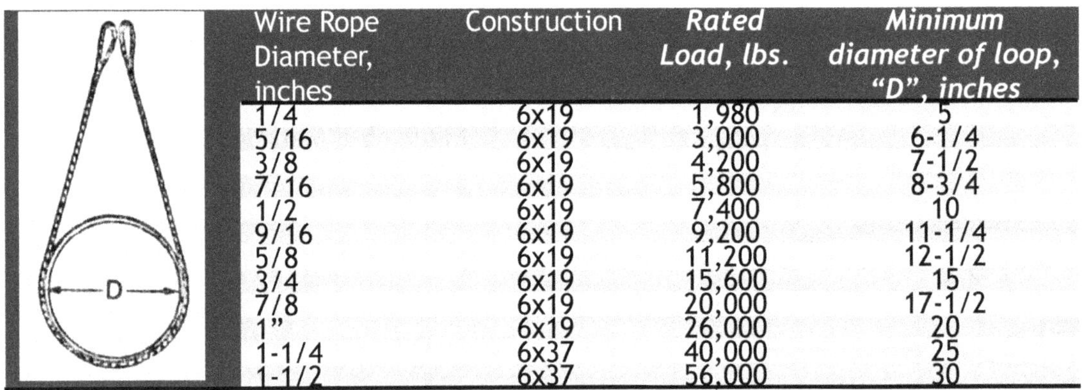

Wire Rope Diameter, inches	Construction	*Rated Load, lbs.*	*Minimum diameter of loop, "D", inches*
1/4	6x19	1,980	5
5/16	6x19	3,000	6-1/4
3/8	6x19	4,200	7-1/2
7/16	6x19	5,800	8-3/4
1/2	6x19	7,400	10
9/16	6x19	9,200	11-1/4
5/8	6x19	11,200	12-1/2
3/4	6x19	15,600	15
7/8	6x19	20,000	17-1/2
1"	6x19	26,000	20
1-1/4	6x37	40,000	25
1-1/2	6x37	56,000	30

Table 5-5 Capacity of Two-Leg Slings (Improved Plow Steel, Fiber Core)

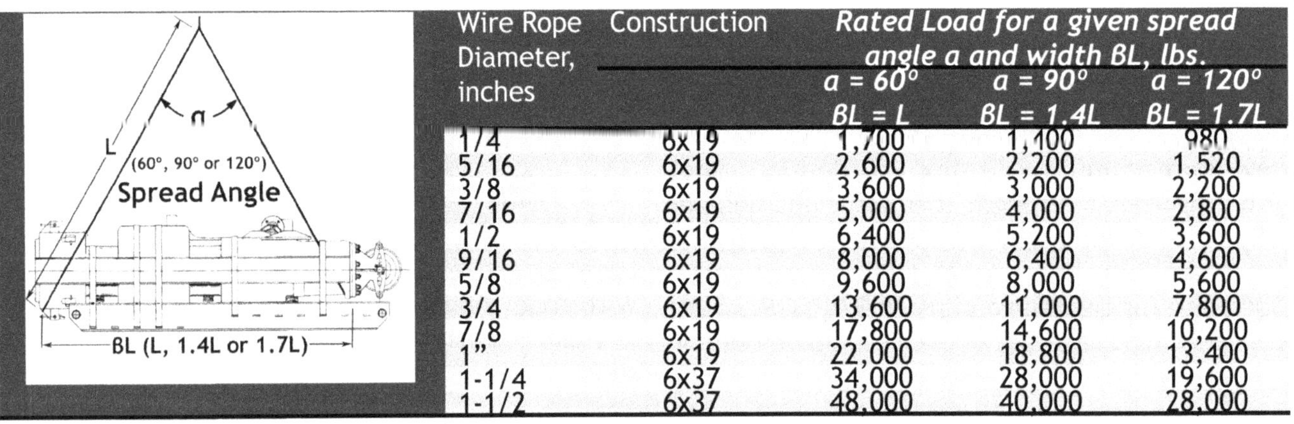

Wire Rope Diameter, inches	Construction	*Rated Load for a given spread angle a and width BL, lbs.*		
		a = 60° *BL = L*	*a = 90°* *BL = 1.4L*	*a = 120°* *BL = 1.7L*
1/4	6x19	1,700	1,400	980
5/16	6x19	2,600	2,200	1,520
3/8	6x19	3,600	3,000	2,200
7/16	6x19	5,000	4,000	2,800
1/2	6x19	6,400	5,200	3,600
9/16	6x19	8,000	6,400	4,600
5/8	6x19	9,600	8,000	5,600
3/4	6x19	13,600	11,000	7,800
7/8	6x19	17,800	14,600	10,200
1"	6x19	22,000	18,800	13,400
1-1/4	6x37	34,000	28,000	19,600
1-1/2	6x37	48,000	40,000	28,000

Table 5-6 Working load limits for common alloy steel chain sizes

Chain Size, inches	Working Load Limit, lbs.
1/4	3,250
3/8	6,600
1/2	11,250
5/8	16,500
3/4	23,000
7/8	28,750

Table 5-7 Capacity of Shackles

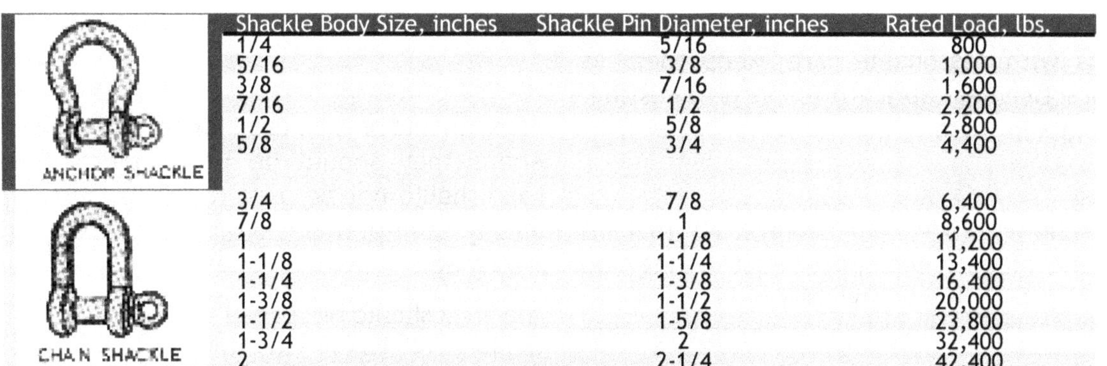

Shackle Body Size, inches	Shackle Pin Diameter, inches	Rated Load, lbs.
1/4	5/16	800
5/16	3/8	1,000
3/8	7/16	1,600
7/16	1/2	2,200
1/2	5/8	2,800
5/8	3/4	4,400
3/4	7/8	6,400
7/8	1	8,600
1	1-1/8	11,200
1-1/8	1-1/4	13,400
1-1/4	1-3/8	16,400
1-3/8	1-1/2	20,000
1-1/2	1-5/8	23,800
1-3/4	2	32,400
2	2-1/4	42,400

Table 5-8 Capacity of Hooks

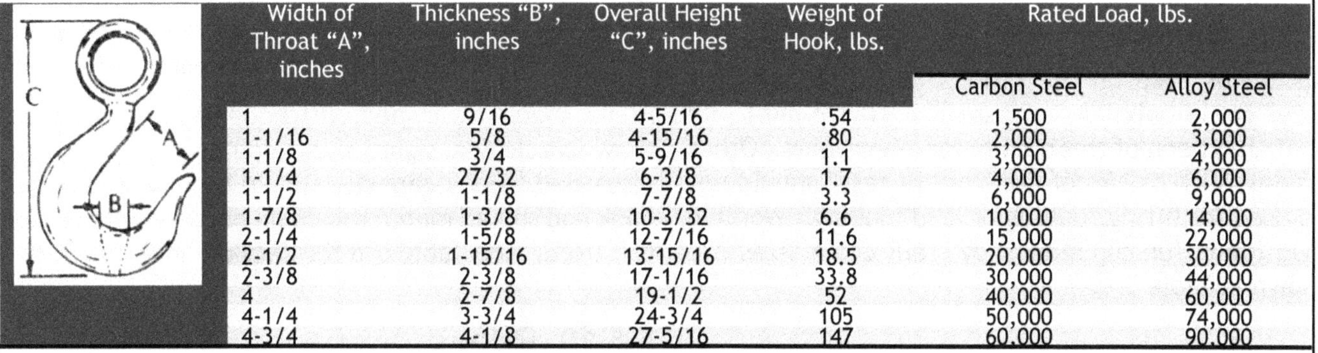

Width of Throat "A", inches	Thickness "B", inches	Overall Height "C", inches	Weight of Hook, lbs.	Rated Load, lbs.	
				Carbon Steel	Alloy Steel
1	9/16	4-5/16	.54	1,500	2,000
1-1/16	5/8	4-15/16	.80	2,000	3,000
1-1/8	3/4	5-9/16	1.1	3,000	4,000
1-1/4	27/32	6-3/8	1.7	4,000	6,000
1-1/2	1-1/8	7-7/8	3.3	6,000	9,000
1-7/8	1-3/8	10-3/32	6.6	10,000	14,000
2-1/4	1-5/8	12-7/16	11.6	15,000	22,000
2-1/2	1-15/16	13-15/16	18.5	20,000	30,000
2-3/8	2-3/8	17-1/16	33.8	30,000	44,000
4	2-7/8	19-1/2	52	40,000	60,000
4-1/4	3-3/4	24-3/4	105	50,000	74,000
4-3/4	4-1/8	27-5/16	147	60,000	90,000

Table 5-9 Rope Strength

Rope Diameter, inches	Manila Rope			Nylon Rope		
	Weight, Lbs Per 100 Ft	Minimum Breaking Strength, Pounds	Rated Capacity, Pounds (F_s=5)	Weight, Pounds Per 100 Ft	Minimum Breaking Strength, Pounds	Rated Capacity Pounds, (F_s=9)
3/16	1.5	450	90	1.2	900	100
1/4	2.0	600	120	1.5	1,650	180
5/16	2.9	1,000	200	2.5	2,500	280
3/8	4.1	1,350	270	3.6	3,700	400
1/2	7.5	2,650	550	6.5	6,080	700
5/8	13.3	4,400	900	10.5	9,880	1,100
3/4	16.7	5,400	1,100	14.5	13,490	1,500
7/8	22.5	7,700	1,500	20.0	19,000	2,100
1"	27.0	9,000	1,800	26.0	23,750	2,600
1-1/4	41.7	13,500	2,700	40.0	35,625	4,000
1-1/2	59.9	18,500	3,700	55.0	50,350	5,600
2"	107.5	31,000	6,200	95.0	87,400	9,700

Table 5-10 Rated Capacity of Rope Slings

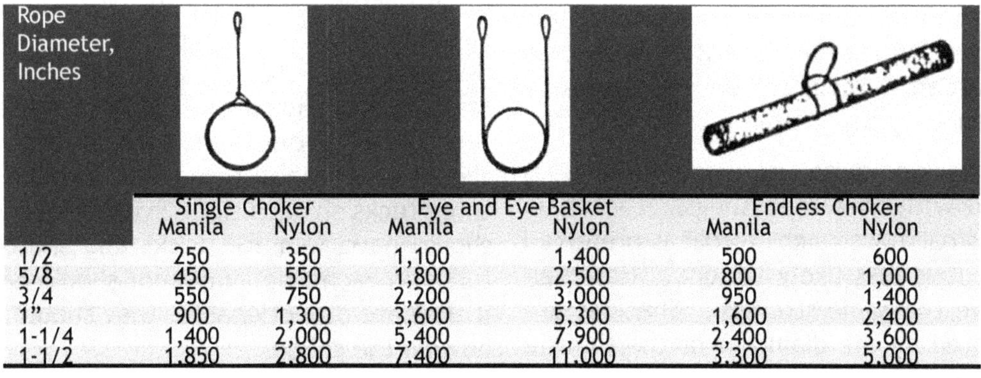

Rope Diameter, Inches	Single Choker		Eye and Eye Basket		Endless Choker	
	Manila	Nylon	Manila	Nylon	Manila	Nylon
1/2	250	350	1,100	1,400	500	600
5/8	450	550	1,800	2,500	800	1,000
3/4	550	750	2,200	3,000	950	1,400
1"	900	1,300	3,600	5,300	1,600	2,400
1-1/4	1,400	2,000	5,400	7,900	2,400	3,600
1-1/2	1,850	2,800	7,400	11,000	3,300	5,000

any sling with six or more closely spaced broken wires should be discarded. Slings are inexpensive compared with the cost of one minor accident. Treat slings with reasonable care; keep them as clean as practicable; lubricate to prevent rusting; don't tie knots in them.

A two-leg sling with 3/4" (19.1 mm) x 20' (6.1 m) legs, one end of each leg spliced into a ring and opposite ends with hooks should be available for handling mats. Slings should not be made with wire rope clips.

5.5.8. Chains

All chains used on the project should be of alloy steel grade only. A standard 5/8" (15.9 mm) pile chain has been of this quality for many years. Other grades, such as "Proof Coil" or "Hi-test" should not be used for lifting, pulling or load binding. Working load limits for common alloy steel chain sizes are shown in Table 5-6.

Chains should be used with links, hooks and rings that develop the full strength of the chain. Chain should be inspected frequently and discarded when it shows obvious wear. A wear of 1/8" (3.2 mm) at any point in the 5/8" (15.9 mm) pile chain is cause for rejection. Chains should have a permanently affixed tag stating size, grade, rated capacity and manufacture.

5.5.9. Shackles and Hooks

Only forged, heat-treated shackles and hooks should be used.

The screw pin type of shackle is to be preferred for most of the work. The pin can easily be tightened or released with either a wrench or a screwdriver through the eye. If the pin is lost, the shackle should be discarded. A common bolt is not a safe substitute for an alloy shackle pin.

The anchor shackle, with its wide throat, is commonly used with wire rope slings. The chain shackle with its narrow straight throat is preferred with chains as it is less likely to kink. However, both styles have the same strength. Shackles are usually designated by the diameter of the shackle body and not the pin size, which is always slightly larger. The rated working loads for forged heat-treated shackles, unless otherwise marked on the shackle, are shown in Table 5-7.

Hooks should only be used for temporary connections where they should be under close observation. They should not be used for drills, jets, siphons or similar semi-permanent attachments; a shackle is preferred.

The hook should be turned so that the pull of the line or chain is always into the hook. Wherever there is a possibility that the load may be momentarily slacked (lowering a bucket that can catch on a ledge, for example), a safety hook should be used or the hook should be moused with wire.

Hooks should be inspected frequently and any that show cracks or signs of straightening should be discarded. Hook capacity should be marked on the hook and should match the chain or cable with which it is used. See Table 5-8 for data on hook capacity.

5.5.10. Rope

Rope is generally preferred for hand lines, where a wire rope sling might slip or damage the material, or where an extra flexible sling is needed for handling small objects.

Manila rope is not as strong as nylon and will deteriorate more rapidly, particularly if stored when wet. However, nylon has a smoother surface and tends to slip more easily; nylon rope stretches and chokers may be hard to loosen. The stretch in a long nylon rope may be very hazardous if the load is suddenly released. Personnel should be warned never to stand in line with or close alongside any rope under heavy strain, particularly nylon.

All ropes are easily cut or abraded and they should be inspected frequently for wear and tear. Rated capacity of rope and rope slings is shown in Table 5-9 and Table 5-10. Slings should usually be bought ready-made. However, if project-made, be certain the splicer is qualified. Manila rope eye splices should have at least three full tucks and short splices three tucks on each side. Synthetic rope eye splices should have four full tucks and short splices four tucks on each side. Tail ends should not be cut off; if they are objectionable, they should be tucked in beyond the splice.

5.5.11. Cutting and Welding

Cutting and welding can be done safely with a few simple precautions. Carelessness, however, can lead to an accident of disastrous proportions - loss of eyesight, severe injury or significant property damage.

The use of a helmet by welders is obviously necessary. Frequently overlooked is the danger to other workers (or pedestrians) who may be accidentally exposed. Portable shields around the point or arc are the most practical way to protect others. When continuous welding is being done, employees who may be accidentally exposed should be provided with shade 2 spectacles with side shields.

People cutting off piles should protect their eyes not only from flying particles but also from the glare of the white hot metal by wearing goggles with shade 3 lenses. People handling cut-offs should wear heavy gloves. Cut-off pieces should not be left; another worker might pick up the hot end.

Fire is an ever-present danger when welding or cutting operations are in progress. Particular caution should be used in working in or near buildings, in deep pier holes or in plants where oil or other flammables are handled or stored.

It is preferable that objects to be welded or cut be moved to a safe location or that all combustible materials be removed from the working area. If this is not practicable:

1. Have a 6# 2A-10BC dry chemical or equivalent fire extinguisher handy and know how to use it. If there is a substantial amount of combustible material, have a fire hose or additional extinguishers available.
2. Cover combustibles, cracks and floor or wall openings with fire resistant covers.
3. Provide a firewatcher, if needed; this includes the far side of a wall when metal may conduct heat through it.
4. If working in a plant, notify the plant Safety Department and get a permit, if one if required.
5. Recheck work area for smoldering material one-half hour after work is completed and again at the end of the day.

Gasoline and other flammable vapors tend to collect in low places. This is a possible hazard when cutting off or building up piles in pier holes. The danger is greatest in plants where flammable liquids may be mixed with ground water. If you have any question, ask the plant safety engineer to check out the area before starting. Another possible source of gas is from organic material when driving piles in marshy areas. If there are a limited number of piles to be cut off, they can be cut with a non-sparking hammer and chisel.

A compressed gas cylinder will take off like a rocket if the valve is broken off while it is in a horizontal position. Such cylinders have force enough to destroy anything in their path including brick walls. Always keep gas cylinders secured in an upright position, except for short periods when being moved. Keep valve covers in place wherever the cylinder is not in active use, particularly when it should be moved.

A cylinder storage area, accessible to trucks, should be provided with a firm level base and a chain or other suitable means of holding cylinders upright. When cylinders are emptied, the valves should be shut and the valve caps replaced.

Inspect cylinders and valves for any sign of leak, damage or defect. Defective cylinders should be removed from the work area.

Figure 5-6 Keep Compressed Gas Bottles Upright

Caution:

When working from oxygen and acetylene pipelines connected to manifolds and generators always follow directions on setting up, taking down and safety precautions applicable to this equipment.

Acetylene cylinders should also be kept in an upright position. If the cylinder is used in a horizontal position, liquid acetone may pass through the torch and then suddenly give up 425 times its volume of acetylene when exposed to atmosphere.

A difficult but necessary precaution is to keep

cylinders in a suitable cylinder cart. If the ground is too rough to be negotiated with the cart, the regulators should be removed, the valve caps replaced and the cylinders and cart carried separately to the new work area. An extra hose will minimize the frequency of moves.

When cylinders are lifted or moved, valve protection caps should be in place. When multiple cylinders are handled, suitable cradles should be available so they can be handled upright; never try to pick up two or more cylinders with a choker. Cylinders should not be dropped or allowed to strike each other violently. When moved by truck, they should be secured in a vertical position. Never put a lifting cable or bar through the holes in a valve cap. Cylinders should never be exposed to excessive heat or open flame, do not permit a welder to strike an arc on a cylinder. Cylinders should never be used as rollers.

Welding and burning equipment should be maintained in good, working order. Hoses, torches and regulators should be inspected at start of each shift; damaged, defective or leaking equipment should be replaced. Do not attempt to salvage cut hose lines with tape; discard them. Damaged torch tips can cause flashback; they should be cleaned with brass wire or replaced. Torches should be lighted with friction lighters.

Welding electrode holders and cables should be fully insulated, in good condition and of adequate capacity. Ground leads should be attached as near as practicable to the work; if remotely attached, make certain that an adequate, high capacity electrical circuit exists. No welding should be done on drums, containers, tanks or pipes unless they are filled with water or thoroughly cleaned of flammable substances, ventilated and tested.

No welding or cutting should be done in confined spaces without adequate ventilation. Before welding or cutting metal with a plated or preservative coating, test should be performed to determine its flammability or tendency to give off noxious gases. When welding or cutting overhead, make sure people and material below are protected from sparks and molten metal.

5.5.12. Excavation

Excavating[9] is recognized as one of the most hazardous construction operations, as can be seen in Figure 5-7. OSHA recently revised Subpart P, Excavations, of 29 CFR 1926.650, .651, and .652 to make the standard easier to understand, permit the use of performance criteria where possible, and provide construction employers with options when classifying soil and selecting employee protection methods. The following is a summary of those recommendations, along with other useful information regarding excavations.

Figure 5-7 Keep Rig a Safe Distance Back from Excavation

5.5.12.1. OSHA Soil Categories

OSHA categorizes soil and rock deposits into four types, A through D, as follows:

B. STABLE ROCK is natural solid mineral matter that can be excavated with vertical sides and remain intact while exposed. It is usually identified by a rock name such as granite or sandstone. Determining whether a deposit is of this type may be difficult unless it is known whether cracks exist and whether or not the cracks run into or away from the excavation.

C. TYPE A SOILS are cohesive soils with an unconfined compressive strength of 3 ksf (144 kPa) or greater. Examples of Type A cohesive soils are often: clay, silty clay, sandy clay, clay loam and, in some cases, silty clay loam and sandy clay loam. (No soil is Type A if it is fissured, is subject to vibration of any type, has previously been disturbed, is part of a sloped, layered system where the layers dip into the excavation on a slope of 4 horizontal to 1 vertical (4H:1V) or greater, or has seeping water.

D. TYPE B SOILS are cohesive soils with an unconfined compressive strength greater than 1 ksf (48 kPa) but less than 3 ksf (144 kPa). Examples of other Type B soils are: angular gravel; silt; silt loam; previously disturbed soils unless otherwise classified as Type C; soils that meet the unconfined compressive strength or cementation requirements of Type A soils but are fissured or subject to vibration; dry unstable rock; and layered systems sloping into the trench at a slope less than 4H:1V (only if the material would be classified as a Type B soil).

E. TYPE C SOILS are cohesive soils with an unconfined compressive strength of 1 ksf (48 kPa) or less. Other Type C soils include granular soils such as gravel, sand and loamy sand, submerged soil, soil from which water is freely seeping, and submerged rock that is not stable. Also included in this classification is material in a sloped, layered system where the layers dip into the excavation or have a slope of four horizontal to one vertical (4H:1V) or greater.

F. LAYERED GEOLOGICAL STRATA. Where soils are configured in layers, i.e., where a layered geologic structure exists, the soil should be classified on the basis of the soil classification of the weakest soil layer. Each layer may be classified individually if a more stable layer lies below a less stable layer, i.e., where a Type C soil rests on top of stable rock.

5.5.12.2. Shoring Types

Shoring is the provision of a support system for trench faces used to prevent movement of soil, underground utilities, roadways, and foundations. Shoring or shielding is used when the location or depth of the cut makes sloping back to the maximum allowable slope impractical. Shoring systems consist of posts, wales, struts, and sheeting. There are three basic types of shoring, timber, aluminum hydraulic and sheet piled.

5.5.12.2.1. Timber Shoring

Figure 5-8 shows a typical timber shoring arrangement.

Figure 5-8 Timber Shoring

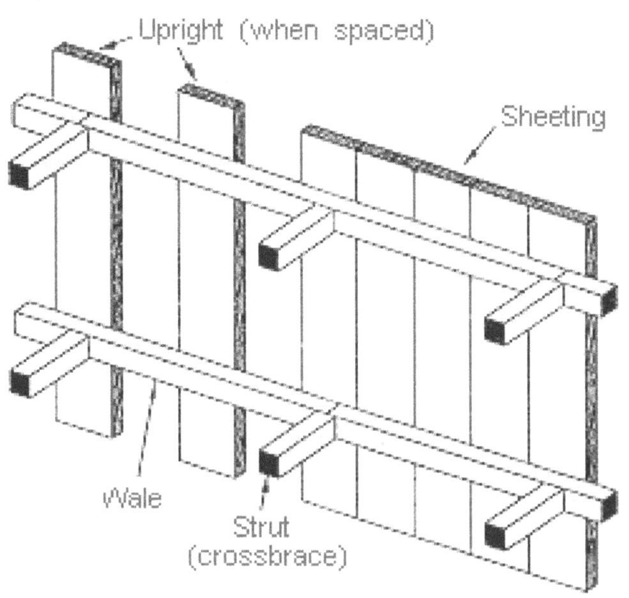

5.5.12.2.2. Aluminum Hydraulic Shorting

The trend today is toward the use of hydraulic shoring, a prefabricated strut and/or wale system manufactured of aluminum or steel. Hydraulic shoring provides a critical safety advantage over timber shoring because workers do not have to enter the trench to install or remove hydraulic shoring. Other advantages of most hydraulic systems are that they:

- Are light enough to be installed by one worker;
- Are gauge-regulated to ensure even distribution of pressure along the trench line;
- Can have their trench faces "preloaded" to use the soil's natural strength to prevent movement; and
- Can be adapted easily to various trench depths and widths.

All shoring should be installed from the top down and removed from the bottom up. Hydraulic shoring should be checked at least once per shift for leaking hoses and/or cylinders, broken connections, cracked nipples, bent bases, and any other damaged or defective parts.

Figure 5-9 Typical Aluminum Hydraulic Shoring Installations

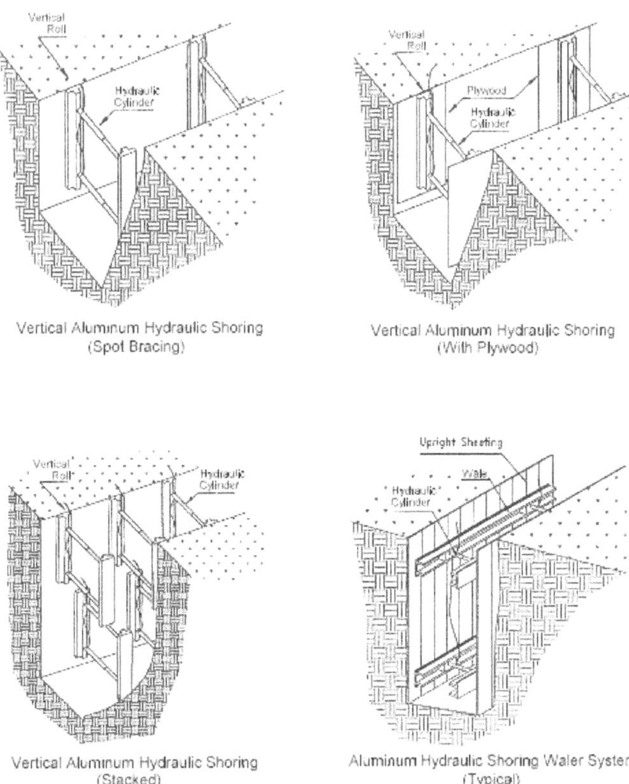

Figure 5-10 Typical Sheeted Trench

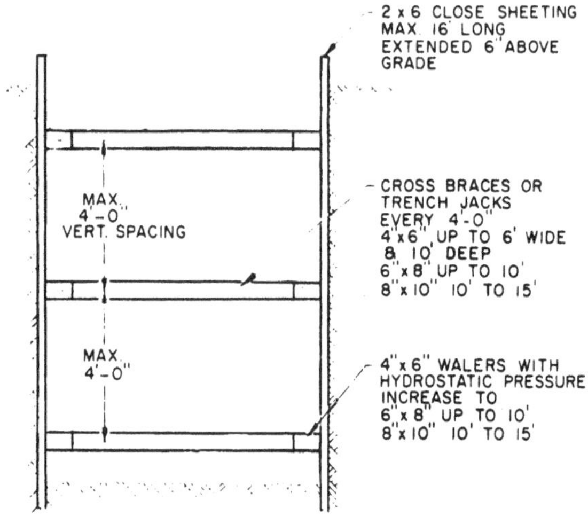

5.5.12.2.3. Sheeted Trenches

If sloping is not practical, or if you are excavating below the footing of an adjacent structure, the excavation will have to be sheeted and braced. The sheeting should be designed by a competent engineer whose instructions are carefully followed with regard to the steps in which the excavation, sheeting and bracing are done.

Simple trench excavations, up to 15' (4.6 m) deep, where there are no complications, may be sheeted as shown in Figure 5-10. Sheeting should project 6" (152.4 mm) above grade, or a toe board provided around holes, to prevent soil, tools, etc., from falling or being kicked into the excavation. When people are working in trenches 3' (914 mm) deep or more, an adequate exit ladder should be available within 25' (7.62 m). Walers and braces should be toe nailed or otherwise secured to prevent slippage.

Before starting any excavation, determine what underground utilities are in the area. Consult the plans, look around for manholes, valve boxes or warning signs, and consult plant or municipal underground utility maps. If in doubt, contact the utility involved.

When excavating near adjacent structures, sheeting should be sufficiently tight and advanced ahead of excavation as may be necessary to prevent loss of ground under footings.

Even though an excavation is sheeted, do not assume that you can place additional load, such as a pile driver, near the edge of the slope unless you can determine that it is safe. Keep excavated material well back from the edge of an excavation.

5.5.12.2.4. Other Types of Shoring

- Pneumatic shoring works in a manner similar to hydraulic shoring. The primary difference is that pneumatic shoring uses air pressure in place of hydraulic pressure. A disadvantage to the use of pneumatic shoring is that an air compressor should be on site.
- Screw Jacks. Screw jack systems differ from hydraulic and pneumatic systems in that the struts of a screw jack system should be adjusted manually. This creates a hazard because the worker is required to be in the trench in order to adjust the strut. In addition, uniform "preloading" cannot be achieved with screw jacks, and their weight creates handling difficulties.

- Single-Cylinder Hydraulic Shores. Shores of this type are generally used in a water system, as an assist to timber shoring systems, and in shallow trenches where face stability is required.
- Underpinning. This process involves stabilizing adjacent structures, foundations, and other intrusions that may have an impact on the excavation. As the term indicates, underpinning is a procedure in which the foundation is physically reinforced. Underpinning should be conducted only under the direction and with the approval of a registered professional engineer and carefully followed.

5.5.12.3. Shielding Types

5.5.12.3.1. Trench Boxes

These are different from shoring because, instead of shoring up or otherwise supporting the trench face, they are intended primarily to protect workers from cave-ins and similar incidents. The excavated area between the outside of the trench box and the face of the trench should be as small as possible. The space between the trench boxes and the excavation side are backfilled to prevent lateral movement of the box. Shields should not be subjected to loads exceeding those that the system was designed to withstand.

Figure 5-11 Trench Shield

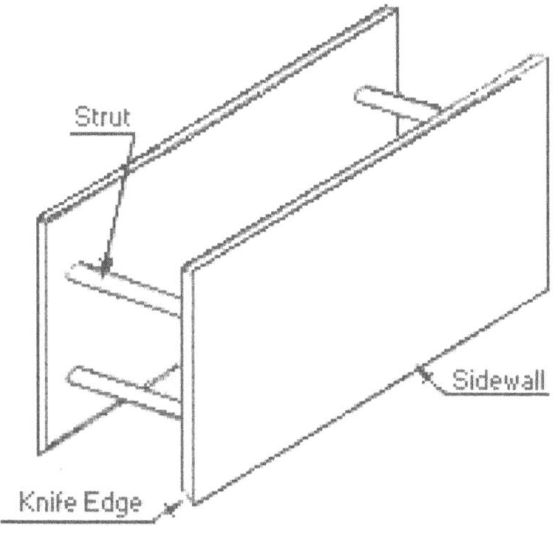

Figure 5-12 Trench Shield, Stacked

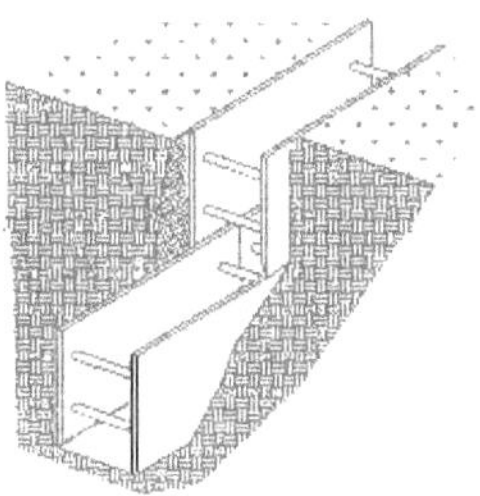

5.5.12.3.2. Combined Use

Trench boxes are generally used in open areas, but they also may be used in combination with sloping and benching. The box should extend at least 18 in (0.45 m) above the surrounding area if it is sloping toward excavation. This can be accomplished by providing a benched area adjacent to the box.

Earth excavation to a depth of 2 ft (0.61 m) below the shield is permitted, but only if the shield is designed to resist the forces calculated for the full depth of the trench and there are no indications while the trench is open of possible loss of soil from behind or below the bottom of the support system. Conditions of this type require observation on the effects of bulging, heaving, and boiling as well as surcharging, vibration, adjacent structures, etc., on excavating below the bottom of a shield. Careful visual inspection of the conditions mentioned above is the primary and most prudent approach to hazard identification and control.

5.5.12.4. Sloping And Benching

5.5.12.4.1. Sloping

Maximum allowable slopes for excavations less than 20 ft (6.09 m) based on soil type and angle to the horizontal are as follows:

Table 5-11 Allowable Slopes[10]

Soil type	Height/Depth ratio	Slope angle
Stable Rock	Vertical	90°
Type A	3/4:1	53°
Type B	1:1	45°
Type C	1 1/2:1	34°
Type A (short-term)	1/2:1	63°

5.5.12.4.2. Benching

There are two basic types of benching, simple and multiple. The type of soil determines the horizontal to vertical ratio of the benched side.

As a general rule, the bottom vertical height of the trench should not exceed 4 ft (1.2 m) for the first bench. Subsequent benches may be up to a maximum of 5 ft (1.5 m) vertical in Type A soil and 4

Figure 5-13 Slope And Shield Configurations

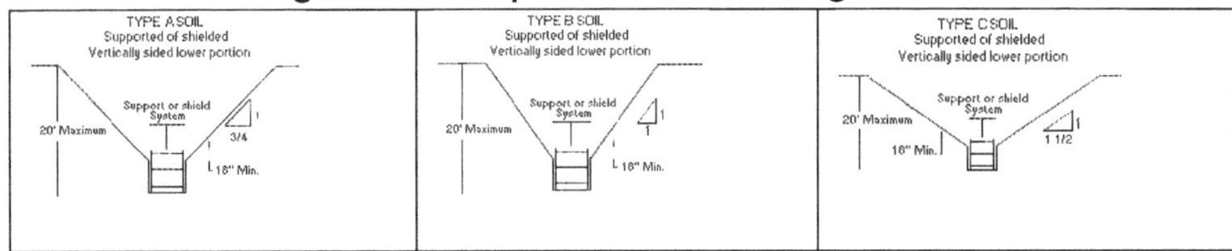

Figure 5-14 Slope Configurations: Excavations In Layered Soils

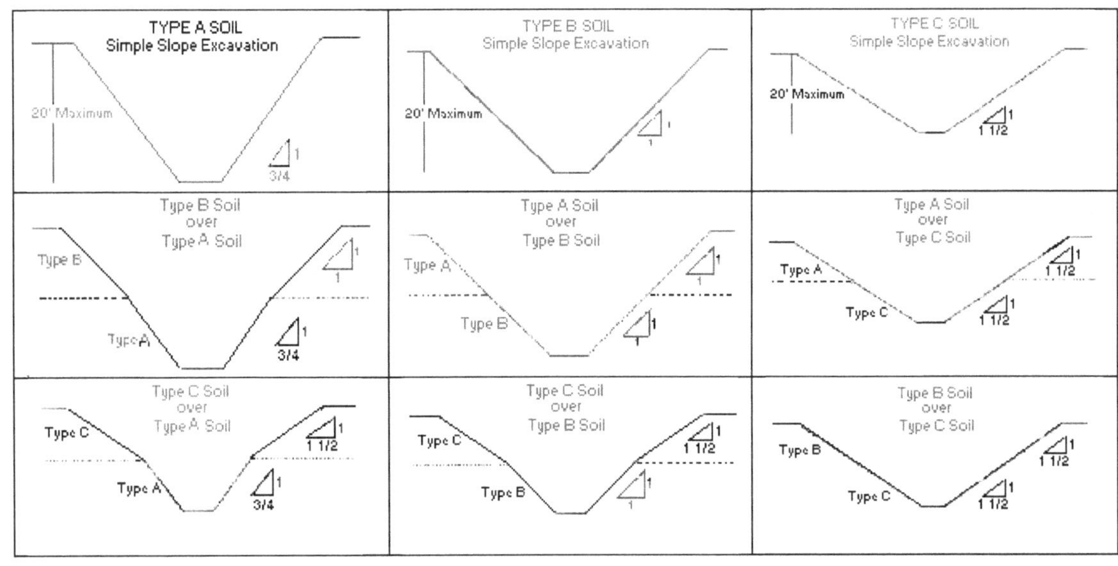

Figure 5-15 Excavations Made In Type A Soil

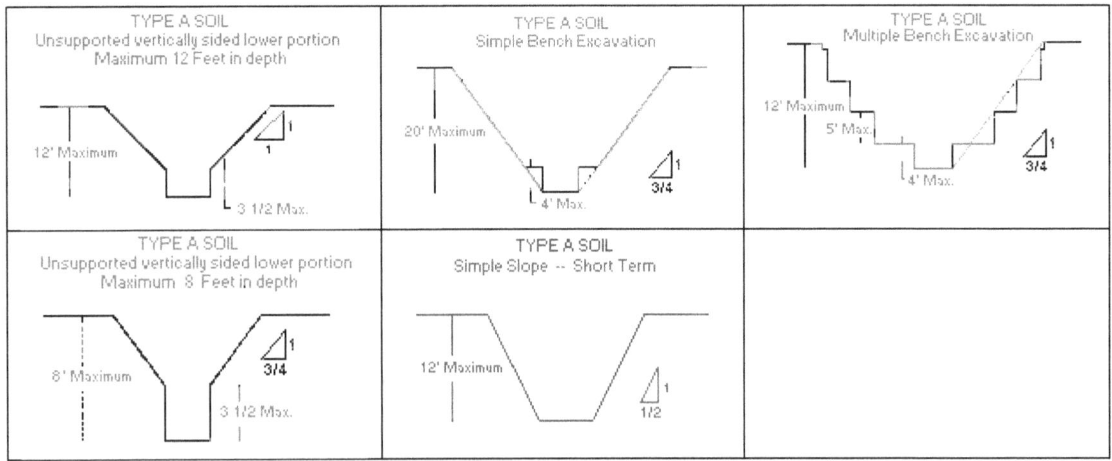

ft (1.2 m) in Type B soil to a total trench depth of 20 ft (6.0 m). All subsequent benches should be below the maximum allowable slope for that soil type. For Type B soil the trench excavation is permitted in cohesive soil only.

Figure 5-16 Excavations Made In Type B Soil

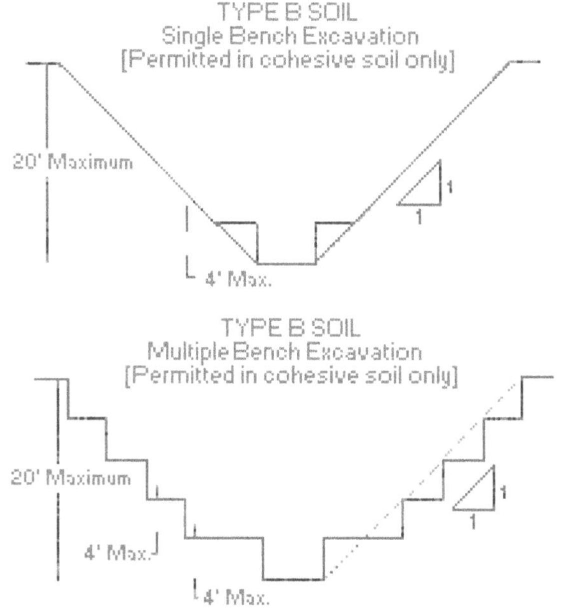

5.5.12.5. Spoil

5.5.12.5.1. Temporary Spoil

Temporary spoil should be placed no closer than 2 ft (610 mm) from the surface edge of the excavation, measured from the nearest base of the spoil to the cut. This distance should not be measured from the crown of the spoil deposit. This distance requirement ensures that loose rock or soil from the temporary spoil will not fall on employees in the trench.

Spoil should be placed so that it channels rainwater and other run-off water away from the excavation. Spoil should be placed so that it cannot accidentally run, slide, or fall back into the excavation.

Figure 5-17 Temporary Spoil

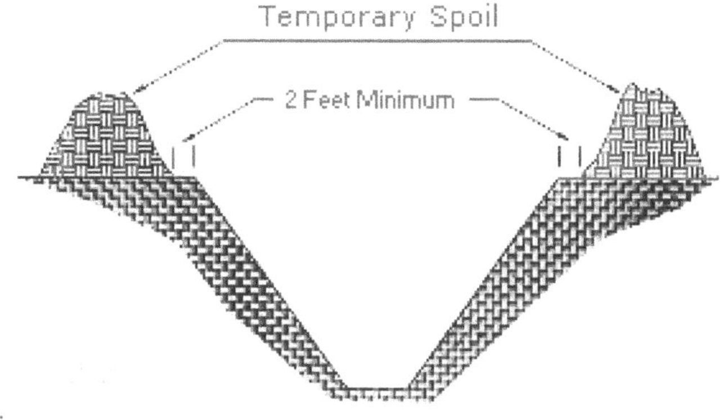

5.5.12.5.2. Permanent Spoil

Permanent spoil should be placed at some distance from the excavation. Permanent spoil is often created where underpasses are built or utilities are buried. The improper placement of permanent spoil, i.e. insufficient distance from the working excavation, can cause an excavation to be out of compliance with the horizontal-to-vertical ratio requirement for a particular excavation. This can usually be determined through visual observation. Permanent spoil can change undisturbed soil to disturbed soil and dramatically alter slope requirements.

5.5.12.6. Special Health And Safety Considerations

A. COMPETENT PERSON. The designated competent person should be able to demonstrate the following:

  a. Training, experience, and knowledge of:

    i. Soil analysis;

    ii. Use of protective systems; and

    iii. Requirements of 29 CFR Part 1926 Subpart P.

  b. Ability to detect:

    i. Conditions that could result in cave-ins;

    ii. Failures in protective systems;

    iii. Hazardous atmospheres; and

iv. Other hazards including those associated with confined spaces.

c. Authority to take prompt corrective measures to eliminate existing and predictable hazards and to stop work when required.

B. SURFACE CROSSING OF TRENCHES. When personnel on the project must cross excavations, provide adequate walkways to discourage jumping across trenches. Such crossings are permitted only under the following conditions:

a. Vehicle crossings should be designed by and installed under the supervision of a registered professional engineer.

b. Walkways or bridges should be provided for foot traffic. These structures shall:

i. Have a safety factor of 4;

ii. Have a minimum clear width of 20 in (0.51 m);

iii. Be fitted with standard rails; and

iv. Extend a minimum of 24 in (.61 m) past the surface edge of the trench.

C. INGRESS AND EGRESS. Access to and exit from the trench require the following conditions:

a. Trenches 4 ft or more in depth should be provided with a fixed means of egress.

b. Spacing between ladders or other means of egress should be such that a worker will not have to travel more than 25 ft laterally to the nearest means of egress.

c. Ladders should be secured and extend a minimum of 36 in (0.9 m) above the landing.

d. Metal ladders should be used with caution, particularly when electric utilities are present.

D. EXPOSURE TO VEHICLES. Procedures to protect employees from being injured or killed by vehicle traffic include:

a. Providing employees with and requiring them to wear warning vests or other suitable garments marked with or made of reflectorized or high-visibility materials.

b. Requiring a designated, trained flag person along with signs, signals, and barricades when necessary.

c. Any excavation adjacent to a walkway or road should be suitable barricaded and lighted at night.

d. When excavations are made on the street or in other areas accessible to the public, provision should be made for security guard service as needed to insure that lights are maintained. Electronic flashers, which can be rented in most areas, are generally the most practical type.

E. EXPOSURE TO FALLING LOADS. Employees should be protected from loads or objects falling from lifting or digging equipment. Procedures designed to ensure their protection include:

a. Employees are not permitted to work under raised loads.

b. Employees are required to stand away from equipment that is being loaded or unloaded.

c. Equipment operators or truck drivers may stay in their equipment during loading and unloading if the equipment is properly equipped with a cab shield or adequate canopy.

F. WARNING SYSTEMS FOR MOBILE EQUIPMENT. The following steps should be taken to prevent vehicles from accidentally falling into the trench:

a. Barricades should be installed where necessary.

b. Hand or mechanical signals should be used as required.

c. Stop logs should be installed if there is a danger of vehicles falling into the trench.

d. Spotters should be provided with highly visible jackets and kept out from behind trucks.

e. Everyone must keep out from under buckets; unless trucks are provided with an adequate safety shield, drivers should leave their cabs while trucks are loaded with power equipment.

f. Soil should be graded away from the excavation; this will assist in vehicle control and channeling of run-off water.

G. HAZARDOUS ATMOSPHERES AND CONFINED SPACES. Employees shall not be permitted to work in hazardous and/or toxic atmospheres. Such atmospheres include those with:

a. Less than 19.5% or more than 23.5% oxygen;

b. A combustible gas concentration greater than 20% of the lower flammable limit; and

c. Concentrations of hazardous substances that exceed those specified in the Threshold Limit Values for Airborne Contaminants established by the ACGIH (American Conference of Governmental Industrial Hygienists).

d. All operations involving such atmospheres should be conducted in accordance with OSHA requirements for occupational health and environmental controls (see Subpart D of 29 CPR 1926) for personal protective equipment and for lifesaving equipment (see Subpart E, 29 CFR 1926). Engineering controls (e.g., ventilation) and respiratory protection may be required.

e. When testing for atmospheric contaminants, the following should be considered:

i. Testing should be conducted before employees enter the trench and should be done regularly to ensure that the trench remains safe.

ii. The frequency of testing should be increased if equipment is operating in the trench.

iii. Testing frequency should also be increased if welding, cutting, or burning is done in the trench.

iv. Employees required to wear respiratory protection should be trained, fit-tested, and enrolled in a respiratory protection program. Some trenches qualify as confined spaces. When this occurs, compliance with the Confined Space Standard is also required.

H. EMERGENCY RESCUE EQUIPMENT. Emergency rescue equipment is required when a hazardous atmosphere exists or can reasonably be expected to exist. Requirements are as follows:

a. Respirators should be of the type suitable for the exposure. Employees should be trained in their use and a respirator program should be instituted.

b. Attended (at all times) lifelines should be provided when employees enter bell-bottom pier holes, deep confined spaces, or other similar hazards.

c. Employees who enter confined spaces should be trained.

I. STANDING WATER AND WATER ACCUMULATION. Methods for controlling standing water and water accumulation should be provided and should consist of the following if employees are permitted to work in the excavation:

a. Use of special support or shield systems approved by a registered professional engineer.

b. Water removal equipment, i.e. well pointing, used and monitored by a competent person.

c. Safety harnesses and lifelines used in

conformance with 29 CFR 1926.104.

d. Surface water diverted away from the trench.

e. Employees removed from the trench during rainstorms.

f. Trenches carefully inspected by a competent person after each rain and before employees are permitted to re-enter the trench.

J. INSPECTIONS. Inspections shall be made by a competent person and should be documented. The following guide specifies the frequency and conditions requiring inspections:

a. Daily and before the start of each shift;

b. As dictated by the work being done in the trench;

c. After every rainstorm;

d. After other events that could increase hazards, e.g. snowstorm, windstorm, thaw, earthquake, etc.;

e. When fissures, tension cracks, sloughing, undercutting, water seepage, bulging at the bottom, or other similar conditions occur;

f. When there is a change in the size, location, or placement of the spoil pile; and

g. When there is any indication of change or movement in adjacent structures.

5.5.13. Pile Driver Assembly

The best of equipment will not be safe if improperly used. It is wise to apply more conservative limitations on the use of cranes as pile drivers than might be allowed by ordinary rating charts. Following the procedure given for using the capacity charts, determine the maximum safe radius at which your rig can be operated. This figure may have to be modified when project conditions are less favorable than specified.

When using new or unfamiliar equipment, the superintendent should study the appropriate section of the factory-supplied manuals to learn the capabilities and limitations of the particular item. When special equipment is to be used, understand and conditions to be used with the supervisor and be sure he understands how it is to be used and discuss any special precautions that will have to be taken.

Under no circumstances should the superintendent add more counterweight than authorized. While this might seem to improve stability, it can result in overstressing and failure of some structural components or reducing stability in the out-batter position.

In preparation for unloading the rig, the superintendent should assure he has adequate handling equipment, in good working order, with slings and other rigging gear in safe condition. He should select a safe, firm, level site free of overhead power lines or other obstructions at which to erect his or her rig.

Equipment should be lifted with adequate size slings, properly placed and tied down to the transport to prevent shifting during travel. Be sure to use tag lines to prevent loads from swinging out of control. Adequate dunnage and blocking should be available so that loads can be landed safely.

As equipment arrives at the site and is unloaded, it should be given a thorough inspection. Leaders, booms, spotters, crane carbody, gantry, topping lift and other structural items should be carefully examined. Look for cracks, evidence of overstressing, deterioration or damage that would impair functioning. Check platforms, ladders and catwalks to be sure they are not damaged and are clean and safe.

All wire rope, including hoisting lines, pendants, slings, etc., should be inspected as detailed above. All headblocks, fairleaders and other topside fittings should be examined to be sure they are working properly and that nothing is loose. This should be one of the best times for these items to be properly inspected. The boom leader connections and boom tip fairleader should likewise be examined. Be sure that the guards that prevent cable from jumping out of the sheaves are in place.

While the crew is bolting up leaders, etc., the operator has an opportunity to conduct a detailed inspection of the crane or hoist. All controls should be operating properly, adjust clutches and brakes, check operation of swing and travel locks and inspect drum ratchets and pawls.

Figure 5-18 Hand Signals

5.5.14. Erection

Before raising leaders, check for adequate tag lines to control bottom of leaders. A check should be made that no tools, extra bolts, etc., have been left on the boom or leaders to fall when they are raised. The hammer extensions and sliding frame should be cabled securely. Inspect steam, jet and hydraulic hoses; if they are connected to the boom tip, they must have safety chains secured to the boom; all joints should be safetied. The lower ends of hoses, boom safety cables, etc., should be wired to the boom so they will not swing free as the boom is raised.

Check that the center of gravity of the leaders is below the boom point so that they cannot pinwheel or else take other precautions such as counterbalancing. Check that the weight of the leaders, extensions, drill guides, and etc., to be raised is within the safe capacity of the crane; otherwise, a helper crane should be required. Do not attempt to add additional counterweight or tie down the back of the crane as this can cause failure of the gantry and/or topping lift.

Start the raising operation slowly; watch rear track rollers; watch the mat and ground under the front end of the crawlers where most of the load should be concentrated. After raising the leaders, connect the sliding frame to the spotter; connect and adjust boom safety cables so they should be taut at maximum anticipated boom angle. As soon as the hammer is lifted, lower the hammer extensions and cable them off to the hammer. As hoses are connected, be sure safety chains are secured.

5.5.15. Daily Inspections

The superintendent and supervisors, assisted by experienced crewmembers, should inspect all work areas and equipment daily to insure that they are in safe working condition.

1. Rig supervisor, assisted by the lofts person, should visually check all platforms, railings and ladders to assure they are clean, unobstructed and safe.
2. Rig supervisor, assisted by the lofts person, should check boom safety cables, the cable tying extensions to the hammer and safety chains on hoses.
3. The lofts person should be trained to keep alert when aloft for loose bolts or fittings.
4. Rig supervisor, assisted by the front-end-person should visually check hammer, drill and jet hoses and their fittings for leaks or damage. Check safety chains on hose at hammer, drill and jet and at hose joints. Immediate attention should be given to leaks.
5. Rig supervisor, assisted by the front-end-person should examine all slings, chains, hooks, shackles and ropes, as explained above, discarding any not in safe condition.
6. Rig supervisor and front-end-person should routinely inspect mats, timbers, blocking and falsework each time they are handled for signs of excessive wear or damage.
7. The operator, under the supervision of the supervisor, should check all rig controls, at starting time, and again after lunch break, to be certain they are operating properly. Picking the hammer and core a few inches above the ground and holding it momentarily on the brake should test the hammer line brake. Check that all safety devices are functioning properly, all guards on machinery are in place, the cab is not littered with anything that would interfere with his or her controls or pedals and that the view is unobstructed.
8. Each time the engine is shut down the operator, under the supervisor's supervision, should check the gauges and listen for any abnormal air leaks. The pressure should be released and operation of spring applied locks and pawls checked.
9. The operator or fireman, under supervision of the supervisor, should observe the steam piping and boiler for signs of leakage, check that the safety valve is unobstructed and operating properly and that the boiler is otherwise in safe operating condition.
10. Each supervisor should check that every crew member is wearing hard hats, that all people are properly clothed, and that they are using other needed safety equipment.
11. The concrete supervisor should check general project housekeeping, that concrete runways and concrete tools are in safe condition,

that oxygen and acetylene tanks are being properly used and stored, that goggles are being worn for burning operations and that all ramps and ladders leading to work areas are in a safe condition.

5.5.16. Monthly Project Inspections

At least once a month:

- The superintendent should personally supervise the above listed daily inspections and assure himself that they are being performed regularly and that defects are being promptly reported to him.
- An experienced person should go aloft and make a detailed visual inspection of boom, leaders, fairleaders, headblock, drill guides and topping lift for loose bolts and fittings, deformed, cracked or corroded structural members, worn or broken sheaves, worn, cracked or distorted pins, bearings, shafts and rollers and condition of ladders and platforms.
- The operator, under supervision of the supervisor, should make detailed examination of the machine. Structural parts should be examined for evidence of deformation, cracking or corrosion. Controls and control linkages should be inspected for loose pins and bolts, excessive wear or any other interference with free operation. Bearings, pins, shafts, gears, rollers, chains and sprockets should be examined for excessive wear or damage. Brake and clutch lining should be checked for wear and replaced if needed.

The superintendent should check the following:

- The OSHA notice and safety posters are displayed;
- OSHA records are up to date;
- Materials are safely stored;
- General project housekeeping is acceptable;
- Fire extinguishers are in usable condition and in their proper places;
- First-aid kits are available and properly equipped;
- Work habits are in accordance with good safety practice;
- Foremen are instructing their people in safe working habits.

Send a written report of the above inspection, noting any defects and the corrective action taken.

5.5.17. Driver Operation

The supervisor should position himself so that he is in view of the operator and can see the operating area to the rear, sides and aloft at all times. He alone should give signals to the operator or clearly designate who is to give signals when he cannot safely do so. The operator should accept signals from one person only, except emergency signals to stop. Standard hand signals, as shown on Figure 5-18, should be used. Misunderstanding of signals can cause a serious accident.

The supervisor, or signalman, should assure everyone is clear before giving a signal to move, swing, pick, or lower a load. When swinging with a load, it should be held against the leaders or controlled with tag lines. Do not allow loads to swing and strike someone.

When operating in inclement weather or in muddy or slippery areas, walkways and mats should be cleaned, or sanded, if necessary. Remember when swinging loads that people cannot move as quickly or may fall if they have poor footing.

The most hazardous part of the rig operation is moving around the site, particularly on slopes; take special precautions when indicated by project conditions. The supervisor should position himself where he can get the best overall view of the rig. A second person should be on the opposite side. The signalman should observe the leaders and be alert when approaching overhead obstructions. The bottom of the core or follower should be no higher than the bottom of the leaders, except when shelling up. When moving, secure the bottom of the core to prevent it from swinging out. Never attempt to drag a load by moving or swinging with a hoist line attached to the load. If it is necessary to move something, it should be leapfrogged by picking it up and swinging it ahead periodically or dragged by a sling attached to the carbody or under carriage.

The rig should be operated on mats unless your general superintendent otherwise authorizes you.

Keep the rig in the center of the mats. Handle mats with a two-leg sling; don't try to slide mats sideways as it puts severe loads in the boom.

When working near ordinary power distribution lines, it is preferred that they be de-energized and visibly grounded or insulating barriers erected between the lines and the rig. If this is not feasible, extreme care should be exercised. No part of the rig should be allowed to come closer than 20' (6.1 m); where the operator has any difficulty in seeing, a special signalman should be designated to observe clearance. Do not attempt to drag shells or piles from under a power line; the rig should be grounded with a maximum resistance of 25 ohms from rig to ground; only manila tag lines should be used. When working near high-tension electric transmission lines or radio transmitter towers, consult a Safety Engineer for proper procedures to be followed. Consider all power lines "hot" unless you know they have been de-energized and are visibly grounded.

People should not stand or walk under suspended loads and supervisor should plan his or her work so that it will not be necessary to hoist or swing loads over areas where normal work operations are carried on.

When it is necessary to perform maintenance on equipment, the engine should be shut down or power turned off, loads landed, brakes set, chocks or toggles placed to prevent accidental movement, as the situation dictates. No one should enter machinery spaces or crawl under a machine for any purpose until both the supervisor and operator are aware of his or her intentions. When remotely operated equipment should be repaired, the controls should be tagged and locked in an inoperative position. When working on equipment driven by a gas engine with magneto, the ignition circuit should be disconnected to prevent accidental started.

Fuelling should be done with care to prevent fuel spills or overflows on to the engine, exhaust, hot pipes or the boiler platform. The gasoline-starting engine on the air compressor should be fuelled in the morning before firing up the boiler. All gas engines should be shut down before fuelling.

When operating at night, adequate light should be available. People are generally not as alert at night when they have been accustomed to sleeping. When permitted by the labor agreements, it is preferable to keep the same crews on the night shifts as it takes two or three weeks before a person becomes accustomed to the shift in working and sleeping hours. On foreign assignments, supervisors and operators may require several days before they are acclimated to a change in time.

5.5.18. Driving Piles

Pile driving presents some special safety hazards, but adherence to common sense and general safety rules supplemented by knowledge of pile driving problems can result in safe working conditions.

Piles should be picked up with proper type slings or other lifting equipment and people should be properly trained in their use. Piles should be picked up from the side away from other operations and everyone kept clear when piles are hoisted into the leads. Shells and wood piles can be picked with the standard pile chain. Slings used to pick other piles should have adequate capacity; however, if they are intended for larger pile than is being picked up, they may be so stiff that they will not grip the pile. For light piles, manila rope slings are preferred over wire rope slings; consult the sling tables. Handling and driving creosoted piles requires protection of eyes and skin with goggles, rubber gloves and proper clothing.

When jetting or pre-excavating for piles, care should be exercised not to undermine the rig or adjacent structures. Augers should be cleaned of clods of earth and rocks as they are hoisted out of the ground, lest these fall and strike someone. Holes are normally pre-excavated or jetted one at a time just before driving. If holes should be made in advance, be certain that they are covered or barricaded to prevent someone from falling into them.

All employees, but the lofts person in particular, should be trained to keep out from under the hammer and core at all times. Should some circumstance arise when it is necessary to work under the hammer, a 12" (304.8 mm) x 12" (304.8 mm) timber should be placed across a platform, the hammer landed on it and either the dogs or the brake set.

When changing capblocks or hammer cushion, the core sling is released on both sides, the hammer raised clear and the drum brake or dog set before placing under the hammer. People on the ground should keep away from the leads when the capblock or hammer cushion is being changed. Whenever possible, the hammer cushion should be changed with the hammer near the ground. Hammer cushion plates, new or used, should not be left on leader platforms.

The operator should keep the hammer under control during driving by maintaining slight pressure on his or her hammer drum brake. He should be prepared to shut off the hammer immediately should the pile break or start running into the ground under weight of the hammer. When lifting the hammer at the conclusion of driving, the signalman must watch the hammer hose to be certain that it does not catch on anything. When the hose loops down below the leaders, a person should hold it out away from the leaders, while the hammer is raised, until the hose is clear.

The driving of batter piles involves special procedures that should be fully understood. Pulling piles and stuck cores require special precautions.

Pile driving may raise questions concerning possible effects on nearby structures or utilities. While experience shows that damage to structures is extremely rare, you should be alert to possibility of claims. When driving close to buildings that are in poor or damaged condition, arrange for a preconstruction condition survey.

Underground utilities are vulnerable to damage by pile driving. If the location of utilities is not shown on the plans, it should be established before driving begins. Important underground facilities such as gas lines, water mains and telephone ducts should be uncovered, if they are near pile locations.

5.5.19. Concreting and Cutting Off Pipe Piles

Closed ended pipe piles should be inspected by reflecting sunlight with a mirror, by flashlight or by lowering a bulb with a wire guard. The use of lighted paper to inspect piles is hazardous. The presence of flammable gas should be anticipated when driving in organic soils or in plants where flammable liquids are made or stored. Before cutting the piles off with a torch when flammable gases are encountered, drop a piece of lighted oil-soaked rag into the pile, keeping head and body away from the head of the pile to safely ignite any gas present.

When piles are driven or cut off so they are nearly flush with the ground surface, the heads should be covered or protected to prevent someone from stepping into the pile and to keep foreign material out of the pile, if concrete does not follow immediately. When such piles are left overnight, heads should be secured so children cannot remove them.

When cutting off piles with a torch, follow precautions given in "Cutting and Welding" above. Workers should wear goggles and keep their head, arms and body away from the pile. Short cut off pieces should be stored on end, hot end down. Workers handling cut offs should always wear gloves. Workers cutting creosoted piles should wear clear goggles and rubber gloves.

When dewatering piles with a steam siphon make certain that hoses are in good condition and unobstructed and fittings are tight. A quick opening valve should be provided at the rig to operate the siphon. When the siphon is being operated, a person must stay at the valve. When the steam is turned on, steam or water should come out the discharge line. If the discharge line is plugged, steam will mix with the water and blow into (and possibly out of) the pile. Workers should not look down the pile, while using the siphon.

Figure 5-19 Observe Safety Rules in Handling Siphon

Pile driving and concrete operations should be coordinated, so it will not be necessary to drag piles or swing the rig over the area when concreting is in progress. The principal cause of concreting accidents is the use of narrow or unstable runways. The minimum safe runway for a wheelbarrow is two cleated 2 x 12's supported every 3' or two cleated 3 x 12's supported every 6'. When runways are over 3' above ground, duckboards should be used.

Cement is irritating to the skin and concrete workers should be given rubber-coated gloves. When dry cement is handled in bags, workers should keep their arms and neck covered as the dust can work into open pores in warm weather, causing serious burns.

5.5.20. Small Tools

Small tools should be inspected frequently for split handles, loose or mushroomed heads, sprung jaws, dull points, or other defects. Damaged tools should be discarded or returned to the tool room for repair. Cutting tools are safest when sharp.

Make sure that each person has the proper tools to safely perform their job. Do not permit workers to misuse tools; a shovel handle should not be used as a crowbar, for example. It is not only liable to damage the tool but may injure the user.

All powered tools should also be examined for defects before being used. Any tool not in proper working order should be removed from service. Workers chipping or grinding should wear goggles or a facemask.

Jacks, chain hoists and come alongs should be checked to determine if they are safe. Such tools should not be used beyond their rated capacity; do not allow use of "cheaters." Bars should be removed from ratchet jacks, when not being used.

5.5.21. Caissons

When not being worked on, caissons should be covered or adequately barricaded. The head of the casing should extend at least 8" (203.2 mm) above ground level to prevent anything from being kicked into the hole.

No one should be permitted to enter a drilled hole unless:

1. A casing of adequate strength has been installed to prevent its collapse.
2. The air at the bottom of the hole has been checked and found safe.
3. There is adequate ventilation.
4. The person is equipped with a shoulder harness with a separate lifeline, which is individually manned, and is independent of any line used to remove materials.

All lines and hoisting equipment should be inspected frequently to insure that it is in first class condition. Only safety hooks or shackles should be used for attaching buckets. No smoking should be permitted in or around caissons.

5.5.22. Electrical

All electric tools either should be of the approved double insulated type or properly grounded. Grounds should have a continuous circuit of low resistance and be adequate in capacity, generally the same size as the current carrying wires. Cables to portable tools should be three-wire type with grounding plugs. The grounding wire should be non-current-carrying and be green in color.

Electrical equipment and tools should be inspected frequently for defective or frayed cables. Check particularly at points where cables enter tool housings and switch boxes; if loose, they should be repaired immediately. At any sign of a short circuit, determine and correct the cause. Under no circumstances should you continue to use the device by substituting a larger fuse or by bypassing the overcurrent device. Very small currents, particularly in wet or damp locations, can be fatal.

In accordance with the National Electrical Code, competent personnel should install temporary wiring and lighting. Temporary light bulbs should be guarded unless they are in deeply recessed fixtures. Light cords should be of the heavy-duty type "S," with concealed contact type grounding connectors, suitable for rough use. Cables should be free of splices other than of the vulcanized type and cables should be located and supported to minimize accidental damage.

5.5.23. Vehicles

All vehicles, whether operated on the highway or not, should have adequate service and parking brakes, brake lights, horn, rear view mirrors, windshield wiper, windows of clear un-cracked safety glass, fire extinguisher and seat belts for each seat. If operated at night, the vehicle should also have two good headlights and two operable taillights. Vehicles on which the operator cannot see clearly to the rear must have a back-up alarm. Vehicles operated on public highways must have marker lamps, reflectors, flares or other emergency lights.

Employees must ride seated in the cab; they must not ride on top of loads or running boards. Trucks should not be overloaded. Loads should be suitably secured to prevent dangerous shifting. Vehicles should be operated in accord with traffic laws or local regulations and in a manner not to endanger others.

All persons who operate company motor vehicles should be in possession of a valid operator's permit for the class of vehicle operated. You should record on the individual's employment application his or her driver's permit number, the date of expiration, where issued, and any qualifications. Drivers should be 21 years of age, read and write adequately to understand how to load, secure and safely operate their vehicle and have no physical impairment that would interfere with safe operation.

Vehicles should be dispatched and operated in accord with traffic laws and regulations, with due regard for the safety of employees and general public. No driver should be permitted to operate a company vehicle under the influence of liquor or drugs or be permitted to have liquor (including beer or wine), drugs or "pep pills" in his or her possession while on duty. Under U.S. regulations, no driver may consume any liquor (including beer and wine) within 4 hours before going on duty and you may not permit a driver to be on duty if he appears to have consumed intoxicating liquor within the preceding 4 hours. No driver should be allowed to operate in excess of 10 hours, within 8 hours off duty or when his or her alertness is impaired through fatigue or illness.

When operating trucks on interstate highways in the United States, the superintendent should understand the provisions of the Department of Transportation Regulations relating to "Qualification of Drivers." In addition to the obvious rules set out above, the driver must give you copies of a medical certificate, a road-test certificate and certificate of written test, all showing that he qualified to operate the vehicle involved.

When the driver is employed on other than an intermittent, casual or occasional basis, he must furnish, in addition to the above, all addresses where he has lived for the past 3 years; a list of all un-expired driver's permits with numbers and expiration dates; the nature and type of past experience; a list of all accidents within the past 3 years, with details; a list of all moving violations within the past 3 years; information in detail concerning any revocation of suspension of his or her license; a list of all employers and reasons for leaving; and a certificate that the information submitted is true and complete. You must verify the above information by inquiry or investigation.

5.5.24. Marine Work

All marine equipment shall meet appropriate U.S. Coast Guard or equivalent regulations. Only properly licensed personnel should operate towboats and launches. Barges and boats should be equipped with Coast Guard approved lights, bells, whistles, life preservers, fire extinguishers and other equipment appropriate to their class and type of operation. No work should be undertaken in a navigable waterway until a permit has been secured from the U.S. Army Corps of Engineers or equivalent local authority, and appropriate notices given to marine interests.

Equipment located in navigable waters should be well lighted at night, in accordance with C.G. regulations, and securely moored to prevent it going adrift and becoming a menace to navigation. Be prepared for the possibility of a storm coming up during the night or over a weekend. A person with adequate marine experience, and a knowledge of local tides, currents and weather, should select the size, type, location, and scope of anchors and lines.

Figure 5-20 Proper Anchoring and Mooring is Essential

All employees working on the decks of barges, on

piers or other places where there is any danger of falling overboard must wear an approved buoyant vest. A Coast Guard approved ring buoy, with at least 90 feet of line, shall be maintained on each barge and on all exposed work areas, not over 200 apart. Ladders should be provided, extending from decks of barges and other work surfaces to the water and a skiff shall be immediately available for rescue purposes.

Suitable access ramps and gangways for equipment and people shall be provided between barges and docks and between barges. Decks and passageways should be kept clear. Safe passageway should be provided over or around deck loads. Where there is less than a 3-foot clear walkway outside of deckhouses, loads or bulkheads, safety grab rails should be provided.

5.5.25. Reporting Accidents

Workers should be instructed to report all accidents, no matter how minor. Prompt reporting can often prevent a minor accident from developing into a serious case.

The superintendent must make a written report of all accidents to the employees which require medical attention or when the employee cannot continue working at his or her normal task. Be particularly alert to report and seek medical help for falls or strains that may involve back or internal injuries and any head injuries, no matter how slight they appear.

All accidents or alleged accidents involving members of the public, employees of the owner or other contractors or the property of others, should be reported in writing, no matter how slight.

In all of the above cases, the original should be submitted to the nearest office of the insurance company, a copy to your general superintendent and other copies as required by your company. In case of a serious or fatal accident, an immediate report should be made by telephone, followed by a written report.

Any fatal accident or any accident that results in the hospitalization of five or more employees must also be reported, within 48 hours, to the nearest regional office of the Occupational Safety and Health Administration, U.S. Department of Labor.

Many local authorities have regulations regarding the reporting of fatal or serious accidents; consult your general superintendent or local insurance company office.

5.5.26. OSHA Records

The project superintendent must maintain a log of all reportable accidents and injuries on OSHA Form 100. You should keep this log up to date and available for examination by Federal or State representatives. At the end of the project, the log should be included with other valuable project records and sent to District or Regional Records Center, as instructed.

At the end of each calendar year, you should summarize the data from Form 100 on OSHA Form 102, "Summary Occupational Injuries and Illnesses," and post it in a place accessible to the employees. Instructions for both forms are included in the OSHA booklet.

5.5.27. Loss or Damage Of Company Equipment

Each superintendent is responsible for notifying his or her general superintendent of any loss or damage of Company owned equipment or other property when the loss exceeds US$100. The report should include description of the equipment, manufacturer's serial number, extent of damage or loss, circumstances surrounding the damage or loss and, if damaged, location of the item and recommended disposition. The report may ordinarily be made in your daily narrative. If a major item of equipment is involved, (including all rig capsizes) or if the loss may exceed US$10,000, report should be made to your general superintendent by telephone fax, or email, followed by a written report. If the property is believed stolen, report should also be made to local police.

5.5.28. Insurance

The original of all accident reports should be sent to the nearest office of the insurance carrier. Include any bills for first aid or emergency treatment; beyond this, you should not incur any expense or make any commitments or statements. The handling of claims should be left to the insurance adjusters who are experienced in this work.

When operating in states where you must deal with a State Insurance Fund, the state fund can provide you with the proper forms in event of injury to an employee. For all other accidents, you should deal with your insurance carrier.

When you employ any sub-contractor, you should secure copies of their insurance certificates before work starts. The standard order form provides that the subcontractor agree to indemnify and save us harmless from all claims for bodily injury and property damage that may arise out of his or her work, under contract with us. Such a provision should be included in any other type of contract. Be careful to avoid informal arrangements, for an employee to handle shells with his or her own truck, for example.

In event of any property damage accident, do not make any repairs except what may be necessary to prevent further loss.

5.6. Welding and Cutting

Welding is often an essential part of pile driving operations. Although most people involved take it for granted, it is both a science and an art that requires forethought in configuring a weld and skill in executing it.

When welds are performed from a drawing, they should adhere to the specifications therein. Figure 5-21 shows the welding symbols used on drawings.

Figure 5-21 Standard Arc and Gas Welding Symbols

1. BASIC SYMBOLS

TYPE OF WELD							
BEAD	FILLET	PLUG OR SLOT	GROOVE				
			SQUARE	V	BEVEL	U	J

2. SUPPLEMENTARY SYMBOLS

WELD ALL AROUND	FIELD WELD	CONTOUR	
		FLUSH	CONVEX

3. LOCATION OF ELEMENTS OF A WELDING SYMBOL

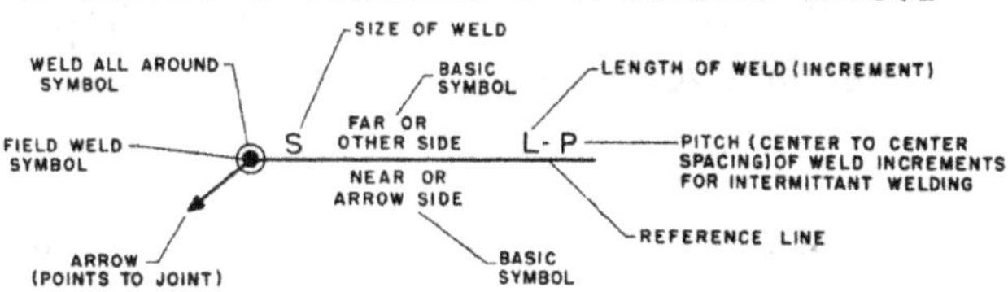

5.6.1. Electric Welding

Only a thoroughly trained and competent operator should do electric welding. It is best to hire an established and reputable welding contractor unless a large amount of welding is to be done, in which case it may pay to hire your own people.

5.6.2. Acetylene Welding and Cutting

5.6.2.1. Procedure

A competent mechanic is one who knows not only the technical aspects of his or her trade but also the safety precautions. The mechanic should be able to recognize hazards quickly and know how to avoid or correct them. Always remember that a careless operator is a hazard not only to himself but also to everyone near him.

It is therefore vitally important that the operator of welding equipment become thoroughly familiar with the Safety Rules given herewith. As an aid to understanding the terms used in these rules, Figure 5-22 show an oxyacetylene welding outfit.

Figure 5-22 Oxyacetylene Welding Outfit

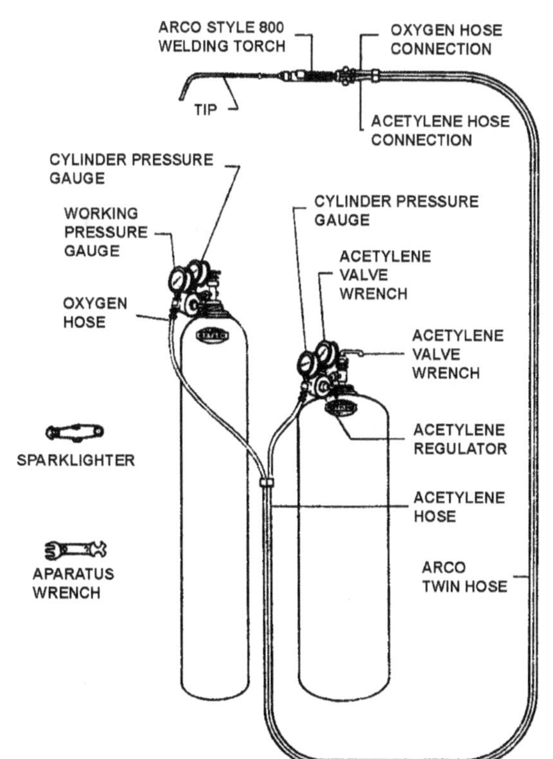

5.6.2.2. Safety Rules

5.6.2.2.1. Oxygen

1. Always refer to oxygen by its full name "oxygen" and not by the word "air."
2. Do not store oxygen and acetylene cylinders together. They should be separately grouped.
3. Never use oxygen near flammable materials, especially grease, oil or any substance likely to cause or accelerate fire. Oxygen itself is flammable and does support combustion.
4. Never permit oil or grease to come into contact with oxygen cylinders - valves - regulators - hose or fittings. Do not handle oxygen cylinders with oily hands or oily gloves.
5. Never use oxygen regulators, hose or other pieces of apparatus with any other gases.
6. Open oxygen cylinder valve fully when in use.
7. Never attempt to mix any other gases in an oxygen cylinder.
8. Be certain that cylinder valve is tightly closed before handling cylinder.
9. Never use oxygen from cylinders without a suitable regulator attached to the cylinder valve.
10. Never tamper with nor attempt to repair oxygen cylinder valves.

5.6.2.2.2. Acetylene

1. Call acetylene by its full name "acetylene" and not by the word "gas." Acetylene is far different from city or furnace gas.
2. Acetylene cylinders should be used and stored in an upright position.
3. Keep sparks, flames and heat away from acetylene cylinders.
4. Never use acetylene from cylinders without reducing the pressure through a suitable regulator attached to the cylinder valve.
5. Turn the acetylene valve so that the valve outlet will point away from the oxygen cylinder.
6. When opening an acetylene cylinder, turn key about one full turn.
7. Acetylene cylinder key for opening cylinder valve should be kept on valve stem while cylinder is in use so that the acetylene cylinder may be quickly turned off in an emergency.
8. Never use acetylene regulators, hose or other pieces of apparatus with any other gases.
9. Never attempt to either transfer acetylene from one cylinder to another or to refill an acetylene cylinder, or to mix any other gas or gases in an acetylene cylinder.
10. When returning empty cylinders make certain that valves are closed tightly to prevent escape of residual acetylene.
11. Should a leak occur in an acetylene cylinder take cylinder out in the open air, keeping well away from fires or open lights. Notify the manufacturer at once.
12. Never use acetylene at pressures in excess of 15 psi; the use of higher pressures is prohibited by all insurance authorities and by law in many localities.

5.6.2.2.3. General Precautions

1. Don't permit anyone to strike an arc on a compressed gas cylinder.
2. Don't weld near inflammable or combustible materials.
3. Don't weld on containers which have held combustible or inflammable materials without first exercising the proper precautions recommended by the American Welding Society.
4. Don't weld in confined spaces without adequate ventilation.
5. Don't pick up hot objects.
6. Don't do any chipping or grinding without suitable goggles.
7. Don't move individual cylinders without the cap over the cylinder valve.
8. Don't drop or abuse cylinders in any other way.

9. Make certain that cylinders are well fastened in their stations so that they will not fall.
10. Don't use a hammer or wrench to open cylinder valves.
11. Don't use while lead, oil or grease, or other pipe fitting compounds to make joints.
12. Never force connections, which do not fit.
13. Never tamper with fuse plugs.
14. Always protect the hose from being trampled on or run over. Avoid tangles and kinks. Don't leave the hose so that it can be tripped over because a connection may be pulled off, or worse still, the cylinders and equipment may be pulled over.
15. Protect the hose from flying sparks, hot slag, hot objects and open flame.
16. Don't allow hose to come in contact with oil or grease; these deteriorate the rubber and constitute a hazard with oxygen.
17. Be sure that the connections between the regulators adaptors and cylinder valves are tight. Escaping acetylene can generally be detected by the odor. Test with soapy water, never with an open flame.
18. Don't use matches for lighting torches; hand burns may result. Use friction lighters, stationary pilot flames or some other suitable source of ignition. Do not light torches from hot work in a pocket or small confined space. Don't attempt to relight a torch that has "blown" out without first closing both torch valves and relighting in the proper manner.
19. Don't hang a torch with its hose on regulators or cylinder valves.
20. Don't cut material in such a position as will permit sparks, hot metal or the severed section to fall on the cylinder, hose, legs or feet.
21. When welding or cutting is to be stopped temporarily, release the pressure adjusting screws of the regulators by turning them to the left.
22. When the welding or cutting is to be stopped for a long time (during lunch hour or overnight), close the cylinder valves and then release all gas pressures from the regulators by opening the torch valves momentarily. Close the torch valves and release the pressure adjusting screws. If the equipment is to be taken down, make certain that all gas pressures are released from the regulators and that the pressure adjusting screws are turned to the left until free.

5.6.3. Setting up the Apparatus

When the foregoing safety precautions have been thoroughly mastered, set up the welding equipment. The procedure given herewith should be followed without variation. The established procedure as outlined here is based on years of experience and assures practical safety to the operator and the apparatus.

- If the cutting torch is to be used instead of the welding torch, proceed as directed through Step 5; then connect the oxygen hose to the oxygen connection nipple of the torch, and connect the acetylene hose to the torch acetylene valve.
- Check that the seating surface of the cutting tip and the cutting torch head are clean and free from scratches or nicks; unscrew the tip nut and insert a cutting tip in the torch head, then slip the tip nut over the tip and screw into the head snugly.
- Open the cutting oxygen valve and adjust the oxygen regulator to the required working pressure according to the cutting table shown on following page. Close the oxygen valve. Open acetylene valve and adjust acetylene regulator to the required working pressure.
- Open the acetylene valve and light the pure acetylene flame using an approved type of lighter. Follow by opening the preheat oxygen valve until an approximate neutral flame is obtained.
- Open cutting oxygen valve to observe its proper operation. At the same time adjust preheat flames to neutral with cutting oxygen on.

Gas pressures for Airco tips are shown in Table 5-12 (for cutting) and Table 5-13 (for welding). Pressures and consumptions shown are fore separable tips with appropriate mixers. Operating pressures for

Figure 5-23 Procedure for Setting Up Apparatus

Fig. 1. "Cracking the cylinder valve." To crack the cylinder valve, open each valve for an instant to blow dirt out of nozzles. Wipe off the connection seat with a clean cloth. Don't stand in front of valves when opening them.

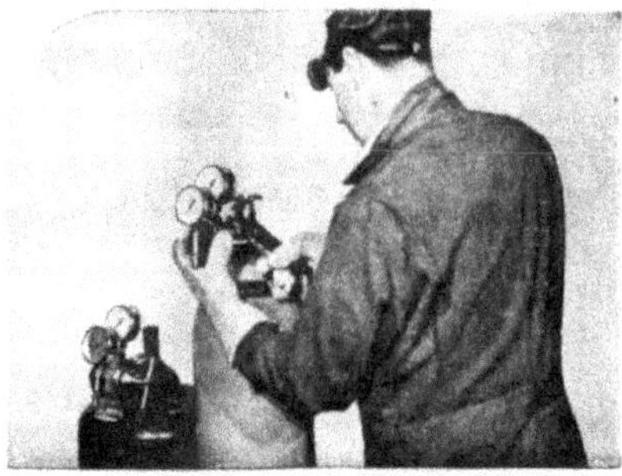

Fig. 2. Attaching the pressure regulators. To attach the regulators, connect the acetylene regulator to the acetylene cylinder and the oxygen regulator to the oxygen cylinder. Screw the nuts up tightly with a close-fitting wrench.

Fig. 3. Connecting the hose to the regulators. To connect the hose to the regulators, connect the black or green hose to the oxygen regulator and the red hose to the acetylene regulator. Screw the nuts up tightly.

Fig. 4. Opening the cylinder valves. To open the cylinder valves, release regulator screws and open valves slowly. Never open cylinder valves until regulator screws have been released.

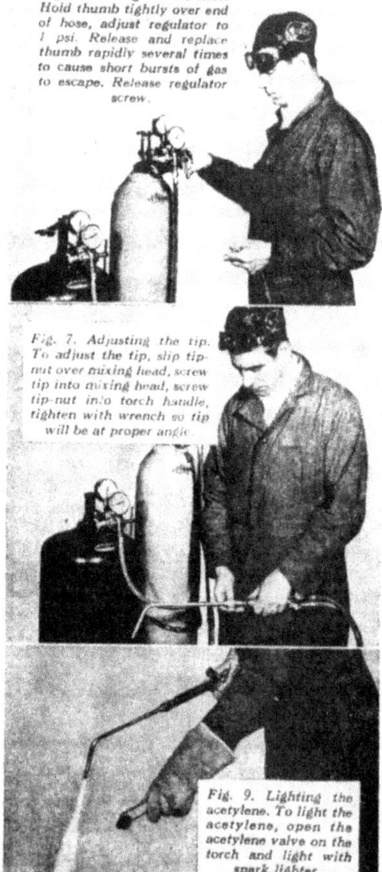

Fig. 5. Blowing out the hose. Hold thumb tightly over end of hose, adjust regulator to 1 psi. Release and replace thumb rapidly several times to cause short bursts of gas to escape. Release regulator screw.

Fig. 7. Adjusting the tip. To adjust the tip, slip tip-nut over mixing head, screw tip into mixing head, screw tip-nut into torch handle, tighten with wrench so tip will be at proper angle.

Fig. 9. Lighting the acetylene. To light the acetylene, open the acetylene valve on the torch and light with spark lighter.

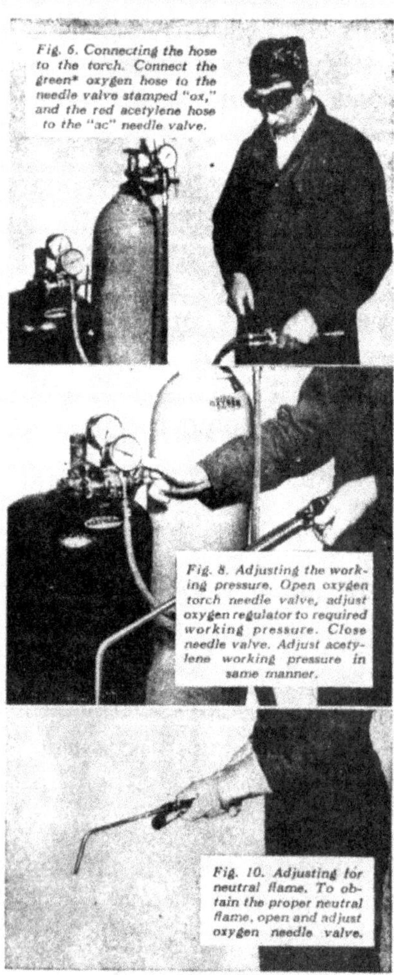

Fig. 6. Connecting the hose to the torch. Connect the green oxygen hose to the needle valve stamped "ox," and the red acetylene hose to the "ac" needle valve.*

Fig. 8. Adjusting the working pressure. Open oxygen torch needle valve, adjust oxygen regulator to required working pressure. Close needle valve. Adjust acetylene working pressure in same manner.

Fig. 10. Adjusting for neutral flame. To obtain the proper neutral flame, open and adjust oxygen needle valve.

tip-mixer assemblies to secure normal flow should be somewhat higher for the smaller sizes. Gas pressures are for hose lengths up to 25'. Increase pressure for longer lengths of hose.

Table 5-12 Approximate Gas Pressures for Airco Tips for Cutting

Thickness, Inches	Size No.	Oxygen Pressure, psi	Acetylene Pressure, psi	Hand Cutting Speed, In./Min.	Machine Cutting Speed, In./Min.
1/4	0	30	3	16-18	20
3/8	1	30	3	14.5-16.5	19
1/2	1	40	3	12-14.5	17
3/4	2	40	3	12-14.5	15
1	2	50	3	8.5-11.5	14
1-1/2	3	45	3	6.0-7.5	12
2	3	50	3	5.5-7.0	10
2-1/2	4	50	3	5.7-6.7	9
3	5	45	4	5.0-6.5	8
4	5	60	4	4.0-5.0	7
5	6	50	5	3.5-4.5	6
6	6	55	5	3.0-4.0	5

Table 5-13 Approximate Gas Pressures For Airco Tips for Welding

Tip No.	Thickness Of Metals, inches	Oxygen Pressure, psi (Approx.)	Acetylene Pressure, psi (Approx.)	Acetylene Consumption, ft^3/hr (Approx.)
00	1/64	1	1	0.5
0	1/32	1	1	1
1	1/16	1	1	2
2	3/32	2	2	5
3	1/8	3	3	9
4	3/16	4	4	16
5	1/4	5	5	25
6	5/16	6	6	30
7	3/8	7	7	40
8	1/2	7	7	60
9	5/8	7-1/2	7-1/2	70

5.6.4. Flame Adjustment

The last step in setting up the apparatus is adjusting the flame. Starting out with acetylene burning in air, gradually turn on the oxygen. As the oxygen is brought in and before the two cones become one, a "Carburizing" flame exists. If used in welding, this will cause a boiling condition in the molten metal, and the finished bead will have a spongy appearance (porosity).

When the two inner cones become one, the flame is "Neutral." This is the proper flame for most welding operations.

As more oxygen is admitted to the tip, an "Oxidizing" flame is obtained. This also will produce a boiling effect in the molten steel and the finished bead should be covered with an irregular deposit of scale. When this scale is removed, it should be noted that the weld is excessively porous; also, the steel that has been melted should be very hard.

In some cases, these apparently undesirable carburizing and oxidizing flames may be put to good use, but for most welding a neutral flame should be used. The neutral flame plus proper manipulation will make for a clean, bright, smooth weld.

5.6.5. Taking down the Apparatus

It is equally as important that a definite procedure be followed when taking down the apparatus. The following series of operations represent the accepted method and should be followed exactly.

1. Close acetylene valve on torch.
2. Close oxygen valve on torch.
3. Close acetylene cylinder valve.
4. Close oxygen cylinder valve.
5. Open the torch acetylene valve, with the oxygen valve closed, to drain the line; release the adjusting screw on the acetylene regulator, then promptly close torch acetylene valve.
6. Open the torch oxygen valve, with acetylene valve closed, to drain the line; release the adjusting screw of the oxygen regulator, then promptly close the torch oxygen valve.
7. Flashbacks may result from mixed gases in either oxygen or acetylene hose. By following these rules acetylene will neither get over into the oxygen line nor oxygen over into the acetylene line when starting up or shutting down. Flashbacks should be eliminated if the torch flame is properly adjusted and maintained, with the correct pressures, and tip outlets kept free from obstruction.
8. Regulators should be removed before moving cylinders. Replace cap over cylinder valve when empty and mark the cylinder "MT."

5.6.6. Care of Tip

If the inner cone has a tendency to curve, the tip should be cleaned. Be sure to keep tip clean at all times. Restrictions of the orifice will alter the flame and cause a lessening in the heat of the flame. Excess amounts of slag adhering to the outside of the tip are equally as detrimental to efficient operation of the tip. Frequent cleaning of the tip, particularly when working in confined areas, is very important. Proper regard for and care of the tip is as important as the care of the rest of the equipment, and is as necessary for good welding as proper welding technique.

5.6.7. Principles of Oxyacetylene Cutting

When iron or steel is heated to a temperature of around 1600° F, it will burn when brought into contact with oxygen. If the oxygen comes only from the surrounding air, the combustion occurs only on the surface of the metal, and the heated iron or steel merely turns black. However, if a blast of pure oxygen is directed at the hot metal, the metal should begin to burn through.

In oxyacetylene cutting, the metal is first preheated by mixing acetylene with the preheating oxygen and using the resulting flame to heat the metal to the ignition temperature. When this point is reached, the cutting oxygen is turned on, and the stream of pure oxygen is directed against the metal. This ignites the iron or steel at the surface and starts the cut.

Oxyacetylene cutting is a progressive action. The burning of the surface material releases heat, which raises the temperature of the metal immediately below the surface and melts it. The cutting oxygen then ignites this molten material, which in turn releases heat to the metal below it, and the action progresses downward until all the metal has been cut. Oxides are molten at cutting temperatures and are blown out of the cut by the stream of cutting oxygen.

The cutting torch (or attachment) is constructed as shown in Figure 5-24. The acetylene and preheating oxygen are mixed and exit the torch through small orifices around the edge of the tip, while the cutting oxygen flows down through the center orifice.

Figure 5-24 Starting a Cut

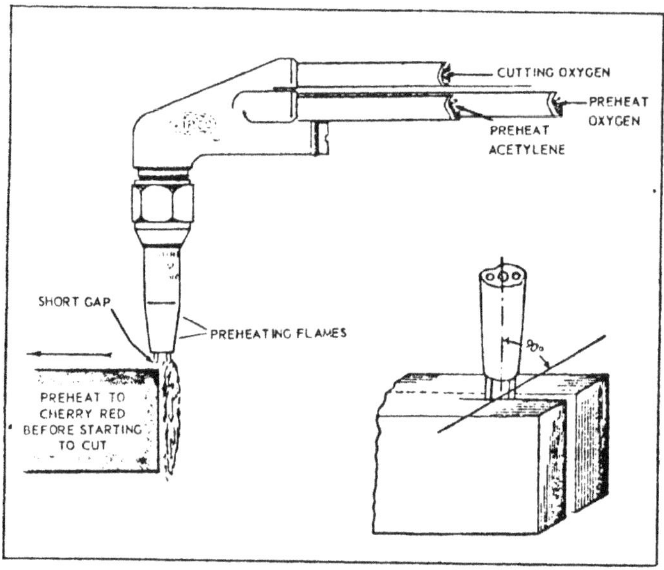

5.6.8. Cutting Steel

When cutting steel, a neutral oxyacetylene flame, is used for preheating (see Figure 5-23.) The cut is usually started at the edge of the piece, as shown in Figure 5-24. Hold the cutting torch lightly but steadily so that the ends of the preheating flame cones are about 1/8" (3.2 mm) above the surface of the metal. When a spot of metal at the top of the edge has been heated to a cherry red, the cutting oxygen is turned on, and the cutting begins.

When starting a cut in the metal away from the edge, a hole first has to be pierced in the metal to start the cut, and this is done by heating a spot of the metal and burning a hole all the way through.

After the spot of metal is heated to a cherry red, raise the torch about 1/2" (12.7 mm) above the normal position for cutting and slowly turn on the cutting oxygen. As soon as the hole is burned all the way through the metal, lower the torch to the normal height above the work and start the cut.

When piercing a hole, be careful to avoid the possibility of slag blowing back and plugging the cutting tip.

When cutting, hold the torch steadily, but not too tightly. For right hand operators, the torch should be held by the left hand a few inches back of the head of the torch, while the right hand holds the handle in position to operate the cutting oxygen valve lever.

Look down into the cut as the cutting progresses

and make sure that the flow of slag is clear and not blocked. Move the torch in a straight line to keep a straight cut, and move the torch at a uniform speed just fast enough to match the speed at which the oxygen is cutting through the metal.

5.6.9. Cutting Cast Iron

Cast iron is much more difficult to cut than steel. When cutting steel, the oxygen combines with the metal and starts the cut when the steel has been heated to a cherry red. In cast iron cutting, however, the temperature of the metal should be almost at the melting point before the cut can be started and continued.

From this, the preheating flame for cutting cast iron is adjusted to be strongly carburizing. The streamer of excess acetylene is adjusted as shown in part (a) of Figure 5-25, and the cutting tip is held somewhat farther away from the metal than in steel cutting. The carburizing flame delivers more acetylene into the cut, thus developing more heat and increasing the temperature of the cast iron.

Figure 5-25 shows the procedure for cutting cast iron. Part (f) of Figure 5-25 shows the use of a rod to aid in cutting. Steel rods are used to assist in making the cut. These rods are inserted in the cut and are consumed by the cutting oxygen, thus greatly increasing the heat along the line of the cut.

Figure 5-25 Procedure for Cutting Cast Iron

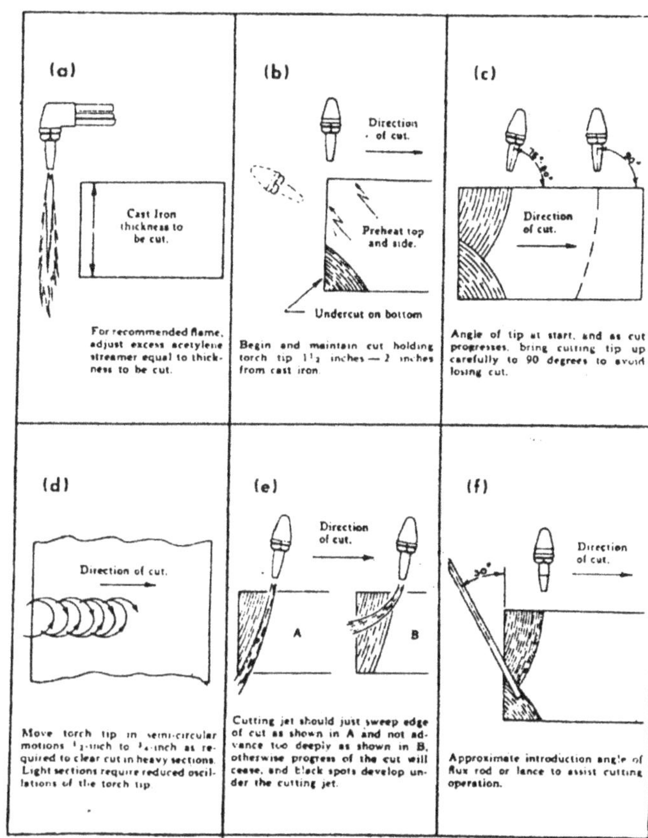

5.7. Moving Around the Site

5.7.1. Rig Capabilities

Capacity tables are published showing the safe operating radius for various rigs. Rig capabilities are always based on the understanding that the hammer and core should be carried as low as practicable when moving and that the core should be restrained with a short sling to prevent it swinging out.

When batter piles are involved, the superintendent must understand the rig's ability to drive in, out, side or combination batters. When more than one rig is employed, the capabilities of each should be used to best advantage.

5.7.2. Prioritizing Work Tasks and Establishing Sequence

One must know the needs of the client, and requirements of the contract, as far as the order in which the work is to be done. When no significant extra cost is involved, one will normally perform the work in the order most convenient to the client and other contractors working at the site. This is very often fixed by the schedule of other work, such as excavation, pumping, and installation of well points, as well as follow-up work such as steel erection and equipment installation.

5.7.3. Planning Rig Moves

The basic consideration is how to reach all the piles within the safe operating radius of the rig with a minimum of moving. It will frequently help if a scale template of the driver is moved about the pile plan on which you have plotted obstructions or areas to be avoided. This may show that you will have to move by a different route to provide adequate tail swing, for example. Batter piles may have to be driven from a particular position and consideration should be given to the elevation of the rig in relation to cut-off.

A second basic consideration is to drive and move in a way that leaves the driven piles uncovered for concreting and the concreting area undisturbed by dragging shells or other rig activities. Check concreting radius and plan driving sequence so there should be no delay in concreting. When two or more rigs are used, they should be kept well apart except where it may be necessary to work close to jet pumps, water supply lines, pile yards or shell racks. In any event, rigs should not be allowed to come within 10' (3.05 m) of each other, when both are at maximum operating radius, without clear understanding between supervisors as to how each rig will maneuver.

Another basic consideration is making sure that the rig can be set up and dismantled near the starting and ending points. Avoid a route that requires dismantling the rig to get under an obstruction, if it can be economically by-passed. When working in an excavation, be certain you will have swing clearance both front and back. If there is less than 50' (15.2 m) between excavation slopes, you will have problems.

The sequence of driving piles should be planned to start near other buildings and drive away from them. Ground heave is sometimes severe enough to lift the rig out of plumb, and should be a consideration in moving.

5.7.4. Unstable Ground

Moving over unstable ground is by far the most hazardous part of the rig operation. Pile drivers are unique in that they move constantly with their load and the center of gravity is high in the air. In this case, it does not take much out-of-plumbness to put the rig in a critical position.

Avoid areas where the ground is soft or where there is uncompacted fill, areas over old sewers or underground structures that will not support the rig, or areas adjacent to excavations. A good rule of thumb is to keep the nearest crawler no closer to the edge of a slope than two feet per foot of height, measured from the toe of the slope.

It is good policy to operate the rigs on mats. In addition, you should consider whether a single layer of mats is adequate. In unusually soft areas, two layers of mats may be necessary. This is particularly important when making square turn; otherwise, almost the entire weight of the rig may bear on a single mat.

Accidents have occurred due to working too close to an excavation. The danger should not be overlooked even with apparently hard to stable materials as these may fail suddenly without warning. The vibration of pile driving can cause the safest looking slope to slough. The general rule is repeated that the edge of the cats should preferably be kept back two feet from the toe of a slope or excavation for each foot of height. When the soil is very soft or loose, even this guidance may not be enough and when in doubt, consult your general superintendent.

When it is necessary to work close to the edge of an excavation, the slope should be sheeted and braced or the mats supported by stacks of mats, cribbing, posts or false-work piles to carry the weight of the rig to stable ground.

The superintendent should observe trucks and other equipment moving around the site to aid in spotting soft areas. When crossing over an old sewer or freshly back-filled trench, the simplest precaution is usually double matting, if the unstable area is not too wide. If there is possibility of damage, special precautions may be necessary and the situation should be discussed with your general superintendent well in advance. Move across unstable areas as promptly as possible. Don't sit on them and drive piles, if it can be avoided.

When pre-excavated holes are close together, it is

best to have the entire area excavated to a single average depth rather than dig individual pier holes. If pier holes are spread apart, the modern rigs usually have sufficient reach to keep back a safe distance from the edge of the holes. Sometimes it should be possible to schedule the sequence of excavation to avoid having to work over pre-excavated holes.

One frequently has to operate the rigs on recently placed uncompacted fills. When the fill is not thoroughly compacted, use mats and back off periodically and level up as the fill compacts. If the fill is thin, and has been placed over a very soft material, there is a possibility of punching through the fill due to failure of the underlying soft material. In addition, fills are usually less thoroughly compacted near the edges than in the center and one should be particularly cautious in working in such areas.

When working near the top of a slope that has been sheeted, keep well back from the edge unless one can determine that the sheeting has designed to support the rig. Consult your general superintendent in determining the best approach to this problem.

5.7.5. Cribbing and Trestles

When it is necessary to operate the rig over an excavation not over two feet deep, it is customary to simply fill it with mats. However, when the excavation gets deeper, cribbing may have to be used for support. Cribs are built out of 12" (304.8 mm) x 12" (304.8 mm) x 6' (1.82 m) blocking provide mat support every 4' (1.22 m).

The top layer of cribbing should consist of mats to span between cribs and provide a level surface on which to lay the crane's mats. The cribs should be founded on planking or on additional mats to spread the load. A typical cribbing support for a 100-ton crane is shown in Figure 5-26. It is sometimes practical to tie cribs together so they can be handled as a unit.

When an excavation is over six or eight feet deep, consideration should be given to spanning it with a steel beam trestle, as shown on Figure 5-27. Suggested spans for various sizes of beams are shown for a 100-ton (690 kN) driver. It is important that the ends of the beams be supported on a sill placed well back from the edge of the excavation. If the span is of any length, one or more intermediate cribs or trestle bents of posts or piles will have to be used.

All timber used for cribbing and trestling should be construction grade oak, Douglas fir or equivalent in first class condition. Because of the many variations in rigs and project conditions, you are urged to discuss your cribbing and trestling problems with your superintendent.

Another method of spanning an excavation is to mount the rig on a gantry. This is particularly useful when driving in a long excavation such as a subway tunnel or sewer excavation. Several different gantries have been designed to meet varied project conditions.

5.7.6. Clearing Obstacles

Overhead obstructions will frequently be unavoidable. You are particularly cautioned when working near high-tension lines. If possible, they should be moved or de-energized while you are working in the area. The superintendent must consider all lines "hot" unless he knows they have been de-energized or are visibly grounded. Generally, you should not permit the leaders (or any other part of the rig) to come closer than 20' (6.1 m) horizontally or 10' (3.05 m) vertically to a high-tension line, if it can be avoided. The working limitations may be specified by law or covered by the specifications. You should be familiar with the public and utility regulations that apply in the area concerned. A hazard also exists when working near radio transmitting towers; consult an experienced safety engineer.

When overhead obstructions cannot be moved, the rig may have to be dismantled to get beneath them. In some cases, it may be possible to remove core and hammer, boom down and retract or fold the spotter sufficiently to clear the obstruction. When going under overhead power lines, they should be cleared by a minimum of 10' (3 m). The exact height of the lines should be known accurately as the sag in long spans can vary considerably with the temperature and the height of a lowered boom will vary considerably if the rig rocks.

Underground obstructions present several problems to us. Large sewers and vaults can not only fail to support the rig but also can be severely damaged as well. Smaller underground facilities, such as cable ducts can be damaged by the weight of the rig if they are close to the surface.

Above ground obstructions nearer than 2' (610 mm)

Figure 5-26 Cribbing

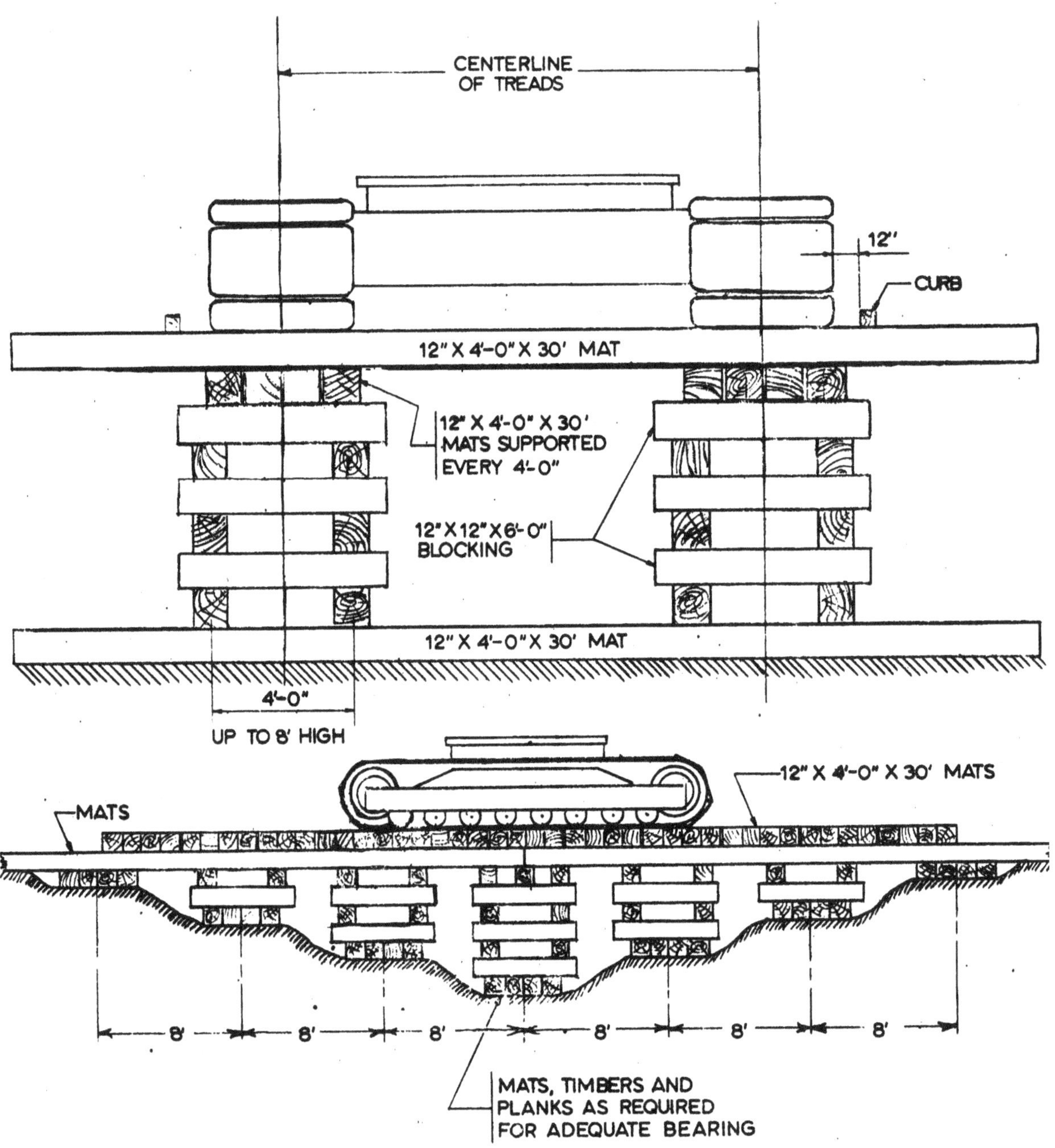

Figure 5-27 Trestling

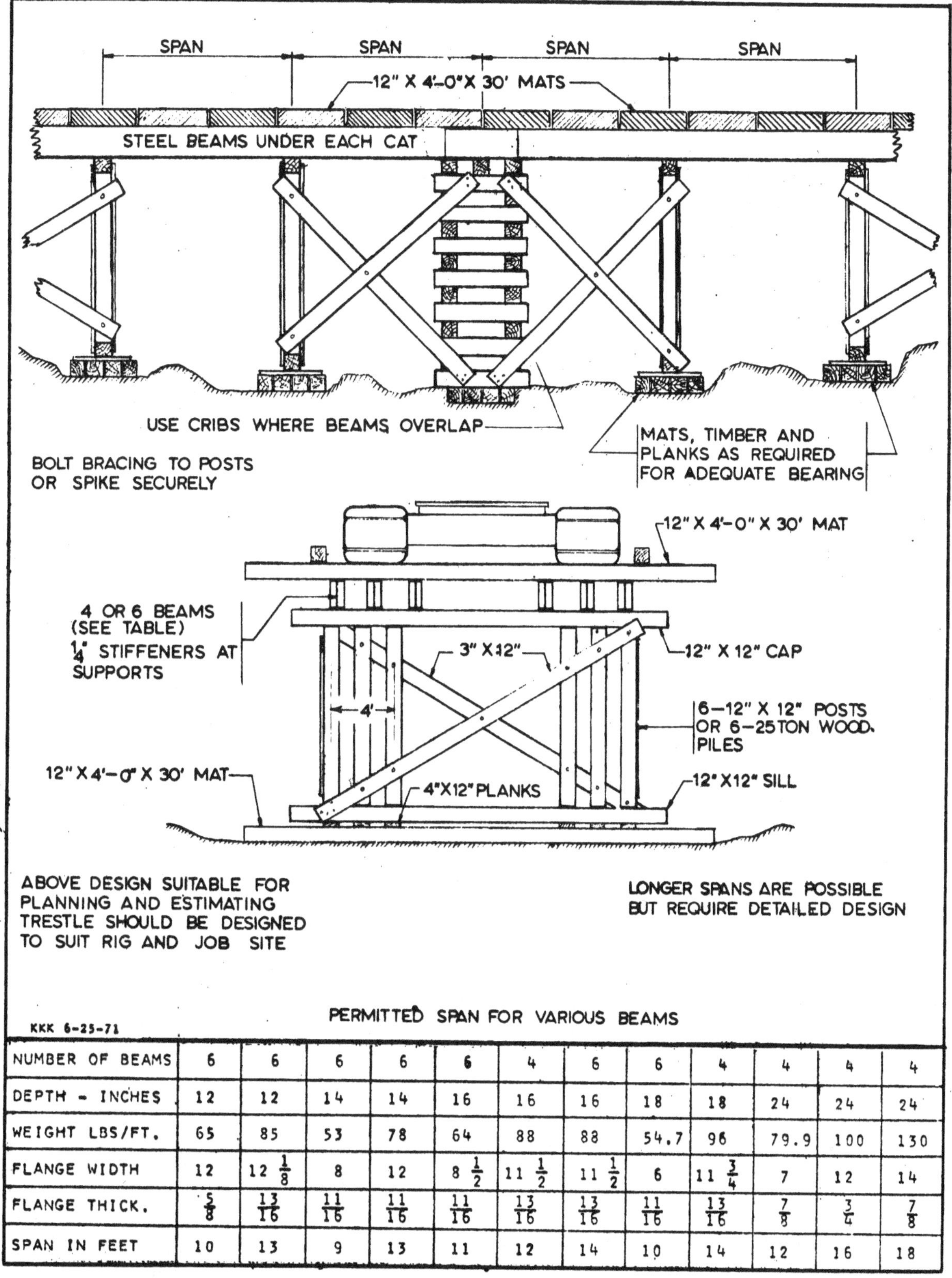

PERMITTED SPAN FOR VARIOUS BEAMS

NUMBER OF BEAMS	6	6	6	6	6	4	6	6	4	4	4	4
DEPTH - INCHES	12	12	14	14	16	16	16	18	18	24	24	24
WEIGHT LBS/FT.	65	85	53	78	64	88	88	54.7	96	79.9	100	130
FLANGE WIDTH	12	12 1/8	8	12	8 1/2	11 1/2	11 1/2	6	11 3/4	7	12	14
FLANGE THICK.	5/8	13/16	11/16	11/16	11/16	13/16	13/16	11/16	13/16	7/8	3/4	7/8
SPAN IN FEET	10	13	9	13	11	12	14	10	14	12	16	18

to the center of the pile will present a problem in hammer clearance. When the rig is equipped with drill, check clearance of left side of leaders as well.

5.7.7. Ramps

Modern cranes have the power to climb slopes as steep as one foot rise in three feet and have travel locks to hold them on such grades. However, they may not be able to get sufficient traction to pull up or stay on such a grade, if the surface is slippery or unstable.

The rig capacity tables are based on driving and shelling up (hammer at top of leaders) with the rig level, that is on a slope not to exceed 1:100; with extra caution, this can be extended to a 1:30 slope, if one does not swing while the hammer is raised to shell up. By removing the core and lowering the hammer to the bottom of the leaders, one can safely negotiate slopes of 1:10. Beyond this, one may have problems in swinging and getting sufficient traction and it may be necessary to lower the leaders.[11]

Traveling on ramps is inherently hazardous and requires good coordination between supervisor and operator. The move up or down the ramp should be planned so that it will not be necessary to sop, swing, or change direction, if this is feasible; swing locks should be engaged. Moving on ramps is preferably done in dry weather. If the ramp is icy or muddy, it should be cleaned and sanded.

When traveling on a ramp, 12" (304.8 mm) x 12" (304.8 mm) chucks should be kept on the downhill side of each track. Travel locks should be checked before entering on the ramp and the operator should be prepared to engage them quickly. If the rig is so equipped, the travel lock should be in ratchet position when going uphill. Keep the leaders facing uphill. Boom down, if necessary, to make the rig heavy enough on the front end when level that it will balance on the kingpin when the rig is on the slope. Hold the hammer on the brake so that it can be lowered to the ground to serve as an anchor, if necessary.

Most ramp accidents have occurred when operating the travel clutches and travel locks to steer the rig. Curved ramps should be avoided, when possible. If it is necessary to cut the rig on a ramp, the cats should be securely chocked before any attempt is made to operate travel clutches or locks. On steep ramps, the added precaution should be taken to insert a jack bar in the sprocket holes or place a piece of rail between one sprocket and a track pad. Cut the rig by locking one cat, powering the other cat uphill and rechecking before re-engaging the other travel clutch.

A critical point in operating on a ramp is when the rig breaks over at the top of the ramp. Changes in slope should be made as gradually as possible. Work should be planned so that the ramp can be negotiated in one continuous operation; if this is not possible, the rig should be securely chocked when left unattended. When going up or down ramps 1:4 or steeper or when the ramp is slippery, a cable should be connected between the rig to the top of the ramp to assist the rig in moving up or to control its descent. The cable should be attached to a large bulldozer, another crane or to a secure deadman.

When it is necessary to drive piles from a ramp, wedges can be cut from heavy timbers to fit under each track, to temporarily level the rig for driving.

5.8. Equipment Operation

Successful pile driving is dependent upon both the condition of the equipment and the operating technique used. Failure to observe the proper equipment procedures can result in loss of energy imparted to the pile failure to reach the desired penetration depth, property damage, personal injury, and equipment breakdowns, all of which lead to downtime, project delays, and contract disputes.

5.8.1. Condition of Equipment

Equipment should be maintained and operated in accordance with the manufacturer's recommendations. This includes proper preparation of the equipment, correct set-up before driving, good operation during driving, and proper transport and storage procedures.

5.8.1.1. Hammer

The hammer is the element that directly interfaces with the pile. Thus, it transmits and receives the most vibration and impact during driving. Although most pile drivers in use today are relatively simple machines, attention to their basic maintenance is still important. It is also essential that personnel with the proper training and experience should operate this equipment, in common with most construction equipment.

5.8.1.1.1. Impact Hammers

The most important element in the efficient operation of impact hammers is proper lubrication. All wear surfaces should be kept lubricated by the means provided for either on the machine or from the outside. Typical lubrication points for an air/steam hammer are shown in Figure 5-28. Diesel hammers vary widely in their lubrication systems, but the entire cylinder assembly should be lubricated for its full length, whether by grease or oil. Hydraulic impact hammers are similar to air/steam hammers except for the hydraulic system and cylinder(s). In these cases, their maintenance is more like that of a vibratory hammer. In addition, due to the high loads during ram deceleration and pile rebound, impact hammers are subject to dynamic stresses. All fasteners should be kept tight, and parts that are worn, cracked, or broken should be replaced.

Figure 5-28 Lubrication Points, Air/Steam Hammer

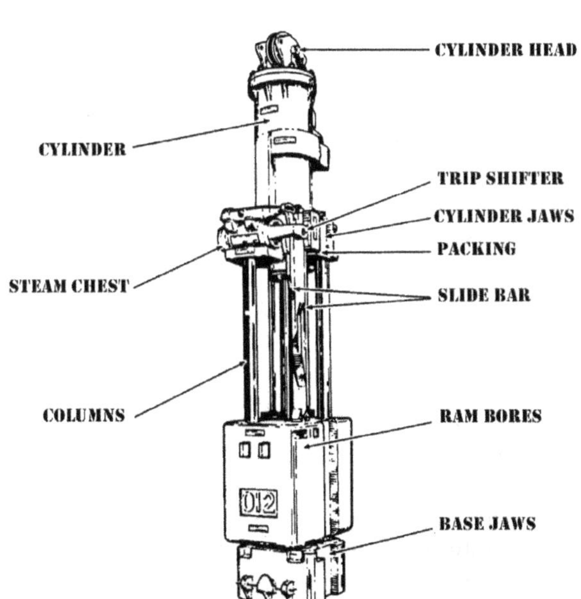

5.8.1.1.2. Vibratory Hammers

A chart of basic items to check on with vibratory hammers is shown in Table 5-14.

Table 5-14 Do's and Do Not's of Vibratory Hammers

Things that always need to be done	Things that never need to be done
Always store oily rags in containers. If these get into a hydraulic system, you will have a mess.	Never allow unauthorized or unqualified people to operate, maintain, or come within thirty (30) meters of the equipment.
Always remove all tools from unit before starting.	Never allow anyone to stand directly under or within at least three (3) meters of the hammer or pile being driven during operation. Failure to do so could result in injury or death by being struck by falling parts, rocks, or dirt on the hammer.
Always be sure that, with hydraulic systems, all pressure is out of the system and that all pressure gauges read zero before you start working on the hydraulics of the system. The high-pressure fluid in hydraulic lines can be very dangerous if released.	Never operate the power pack's engine in a closed area. The breathing of the fumes can be fatal.
Always make sure that you make any hose fittings or connections very tight when you reassemble them. Failure to do so can result in the hoses coming loose, resulting in hoses flying around, hydraulic fluid spraying everywhere, and injury or death.	Never smoke or use open flame when servicing batteries. Proper ventilation is necessary when charging batteries. On units with a power pack enclosure, all of the doors of the unit should be open during battery charging.
Always make sure that electrical systems are properly grounded during operation. Also, make sure that they are not connected to any power source in any way and that there is no voltage of any kind in the system before servicing it.	Never smoke when filling fuel tank or hydraulic reservoir, or for that matter while anywhere near exciter, hoses, or power pack. Diesel fuel, gasoline, and hydraulic fluid are all very flammable.
Always make sure that electrical connections and wiring are tight and completely insulated to prevent shock if accidentally touched. This is especially important in waterfront or marine situations, in which case uninsulated wire can result in electrocution.	Never adjust or repair the unit while it is in operation, except with the main motor and clamp controls provided for that purpose. If you need to make any other adjustments, shut the entire system down first.
Always be sure to wear gloves and protective clothing while working on any part of the system, or even better to wait until the system has cooled down. Hydraulic components, electrical wiring and switchgear, and the engine get very hot during operation.	Never attempt to operate the engine with the governor linkage disconnected.
Always make sure that the pile is firmly gripped by the jaws when clamping.	Never store flammable liquids near the engine.
	Never unclamp the exciter from the pile when there is any line pull on the suspension or when the hammer is still vibrating.

5.8.1.2. Driving Cap

The driving cap transmits the impact force from the hammer to the pile. Its proper setup and maintenance are central to successful pile driving.

5.8.1.3. Cushion

Both hammer and pile cushion material should be configured and installed so that there is adequate cushioning for the ram and the pile. Materials which are too hard, or less than complete cushion stacks should be avoided; cushions which are added piece by piece during driving should also be avoided. Cushion material is an expendable item; it degenerates during its life with heat and plastic deformation; it is essential that it be replaced when it is burnt or deformed to its compressive limit. Cushion whose stack height is 75% of the original, uncompressed height should be replaced. Also, when piling are designed using the wave equation analysis, it is essential that the cushion configuration assumed during the wave equation analysis be a realistic one in the field and that the configuration assumed during the analysis be replicated in the field. This is especially important with a concrete pile cushion; inadequate cushion may lead to tension cracking during driving and excessive material may lead to unanticipated energy losses. When used, a new pile cushion should be used with every pile.

5.8.1.4. Anvils and Helmets

Anvils and helmets should be able to withstand the high loads and stresses of impact pile driving. They should be inspected thoroughly before use. Any helmet or anvil that has any cracks, broken pieces, or thin wall sections that might break in use should be rejected before driving begins. During driving, the cap should be inspected frequently for developing cracks or broken or missing pieces, and removed from the project if any of these is discovered during inspection.

5.8.1.5. Mandrel

Mandrels also transmit high loads. They should be inspected during use for any cracks or other mechanical failures, and repairs made before driving is resumed.

5.8.2. Hammer Operation

There are several important performance parameters to monitor during pile driving.

5.8.2.1. Hammer Stroke

The single most important parameter to monitor is the stroke of the ram.

5.8.2.1.1. Single Acting Hammers

The down stroke of a single acting hammer it its only source of impact energy. Therefore, it is essential that its stroke be what the piling being driven requires, whether that is a full or partial stroke. This stroke should be achieved when the pile is at refusal; some hammers (especially diesel hammers) cannot achieve full stroke without the pile energy rebound of refusal driving.

5.8.2.1.2. Other Impact Hammers

Although the stroke is not as important with hammers that have assistance on the down-stroke, it is important to establish what stroke is required for the particular project and monitor the stroke by the appropriate means.

5.8.2.2. Monitoring Of Stroke

If the ram is visible from the outside of the hammer, then visual monitoring of the stroke is possible. With open type air/steam hammers, as the ram moves it wipes most of the lubricant off the columns. The length of this wiped area will determine the stroke of the hammer. With open-end diesel hammers, a striped pole can be mounted on top of the hammer. As the ram rises, the stroke can be determined by comparing the maximum height of the ram with the highest point of the ram rising. A Saximeter can also be used for this purpose.

5.8.2.3. Speed

Although the hammer speed is a general indication of the performance of the hammer, it can be used to determine the energy output of the hammers only for single-acting diesel hammers. This is because there are simply too many variables influencing the motion of the ram, especially the variations in energy rebound from the pile. With single-acting diesel hammers, devices exist that can estimate the energy of the hammer based on its blow rate, using either a stopwatch or a Saximeter™.

5.8.2.4. Pressure

With external combustion hammers, the pressure of the operating fluid is an important variable in the energy output of the hammers. With single-acting hammers, although no direct correlation can be made between pressure and output energy, monitoring the pressure can be a useful tool in diagnosing pressure of the motive fluid. For proper readings, any pressure readings should be taken at the hammer by a pressure gauge at the hammer or by remote sensor.

5.8.2.5. Direct Energy Monitoring

If the ram is not visible, or for more complete information on the energy output of the hammer, devices exist that can determine the energy output of the hammer, either from the hammer itself or after the stress wave has passed through the hammer cushion and drive cap. These can be either obtained from the manufacturer or third-party testing firms and consultants. It is very important that these devices be properly installed, calibrated, and operated by experienced and trained personnel who are very familiar with their use; incorrect installation, calibration, or operation can result in erroneous readings. Complete reports on the use of these devices and interpreted results should be required if they are used.

5.8.3. Hydraulics

5.8.3.1. Maintenance

Most hydraulic difficulties are the result of dirt entering the system. Since dirt and dust are ever-present on the jobs, it requires special effort to keep the oil from becoming contaminated. The following precautions should be observed:

1. Clean off the tops of oil barrels and other containers before they are opened. Keep them horizontal and tightly closed when not being used.
2. Keep the fill cap securely fastened on the reservoir; clean around the fill-cap before removing it, lest some dirt fall in to the tank. Make sure that funnels and other oil handling equipment are perfectly clean.
3. Hydraulic reservoirs should be opened and cleaned, at least once every six months. All sludge should be removed, the tank flushed with kerosene, the strainer replaced and the system refilled with fresh oil.
4. Filters should be observed daily and elements replaced at least once a month. It should be routinely done at the beginning of each project unless it is known that the elements are new. Spare filter elements should be kept on the project. Under severe service conditions, filters should be examined frequently, particularly at the start of a project.
5. When a pump or motor falls in use, it will probably have deposited fine metal particles in the oil. In such cases the system should be completely drained, tank cleaned, filter elements replaced and the system filled with fresh oil.
6. The ends of hoses and hose connections should be capped when not in use. Before reconnecting, inspect fittings to be certain they are free of dust and dirt.
7. If hoses or fittings have been left open, they should be thoroughly flushed first with kerosene and then with oil, before being used.

If dirt is kept out, the system will usually operate with little trouble. The following other maintenance items should be considered:

1. Inspect hydraulic hoses for physical damage before using. Any hose that has wire braid showing, has bubbles or which has been crushed should be discarded. Hydraulic oil under pressure has great force and can cause injury should a hose spring a leak.
2. Examine all hydraulic fittings periodically and tighten at any sign of a leak.
3. The temperature of the hydraulic oil should preferably be kept below 150°F but never be allowed to go above 200°F, otherwise it may break down. If oil has a burned smell or color when compared to fresh oil, it is an indication that it needs to be replaced. Some power packs are equipped with "Thermomarkers" which give an indication of maximum operating temperature.
4. Drilling is hard on hydraulic components, particularly the pumps, due to the shock of hitting hard strata and stalling the drill. The good operator advances the auger at a steady rate with a minimum of stalling. He does this by keeping hydraulic pressures at a

range of 60% to 80% or relief pressure.

5. If the pressure gauge indicates that system pressures are above authorized maximum pressure (2500 psi (172 bar) on low pressure or 5000 psi (345 bar) on high pressure), relief valves require resetting downward or gauge should be checked. If there are no relief valves on the system the pump may require adjustment.
6. When drilling, if the maximum attainable pressure drops to less than 80% of authorized maximum pressure, it may indicate too low a setting of the relief valve, pump or motor wear, insufficient oil, or leak in the system. Pumps and motors should be replaced and returned for repair, if necessary.
7. Keep the hydraulic reservoir filled, with equipment shut down and large cylinders retracted.
8. Protect all hydraulic components from damage, particularly hoses and fittings in the front of the rig where they can be hit by a mat or a pile.
9. Listen for noisy pumps and motors that indicate trapped air, cavitation or wear.
10. When starting up on a new project, connect drill or hammer supply and return hoses together and circulate the oil before actually running the drill.
11. Bearings and couplings should be greased once a week when units are in active use. Check and refill air line lubricators. Occasionally apply a few drops of engine oil to valve operating levers and linkages to keep them operating freely.

5.8.3.2. Trouble Shooting

In spite of good maintenance practices, hydraulic systems will just not work sometimes. Some of the common things to review, before you seek help:

1. Is there sufficient oil in the system?
2. Is the oil clean and in good condition; you can't operate with sludge.
3. Is there free flow of oil to the pump intake? Disconnect suction line and check; if insufficient flow, check for obstruction or dirty tank or strainer. In below zero weather, the oil may be too viscous. Is suction airtight?
4. Check pump to be sure it is rotating and that it is rotating in the proper direction; if uncertain, disconnect inlet and outlet hoses and rotate the pump by hand. Check pump openings to determine which is suction and which is discharge.
5. Check the condition of filters, replace elements if dirty, and look for obstructions in the inlet or outlet.
6. Disconnect hoses to drill and spotter, operate control valve in each direction; if there is no flow, then look for obstruction in valves or lines, non-working valve or improperly adjusted valve operating mechanism. Check the drain hoses for excessive flow.
7. If oil flows, but pressure is inadequate to operate components, check relief valve setting; check for continuous flow through relief valve due to stuck valve or broken relief valve spring; if these are O.K., then the pump or motors may be worn and need replacement; gauges may be faulty.

5.8.3.3. Handling and Storage

All hydraulic equipment should be dissembled with care and all openings capped and covered, with covers secured. All projecting fittings and parts should be protected from physical damage.

Unless the oil has been very recently changed, the end of a project is the best time to drain the system, install new filter elements, fill with clean oil, and operate for ten minutes. After this is completed, put on a tag describing what you have done.

Advise your general superintendent of any difficulty you have had and any components you believe need repair or replacement.

When storing, always leave hydraulic units full of oil to prevent rusting of interior parts and coat all exposed machined surfaces with grease. Hydraulic equipment should be stored under shelter. If this is not practical, vulnerable parts should be shrouded with plastic or tarpaulins, securely wired down.

Equipment should be stored on blocking off the ground and located in an area not likely to flood. In-line power packs should be stored upright to keep pumps flooded; they may be shipped lying down.

Store all units so they are accessible for inspection, inventory, repair or modification.

5.8.4. Pile Stakes

Even if pile stakes are not your responsibility, you will save yourself delays and headaches if you assure the layout is done properly.

Normally the survey crew will lay out the column lines and erect batter boards or targets outside the work area. You should caution your crews to use care when working around batter boards, benchmarks and monuments.

The survey crew locates the centerlines of each foundation and then places a stake for each pile. As foundation layouts are frequently repeated, time may be saved on a large project by using a layout template for each type foundation. The heads of the pile stakes should be painted with some brilliant, distinctive color, and driven down so to be flush with the ground. If a centerline stake is used, and it is not a pile location, it should be removed after the pile stakes are driven, lest it be mistaken for a pile stake.

The driving sequence should be discussed with the survey crew so that the stakes can be installed well in advance of driving. It will cause a significant slowdown of the operations if a survey crew is working near the rig or between the rig and the shell rack, for example.

While it is not possible to check the location of the pile stakes in detail, the superintendent, supervisor and inspector should review the pile stakes to be on the alert for any layout errors. The survey crew is more likely to make an error of several feet than one of a fraction of an inch.

Before starting to drive any pile, the supervisor should uncover or locate all the stakes and make sure that they agree in number and general appearance to what is shown on the pile plan, a copy of which should be available to him.

Once the project is staked, the location tolerances of the pile driving project become important. Typical tolerance of lateral deviation from the specified location at the head is not more than 100 mm (4") horizontally, except for bridge abutments, where the deviation should not exceed 25 mm (1"). A deviation of ± 1 inch from the specified cut-off elevation is reasonable. Sloping land surfaces may require adjustment of the pile location if the surface varies from the reference plane used in the plans to depict pile locations.

5.8.5. Mandrel Driven Piles (Shells)

For efficiency, sufficient shells should be kept on hand to allow for variation in delivery time and changes in driving rate. Having too many shells, however, can require extra handling and may prove embarrassing and costly, should gauge requirements, pile lengths, etc., change unexpectedly.

5.8.5.1. Shell Up

In order to place the shell over the mandrel for the first pile, it may be necessary to screw some of the shells together in the leaders. To save time you should assemble as many shells together as you have clearance between the bottom of the core and the ground, with the hammer at the top of the leaders.

On subsequent piles, the string of assembled shells can be lowered into a previously driven pile. They will stick up above the head of the driven pile a distance equal to the length of the two longest shells in the assembly.

When it is necessary to screw or unscrew shell joints in leaders with any frequency, it is done by taking several wraps around the shell with a 1" (25.4 mm) manila line and pulling on it with a tugger hoist or #1 line. An old piece of 1" (25.4 mm) rope looped into the head of the previously driven shell will prevent the lower sections from turning. Screwing shells in the leaders is time consuming and should be avoided, if possible.

A special problem is encountered when driving long straight-sided shells with mandrel when sufficient leader height is unavailable. Since the shell cannot be dropped into a previously driven pile, it may be necessary to provide "doodle" holes by augering or driving pipe to provide the additional shell up space.

If the required driving resistance is not reached when the head of the shell reaches ground level, the core should be withdrawn and additional shell added. Conversely, if the pile is stopped before the shell clears the bottom of the leaders, it may be necessary to unscrew or burn off the excess shell before the rig can swing clear to drive the next pile.

It is preferable to move the rig to drive in another pier while the concrete crew removes the excess shell, if possible.

If pile lengths are erratic, and it is not possible to shell up with the correct length shell, it is better to have too much shell rather than too little so the core does not have to be removed before driving is complete.

5.8.6. Sequence of Driving

One usually backs away from the driven piles, having driven all one can reach within the safe operating radius of the rig.

Particular attention should be paid to isolated piles, and the superintendent should always check personally to assure these are not missed.

Pile driving frequently causes compaction and lateral movement of adjacent soil. For this reason, the driving of a multiple pile foundation should start in the center and work outward or from one side and work toward the opposite side. When driving near a high slope or structure, you should start by driving the piles nearest the slope or structure and work away from it. Leave such piles with a high shell on them if there is any chance of the slope caving.

When driving a group of piles for testing, the proper sequence will vary with soil conditions.

5.8.7. Driving

Driving progress usually controls the overall progress of a pile project. If you are behind schedule, it should be costly to catch up. If the completion date is fixed, additional rigs should be mobilized, overtime worked or additional shifts added, all of which will adversely affect profits.

The steady progress that is needed to keep on schedule requires constant advanced thinking on the part of the superintendent to insure that the driver consistently operates. In addition, when progress does not keep up to schedule, it should be recognized early and discussed with your general superintendent so that corrective measures can be taken. The further you get behind, the more difficult it becomes to catch up. Driving progress can often be improved by pre-excavating or jetting, but this may require approval of the owner after the project has started unless allowed in the specifications.

Proper hammer operation and driving progress are dependent on adequate steam or hydraulic pressure. The superintendent must assure himself that the required pressure is available and that boiler capacity is adequate for sustained operation. This is particularly true when using differential hammers.

The size of hammer or minimum rated energy will often be specified. On larger or more complex jobs, the company frequently runs wave equation analysis on the hammer, hammer cushion, pile/soil system to determine what hammer will install the pile to the desired resistance most economically, without exceeding permissible driving stresses in core or pile.

As pile loads become larger, it becomes increasingly important that piles be kept within the dimensional limits specified. The slope from vertical alignment is typically not more than 0.25 inch per foot of length (2%) for large pile groups. The superintendent must see from the start that the rig foremen understand the importance of this, as it is easy to become lax.

A supervisor should check pile location after core is set, from adjacent stakes or from offset stakes, made from short pieces of 3/8" (9.5 mm) rod with handle bent at the top. This is particularly important when the ground is heaving. It is unfortunate not to discover that piles are off location until they start setting the forms for the pile caps. When piles are off location, a pile cap redesign may be required, and all too often at the expense of the contractor.

5.8.7.1. Pile Driving Records

A complete pile driving record should be prepared for each pile driven to its embedment depth. These records provide important data concerning the nature of the soils that are penetrated, provide evidence of any soil disturbance that is a result of driving, and indicate the penetration resistance. The penetration resistance, pile driving analyzer data recorded during driving of indicator and some of the production piles, and recorded later during restrikes, can be analyzed to assess the driving stresses, capacity, and quality of the pile.

5.8.7.1.1. Impact Hammers

The make, model, energy, stroke, pile, and hammer cushions should all be recorded on the header portion of the pile record. The remaining portion of the pile driving record is used to record the penetration resistance (blows applied to the pile within a unit distance such as 1'.)

A wave equation analysis should be performed if blow counts that are recorded are not similar to those observed during driving and restrike of the indicator piles at the start or prior to construction.

Blow counts should be plotted as a function of depth to determine depths where the soil may be strongest and to indicate if the pile had been driven to the proper embedment depth and bearing stratum. One good setup to make the recording of pile penetration easier is shown in Figure 5-29.

Figure 5-29 Indicator Device for Field Use

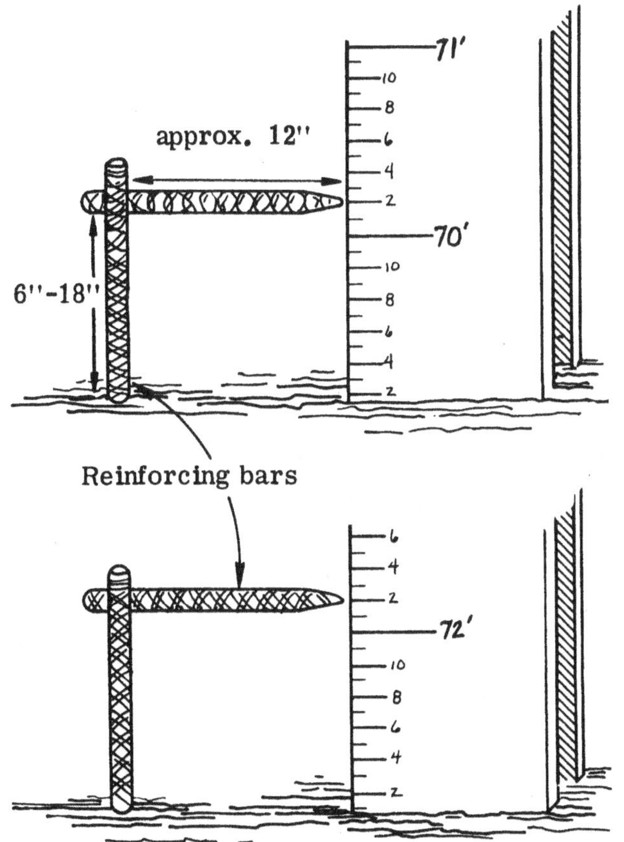

The average blow count should be computed for each pile and the average of each pile plotted versus time. This plot will indicate if the penetration resistance is changing as additional piles are installed. Increases in the penetration resistance with time may indicate that the soil is becoming stronger, perhaps from soil densification due to driving. Increasing penetration resistance with time may also indicate a temporary drop in pore water pressure because of the disturbance of dense sands or stiff clays. Decreases in the penetration resistance with time may indicate driving into weaker soil or perhaps the temporary build-up of excess pore pressures if the soils are loose and fine-grained with low permeability.

Abrupt changes in penetration resistance should be investigated to determine the cause. Abrupt changes can indicate a variety of problems that will cause the pile to be rejected. The pile should be replaced if driving damaged the pile. The installation equipment may have to be changed or a different type of pile and pile length may need to be installed.

5.8.7.1.2. Vibratory Drivers

The make, model, weight, dynamic force, frequency or range of frequencies, maximum eccentric moment, and clamping method should be recorded on the form. The remaining portion of the form should be used to record the depth, the time to reach a given depth, and the frequency to reach the depth.

The ultimate pile capacity cannot yet be determined from vibratory driver records. An impact hammer is required to drive the pile to assess capacity at the embedded depth.

A static load test is typically required to verify that vibratory driven piles have adequate pile capacity. Specifications should require a static load test for each type of pile and soil.

Steel H-piles and pipe piles can be easily clamped to a vibratory driver. Concrete and timber piles require special clamps for vibratory driving.

The vibratory driver is nearly always the most efficient tool for extracting a pile and can be used for steel, wood, or concrete. The vibratory driver first breaks the pile loose from the soil and then extracts the pile. A line attached directly to the pile can aid extraction when the pile is too heavy for the driver.

The dynamic force should be sufficient to maintain adequate amplitude to continue penetration of the soil. This is especially important if the soil is clay.

5.8.7.1.3. Pile Driving Analyzer (PDA)

The pile driving analyzer is recommended during

pile installation or during some part of the pile installation operation. It is discussed in detail in Chapter 6.

5.8.8. Prevention of Pile Damage during Driving

Damage can be prevented by not overstressing (overdriving) of the pile during driving. Overstressing occurs when driving stresses exceed the allowable limits (see Chapter 4).

5.8.8.1. Pile Damage

Overstressing is frequently attributed to using a hammer too large for the pile being driven, misalignment of the pile and hammer, material failure within the drive cap such as loss or insufficient cushion, or malfunction of the equipment.

5.8.8.1.1. General Indicators of Damage

The pile is probably damaged if any of the indicators given in Table 5-15 are observed.

- Voided piles that can be inspected after driving may exhibit damage from overdriving (crushing or distortion) the pile.
- Any piles suspected of being damaged should be promptly evaluated to determine its effect on the overall foundation design.
- Repetitive damage to piles may require modification of the installation equipment or installation method.

Table 5-15 Indicators of Pile Damage.

General Indicators of Damage	
Observation	**Pile Damage**
A sudden increase, decrease, or irregular penetration resistance	Pile breakage if the soil formation has not changed
Drifting of the pile off location as observed at the ground surface	Pile breakage
A sudden lateral snap of the pile head.	Pile breakage from bending
Damage observed at the head	Similar damage may occur at the toe
Two or more cycles of decreasing and increasing penetration resistance	Progressive crushing of the toe.
Vibration and/or noise of previously driven piles	Interference with the pile being driven

Types of Observed Damage	
Type of Pile	**Damage**
Timber	Brooming at the toe or head, splintering, or breaks
Concrete	Spalling, transverse cracks
Steel	Bending, reduction in cross-section

5.8.8.1.1. Prevention of Damage

Determining how much driving the pile can tolerate can prevent overstressing the pile. This is accomplished by completing a wave equation analysis to determine driving stresses. These stresses and corresponding penetration resistances should be avoided during pile installation. Experience on how hard the pile can be driven in the field without damage can also assist in avoiding damage.

5.8.8.2. Good Driving Practice

5.8.8.2.1. Driving Limitations

- The piles should be properly aligned with the pile hammer and the driving energy limited to maintain driving stresses below allowable limits.
- Cushions should be maintained at the thickness determined from results of driving of indicator piles with the PDA.
- Timber piles are especially vulnerable to brooming and splitting near the pile head or toe when driven. Driving should stop immediately when hard driving is encountered as indicated by a sudden increase in the penetration resistance or if the pile is observed to drift off location. A vibratory driver can be substituted for an impact hammer to reduce potential damage from hard driving.
- The driveability of concrete piles is often limited by the tensile strength of the pile, especially long prestressed concrete piles. Concrete piles should be driven initially with low energy that is about 50% of the maximum energy to be applied to the pile. The maximum driving energy is applied when the pile is nearly at the embedment depth. This guidance therefore requires that concrete piles should be driven with hammers of variable energy to safeguard the

integrity of the pile.

- Steel piles, especially open-end pipe and H-piles, are capable of hard driving. Pile toes may be protected with driving shoes, but prolonged hard driving should be avoided.
- Piles suspected of being damaged should be extracted and replaced. Vibratory drivers are efficient extractors of piles if the pile can be clamped to the driver.
- Pile driving should cease immediately when sudden increases in penetration resistance are observed to avoid overdriving. Prolonged hard driving should be avoided.

5.8.8.2.2. Restrike of Selected Piles

- Selected piles should be restruck while monitored by a pile driving analyzer to measure the quality of the driven piles and to assess pile capacity.
- Piles that should be restruck include those that have some sign of suspect behavior, but not judged serious enough to extract. Examples include penetration resistance that is not as high as expected and where the pile capacity is in doubt, unexplained changes in penetration resistance, penetration resistance not as high as previously driven piles, and piles with some damage at the pile head.
- Driving stresses evaluated by the analysis during restrike will not exceed the yield or ultimate strength of the yield or ultimate strength of the pile material. This provides a slight margin of safety against damage because the duration of the peak of the driving stresses is short.
- The quality of restruck piles should be evaluated by the PDA engineer and made a part of the quality assurance record.

5.8.9. Moving Between Piles

When piles are closely spaced, the supervisor may be able to drive several piles merely by swinging and, perhaps, changing driving radius, without actually moving the rig. He should understand the layout and plan several piles ahead just how he will reach each pile, with minimum of moving.

The ground ahead of the rig should be prepared so that it should be level and unobstructed when ready to move. If additional cribbing, planks or mats should be required, they should be available before they are actually needed, so as not to delay the rig. If leveling is required, the driver crew should not do it.

Except when urgently required by the pile layout, the supervisor should not be allowed to drive by moving the rig forward so that driven piles will subsequently be under the mats and inaccessible for concreting.

The vibration from pile driving may cause subsidence of what appears to be firm ground. Mats should normally be used under the rig unless your general superintendent authorizes you.

The rig should be kept in the center of the mats and as level as practicable. A level rig makes it easier to drive plumb piles with minimum stress in the equipment and with maximum safety.

Mats should be handled by breaking the suction at one end, lifting them vertically with two hook slings, and setting in place. Avoid trying to slide the mats sideways as it puts severe loads on the boom. Mats should be handled carefully to avoid striking the leaders hard enough to damage leader columns and bracing.

5.8.10. Batter Piles

5.8.10.1. General

The determination of batter pile capability is a complex engineering problem. The usual crane capacity tables are of little help and reliance on rules of thumb like "keeping the hammer below the boom tip", can be very risky. There are so many possible combinations of leaders, hammers, cores, piles, drills, jets, type and degree of batter, etc., that it is impractical to reduce this data to a table.

The piles should be picked up or the core shelled-up with the leaders plumb. The boom length specified for batter capabilities is normally the safety length; do not assume that a shorter boom is safer.

In addition to stability considerations in the batter position, the rig should be checked for:

- Safety in picking up the pile (or shelling up),

- Stability after the pile is landed,
- Ability to properly handle drill or jet,
- Ability to raise the hammer after driving,
- That the leaders are strong enough when the batter position and that the extensions are long and strong enough to hold the hammer in the leaders in the batter position and to support the hammer when driving below the leaders.

Since the rig is often loaded to its capacity in driving batter piles, a firm footing for the rig is particularly important. The rig should be kept a safe distance back from the edge of any excavation, should be kept level, and, unless otherwise authorized, should be on mats.

Batter pile driving is likely to impose particularly severe loads in the topping lift supporting the boom. The pendant cables, boom cable, topping lift blocks and connections should be carefully inspected before starting work. The boom hoist and boom hoist brakes should be properly adjusted and the operator should have the boom fully under control when booming down.

Alignment is also more difficult to maintain during batter driving. One way to maintain proper batter angle is to use the set-up shown in Figure 5-30. As with plumb piles, a typical alignment specification is 1/4" deviation per lineal foot of pile (2%).

Figure 5-30 Apparatus for Checking Alignment of Batter Piles

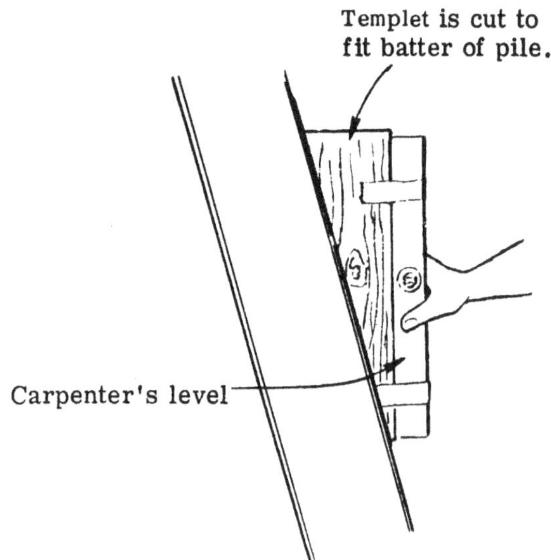

5.8.10.2. In Batter Piles

1. Keep the hammer and core or pile as low as possible (with the pile toe near the ground) when battering the leads. The hammer should be held on the brakes and not dogged.
2. Use a pile guide or cable to hold the core or pile in the leads and do not let the core or pile swing free while the leads are being battered.
3. When the core is withdrawn from a driver shell, do not let the toe swing out as it comes out of the shell. This can be avoided by booming up as the toe of the core nears the top of the shell or by the use of a pile guide or timber placed at the bottom of the leaders to restrain the core.
4. Unless the batter capabilities give specific instructions to the contrary, the spotter should be retracted as far as it will go and locked in place before booming down.
5. Keep the bottom of any drill or jet no higher than the bottom of the leaders.
6. The most dangerous moment is when the leads approach maximum batter. If the rig is not sufficiently counterweighted, it will tend to tip forward. If there is any concern, the stability can be checked by resting the toe of the core or pile on a mat or timber and booming down. When the desired batter is reached, try easing the core up, watching the rear track rollers for a tendency to lift.

Figure 5-31 In and Out Batter

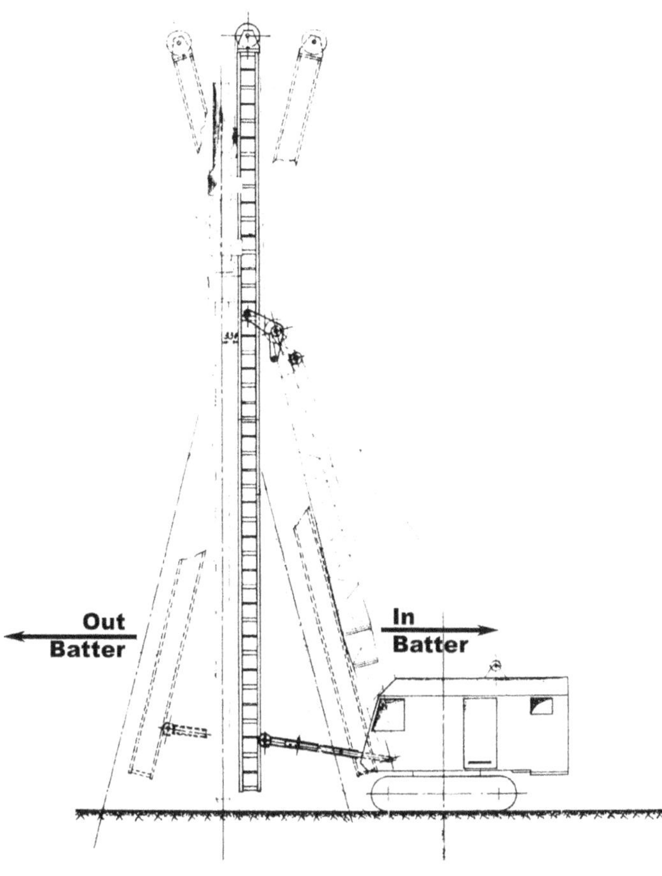

5.8.10.3. Out-Batter Piles

1. Keep the hammer and core or pile as low as possible when battering the leads, with the pile toe near the ground.
2. The spotter should normally be extended as far as it will go and locked in place before booming up, unless specific instructions have been received to the contrary. When using a very heavy hammer to drive a long, relatively light batter pile, for example, instructions may be given to boom up before extending the spotter.
3. The length of the boom safety guy cables should be adjusted, before the pile or core is landed, so that they are just tight when the leaders are in the required batter position. When more than one boom position is required on a project, adjust the cables for the highest boom position, unless otherwise instructed. The boom safety cables should be attached to a fixed spotter or safety beams.
4. Keep the bottom of any drill or jet as low as possible but no higher than the bottom of the leaders.
5. The most dangerous moment is when the core or pile touches the ground with the leads battered. If the rig is not correctly balanced, and safety cables adjusted, either the boom or the entire rig may go over backward as soon as the hammer line is slacked. Make sure the operator is told to release the brake slowly and cautiously especially on the first batter pile, watch for signs of lifting at the front track rollers, and watch the topping lift cables for slack to detect any tendency of the boom to go over backward.
6. When the core is withdrawn from the driven shell, it may be desirable to boom down as the toe of the core nears the top of the shell.

5.8.10.4. Side Batter Piles

1. Side batter piles should not be driven with a commercial boom due to the likelihood of boom failure. A Raymond-type or other reinforced boom should always be used to drive side batter piles.
2. No attempt should be made to pull a stuck core on a side batter. Even with the Raymond-type booms, modern cranes can exert such a great line pull that the boom can be easily buckled without warning.
3. If the pile is a pure side batter, the spotter should be positioned so that the leaders are at the maximum radius authorized for plumb piles, to minimize possibility of buckling the boom. For a combination side-and-in batter, the spotter should be retracted. For a combination side-and-out batter, the spotter should be extended. Combination batters may be used to reach the pile from the most advantageous rig position. They also permit a steeper batter than can be secured by in and out or side batter alone.

Figure 5-32 Side Batter

5.8.11. Complications

Driving of the production piles may lead to complications that could compromise the capacity of the foundation and interfere with the operation of adjacent structures. Penetration resistances that are either too low or too high when driven to the embedment depth, ground movement, and environmental problems can cause these complications.

5.8.11.1. Problems with Low Resistance

The pile foundation will not have sufficient capacity if the penetration resistances are too low. Abrupt unexpected reductions in the penetration resistance may indicate a damage pile. Other causes of low penetration resistance include soil freeze, soils with low strength, and driving into a landfill. Materials in landfills will usually be poorly compacted.

5.8.11.2. Driving Of Initial Piles

If the first production piles driven in a group have penetration resistances that are too low, then one must first check that the hammer and driving energy are adequate and similar to those used for driving nearby indicator piles prior to or at the start of construction. PDA data should be obtained if the equipment is available, and driving stresses, capacity, and pile quality should be checked. PDA equipment should be available during the driving of the first piles and used to assist the installation.

If the pile capacity is not adequate at the embedment depth as indicated by low penetration resistance or the PDA, but the pile is not damaged and driving stresses are satisfactory, several additional piles may be driven as specified in the contract documents. With granular soils, the first piles driven in a group often have penetration resistances that are lower than those driven later, since driving tends to compact granular soils and increase the penetration resistance.

Plot the average penetration resistances of the piles that were driven as a function of time to see if the penetration resistances had increased with time to see if the penetration resistances had increased with time and have reached a satisfactory level. The first piles that were driven in a group should be restruck after driving additional piles to see if the penetration resistances and pile capacity have increased to a satisfactory level.

If the penetration resistances and pile capacity have not increased to an adequate level, then other options should be considered. Such options are driving the piles to a deeper embedment depth, driving additional piles to reduce the spacing between the piles and increase the number of piles in the group. A cost and efficiency comparison should be made between these options, and the most cost effective option should be selected first.

5.8.11.3. Piles Driven Later in Production

If piles driven later in a group do not have sufficient penetration resistance, then the cause may be from the effects of soil freeze. Solis that exhibit freeze include loose to medium dense sands, silts, and clays.

If the penetration resistance of the pile is low when driven, the pile should be restruck after a 1-, 2-, or 7-day delay to determine if soil freeze effects are present, and a 30-day delay may be necessary to insure complete soil freeze. The penetration resistance of restruck piles should be plotted as a function of time to determine the rate of increase in penetration resistance. Adequate capacity may develop after a sufficient delay following pile installation.

If the penetration resistance is low, although

driving energy is adequate, there is no evidence of pile damage. If restrike does not provide larger blow counts, then the soils may be weaker than anticipated from results of the exploration program. Either driving of additional piles or driving to a deeper embedment depth are possible options.

If the penetration resistance suddenly drops and pile damage is not suspected, the pile toe may have penetrated a subsurface void. The penetration resistance should suddenly increase when the bottom of the void is reached. Driving piles at nearby locations will aid the determination of the lateral extent of this void. A possible complication with subsurface voids is that they may become larger with time and could eventually compromise the integrity of the foundation. Subsurface voids can be filled with a cement-bentonite grout when they are found; this should only be done with the approval of the designer. Pile installation should continue at another location of the site until the grout has had a chance to set. If the void in question is an underground cavern (as can occur in limestone), grouting may not be practical because of the size of the void.

5.8.11.4. Problems with High Resistance

Encountered blow count resistance that is too high may be caused by inadequate driving equipment. If the driving equipment, driving energy, and piles are found adequate, then soil relaxation may be a cause of the problem. Relaxation may occur in very dense, fine, submerged sand, inorganic silt or stiff, fissured clay or by driving a toe bearing pile into friable shale or a clay stone. Obstructions such as cobbles or boulders, an unexpected dense or stiff soil, large bodies in landfills, interference with adjacent piles, or bedrock may also cause high penetration resistances.

5.8.11.5. Driving Equipment and Piles

- Driving equipment that does not have an adequate energy to install the piles will cause high penetration resistances with little or no set of the pile.
- The size and type of pile should be checked to be sure that the proper pile is being installed to support the structure and that has sufficient driveability requirements.
- Contract specifications commonly indicate the minimum driving energy required for the work. The driving energy should exceed the minimum required energy.
- If the PDA indicates that driving stresses are low, efficiency of the hammer is adequate, and pile quality is adequate, energy of the hammer should be increased and driving continued. If the efficiency of the hammer is low, then driving equipment requires maintenance.
- If the PDA indicates that driving stresses are high and near the allowable limits and hammer efficiency and energy are adequate, then soil relaxation or obstructions may exist. If soil relaxation or obstructions such as boulders or cobbles are not found, then the soil through which the pile is being driven may be of a higher strength than anticipated may be required to reach the design embedment depth. Soil data recorded during the investigation should be checked to determine if the designers had anticipated high-strength soil at depths above the design embedment depth. Adequate bearing capacity may be obtainable at a lesser embedment reducing the required length of the pile. Jetting the preboring are also options that should be considered in such cases.

5.8.11.6. Relaxation

Penetration resistances that continue to increase as the pile is driven may be caused by soil that increases in strength with increasing depth. Driving may also reduce the pore water pressures and cause the soil strength to increase in response to driving.

Small displacement piles such as H-piles or open-end pipe piles are recommended if relaxation effects are significant. Open-end pipe piles can be cleaned periodically during driving to minimize soil displacement. The pipe pile, after it had been cleaned, can also be filled with an internal core of reinforcement steel and concrete to achieve adequate pile capacity if necessary.

Driving of the pile should continue until driving stresses reach the allowable limit or until the pile is driven to the design embedment depth. The pile should be restruck after 1 or 2 days, even if the pile had been driven to the embedment depth, to determine if the soil will relax and the soil strength should be reduced following equalization of pore pressures. If the penetration resistance had decreased following the delay, driving can continue until the pile is driven to the design embedment depth or until blow counts become excessively high.

Temporary soil strength gain caused by relaxation can exceed the capacity of the driving system to install the pile. Delays of 1 or more days may be required to drive the pile to the design embedment depth if the capability of the driving system is not increased. The driving energy should be increased until driving stresses reach the allowable limit to optimize driving efficiency. Increasing the section of H-piles or wall thickness of steel open-end piles may also increase the capacity of the pile and avoid excessive overstresses.

Several production piles should be restruck after a long-term delay such as five or ten days to verify that long-term equalization of pore pressures will not reduce the penetration resistance and pile capacity to excessively low levels. The decrease in penetration resistance recorded during the restrikes should be plotted with time to determine the rate that relaxation occurs. If the decrease in penetration resistance is negligible after 5 or 10 days, then no later restrikes should be necessary.

5.8.11.7. Obstructions

Obstructions such as cobbles, boulders, and large bodies in landfills will cause significant increases in the penetration resistance leading to refusal of the pile to be driven further and/or the pile can be deflected by the obstruction. The pile should be damaged if driving stresses exceed the allowable limits.

Obstructions near the ground surface are evident by drift of the pile or by abrupt increases in the penetration resistance. Deep obstructions are indicated by an abrupt increase in the penetration resistance or by a significant difference between toe elevations of adjacent piles when the pile had been driven to refusal.

Deep obstructions can deflect the pile without any apparent effect to the exposed portion of the pile above the ground surface. If the deflection breaks the pile, then the penetration resistance may be decreased abruptly. If the deflection caused by the obstruction bends the pile, then the pile capacity may not be adequate. Steel piles can bend through large angles approaching 180° so that the toe is driven toward the ground surface and may even break the surface.

Obstructions may be removed by assisted installation or may be pierced with a spud. A spud is a mandrel, heavy steel pipe or H-pile section driven to provide a pilot hole. The spud is withdrawn and the pile inserted into the hole and driven to the embedment depth.

A production pile may also be pulled when difficult driving is encountered and redriven. However, shaft friction may be reduced causing these piles to be unacceptable if they are friction piles.

5.8.11.8. Problems with Ground Movement

Displacement piles such as timber, precast concrete, and closed-end pipe piles displace the ground as they are driven to the embedment depth. Driving these piles in dense sands or saturated cohesive soils will possibly cause heave of the ground surface around the driven pile.

5.8.11.8.1. Pile Heave

Immediately after driving, an accurate location and elevation should be taken on each pile, referenced to permanent markers outside the area that may be affected by further driving. If lateral movement is noted, it should be brought to the attention of the engineer. If upward pile heave of more than 6 mm (1/4") is noted, the pile should be redriven as necessary to assure full design capacity. Piles that heave should be redriven to their original embedment elevation, but redriving should not begin until all heave had occurred. The penetration resistance required to redrive the piles to the embedment elevation should be recorded and, any significant change in the penetration resistance should be noted.

5.8.11.8.2. Heave of Adjacent Structures

Increases in surface elevation or ground heave exceeding 13 mm (1/2") may cause damage to adjacent structures. Therefore, the effect of ground heave from pile installation should be observed on adjacent structures.

The increase in the surface elevation from ground heave should be measured and recorded. Methods of controlling ground heave include pre-excavation, increasing pile spacing, or by installation of low displacement piles.

5.8.11.9. Environmental Conditions and Requirements

All environmental requirements should be fully addressed prior to construction and necessary approvals fully documented.

5.8.11.10. Vibrations and Noise

Pile driving operations produce considerable noise that may adversely influence the well being of inhabitants of the area.

Vibrations can damage nearby structures. Structures adjacent to the construction should be monitored to assess their integrity prior to and during construction. An assessment of the condition of adjacent structures should be conducted prior to construction to determine the initial condition of the structures.

Noise can interfere with the quality of life and may cause hearing loss. High noise levels that are part of an impact hammer pile driving operation can and frequently do prevent installation of driven piles where people can be adversely affected. People within range of damaging noise levels should be protected with safety measures. Vibratory drivers produce much lower noise levels than impact hammers and may be used where noise can be harmful.

5.8.11.11. Hazardous Materials and Gases

Some construction sites may be located in areas that had been used for landfills, storage depots that may contain natural gases.

The environmental requirements for the site should be fully addressed prior to construction. This is an important reason for thorough exploration prior to construction. If hazardous materials or gases are suspected during pile installation, then the construction activity should be delayed until an environmental assessment can be completed. Hazardous materials or gases can become evident when piles are extracted or when the soil is excavated.

5.8.11.12. Planned Delays

There are certain necessary activities that interfere with the rig's normal operation. However, if these are conducted in a planned fashion, the loss of production should be minimized.

Some delays can be anticipated, and planned for, are the performance of load tests, fuelling, lubrication and planned maintenance, changing equipment (one size hammer to another, for example), changing hammer cushions, safety talks, moving equipment, and inspections by public authorities.

Some things that can be done to properly plan for delays:

1. Have as many things done at the same time as possible. If you have to shut down until load tests are completed, use the time to make necessary repairs, alterations and adjustment to equipment; have the needed parts on hand.
2. Schedule routine activities, like fuelling, for periods that will not interfere with driving, such as lunchtime, before starting in the morning or after quitting time.
3. Operator should inspect, repair and lubricate the rig during downtime or when the crew is working independently of the rig.
4. If the rig should be stopped for fuelling, use this time for safety talks, lubrication, inspection, moving hoses, etc.
5. Use steam up time for lubrication and maintenance checks.
6. Schedule one or two people for overtime when necessary to avoid delays to the entire crew.

5.8.11.13. Weather

More time is lost due to weather than any other single cause. The work is affected to varying degrees by rain, snow, freezing, flooding, unusually high or low temperature, and wind.

Rain is the most serious delay problem, since it not only makes it unpleasant to work but also may slow operations and make working more difficult. It will sometimes help to provide crews with raingear. Temporary shelter can be provided for pile yard welders and similar operations where it is practical.

When haul roads are in poor condition, rain will turn potholes into quagmires and may stop delivery of concrete and pile material for days. Flooding of excavations can be anticipated and the superintendent should make certain that adequate pumps are on hand and operated so as not to delay the driver.

In areas where significant snow can be expected, arrangements should be made to plough access roads and parking areas. The superintendent should be able to contact the labor crew early to shovel snow when necessary.

Cold and freezing weather need not affect work if properly anticipated. Construction crews have often worked for extended periods in below freezing temperatures. Cooling systems should be protected with anti-freeze, changes made to proper winter lubricants and starting batteries and ignition systems should be in good condition.

Extremely high temperatures present overheating problems for cooling and hydraulic systems. Radiators and oil coolers should be clean and unobstructed, water levels should be correct and fan belts properly tensioned. Oil and oil filters may require replacement that is more frequent. If engines overheat, operators should be cautioned not to run wide open for extended periods; if clutches overheat, they are being slipped excessively. Watch batters for overcharging. Personnel should be supplied with plentiful quantities of cool drinking water and salt tablets. The engine fan should be changed so that the operator has fresh air drawn into the radiator. Working hours may have to be adjusted to avoid the hottest part of the day.

The equipment is designed to operate in any normal windstorm in which personnel can function efficiently and safely. When operations are suspended because of a predicted hurricane, the core should be left in a pile or removed from the leaders, the hammer lowered to the bottom of the leaders and the swing and travel locks engaged. All loose items should be removed and stored inside the rig or in the toolboxes. Hoses and similar items should be removed or wired securely to the rig. All enclosures should be securely fastened and, in extreme cases, windows boarded up with plywood to protect them from flying objects. Shanties, tool houses, stacks of shells, etc. should be staked and tied down.

When weather forces a shut down of the driver and hourly people are laid off, it is frequently possible to use straight time people to make repairs or adjustments under cover or wait until the weather improves.

When the crew should be paid show-up time but can't work because of weather, the time may be used for safety talks or other instructions.

5.8.11.14. Other Delays

Breakdown of equipment can result in substantial loss of productive time. Although it can never be eliminated, it can be measurably reduced by conscientious day-to-day lubrication, inspection and maintenance, prompt repair before the equipment breaks down or at least having needed parts on hand, in case something does break down.

While delays in delivery of material may be due to causes beyond the superintendent's control, they should not be the result of failure to anticipate needs.

Delays caused by others, such as failure to provide plans or pile stakes, failure to give instructions as required by the specifications, failure to provide access to pile locations, etc., can be minimized by close coordination of the work with the owner and general contractor. One cannot hold them responsible for such delays unless they are provided proper notice.

When you foresee that one should be faced with a significant delay due to the action or lack of action by others, you should be certain that the potential cost involved is made known to responsible members of the client's staff. It should also be understood that reimbursement is expected for the costs during any delay that occurs. Notations regarding the delay should be made in the pile report.

Extra moves occasioned by missing piles, driving in the wrong place, etc., even though the fault of others, are best avoided by some of the steps previously mentioned. Even when one is paid for such moves, it may be difficult to recover all costs; such things sour contractor-owner relationships.

Accidents and injuries not only cause direct losses

but also often result in loss production. They can be minimized by the proper instruction of all concerned in safe practices and constant attention by the superintendent to assure such safe practices are followed. It is not the intention of safety rules to limit production but rather to insure that it will continue. When you see an unsafe practice, use the opportunity to explain how the work can be performed.

As with other delays, the superintendent should have plans for what the crew should do in event of an unplanned delay. The operator and fireman can use the time profitably for adjustments, inspection, changing oil and oil filters, flushing radiators, washing boilers, etc. The driver crew can be used for inspections, lubrication of fairleaders and headblock, cleaning the rig, making up shells, etc. Where the delay is of significant duration, the hourly paid people can be laid off.

5.8.12. Extracting Piles

It is frequently necessary to pull piles that were driven for temporary structures that obstruct the driving of a new foundation, that are damaged during driving or that are driven out of position. The resistance offered will depend on the length of the pile, the type of soil, how long they have been in place, how hard they were driven, and the extent of any damage during driving. Piles driven in granular soils will generally pull easier than those driven in clays.

While sheet piles are generally driven in pairs, they can be extracted as single piles to facilitate handling, storage and rehabilitation. With cofferdams, when removal is necessary it should be removed in approximately the reverse order of driving. Once the cofferdam is flooded to the level of the outside water, any berms are removed then the fill material inside the cell is clammed out to the water line. If there is no overburden outside of the cell, depositing this fill next to the cells will help balance the soil pressure. When the interlocks of sheet piles are damaged or when they have been in long enough to corrode, the sheets should be extremely hard to pull.

There are four basic methods of extracting piles: direct pulling by the crane or excavator, impact extraction, impact-vibration extraction, and vibratory extraction.

5.8.12.1. Crane Extraction

Putting a sling around short piles and attaching it to the hammer or to a crane reeved several parts is often effective in extracting piles. In difficult cases, it may be necessary to set up a gin pole, pulling mast, or A-frame with multiple parts of cable. The crane can set the A-frame over the pile to be pulled, attach the lower A-frame block to the pile and attach its hoist line to the free end of the A-frame cable. A small crane can develop a pull of several hundred tons. The A-frame must have good support; sometimes it can be rested on unpulled piles.

If you are not successful in pulling a pile, try working on one nearby. Removing the second pile will sometimes relax ground pressures and make it easier to pull nearby piles. Do not attempt to pull a batter pile from the side or to try to rock a pile by swinging a crane boom back and forth. You can easily buckle the boom. When pulling piles, attach the #1 (or ship line) to an adjacent piles to prevent the rig going over backwards in case the hoist line breaks or the pile suddenly comes loose.

5.8.12.2. Impact Extractor

Although it is a mostly obsolete technology, some mention should be made of the impact extraction of piles. Such an extractor is shown in Figure 5-33. The extractor has a ram which impacts upward on an anvil, which is connected by sidebars or cables to the bottom of the extractor. This is in turn connected to the pile either by pins or by a wedge action clamp. The impact action of the extractor breaks the soil friction, allowing the pile to be pulled out of the ground by the crane. Some impact hammers can be rigged to act as extractors as well.

Before operating the extractor, take the maximum pull that can safely be made by the crane. As soon as the pile starts to pull easily, shut off the extractor. Keep the hoisting line vertically over the pile. A clamp on the bottom of the extractor can be attached by a cable sling around the pile but this should be less effective, due to the give in the cable.

Figure 5-33 Impact Extractor

Figure 5-34 Impact-Vibration Extractor

5.8.12.3. Impact-Vibration Extraction

Impact-vibration hammers have been designed for and used with steel sheet piling in soils that have large toe bearings and thus are not suitable to vibratory driving. Hammers that are used for both driving and extraction are able to apply impacts both upward and downward. Impact-vibration hammers for sheet piles generally are equipped with clamps instead of an inertial frame. Such an extractor is shown in Figure 5-34.

5.8.12.4. Vibratory Extraction

If extracting with a vibratory hammer, the crane must exert a net pull on the hammer-pile system. This will cause the pile to move upward. When extracting, in general the best procedure is to start the exciter without crane pull, allowing it to come up to speed, loosen the soil and to drive a little. Once this loosening has taken place, extraction is easier. It is very important that the crane not exert an upward pull greater than the rated capacity of the exciter's suspension. This notwithstanding, the crane capacity should be adequate for the project. The pulling capacity of the crane should be at least twice that of the combined hammer and pile weight for piles that have been in the ground for a short time, and 3-4 times that when they have been in the ground for an extended period.

In the case of rusted interlocks on steel sheet piling, an impact hammer such as a diesel can be used on alternate sheets to initially break the interlocks when the vibratory hammer is unable to. Simply use the impact hammer to drive every other sheet a few inches then lay the impact down and begin extracting with the vibratory hammer. In extreme cases, the impact is used to drive one sheet while simultaneously the vibratory hammer is extracting the adjacent sheet. This requires two cranes and would be the last resort.

An example of a vibratory hammer pulling sheet piling is shown in Figure 5-35.

Figure 5-35 Vibratory Hammer Extracting Sheet Piling

5.8.12.5. Special Considerations

Jetting or augering around a pile to be pulled should usually make piles easier to pull. This is particularly effective when pulling old precast piles in marine structures. Because of their weight, large precast piles cannot usually be removed with extractors. A pile that cannot be pulled at once may be loosened if a strain is left on it overnight. Driving piles down a few more inches and then pulling can loosen them sometimes. Higher capacity cables may have to be used or the piles jacked out. When all else fails, you can bore down alongside the pile and dynamite it loose or at least cut it off low enough below ground so it will not be any problem. When a number of piles have to be extracted, and ordinary methods will not work, special hydraulic pulling rigs have been used successfully.

5.8.13. Shift Work

While it is generally preferable to use additional rigs rather than additional shifts, the latter may be required to meet schedules. The two principal problems experienced when adding shifts are providing adequate light and coordination between shifts.

For efficient night work, one should have about 5000 watts of light per rig. The lighting system should consist of four light towers, at least 12' (3.66 m) high, each with two 500-watt floodlights and two 150-watt reflector lamps. These should be located around the rig outside the operating area. An example of such lights are shown in Figure 5-36. The individual lamps can be aimed for best overall lighting but so they don't blind the rig operator; some should illuminate the leaders. Light towers should be skid or trailer mounted to facilitate moving. Sufficient 3-conductor type S cable, with grounding connections, should be provided to connect lights to the power source. For runs up to 100' (30.5 m) #14 wire should be adequate; #12 wire cable should be used for longer runs.

Figure 5-36 Light Trailer

The ground light towers should be supplemented by

lights on the rig to the extent power is available. At least some 12-volt lights can be operated using automotive or marine type fixtures. Mount lights on corners of the cab, at the back of the cab and toward the drums and boom so the operator can see the cables. Additional lights can be mounted on the boom, located so that they will not be knocked off. If the project is spread out, additional lighting may be needed at shell rack, pile unloading areas, walkways, offices, change shanties and tool houses.

Coordinating the work of the two shifts will require extra effort on the part of the superintendent. Responsibility for the problems previously discussed, proper driving sequence, checks on piles driven, maintenance etc., is difficult to fix on one supervisor or the other. There is a tendency for each shift to drive all the easy to get piles and leave the more difficulty or unrewarding tasks for the next shift. The superintendent should schedule the work so that moving and difficult tasks are done in the daylight and driving or other routine work done at night.

Some other special problems with shift work are the unavailability of support personnel and services (clerks, inspectors, mechanics, survey crews, transportation, etc.) disturbance of nearby residential area, increased accident exposure due to working in the dark, security of personnel traveling to and from work and protection of company property and personnel automobiles in some areas. Materials should be used at an increased rate so that scheduling will have to be more carefully done. Even though the driver can be operated on two or more shifts, one usually increases production of the pile yard, concrete and cutting-off crews so that all this work can be done on one shift.

5.8.14. Servicing Rig

The consumption of fuel, water and lubricants will vary considerably depending on the type of work. A diesel crane engine may use as little as 15 gallons (57 l) of diesel fuel just idling all day to 80 gallons on a high production project with a hydraulic hammer; 50 gallons (189 l) per day is a safe average. Diesel fuel in similar to #2 Furnace oil but is more carefully filtered for diesel engine operation.

Lubricant consumption, after initial filling, will vary considerably but the following estimates can be used for initial stocking:

Engine Oil (Crankcase and hydraulics)	4 gal (15 l)/wk
Gear Oils	2 gal (7.5 l/wk
Grease	2 lbs (0.9 kg)/wk

Fuel tank, hydraulic system and lube capacities vary, but the following is typical:

Engine Fuel Tank	225 gallons (852 l)
Hydraulic Power Pack	300 gallons (1136 l)
Engine Crankcase	7 gallons (26 l)
Torque Converter	16 gallons (61 l)
Gear Cases (total)	35 gallons (132 l)
Cooling System	18 gallons (68 l)

5.8.15. Maintenance

Maintenance can be ignored only at the risk of a breakdown and a complete halt of production. An alert supervisor will normally find time for lubrication and routine maintenance during the regular workday without interfering with production. On high production jobs it may be necessary to schedule some time for maintenance outside regular work hours.

One task often neglected is lubrication of the crane undercarriage. This is a difficult and dirty job since one has to work in the mud. Although the cranes do not move very much, they sit and rock back and forth in the mud with every blow of the hammer and require lubrication every four hours or as recommended in the lubrication manuals. Similarly, headblock and fairleader sheaves are out of sight and hard to lubricate. The superintendent must assure these tough lubrication jobs are regularly completed.

5.9. Special Aspects of Piling

5.9.1. Pipe Piles

Shorter lengths of pipe can often be unloaded by rolling them off the truck if done carefully to avoid distorting them. For heavier pieces, it is usually better to unload by crane. Pipe stacks should be adequately chocked to prevent accidents. When pipe is delivered to the project in shorter lengths than required for the pile, lengths are spliced by butt-welding, preferably before delivery to the driver, if leader height permits. Particular care should be taken to insure that the two sections have true axial alignment. Pipe should be ordered with the ends beveled; back-up rings will facilitate welding. Pipe is usually unloaded onto racks where it can be rolled into welding rollers that turn the pipe for welding.

Boot plates will normally be welded at the same time. Pipe pile splices and points are discussed in 5.10.3.

When dragging closed end pipe piles to the rig, care should be taken to keep out dirt. Adequate lifting gear should be used; round turns should be made using slings and all slings checked for tightness.

Using too heavy a hammer may damage the pipe without gaining any additional penetration. Long light pipe piles will flex excessively when driven with large hammers, particularly if there is a long unsupported length above ground. This may cause the pile to move off location and may absorb a substantial part of hammer energy that rebounds against the hammer base resulting in excessive hammer maintenance. This difficulty can be minimized by pre-drilling a hole through hard upper layers and using a sliding pile guide (Iron-man) in the leaders at the pipe's midpoint, to prevent excessive bending. The pile guide can be suspended on cables below the hammer or handled on a live line. The latter method will permit the pile guide to be used to center the pipe under the hammer.

Pipe piles are sometimes driven with an internal mandrel. This requires that the pipe length be very accurately cut to a specified amount longer than the mandrel.

A problem with small diameter shell is that concrete falling from the top may wad up and block entry to the small diameter lower lengths. A proper concrete mix and constant attention are required to assure filling to the bottom. A hopper smaller than the pile should be used to center concrete in the shell.

Pipe piles are sometimes driven open-ended, particularly if it is desired to penetrate a hard stratum. They may later be cleaned out with air, water, or a combination of these. Large quantities of air are needed for short periods and auxiliary receivers are normally used to supplement compressors.

Overdriving can damage pipe piles and lighter wall pipes are vulnerable to collapse. This is particularly true when hard driving is encountered before a substantial portion of the pile is below ground. When driving light wall pipe to high blow counts be alert to a sudden drop off in blow count; this requires inspection of the pipe for possible collapse. Lighter wall pipe piles are also subject to a great deal of elastic compression in pile driving; this can result in hammer and cap instability and the cap actually falling out of the leaders.

All reasonable effort should be made to keep pipe waste to a minimum. Ordered lengths should be adjusted, where possible, as driving experience is gained. On large jobs, distribution of pipe to various areas should be carefully planned so that the driver can pick up the shortest adequate length of pipe. Determine the shortest cut-off piece that can be economically salvaged by dividing the cost of welding by the delivered cost per foot of pipe. Start salvaging cut-offs above this length at very beginning of the project. Weld these pieces together or use them to lengthen piles.

5.9.2. H-piles

H-piles look and are tough. They are easy to handle, drive and splice for needed extension. But careless handling and excessive hard driving can damage them. Because of their weight, H-piles always present more of a problem in unloading and handling than do pipe piles or shells. Flanges can be kinked and the corners bent by rough handling. Bent corners can generally be straightened with a sledgehammer. Minor flange kinks can also be straightened. Actual "bends" may require cutting out a section and splicing. If H-piles are stacked, they should be nested or dunnage should be provided between layers, with support points directly above the ones below. Proper stacking of H-piles is shown in Figure 5-37.

Figure 5-37 Proper Stacking of H-piles

Many times H-piles are lifted from a truck with a special slip-on clamp and a bridle sling from a crane. Attach the clamps from approximately one-quarter of the length from each end to equalize the stress. Stack the piles on timbers so they are kept straight. A small hole can be burned in a flange between the upper third and quarter-points for attaching a shackle for lifting the pile into the leads.

It is frequently necessary to splice two or more lengths of H-pile together to obtain the required pile length. If headroom permits, splicing should be done before delivering piles to the driver. If there are more than just a few piles, provide some sort of welding fixture that will keep the beam in line and permit it to be rotated for welding; consult your general superintendent. Splices may be butt-welded or splicing plates may be required, depending on the specifications.

If the resulting piles are very long, the move from the welding yard to the rig may present problems that should be carefully planned in advance. When operating in an open area, one sometimes uses a crawler tractor to drag the beams within reach of the rig. Because of their weight, beams must usually be brought closer to the rig than might be the case with shells.

H-piles with any type of protective coating should be handled with particular care to prevent damage to the coating. Special nylon web slings may have to be used. H-piles up to 2000 lbs. can be handled with wire rope slings; two turns taken around the beam and the sling checked for tightness before the beam is raised more than a few inches above the ground. Lugs may be welded near the top to guard against the sling slipping off. Heavier beams can be handled with a shackle placed in a hold burned in the flange, several feet down from the top.

H-piles can be lifted with a single line if they are not too slender, as shown in Figure 5-38. However, when the length exceeds 60 times the flange width, it may be necessary to use two lines to upend them without permanent bending. Handling piles into the leads should be carefully watched. It is easy, and less dangerous, to allow the lower end to drag across the ground but it may bend the flanges unless they are protected with a pile point. Every pile should be inspected to assure that flanges are straight and ends cut square when they are set in position for driving. Bent flanges will start the pile eccentrically. Long piles may be overstressed or kinked by handling. Use a two-point pickup for longer lengths. For single-line handling, attachment near the upper quarter point, rather than near the end, will help.

Figure 5-38 Single-Point Pickup of an H-pile

When hard surface overburden is encountered, it may be necessary to drill or spud through it before setting the pile. The use of a pile guide at the mid-point will reduce the whip of the pile, minimize chances of buckling, and reduce hammer maintenance.

Since H-piles cannot be inspected after driving, special care should be exercised not to damage them by overdriving. Be particularly watchful for a sudden drop off in blow count that may indicate failure at the toe of the pile.

H-piles are frequently used as soldier beams in conjunction with timber lagging to support slopes during excavation. This application requires extremely accurate pile placement plumbness in all directions. Pre-drilling for soldier beams should always be considered.

Most of what is suggested above to avoid pipe waste is also applicable to H-piles. Although H-pile splices are more costly, this is often partially offset by a higher cost per foot for the H-pile as compared to pipe.

5.9.3. Monotube Piles

Monotubes are handled and driven similar light closed-end pipe piles and the problems encountered are similar. Long piles have straight-sided upper sections and these should be welded to the tapered lower section. When headroom permits, this is preferably done before driving. The extension section is inserted into the lower section, with a 6" (152.4 mm) overlap and a continuous fillet weld placed around the joint. More elaborate welding procedures may be specified in some situations.

5.9.4. Step Taper Piles

Pipe Step Taper piles are normally used where there is insufficient leader height to drive an all shell pile. However, there may be other reasons.

Much of what is said above about handling and driving shells and pipe will apply to driving Pipe Step Taper piles. However, every time you switch from shell to pipe, the core has to be removed from the leaders. To the extent that it can be done without excessive moving and without causing driving problems, driven as many pipe sections as convenient before adding the shell sections.

Normal procedure is to leave the core in the last pile driven, attach the pipe follower and drive all the pipe sections you can reach. You then unhook the pipe follower, pick up the core and drive all the piles to final resistance. When the pipe section encounters little driving resistance, care should be taken to insure that the shell is properly seated on the pipe. The top of the pipe should be flat and not beveled. Three turns of 1/4" (6.4 mm) wire may be tack-welded about 3" (76.2 mm) from the top of the pipe to look the pipe into the PSJ sleeve.

When the pipe can be pushed down to a firm stratum with the weight of the hammer, the male core head may be used in lieu of a pipe pile follower. However, the male core head will stand very little driving when so used. Any attempt to use the male core head to drive pipe will cause excessive wear on the core head and may break off the tendon. The superintendent must weigh whether the few minutes saved changing from core head to follower offsets the higher core maintenance costs.

Occasionally, pipe should be used in place of the lower shell sections, with the core extending to the toe. This is not strictly a pipe step taper and is driven like an all shell pile. In this case, the pipe is accurately cut to length (as instructed) and attached to the lowest shell before shelling up.

5.9.5. Concreting Piles

This section describes the concreting of all tubular piles, including pipe piles, Step-Taper piles, corrugated shell piles, and Monotube piles.

5.9.5.1. Protecting Open Piles

An unfilled pipe or shell pile is most susceptible to damage or to receiving debris while the rig is driving nearby. Cover open piles driven to, or near existing grade as soon driving is ended. Piles 4' (1219 mm) or more above grade may not need to be covered immediately unless augering or other types of pre-excavation exposes them to mud or debris. The vulnerability of unfilled pipe piles is illustrated in Figure 5-39.

Figure 5-39 Open Head Pipe Piles

Covers should be placed on all pipes or shells after they have been cut-off and awaiting concrete, unless it is obvious that they will not be contaminated.

Covers should be firmly attached on all pipes and shells that are left open overnight. Laws in many states require this. Most pile covers fit rather loosely and are not protection against vandalism or surface water rising above the top of the shells. In an emergency, the heads of the shells can be slit vertically and the segments bent over each other to form a cap. Unfilled piles left overnight should be kept at an absolute minimum.

5.9.5.2. Inspection

One of the most important features of cast-in-place piles is their ability to be inspected internally after they are driven. Every pile is usually inspected immediately after driving and should be inspected its full length immediately before concrete is placed. If concreting of piles is delayed for any length of time the driven piles should be inspected periodically. Each pile shell should be clear of all foreign matter and be reasonably dry (less than 4" (101.6 mm) of water) before it is filled with concrete. Each rig should have on hand at all times the following items:

1. Mirror
2. 100' (30.5 m) Tape
3. Sounding line

In addition, each project should have a droplight or powerful battery light when mirrors cannot be used for visual observations.

5.9.5.3. Dewatering and Cleaning

Removal of water and certain foreign matter from open pile shells can be accomplished by various means. A few basic techniques that have been successful follow:

1. Water: Clear water and some types of slurry (soil and water) can be removed by a bailing bucket or steam siphon. It may be necessary to stir up the slurry (using the bailing bucket or siphon) to keep the soil in suspension so that it will come out with the water. Three or four inches of water left in the pile after bailing will not be harmful to the concrete.
2. Soil: May be washed out of the pile by using a 10-foot length of 1" (25.4 mm) or 2" (50.8 mm) pipe connected to the end of a water hose. With the water turned on, the pipe is worked up and down into the soil to get the toe of the pipe to the bottom of the pile. Sufficient water volume and pressure are necessary to carry the soils solids to the head of the pile. The washing action is continued until the water runs clear for several minutes, and then the pile should then be immediately dewatered and concreted. Another method is to use a piston-type bailing bucket (sand pump) as manufactured by Larkin, Rampp or Bucyrus Erie. A small crane or the rig is usually required to handle the weight and length of sand pump.
3. Wood: Wood can often be removed by spearing it with a piece of sharpened material, or by floating the wood out of the pile, then dewatering.
4. Metal Objects: Depending on the shape of the metal object, it may be possible to retrieve it with a hooked-rod. If the object cannot be grasped or hooked, certain types of heavy U-magnets can be lowered in the pile to pick up ferrous metal objects. (Steel or iron).

5.9.5.4. Placing Concrete

5.9.5.4.1. Techniques

Figure 5-40 Concreting Piles

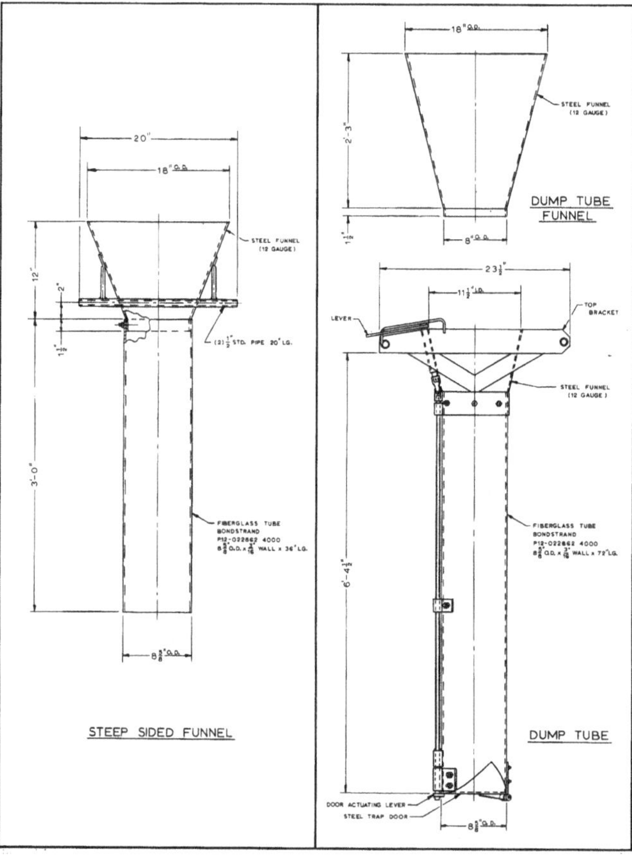

5.9.5.4.1.1. Conventional Concreting Method

- Check the open pile to insure that it is clean and dry (any water in the pile should be less than 4" (101.6 mm) in depth).
- Place steep-sided funnel clearly in the center of the pile so that the spout hangs directly down the center of the pile.
- Place 1/2 ft^3 of flowable grout through the steep-sided funnel.
- Position the truck-chute to the funnel so that it has sufficient slope to move the concrete rapidly.

- Start the pour with the concrete truck discharging as rapidly as possible without interruption.
- When you are near cut-off grade, you may slow down the discharge of the truck and/or remove the funnel to adjust final cut-off.
- Rod the top 6' (1.83 m) of the pile.

5.9.5.4.1.1. Modified Dump-tube Method

- Check the bottom door of the 6' (1.83 m) dump-tube to assure it opens freely and that the latch functions properly.
- Check the open pile to insure that it is clean and dry (any water in the pile should be less than 4" (101.6 mm) in depth).
- Latch the bottom door of the dump-tube and locate the tube directly in the center of the pile.
- Be sure that no obstructions are in the way of the bottom door release level and place the dump tube funnel squarely on the top of the dump-tube.
- Position the truck chute to the funnel so that it has sufficient slope to move the concrete rapidly.
- Place one cubic foot of flowable grout in the dump-tube.
- Inform the transit-mix operator that you must have a rapid, uninterrupted flow of concrete into the dump tube funnel. Rate of flow is to be limited only by the capacity of the truck, the chute or funnel.
- Begin the concrete into the tube. When the concrete fills the tube and reaches the bottom of the funnel, engage the release level and continue to pour from the truck as rapidly as possible without interruption.
- Once the pour is started, it should not be stopped until the concrete level is within 40' (12 m) of the head of the pile. If, for some reason, the pour is interrupted below 40' (12 m), the dump-tube should be removed and relatched and the entire operation repeated.
- When you are satisfied that the concrete level is within 40' (12 m) of the head of the pile you may elect to continue the pour through the dump-tube, or remove the dump-tube if it is more convenient.
- Rod the top 6' (1.83 m) of the pile.
- Thoroughly clean the dump-tube and funnel if immediate re-use is not anticipated.

5.9.5.4.2. Placing Precautions

The following are some general precautions concerning the placing of concrete in general:

1. The slump of the concrete when placed should be between 3" (76.2 mm) and 5" (127 mm). Do not estimate the slump.
2. The recommended maximum size coarse aggregate is 3/4" (19.1 mm); in no case should coarse aggregate be larger than 1" (25.4 mm).
3. Concrete should not be used if it has been mixed more than one hour except in special cases involving set retarding admixtures (see admixtures further in this section). Mixing time starts when the water is first added to the mix. Most transit mixers have part of the mixing water placed on the sand in the rear of the barrel before the coarse aggregate and cement are added. In this case, mixing time starts with the first revolution of the barrel. Concrete that has started its initial set should be discarded. Do not attempt to renew workability lost to initial set by adding water or remixing.
4. Don't place concrete by drop chute over 30' (9.1 m). If the concrete truck chute cannot be placed directly over the pile, use an alternate method such as concrete pump, collection hopper, concrete bucket, buggies or wheelbarrows. Buggies or wheelbarrows, however, cannot be used either with the modified dump-tube method. Wheelbarrows or buggies should not be used in any case without consulting the general superintendent.
5. Concrete chutes should be steep enough so that the concrete will flow by gravity in a continuous stream.
6. The concrete funnel used for concreting all types of piles should be steep-sided.
7. The upper 6' (1.83 m) of each pile should be

Figure 5-41 Concrete Slump Tests

<table>
<tr><td colspan="3">Purpose of test: To determine the consistency of fresh concrete and to check its uniformity from batch to batch. This test is based on ASTM C 143: Standard method of Test for Slump of Portland Cement Concrete.</td></tr>
<tr><td colspan="3">Take two or more representative samples - at regularly spaced intervals - from the middle of the mixer discharge; do not take samples from beginning or end of discharge. Obtain samples within 15 minutes or less. Important: Slump test should be made within 5 minutes after taking samples. Combine samples in a wheelbarrow or appropriate container and remix before making test. Dampen slump cone with water and place it on a flat, level, smooth, moist, non-absorbent, firm surface.</td></tr>
<tr><td>
1. Stand on two-foot pieces of cone to hold it firmly in place during Steps 1 through 4. Fill cone mould 1/3 full by volume (2-1/2" high) with the concrete sample and rod it with 25 strokes using a round, bullet-nosed steel rod of 5/8" diameter x 24" long. Distribute rodding strokes evenly over entire cross section of the concrete by using approximately half the strokes near the perimeter (outer edge) and then progressing spirally toward the center.</td><td>
2. Fill cone 2/3 full by volume (6" or half the height) and again rod 25 times with rod just penetrating into, but not through, the first layer. Distribute strokes evenly as described in Step 1.</td><td>
3. Fill cone to overflowing and again rod 25 times with rod just penetrating into but not through the second layer. Again distribute strokes evenly.</td></tr>
<tr><td>
4. Strike off excess concrete from top of cone with the steel rod, so that the cone is exactly level full. Clean the overflow away from the base of the cone mould.</td><td>
5. Immediately after completion of Step 4, the operation of raising the mould should be performed in 5 to 10 sec. By a steady upward lift with no lateral or torsional motion being imparted to the concrete. The entire operation from the start of the filling through removal of the mould should be carried out without interruption and should be completed within an elapsed time of 2-1/2 minutes.</td><td>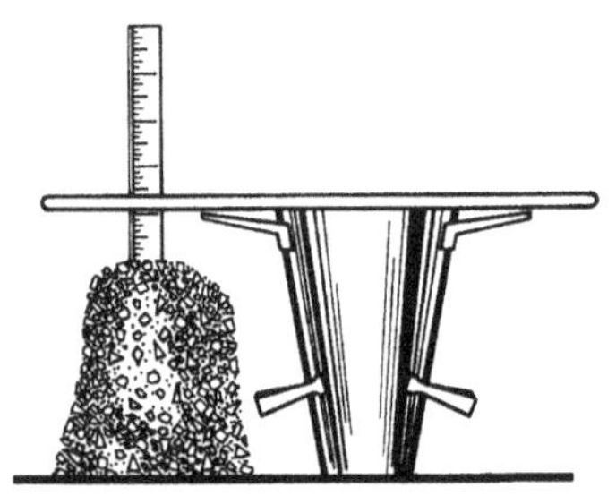
6. Place the steel rod horizontally across the inverted mould, so the rod extends over the slumped concrete. Immediately measure the distance from bottom of the steel rod to the original center of the top of the specimen. This distance, to the nearest 1/4 inch, is the slump of the concrete.</td></tr>
</table>

Figure 5-42 Concrete Test Cylinders

NOTE: For complete procedure, see A.S.T.M. Designations: C31, C94 and C172.

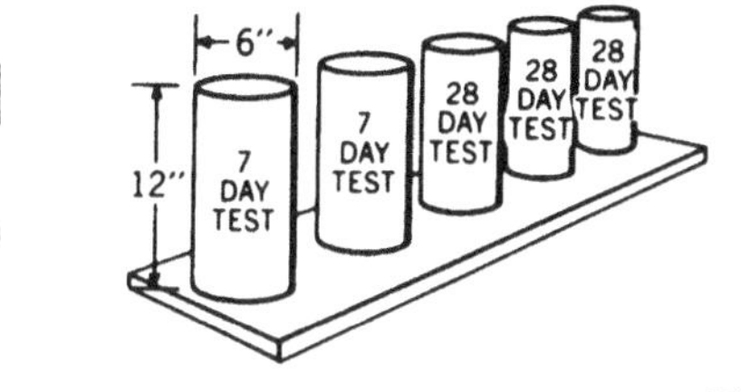

USE ONLY NON-ABSORPTIVE MOLDS

Steel, or parafined paper molds, 6" in diameter by 12" long, with base plates or bottoms, are used for casting concrete cylinders in the field. Before filling, they should be placed on a smooth, firm level surface. Three cylinders should be made for the 28-day test.

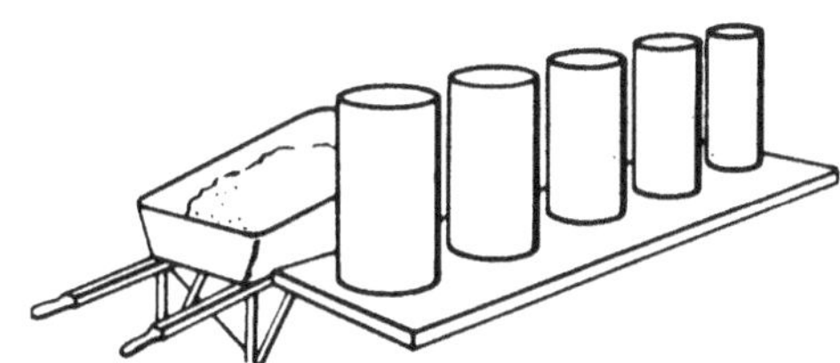

TAKE 3 PART SAMPLE

A sample should be obtained from at least 3 parts of the load. They should be taken directly from the truck or mixer discharge at well distributed points. Before filling the molds, the individual samples should be combined and re-mixed with a shovel in a wheelbarrow, buggy, or metal pan to insure uniformity.

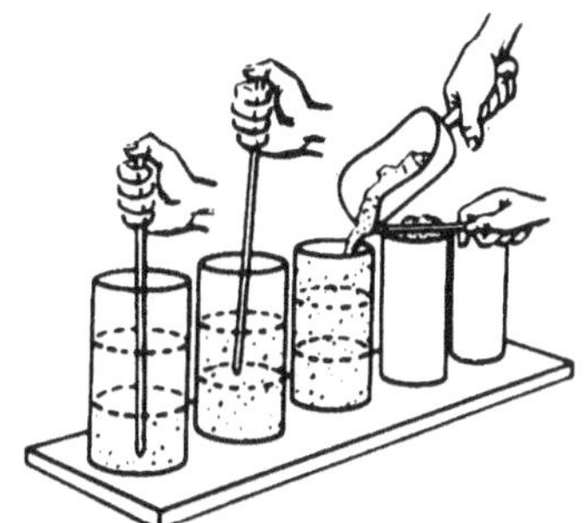

FILL MOLDS IN 3 LAYERS AND ROD EACH LAYER 25 TIMES

Molds should be filled in 3 equal layers, and each layer rodded uniformly 25 times with a 5/8" bullet-pointed rod. When rodding upper layers, the rod should just break through into the layer underneath. All molds should be filled uniformly – that is, place and rod the bottom layer in all, then the second layer, etc. The third layer should contain an excess which can be struck off smooth and level after rodding.

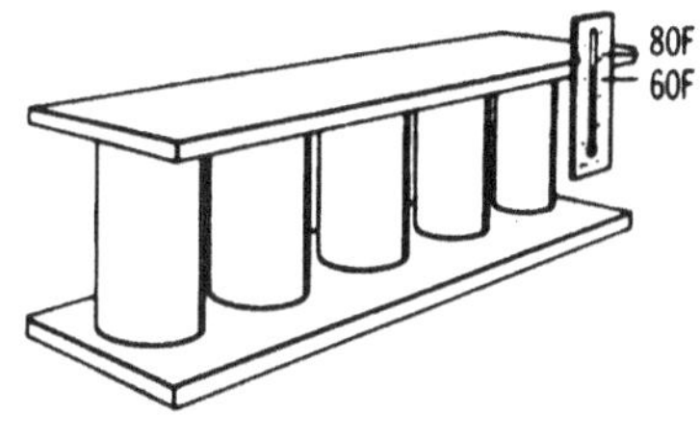

LET CYLINDERS SET AT LEAST 24 HOURS AT BETWEEN 60F AND 80F

Cylinders should be cast in a protected area where temperatures* will not exceed 80F (27C) or fall below 60F (16C). Tops should be covered to prevent loss of moisture and cylinders should not be disturbed or moved for 24 hours. Cylinders left on the job for several days and exposed to hot sun or low temperatures will give sub-standard results. Additional cylinders used for determining when forms may be stripped or when concrete may be put into service should be removed from the molds after 24 hours and then be job-cured adjacent to and under the same conditions as the concrete they represent.

*Temperatures are those designated by the American Society for Testing and Materials.

CURE AND HANDLE CYLINDERS WITH CARE

After 24 hours, cylinders for acceptance tests should be placed in moist curing at 73.4 ± 3F (23 ± 1.7C) or sent to a laboratory for similar standard curing. Careful handling during moving is necessary since cylinders which are allowed to rattle around in a box, or the back of a car, or pick-up, can suffer considerable damage.

IMPORTANT: **Strengths indicative of the quality of concrete can be obtained only if this procedure is closely followed.**

thoroughly rodded.

8. Vibrating is not necessary if the procedures set forth in this section are followed. Excessive vibrating can cause separation of the aggregates in the mix and possibly leave cavities in the concrete, especially when lower slump mixes are used.
9. The volume of concrete poured should be checked periodically with the theoretical volume of the piles. If any discrepancy is noted the cause should be immediately identified.
10. Tools, chutes, funnels, pump lines, etc. should not be made of aluminum. Aluminum reacts chemically with concrete resulting in the release of hydrogen gas and forms deleterious compounds in the concrete.

5.9.5.4.3. Concrete Pumping

When site conditions dictate, a concrete pump may be used to deliver concrete to the pile head. Only pumps having a steady even flow at a rate of 60-yd^3 per hour or more are to be used. The pump should be operated at maximum speed when pumping into the pile especially during the filling of the lower one-third of the pile. The use of a concrete pump does not negate any of the other requirements set forth in this section.

5.9.5.4.4. Admixtures

Concrete admixtures should not be used in cast-in-place piles unless specifically designated in the approved mix design. Check with the concrete supplier to make sure admixtures outside of the specifications have not been furnished.

Set controlling admixtures may sometimes be used to compensate for weather extremes, but should only be used with the express permission of the general superintendent.

5.9.5.5. Concrete Quality - Tests

The quality of concrete placed in a cast-in-place pile is the responsibility of the project superintendent. The superintendent should be assured that the concrete supplier is delivering the mix specified in the proper manner and that the concrete, once reaching the project, is properly handled and placed by the field crews.

For those contracts where the piling contractor has the responsibility for testing, copies of reports should be sent to the main office as well as the project site. Only reputable laboratories should be used.

On projects where laboratory testing is not required by specifications, concrete cylinders should be properly taken and cured. No project should require less than four cylinders. A minimum of four cylinders for each 200 yd^3 (153 m^3) of concrete should be taken. At a minimum, the first batch should be broken at three, 7, 14 and 28 days.

It is essential that test cylinders be properly prepared and cared for in the field.

1. Freshly made cylinders should be immediately placed where they will not be handled for at least 24 hours.
2. Cylinders should be delivered to the laboratory promptly after 24 hours.
3. Cylinders should be protected from extreme heat and cold at all times.
4. Cylinders should be kept moist during the entire time they are on the project.
5. Cylinders should be very carefully handled and well protected whenever they are moved.

If you suspect that a cylinder has been damaged in any way due to heat, cold, improper handling, etc., you should notify the testing lab before the cylinder is tested. Damage may result in erratic test results, which often cannot be explained. It is better to know in advance and discard the cylinder.

5.9.5.6. Protection of Concrete Piles

After concreting, all piles should be protected from extreme heat (above 90° F (32° C)) or cold (below 40° F (4° C)) immediately after placement.

In hot weather, the exposed concrete at the pile head should be covered to prevent rapid loss of moisture. This moisture is essential for the proper curing of the concrete in all kinds of weather. For additional information, see ACI Standard Recommended Practice for Hot Weather Concreting.

In cold weather, heated concrete should be used. In addition, it is necessary to protect any exposed concrete by means of hay, tarpaulins, etc. In extremely cold weather, salamanders, steam (or air) heaters or steam lines may be needed to protect the piles from freezing. Note that if the pile extends above the ground surface, it is necessary to protect all of the exposed portion of the pile and not just the head surface of the pile.

The rate of strength gain for concrete is reduced when it is exposed to cold temperatures. For example, at or near freezing temperatures, early strengths are less than one-quarter of normal and the concrete is exposed to danger of frost damage. The use of Type III cement (Hi-Early) will provide a faster strength gain and thus reduce the exposure of the concrete to damage as well as reduce the time required for protection. If, for example, the temperature is kept at 40°, the time required for Hi-Early to gain service strength is about one-quarter the time required by regular cement. The potential savings should be compared to the higher cost of type III cement. Calcium chloride added to the concrete will hasten the setting time and thus reduce the exposure to frost damage.[12] For winter concreting, it is advisable to consult the local weather bureau for current information on expected day and night temperatures and plan accordingly. Other suggestions are:

1. Use low water content in mix since the more water used, the longer the concrete is exposed to frost damage.
2. Raise temperature of pile if several feet of pile extend above ground.
3. Cover and protect the pile immediately after pouring to retain the heat and take advantage of the natural heat of hydration. Salt hay piled loosely around and on head of the pile is usually adequate.
4. Be sure heat enclosures are tight. Heating units should provide area warmth, not local heat directly on the concrete.
5. Prevent rapid drying out of the concrete surface. Moisture is always necessary for proper curing.
6. Three or four days after pouring, pile heads should be inspected for frost damage. Concrete that is damaged by freezing will have a deep olive drab color and will crumble or chip easily. If any damage is noted, advise the general superintendent immediately. Damaged portions should be removed and replaced.

For additional information, see ACI Standard Recommended Practice for Winter Concreting (ACI-604).

5.9.5.7. Onsite Project Mixing

While project mixing is generally not considered, it should not be completely overlooked. There are occasions on small projects in isolated areas where project mixing may be the most feasible method of casting the piles.

Following is a partial list of items that should be provided for prior to beginning any project mixing operation:

- ✓ An adequate supply of tested sand and aggregate.
- ✓ An adequate supply of clean tested water.
- ✓ A supply of Portland cement, normally Type II.
- ✓ Mix design tests performed on above concrete materials.
- ✓ Adequate storage area for the aggregates and the cement.
- ✓ A mixer of sufficient size to keep up with the expected concreting operation.
- ✓ A method of weigh-batching the dry material into the mixer and of measuring the water in gallons.
- ✓ Proper handling facilities for the mixed concrete.
- ✓ Proper storage facilities for concrete cylinders.

5.9.6. Precast and Prestressed Concrete Piles

A variety of methods have been used for the installation of prestressed concrete piles. These methods differ according to many factors, but all have one common objective: Any prestressed pile should be so installed as to insure the structural integrity of the pile itself, and insure that is will properly resist the imposed design loads.

Installation methods may vary with factors as:

1. Type of pile: bearing, sheet, combined bearing and sheet, tension, fender, trestle, columns, caisson, to name a few
2. Type of soil and intervening substance between soil and structure (water, air)
3. Vertical or batter piles
4. Design purpose, i.e., forces to be resisted
5. Type of structure to be supported
6. Site location and accessibility

5.9.6.1. Prevention of Damage

On some projects prestressed concrete piles cracking and spalling have been experienced. Damage of such concrete piles occurring during driving can be classified as one of three types:

1. Spalling of concrete at the pile head due to high compressive stress
2. Spalling of concrete at the pile toe due to hard driving resistance at the toe
3. Transverse cracking or breaking of the pile due to combination of torsion and reflected tensile stress sometimes accompanied by spalling at the crack

Spalling of concrete at the pile head is due to very high or irregular compressive stress concentrations caused by the following:

1. Insufficient cushioning material between the driving head and the concrete pile will result in very high compressive stresses on impact of the hammer ram.
2. When the head of the pile is not square or perpendicular to the longitudinal axis, the ram impact force should be concentrated on one edge.
3. If the prestressing steel is not cut flush with the end of the pile, the ram impact force may be transmitted to the concrete through the projecting prestressing steel resulting in high stress concentrations in the concrete adjacent to the steel.
4. Lack of adequate spiral reinforcing at the pile head and pile toe may lead to spalling or splitting. In prestressed concrete piles, anchorage of the strands is developed in these areas and transverse tensile stresses are.
5. If the head edges and corners of the concrete pile are not adequately chamfered, they are likely to spall on impact of the ram.

Extremely hard driving resistance at the toe can cause spalling of concrete at the toe of the pile. This type resistance may be encountered when founding the pile toe on very dense sand and hard rock. Compressive stresses when driving on hard rock can theoretically be twice the magnitude of those produced at the head of the pile by the hammer impact.

Under such conditions, overdriving of the pile and, particularly, high ram velocity should be avoided. In the more normal cases, with significant overburden thicknesses overlying the rock, toe stresses will generally be of the same order of magnitude as, but slightly lower than, the head stresses.

Transverse cracking of a pile due to reflected tensile stress is a complex phenomenon. It may occur in the upper end, mid-length, or lower end of the pile. It usually occurs in long piles (50' (15.2 m) or over). It can occur when driving in very soft soil. Although rare, it can also occur when driving resistance is extremely hard or rigid at the toe such as bearing on hard rock.

When the hammer ram strikes the head of a pile or the cushion top, a compressive stress is produced at the head of the pile; this compressive stress travels as a wave down the pile at a velocity of about 13,000-15,000 feet/sec. The intensity of the stress wave depends on the ram (weight, shape, material), the impact velocity, the cushion, the pile (modulus of elasticity, wave velocity), and the soil resistance. Ram weights vary from 1000 to 40,000 lb. Since, in a given pile, the stress wave travels at a constant velocity, the length of the stress wave will depend on the length of time the ram is in contact with cushion or pile head. A heavy ram will stay in contact with the cushion or pile head for a longer time than a light ram, thus producing a longer stress wave. If a ram strikes a thick or soft cushion, it will also stay in contact for a longer period than if it strikes a thin hard cushion, with a resulting decrease in driving stress.

The compressive stress wave traveling down the pile may be reflected from the toe of the pile as either a tensile or a compressive stress, depending on the soil resistance at the toe. If little or no soil resistance is present at the pile toe, the compressive stress wave should be reflected as a tensile stress wave.

The net tensile stress in the pile at any point is the algebraic sum of the compressive stress traveling down the pile and tensile stress traveling up the pile. Whether or not a critical tensile stress to crack the pile will result, depends on the magnitude of the initial compressive stress and the length of the stress wave relative to the pile length. A long stress wave is desirable in order to prevent damaging the pile.

If the toe soil resistance is very hard or firm, the initial compressive stress wave traveling down the pile should be reflected back up the pile also as a compressive stress wave. Tensile stresses will not occur under these conditions until this compressive stress wave is reflected from the free head of the pile back down the pile as a tensile stress wave.

It is possible for critical tensile stress to occur near the pile head in this case; however, internal damping characteristics of the concrete pile and surrounding soil may reduce the magnitude of the reflected tensile stress wave by this time. Such cracking has occurred, however, when driving onto rock with very light hammers.

In summary, the following can cause tensile cracking of prestressed concrete piles:

1. When insufficient cushioning material is used between the pile drivers steel helmet or cap and the concrete pile, a stress wave of high magnitude and of short length is produced, both characteristics being undesirable because of potential pile damage. Use of adequate softwood cushions is frequently the most effective way of reducing driving stresses, reduction in the order of 50% being obtained with new cushions. As the cushion is compressed by hard driving, the intensity of the stress wave increases; therefore, a new cushion for each pile is recommended.
2. Stresses in piles can be reduced by replacing some of the aluminum plates with laminated plastic (Micarta) plates to soften the hammer cushion. Check that the hammer cushion is replaced before it is fully compressed. If spalling of the pile head is experienced, consider increasing the thickness of the cushion at the head of the pile.
3. When a ram at a very high velocity strikes a pile, a stress wave of high magnitude is produced. The stress developed is proportional to the ram velocity.
4. When little or no soil resistance at the toe of long piles (50' (15.2 m) or more in length) is present during driving, critical tensile stresses may occur in the pile. This may occur when driving through a hard layer into a softer layer below, or when the soil at the toe has been weakened by jetting or drilling. Most commonly, these critical tensile stresses occur near the upper third part, but they may also occur at mid-length or lower.
5. When very hard driving resistance is encountered at the toe of piles (50' (15.2 m) or more in length), critical tensile stresses may occur in the upper half of the pile when the tensile stress is reflected from the pile head.
6. When reflected tensile stress is coupled with torsional stresses where torsional stress may be caused by excessive restraint of the pile in the leads, or by too tight a fit between pile caps and pile, thus preventing the pile from rotating freely during driving due either to soil action or rotation of the leads.

Spiral or transverse cracking of concrete piles is usually caused by a combination of torsion and reflected tensile stress. Diagonal tensile stress resulting from a twisting moment applied to the pile can, by itself, cause failure. If reflected tensile stresses occur during driving and they combine with diagonal tensile stress due to torsion, the situation can become even more critical.

The following may cause torsion on the pile:

1. The helmet or pile cap fitting too tightly on the pile, preventing it from rotating slightly due to soil action on the embedded portion of the pile.
2. Excessive restraint of the pile in the leads and rotation of the leads.

5.9.6.2. Driving Heads (Helmets, Cap Blocks)

Piles driven by impact require an adequate driving head to distribute the hammer blow to the head of the pile. This driving head should be axially aligned with the hammer and pile.

In addition to the regular hammer cushion, the head of precast piles should be protected from localized damage, usually with a plywood cushion. The cushion can be made from an inexpensive grade of plywood or laminated 1" (25.4 mm) lumber. A new cushion should be used for each pile unless driving is very easy. The cushion can be centered on the pile by nailing a lath on each side, and tying them with several turns of wire around the pile. A short piece of wire rope over the top will facilitate removal of old cushions from the follower. Alternately, the cushion can be inserted in the helmet and held by two nails driven through holes in the side of the helmet.

The driving accessory should not fit tightly on the pile head as this might cause transfer to moment or torsion. The driving head also holds or retains the cushion block to reduce the shock of the blow and distribute driving force evenly over the pile head, as shown in Figure 5-43.

Figure 5-43 Pile Cushion for Concrete Cylinder Pile

5.9.6.3. Predrilling, Jetting or Spudding

It is customary to pre-drill or jet precast piles not only to save driving time but also to avoid damaging the piles. Predrilling or spudding a starting hole may be helpful if there is material near the surface that tends to deflect the pile. Or it may be necessary to excavate and remove the material before starting pile installation. If a pile begins to shift location, you can drill or jet on the opposite side to bring it back in position. The pile should not be forced back on location as this may induce excessive bending stresses and break the pile.

The same effect of penetrating hard layers with a jet may also be done during driving, using an external or internal jet at the tip. Jetting will replace the water along the pile and reduce the shaft friction in sands and sandy materials.

5.9.6.4. Handling and Transportation

Precast piles may be moved when the compressive strength of the concrete has reached 4,000 psi, but are not to be driven until 5,000 psi strength is attained. Piles in storage should be supported to insure no excessive loads are induced nor a permanent sweep introduced during curing and shortening.Piles are to be lifted or supported only at designated handling points.

Long slender piles will have three, four, or five pickup points, equalizing slings and strong-backs for handling. The pile should be handled and driven in such a manner as to avoid excessive bending stresses, cracking or spalling. Calculations for these stresses, along with the usual pickup point positions, are given in Chapter 4. Typical pickup and support points for concrete piles are shown in Figure 5-44.

Precast concrete piles are generally heavy and should be picked up in the proper manner. If proper care and caution are not used, severe damage can occur. Piles cracked due to mishandling cannot be relied upon for resisting driving tensile stresses that may develop.

To avoid damage in handling, they should be unloaded with a crane or forklift. It is important to understand in advance what pick-up gear (slings, lifting beams, etc.) is required and how it should be attached to the piles. A handling beam for forklift trucks, which has been used with success in handling 12" (305 mm) piles up to 130' (40 m) long, as illustrated in Figure 5-45. The forks should be adjusted so the pile is supported at or on both sides of the specified pick-up points. When slings are used to pick up a precast pile they should be attached as shown in Figure 5-46 at the designated lifting points. Conventionally reinforced piles require more care in handling than do prestressed-precast piles.

Figure 5-44 Typical Pickup and Support Points for Concrete Piles[13]

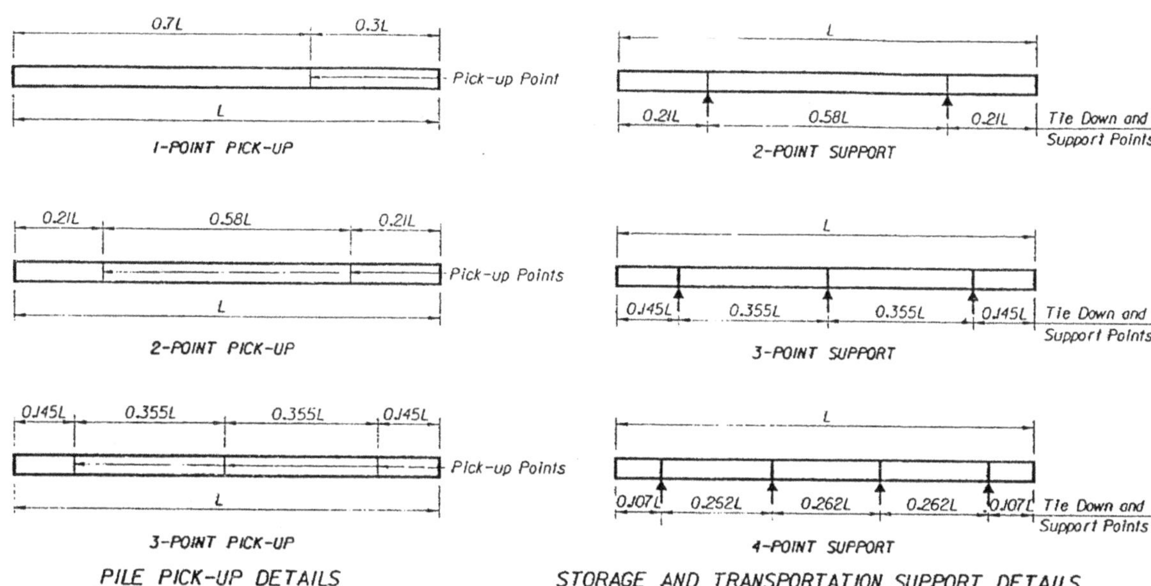

TABLE OF MAXIMUM PILE PICK-UP AND SUPPORT LENGTHS

	Square Pile Size (Inches)						Required Storage and Transportation Detail	Pick-Up Detail
	12	14	18	20	24	30		
Maximum Pile Length (Feet)	48	52	59	62	68	87	2, 3, or 4 point	1 Point
	69	75	85	89	98	124	2, 3, or 4 point	2 Point
	99	107	121	128	140	178	3 or 4 point	3 Point

Figure 5-45 Prestressed Pile Handling Beam

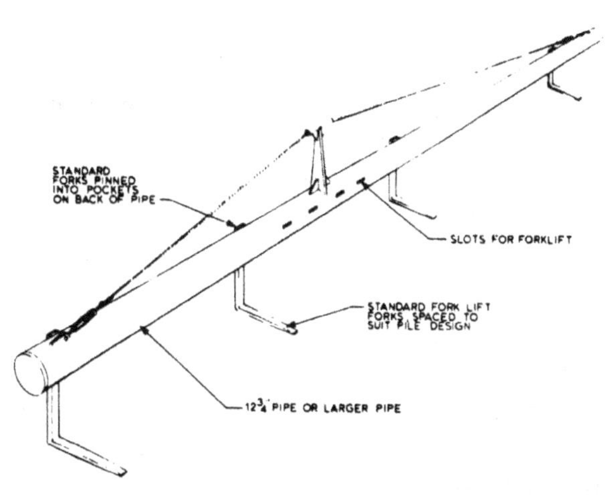

Figure 5-46 Sling Detail

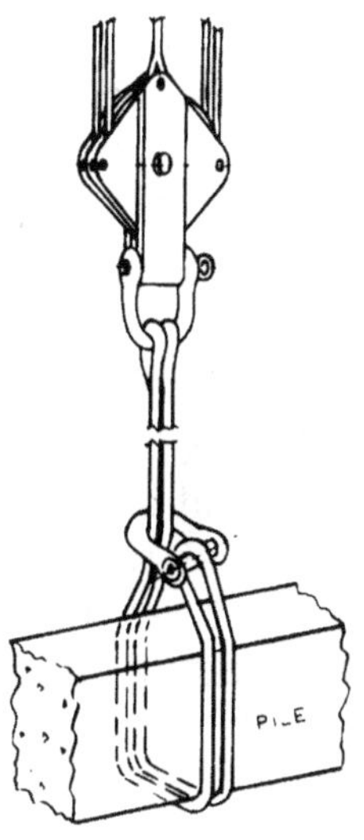

Piles up to approximately 45' can be carried on flat bed trailers. Piles over this length are generally carried on expandable flat bed trailers or telescoping pole trailers. For piles requiring more than two support points, special supports should be used on trucks. These supports should be located so as not to induce bending stress in the pile. Care must be exercised at all times, but especially during turns on highly curved roads. Project access conditions should be reviewed prior to delivery and all obstructions, ruts, holes or dangerous conditions corrected. Concrete pile being hauled is shown in. Figure 5-47.

Figure 5-47 Concrete Piles Hauled on Truck

When piles are stacked or trucked they should be supported at the designated points. Special care is essential when piles are long and require three-point handling. In the field an acceptable procedure is to attach a single-line sheave to line from the crane and connect a slack cable through this to the two lower pick-up points. Another line from the crane picks up the upper end of the pile. Both lines lift the pile clear of the ground, and then the line to the upper end straightens the pile to upright. The line to the lower end can then be released. Sometimes a lifting beam may be needed but prestressing has made this unnecessary in most situations.

Piles should be checked before unloading for cracks and that head end is square. Establish an identification system that will indicate when the pile was cast and what concrete cylinders were made from the concrete in the particular pile. Check specifications for disposition of embedded lifting eyes.

Piles up to about 50' (15.2 m) can usually be picked up with a single line as shown in Figure 5-48. For longer piles, it is usually necessary to have two lifting lines on the rig for upending the piles without overstressing them. Because of their weight, a pile-lifting bracket at the top of the leaders may be required.

Figure 5-48 Single Point Pick-up of Concrete Pile

Stacked piles should have intermediate dunnage supports in vertical alignment.

5.9.6.5. Positioning and Alignment

Correct position can best be assured by accurate setting of the pile. Removal of surface obstructions will aid in attainment of accurate positioning. Where accuracy of position is critical, a template or a predrilled starter hole, or both, can be employed to advantage. The position is largely established when the pile is set. After driving has commenced, attempts to correct position usually results in excessive bending and damage to the pile, and should not be permitted.

As a general statement, proper control of alignment is largely exercised before driving starts. It is almost impossible to correct the alignment after driving has commenced without inducing bending stresses.

Alignment may be aided by checking that the pile is initially vertical or on the project batter, as the case may be. Once the driving starts, the hammer blow should be delivered essentially axially, and excessive sway prevented at the head. The use of fixed leads, often specified, is a means to insure these two steps.

Attempts to correct misalignment by chocking at the base of the leads may introduce excessive bending and damage the piles except at the start of driving.

Long piles should be provided necessary support in the leads. Batter piles should be so supported as to reduce the gravity bending to acceptable limits; use of rollers in the leads is one method. Long slender vertical piles may require "guides" at intervals to prevent buckling under the hammer blow.

When driving a long way below the leader bottom, especially with batter piles, support telescopic leads should be provided to prevent excessive bending and buckling.

After installation in water, the pile should be protected against excessive bending from waves, current, dead weight (in case of batter pile), and accidental impact. Staying and girting should be employed until the pile is finally tied into the structure. Pile heads should be stayed to eliminate bending; this is particularly relevant to batter piles where the head should be lifted to overcome the dead weight. Frequently, when driving in deep water, a batter pile should be stayed before it is released from the hammer.

The heads of piles, even in water, cannot be pulled into position without inducing bending. Even with relatively low pulling forces, many piles have been severely structurally damaged because of the long lever arm available in many water installations. Positive limits should be set by the designer to control the pulling force. Pile heads should be protected from direct impact of the hammer by a cushion head so arranged that strands or bars projecting from the pile head would not be displaced or deformed during driving. The engineer will establish minimum energy per blow of the hammer to be used. Jetting may be permitted, or required where necessary, to reach the desired depth.

5.9.6.6. Good Driving Practices

Some guidelines for good driving practice for prestressed concrete piles can be summarized as follows:

1. Use adequate cushioning material between the driving head and the concrete pile. Three or four inches of wood cushioning material may be adequate for short (50' (15.2 m) or less) piles with moderate toe-soil resistances. Softwood should be used. 6" (152.4 mm), 8" (203.2 mm) or as much as 20" of softwood cushioning material may be required when driving longer piles in very soft soil. When the wood cushioning becomes highly compressed, or chars or burns, it should be replaced. A new cushion should be provided for each pile. If driving is extremely hard, the cushion may have to be replaced during driving of a single pile. Use of an adequate cushion is usually a very economical means of controlling driving stresses. In the past, concern has been expressed that cushioning might reduce the effectiveness of the driving energy transmitted to the pile. Actual experience with concrete piles and recent dynamic wave theory both indicate that normal cushioning, by lengthening the time that the ram is in contact with the head of the pile, may actually increase in some cases the delivered energy. Further, as the pile nears final toe elevation, the cushion usually is substantially compressed. Within practical limits, adequate cushioning does not reduce driving penetration.
2. To reduce driving stresses, use a heavy ram with a low impact velocity (short stroke) to obtain the desired driving energy rather than a light ram with a high impact velocity (large stroke). Driving stresses are proportional to the ram impact velocity.
3. Reduce the ram velocity or stroke during early driving when lower soil resistance is encountered. Anticipate soft driving or at the first sign of soft driving, reduce the ram velocity or stroke to avoid critical tensile stresses. This is very effective when driving long piles through very soft soil.
4. If predrilling or jetting is permitted, ensure that the pile toe is well seated with moderate soil resistance at the toe before the full driving energy is used.
5. Avoid jetting near or below the toe of the pile to avoid reducing toe resistance. In many sands, it is more economical to drive against a greater resistance with the larger hammer and/or higher blow counts, rather than to jet and drive simultaneously.
6. Insure that the driving head fits loosely around the pile head so that the pile may rotate easily within the driving head.
7. Insure that bearing piles are straight and not cambered because of uneven prestress or poor concrete placement during casting. High flexural stresses may result during driving of an initially

bent pile.

8. Insure that the head of the pile is square or perpendicular to the longitudinal axis of the pile, and that no reinforcing strands protrude from the head. Chamfer head edges and corners of pile head.
9. Use adequate spiral reinforcement throughout pile, particularly near the head and toe.
10. The prestress level should be adequate to prevent cracking during transport and handling and, in addition, the values should be adequate to resist reflected tensile stresses. This prestress level has been established empirically from the driving of hundreds of thousands of piles at about 700 to 1200 psi minimum effective prestress after losses. Very short piles have been installed with lower prestress (350 to 400 psi). Where bending in service is a requirement, higher values of prestress, up to 0.2 f'c and even higher have been used without difficulty.

5.9.6.7. Cutting off and Spalling

Precast piles are difficult to extend if driven below cut-off. If they stop short, they are cut off where necessary. Piles driven below cut off grade may be extended as shown on the plans or in a manner approved by the engineer.

Chips and minor spalls which may be created during handling, pitching and driving and which do not impair performance of the pile should be allowed; however, personnel working around concrete piles should not be under these piles as they are lifted or in the leaders, as chips and spalls can cause injury should they fall and strike a person.

5.9.7. Concrete Cylinder Piles

Good practice for square and octagonal prestressed concrete piles also applies to concrete cylinder piles; however, there are some special characteristics of these piles that should to be noted. The piles and the reinforcing of each pile is custom designed to suit of the application. The methods of handling and driving are specified for each project and should be closely followed.

A few larger rigs can handle a cylinder pile, leaders and hammer, installing the pile in a conventional manner. In other cases, the pile has been picked up and set in a drilled hole, or held in place by a template, while the hammer is raised and placed on the pile.

Figure 5-49 Driving 54" (1371.6 mm) Pile

As with other precast piles, special lifting gear is required to prevent overstressing the pile during the pick-up operation. A typical pick-up is illustrated in Figure 5-50. However, keep in mind that the pick-up gear for each pile should be individually designed and may differ from what is shown.

Figure 5-50 Typical Two-Point Lifting Arrangement

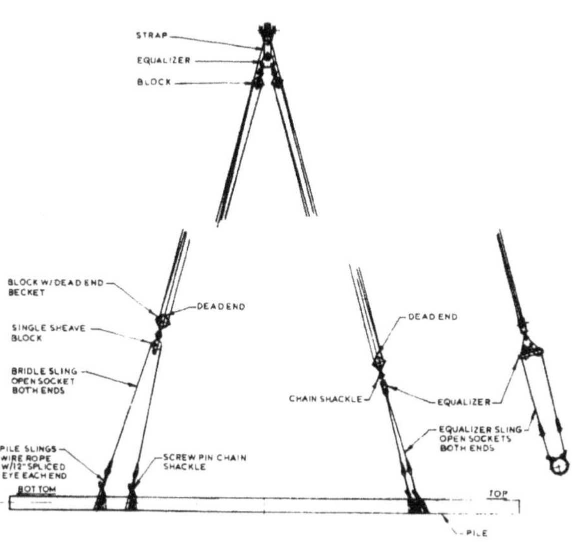

Figure 5-51 Driving Cylinder Piles

Driving is normally performed with large hammers. The values selected are those that will obtain the required penetration in the minimum time without pile damage. To minimize possible damage, a short hammer stroke may be used initially and later lengthened to obtain required final penetration. To reduce driving stresses, a laminated wood cushion is used on head of the pile in addition to the usual hammer cushion.

Figure 5-52 Setting 36" (914.4 mm) Pile in Pre-drilled hole

Cylinder piles are normally assembled from 16' (4.88 m) sections. If there is one shorter section on the end, the short section should be on the bottom of the pile. When driven open end, water or soil may enter the inside. If water and soil fill the inside of the pile and the helmet closes the head of the pile, continued driving can create an internal pressure sufficient to crack the pile. It is necessary to provide escape holes for water or stop and clean out the pile. In a related situation, the internal pressure of the water and soil may be increased by the water hammer effect resulting from driving the pile. The peak pressures resulting from decelerating the water column have been reduced by bubbling air inside the pile during driving.

Figure 5-53 Placing a 54" (1371.6 mm) Pile

Longitudinal cracks due to internal bursting pressure may occur with open-ended cylinder piles. When driving in extremely soft soils, the fluid pressure builds up and a hydraulic ram effect occurs. Providing vents in the walls of the cylinder pile or cleaning or pumping periodically can prevent this. This may also occur when the head is driven below water, in which case adequate venting should be provided in the driving head.

When driving open-ended precast piles in sands, a plug can form and exert a splitting action. This can be prevented during driving with a jet inside but the most practicable remedy appears to be the provision of adequate lateral steel in the form of spirals or ties.

Use of a solid tip will eliminate the splitting problems mentioned, but may not be compatible with other installation requirements. Internal jets can sometimes cause bursting, particularly in a hollow pile with closed toe and head.

If the jet breaks during driving, the water pressure in the hollow pile can cause excessive tensile stresses in the concrete. Vents can prevent this if they are located so as not to plug during driving.

5.9.8. Wood Piles

Woodpiles are relatively light and easily handled. They should be inspected in the cutting areas or at the treatment plant before loading for delivery to the project. While there is little risk of damage in handling woodpiles, a long pile can be broken through carelessness. Woodpiles are normally unloaded by rolling off on to racks or on to the ground. Short skids should be used to keep the piles from rolling under the truck. They can also be unloaded with forklifts. Care should be taken to secure the load while removing the side stakes. One method is pass two slings around the load, one end of each fastened to a stake pocket and the other ends to a tractor. The tractor is moved away from the load, the loose stakes are then removed and the tractor can back up allowing the piles to roll off.

After the piles are unloaded, they should be re-inspected. If they meet specifications, the toes and butts should be squared, headed and chamfered as necessary, the length measured and the length marked on the head. Untreated timber piles need not be peeled of bark unless required by the plans or specifications. Size of pile excludes the bark. Measurement can best be made by taking the circumference on an area without bark and dividing by π to determine the diameter.

Because it is impractical to lengthen woodpiles and because the cut-offs cannot be salvaged, it is particularly important to have the proper length piles available. This can be a problem. If the project is of any size, test piles are usually driven to determine the required length in each area. Actual pile lengths may still vary considerably. Piles of the proper length are usually dragged by tractor from the unloading area to within reach of the driver.

Woodpiles are easily damaged by obstructions or by overdriving. More woodpiles have been damaged by overdriving than have been damaged by all other installation operations. Long woodpiles will spring under hammer blows, if there is a large pile length out of the ground when significant driving resistance is encountered. This cannot only damage the pile but will cause excessive hammer maintenance. The pile should be supported at its mid-point with a sliding pile guide.

If there is a tendency for the piles to split or if there are checks or shakes, the pile can be reinforced, at the toe, head and damaged places with wire or steel straps, similar to those used on crates. Splitting or brooming of the head can be reduced by chamfering the head on a 45° bevel. A steel ring, made from 10-3/4" (273 mm) or 12-3/4" (324 mm) O.D. pipe, 1-1/2" (38.1 mm) to 2" (50.8 mm) long, and driven into the head of the pile will restrain the wood inside the ring from splitting. The rings can often be pried off with a peavie after driving and reused.

When handling and driving creosoted piles, precautions should be taken to avoid burns to the skin and eyes. Workers should wear rubber gloves and be adequately clothed, including long sleeved shirts. Creosote may drip or be squeezed from the pile during driving, particularly when the pile is freshly treated and people around the rig should protect their eyes with clear goggles. The cut-offs of creosoted piles are an equal problem. The crewmember cutting off piles should be provided with hand and eye protection. Do not burn the cut-offs; the smoke is very irritating to lungs and eyes, and many jurisdictions prohibit open burning.

5.10. Pile Points, Splices and Accessories

5.10.1. Introduction

Cast steel point reinforcement and fabricated splicers are common accessories in a well-engineered pile foundation. Points for H-Pile, Pipe, Sheet Piles, Precast Concrete, as well as Timber Piling aid in increasing penetration to dependable bearing as well as minimizing the possibility of damage to the pile generally caused by hard driving.

Pile splicers are routinely used to extend H-Pile, Pipe and Prestressed Concrete Piles to increasingly longer lengths. Both can speed pile installation, aid in improving structural support, and help to reduce costly construction time. Figure 5-54 shows various types of pile points and splicers.

Figure 5-54 Pile Points for Various Types pf Piling

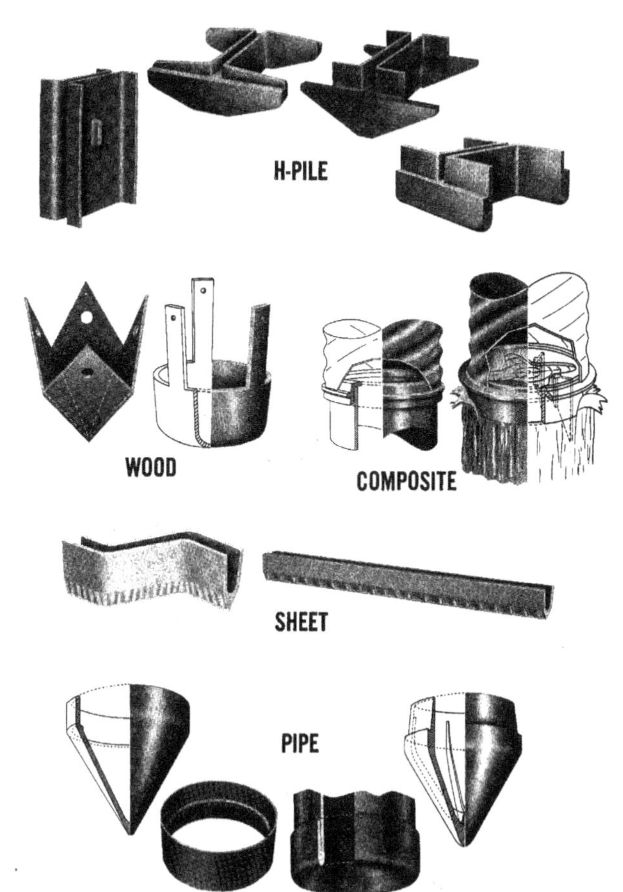

For the most part, accessories for piles are easy to attach. Specifications normally require that welders be certified, however, full certification for all position welding may not be essential. As a result, a capable laborer may install certain points and/or splices.

With increasing heavier loading on pile structures installed in difficult soils combined with the specifying of longer pile lengths the use of protective points and splice accessories are not only cost effective but may also be necessary to insure a properly designed and constructed pile foundation.

5.10.2. H-Piles

5.10.2.1. Points

Driving steel H-piles in soils of moderate grain size is generally easy because of the non-displacement character of the pile. However, problems can arise when driving H-piles through man made fills, very dense gravel or deposits containing rock or even boulders. If left unprotected under these conditions, the pile tip may easily deform to an unacceptable extent and separation of the flanges and web may occur, as shown in Figure 5-55. Point attachments can help prevent these problems. Cast steel point reinforcement for H piles make it practical to drive through compacted fills, medium dense gravels and to some extent penetrate boulder riverbeds. Point attachments are also desirable for end bearing piles on rock particularly on sloping rock surfaces where a toehold on rock is necessary. One-piece points may be attached at the site with a minimum of labor and without rotating the pile.

Figure 5-55 H-Piles Damaged by Underground Obstructions

It is recommended that for a given set of subsurface conditions, pile points be considered equivalent, if they are manufactured from similar materials and by similar fabrication techniques. Minor variations in configuration should be given minimum importance, except in specific subsurface conditions where a certain shape would give a definite advantage.

Several patented driving points are available as shown in Figure 5-56. These points are attached to H-piles with fillet welds along the outside of each flange. Pile points manufactured from cast steel (ASTM A 27) are recommended because of their strength and durability.

Figure 5-56 Configurations of H-Pile Points

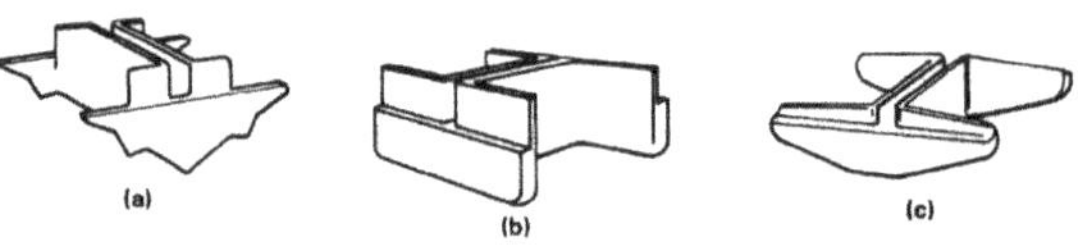

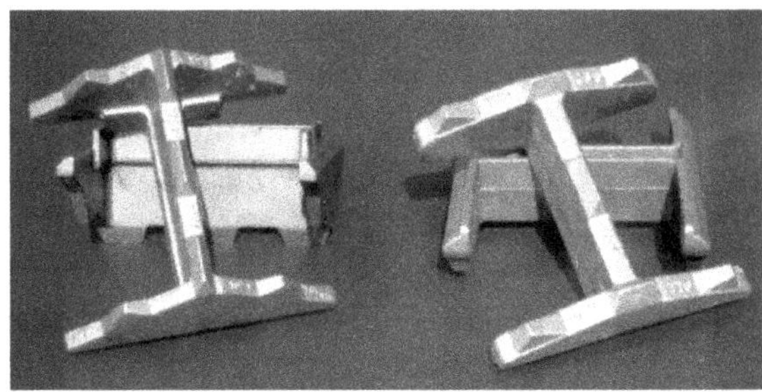

Cast steel points are available in different configurations for differing conditions. In some cases, a slim pattern may penetrate better. In other situations, metal in a more compact shape, while having a broader area, may be better for end bearing piles driven onto rock. This is especially true where driving has been relatively easy through soft soil above the support strata. Some points are designed to increase rock contact to twice the area of the H, halving the unit pressure on rock. Points with integrally cast small "teeth" will help with driving through obstructions and should be utilized when attempting to drive into rock that is sloped in relation to the axis of the pile. The small teeth are designed to cut into the hard strata and aid in increasing the actual bearing capacity of the pile.

Some manufacturers fabricate points from ASTM A 36 rolled steel plates. The plates are cut and welded to form the required shape. Welds used to form these shapes are known to break under hard driving conditions and hence this type of fabricated point is not recommended. H-pile tips reinforced with plates welded to flanges and webs are not recommended because the reinforcement provides neither protection nor increased strength at the critical area of the flange to web connection. In addition, plate or plate and angle tip reinforcement requires handling four or six small pieces, turning the pile one to three times and making up to as much as 270 in. of weld. Much of this weld is in difficult and hard to reach areas between the flanges of the H.

Figure 5-57 Welded Pile Points

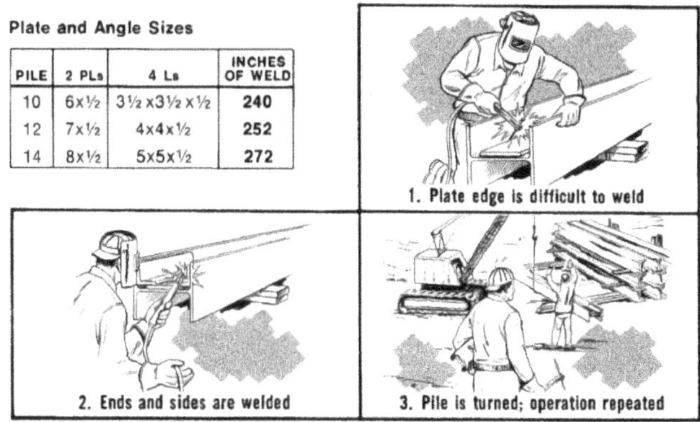

Plate and Angle Sizes

PILE	2 PLs	4 Ls	INCHES OF WELD
10	6x½	3½x3½x½	240
12	7x½	4x4x½	252
14	8x½	5x5x½	272

Plate Sizes

PILE	2 PLs	2 PLs	INCHES OF WELD
10	½x8x9	½x9x10	135
12	½x8½x10	⅝x11x12	154
14	½x9x11½	¾x13x14	182

A. Plate is welded to web of H-Pile
B. Pile turned; flange corners coped
C. Plate added to flange; pile turned
D. Second plate attached to web
E. Crane turns pile for third time; final plate is welded to pile

For most field welding E-60 or E-70 low hydrogen rod is favored. For semi-automatic welding AWS E 70T-G (Lincoln NR 211) flux-cored wire seems suitable for "out-of-position" and ordinary welding.

This one-piece point is made to fit all weights of H in each nominal size. It attaches with a groove weld along the outside of each flange. The point is slipped on the pile and a vertical weld made along each flange. This should be full width of the flange. Welding only the ends overstresses the vulnerable corners). Preferred points will have a pre-bevel on the casting. This assures an even space for a proper weld. Should this pre-bevel not exist it may be necessary to field burn a 45-degree bevel along the outside of each flange of the H.

When preparing the piles for attachment of the point stack the H piles so the ends are accessible with the flanges vertical. (Stack the H-sections with the webs horizontal and the ends accessible to the welder. Alternate piles can be extended beyond the stack in each direction to achieve this.) The cast steel point can then be attached by holding it tight against the H and then making a simple grove weld along each side. By stacking the piles in this manner both sides may be welded, without having to turn the pile. If the cast steel point is not pre-beveled, an equal leg angle can be made as a guide for a straight cut.[14]

A good weld is essential to hold the point reinforcement in position when it strikes tough obstructions or sloping rock at one edge only. Beveling across each flange of the H assures a base for a good weld and is essential for attachment of points. For field welding of cast-steel points use of AWS E60 or E70 welding rod has been found satisfactory.

A definite plan should be made for the stacking of these piles with the pre-attached points. To minimize the need for additional handling, one should stack these piles out of the immediate work area, while still keeping them within reaching distance of the crane.[15]

It is customary to burn a small hole in the flange of the H, about one-fourth of the distance from the top, for a shackle connection for handling. This is by far the safest and quickest means of picking up the H.

5.10.2.2. Splices

Traditionally H-pile splices were made by riveting or bolting. They are now routinely made by full penetration butt-welding or with patented splicers.

Splicers are generally manufactured as a strong unit prefabricated from carefully formed structural plate and accurately spaced to the thickness of web for each weight of H-pile by a heavily welded spacer. Made as directed, a splice makes a member as strong as an unspliced pile. Ends of the splicer are flared in two directions to a pre-determined shape that guides the end of the H into alignment with the next section. Use of the splicer makes the connection quicker and safer; once the H enters the splicer, it is held against dangerous slippage from movement of the crane or a gust of wind. For moderate additions to vertical piles, in some cases the splicer has been used to hold the added length until the welds could be made at the other edges and across the flanges. This releases the driver, or auxiliary-lifting device, at once to continue essential work elsewhere. The close fit holds the pile in alignment. A loose fit would permit wobble and require that the H extensive be held by the crane until it is securely welded. In addition no time is lost welding guides to the H and using wedges for alignment as would be required when splicing an H-Pile without a "factory splice" as the manufactured splice performs this function.

During driving and in service, splices should develop the required strength in compression, bending, tension, shear and torsion at the point of splice. The current AASHTO Bridge Specifications require that a splice must provide the full strength of a pile (for a compression pile full strength in bending is required).

The splicer can be attached in advance to the length of pile to be added. This should be put on the driven length. The attached fabricated splicer will aid in positioning the added length and hold it firmly against dangerous movement. This is especially advantageous when having to drive batter piles or when working off a barge.

A heavy spacer bar is welded to hold the accurately shaped members at about 1/16 in. tolerance over the thickness of any weight of H. This close fit requires that burrs from cutting and other protrusions be removed.

For installation a 7/8 in. x 1-1/8 in. notch is cut in one length only. (Note that the spacer is offset so that only one slot is needed. Be sure that the splicer is turned so it centers on the joint). The notch is cut into the web of the driven section of pile and the splicer is slipped on the pile. Short welds are then made to the flanges near the corners of the splicer. The top section must have flanges chamfered to achieve effective welding. The length of pile to be added is positioned and held while welds across flanges are made. Proprietary H-pile splicers are fabricated from ASTM A 36 steel. This type of splicer has been tested in the laboratory and the results have shown that it provides full strength in bending (as required by the AASHTO Bridge Specifications).

The outside edge of each flange of the H to be added should be beveled to half the thickness of the steel. Irregularities in the contact edge will have a little space for weld metal to penetrate full depth. Make a short fillet weld at each corner of the splicer to the inside of the flange of the H. Weld along the full width of each flange with a minimum 5/16 in. groove weld. Resume driving as soon as the weld cools to "touchable" temperature.

As shown in the sketches, a good weld across each flange and a short weld near each corner of the splicer to the flanges completes the joint. Splices can be placed on the driven length if more convenient for the project.

Tack weld each flange of the sleeve to the pile with a 2-1/2 in. long 5/16 in. fillet weld. This can be most effectively placed just below the flare of the splicer. (The closer it is to the end the greater strength it will have for holding an added length of pile while welding is completed. There is no need for welds across the ends of the splicer.) The splice can be placed on either the driven H-pile or the section to be added. If placed on the driven length, the four short welds can be made at once. If placed on an H on the ground, the piece should be turned or the welds underneath made after the crane raises the pile.

After the next section is set into the splicer and forced to tight contact, a weld is made across the outside of the flanges of the H. Short welds are made near the remaining corners of the splicer.

For any method of splicing, the flanges of the upper section of H should be beveled for welding.

Full penetration butt welds of the H section may be used. Butt welds usually require back-up plates or back gouging. When welding is performed with the pile in the leads, special precautions should be taken to assure that the upper section does not disengage from the driven length, developing a hazardous condition. Piles usually do not line up at a joint due to tolerances and bends so special wedging devices may be needed for alignment prior to the beginning of welding.

Whenever piles are spliced it is essential that the sections be accurately aligned before welding. The pile sections should be checked for straightness on each axis by a straight edge before welding.

If there is trouble in getting the splicer on the pile, check for battered ends and for burrs from cutting or burning. Many splices are made to very close clearance, about 1/16 in. This is to assure a full strength splice and to help hold the added section of H pile securely until it can conveniently be welded. Foreign H piles sometimes have more variations than domestic. A "T" made by cutting a short piece of H-pile can be used in driving the splicer on with a sledge.

Close inspection is essential for plates on webs and especially for all welds across the flanges.

Welding in place without a splice requires exceptional attention, as it is difficult to exactly align and hold the pile sections. There is always a tendency to rush the weld to continue driving or to release a holding crane for other work. A manufactured splicer helps to correct for misalignment or flanges out of square and holds the pile so welding time is minimized.

When factory splices are not utilized, splices may be made by full penetration butt welds across both flanges and the web. To attain full penetration welds, plates should be used to back up the weld or the weld should be back gouged to good metal and filled in. For welding, flanges should be accurately aligned with plates and wedges.

Plates may be welded across the joint at the pile ends, either inside or outside the flanges, and on the web. Flange plates should be 1 in. narrower than the flange space available and the web plates the maximum width that can be welded between the flanges. Each plate should be fillet welded along what should be the vertical edge. Ends of the flanges of the upper length of pile should be beveled and a penetration butt weld made across them. Where plates are used they can sometimes be attached in advance to the upper length of a pile in a way to help position and hold it on the driven length.

A bevel is cut to one half the metal thickness on the section to be added (do not bevel where the hammer will strike). A minimum 5/16 in. groove weld is made along the full contact surface.[16]

5.10.3. Pile Piles

Most pipe piles are installed closed-end, using a sufficiently thick wall pipe that driving can be done on the top of the pipe. Occasionally the wall thickness is as thin as 0.188 in. wall but for moderate driving 0.250 in. or even thicker wall may be necessary. Problems during installation of closed end piles arise when driving through materials containing obstructions. In this case piles may deflect and deviate from their design alignment to an unacceptable extent. In the case of driving open-end pipe piles through very dense materials, the tip of the pile may be deformed. As a result, points attached on closed and open- end piles are used to reduce the possibilities of damage and excessive deflection.

Pipe piles frequently will come to the project with cutting shoes, conical points or flat-plate closures already attached. Sometimes a splicer sleeve will also be welded on. Local work rules may sometimes require that such attachments be made at the project site. If pipe arrives at the jobsite with points and/or splices already attached, assure required material certificates or test reports for all materials (welding as well as the accessory) are supplied.

5.10.3.1. Points

When pipe piles are installed with a closed end, 1/2" to 3/4" inch thick flat plates are sometimes used as a form of tip protection. The suggested diameter of this plate is generally a maximum of 1/2" larger than the actual O.D. of the pipe. Flat plates usually are permitted to extend 1/4" outside the periphery of the pipe and are welded for attachment. When plates are required to be kept flush accurate beveling of the pipe and skilful welding are required.

When steel plate is used around the pipe as reinforcement it should be heat-treated steel. Consult experienced welding authorities on how this should be welded and inspected as driving produces shear on the welds.

End closures may be flat plates or conical points. Generally, conical points have sixty degree configurations and are available with either an inside flange or outside flange. Examples of pipe pile points are shown in Figure 5-58 and Figure 5-59.

Figure 5-58 Open End Pipe Pile Point

Figure 5-59 Closed End Pipe Pile Point

Conical points generally cost more than plate type protection. Conical points should be cast steel meeting the requirements of ASTM a-27 65/35; for tougher conditions A-148 80/40 is preferred. Plates may be A-36 steel and should be thick enough to resist all driving stresses.

Inside-flanged conical points for pipe attach with just a simple weld as driving is in compression. For advance preparation, the pipe is laid on supports and rolled slowly while hand or machine welding is conveniently done.

Outside-flanged conical points can be made for a drive fit so no welding is required. Minor water leaks can be controlled, by swabbing roofing mastic around the joint. Both inside and outside flange types have a level ledge on which the pipe seats and drives.

It should be noted that the inside type have a limitation on the pipe thickness they can accommodate, normally 1/2 inch for popular sizes. When a large quantity is required, and should time be permitting a pattern can be made for almost any size. In the case of the outside flange type, these shoes can be welded or tapered for a drive fit. Pipe wall thickness of outside flange type pipe is generally not of concern, as heavier wall pipe will increase in wall thickness inward, while the outside diameter will remain constant. Points should be inside ribbed for strength with no metal less than 1/2 inch. This thickness should be increased for larger diameter pipes.

Weld beads on spiral (or other) pipe that protrude outside the surface should be ground smooth at ends where drive-fit end-closures or splicers might be adversely affected. (The pipe manufacturers will remove protrusions, if requested).

Pipe in any nominal size has the same outside diameter. Outside flanged points and splicers can be made to one size and splicers can be made to one size and tapered for a drive-fit for all wall thicknesses. Such points and splicers can be attached by friction driving the pipe into them.

Inside diameter of pipe varies with the wall thickness. Fittings are made to match and accommodate about a 1/2 in. wall thickness; they are welded for attachment. This strengthens the pipe against stretching and splitting from driving over the fitting. The inside flange type does not extend beyond the pipe; this may be an advantage in some soils, especially permafrost where soils do not readily deform after driving.

5.10.3.2. Cutting Shoes

Many times, pipe piles are installed open end. These open-end pipe piles are especially susceptible to end damage.

Open-end pipes can be driven through quite difficult strata if the pipe is thick walled and/or an adequate cutting shoe protects the end. Shoes may be a circumferential band of steel around the end of the pipe, a short length of thick circular plate attached to the pile end or a strong cast-alloy steel shoe

When installing open end piles in dense or to rock, the use of cutting shoes will help protect the piles and may make it possible to use thinner wall pipe. Cutting shoes made from cast steel have with a ridge to allow for pile shoe bearing.

Structural plate for cutting shoes is generally specified as 40/50 carbon steel. This will provide a hard cutting edge, however it requires special attention when welding to the pipe and annealing or treatment of the weld area to ensure a strong joint. It can be attached as an additional length of the pipe or as circular reinforcement of the pipe. While these shoes do not have to be watertight, the welds should be strong enough to prevent tilting of the shoe should an obstruction be encountered. Obviously a skilled certified welder is required for the attachment of any "cutting shoe".

5.10.3.3. Splices

Pipe piles can be conveniently be extended with a splice. Manufactured splices are designed either with a taper for a drive fit without welding so no advance preparation is required or a slip on splice requiring welding. If water is a problem, roofing mastic swabbed around the joint probably should be adequate to seal friction splices.

For piles where the splice is not significantly embedded into firm soil or where uplift is expected the splice can be supplied without a taper. It can be slipped on a pile section while horizontal and welded in a convenient position as the pipe is rolled. This joint has substantial lateral resistance and tension capacity. Driving can be done on the splice; the next section is set into it and a quick down hand weld made to complete the joint. Do not bevel the pipe with the splice as it is designed to bear on a square ledge.

The circular shape of pipe makes it practical to use a tapered compression splicer without welding. The design incorporates a "ledge" at the center makes a "drive stopper" at the interface. Newly developed foundry techniques permit the casting of this splicer to a water resistant connection. During the manufacturing process an accurate length is cut, rolled to a circle, flash welded, then stretched to the required size.

When attaching the no weld pipe splicer, the splice is set on a driven length and leveled by tapping with a sledge. The pile driver then picks up the next length of piling and sets it on top of now present splice. Driving continues with very little delay. It is squared with the axis of the pipe while an added section of pile is picked up. As the pile is driven the pipe ends are swaged into the tapered splicer, making a tight friction joint that resists lateral bending of the pile. Filet welding of this splice provides increased bending moment capacity.

Splices for pipe piles may be full-penetration butt welds. Where attachment is made after a length of pile is driven, only the lower end of the upper section should be beveled so hammer contact is on the full area of the steel. A back-up ring, with protruding knock-off pins is available to space and back up the welds.

Full penetration butt welds or fillet welds shown are commonly used for splicing pipe piles.

In some cases a splicer has been used as a driving head on thin-wall pipe piles. For 0.188 in. (thin) wall pipe, for example, the regular tapered sleeve can be driven on with a sledge, compressing the pipe end sufficiently to hold the sleeve in place while the pile is lifted into the leads. This procedure should be used with great care, however, as a falling sleeve could be injurious. Use of a splicer as a driving head has been found in many cases to reduce pipe pile damage during driving.

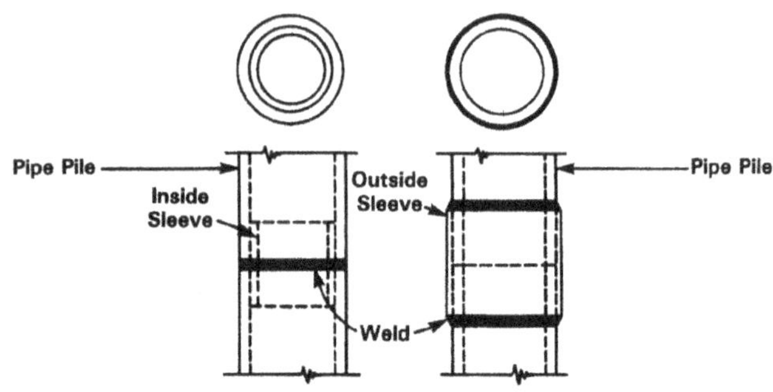

Where full uplift resistance is needed, and for piles in freestanding water, some welding at the joint may be desirable. Some splicers can be supplied without the taper. It can then be slipped on a section of pipe in the shop or on the project and welded in convenient position before it is placed into the leads. Driving can be done on the splice. As sections are added a simple down hand weld can be made in the top of the splicer with very little delay to the driver.

Protrusions on pipe, such as spiral welds or burrs should be ground smooth for about 3 in. from the end of pipe where outside flanged points or splicers are to be used as a friction connection. The pipe can be ordered with this done at the mill. Protrusions can be quickly removed on the project with a power grinder.

5.10.4. Timber Piles

5.10.4.1. Points and Boots

The potential problems associated with driving timber piles are splitting and brooming of the pile tip and driving end, splitting or bowing of the pile body, and pile breakage during driving. To avoid some of these problems, protective attachments at the pile tip and pile head can be used.

The trend toward heavier hammers and heavier design loading may mean that timber should be subjected to damaging blows when the pile reaches obstructions or end bearing. As a result, timber piles can be protected by a metal boot or point. The point shown in the picture covers the entire pile tip without the need for trimming the pile tip.

Tip protection by a boot or shaped point are recommended with banding near the top to alleviate splitting. The boot is preferred as it attaches quickly and utilizes the full area of the wood for load carrying. Heavy steel banding, 1-1/4 in. wide and .031 in. thick at 1.5 to 2 ft. below the top of the pile is recommended.

Attachment of points (as opposed to a "boot") for timber piles requires that the tip be accurately cut to shape. Otherwise the point has limited value, as bearing is uneven. If not straight, it may deflect the pile.[17]

Both boots and points are available in several sizes. As a result, a size can generally be selected that can be placed on the pile with little trimming of the pile tip.

5.10.4.2. Splices

Although timber piles have been spliced, the practice is not encouraged. Timber pile splices are generally undesirable as it is virtually impossible to develop the full bending strength of the piling through simple splices. However, there are obviously project conditions encountered where the need for splicing will arise.

When timber piles have been spliced prior to driving, they are generally bolted on with the use of timbers or steel plate. Bolt holes should be drilled at least 1/16 in. less than bolt diameter and through the entire assembly so they will fit tightly. If the splice should be above permanent ground water special provision should be made for wood preservation. This type of splice is seldom used.

A more common splice for timber is a pipe with interior wedges, installed after a length of timber is driven. The pile head is shaped to the pipe splice size and an additional length added. Again, provision should be made for preservation of the wood if the cut ends are above permanent ground water level.

Timber piles require care in splicing, partly because of the difference in size between the butt of the driven section and the tip of the added section. Bolted timber lagging requires accurate shaping of both sections of pile and tight attachment.

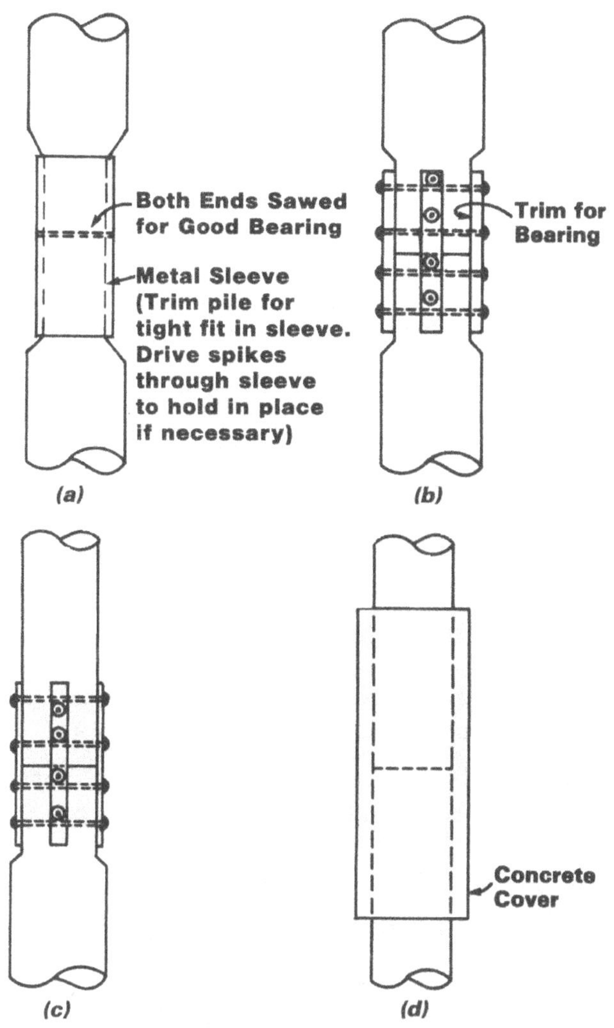

Splicing timber piles should be avoided if practical. Splicing can be done but is expensive and time consuming. A better solution to extension might be a composite pile.

5.10.5. Concrete Piles

Points and splices for concrete piles come in a wide variety of configurations. One thing that should be kept in mind is that precast and prestressed concrete piles are spliced and pointed in different ways; a method that would work with precast concrete piles may not be applicable to prestressed piles, and vice versa. The type of point or splice used also depends on the locality; points and splices used in one area of the world may not be available in another.

5.10.5.1. Points

The tips of concrete piles may be crushed in compression in hard driving. For hard driving conditions, or for end bearing on rock, special steel points can be attached to the tip. A common type of tip protection is a structural H section at the toe of the pile.[18] Prestressed concrete piles can have the toe protected with an H-Pile Point as a means for penetrating difficult strata or extending the length of the pile.[19] Several methods have proved satisfactory for attaching the extension. An example is shown in Figure 5-60.

Figure 5-60 H-Pile Points on Concrete Piles

Cast iron or steel shoes or the "Oslo Point" have also been used frequently. The characteristics of the "Oslo Point" are such that it can be chiseled into any type of rock to ensure proper seating. This detail is shown in Figure 5-61. These points are especially useful with pretensioned spun concrete piles.

Figure 5-61 Cast Steel Shoe and Oslo Point

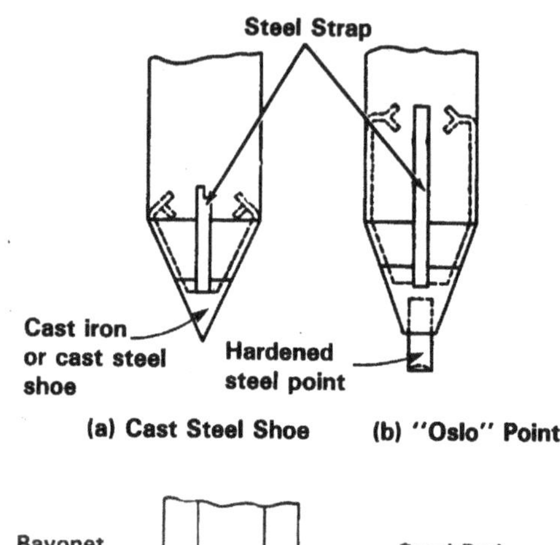

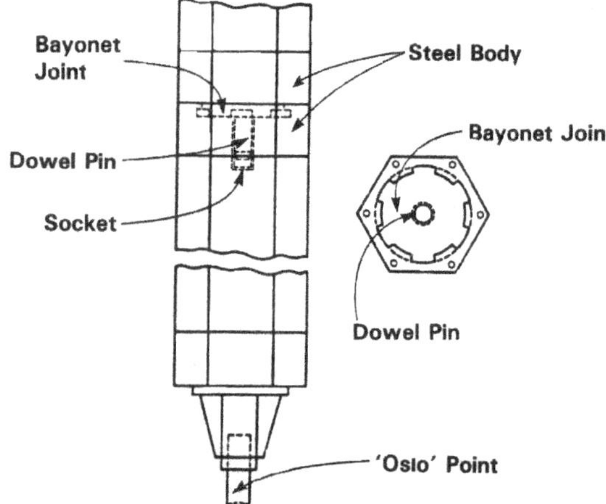

Other cast steel point protection is available. Most are of European design. Most of these devices are planned for attachment in the concrete casting bed. Attaching the point may be complicated due to having to work around the prestressing strands, all of which are continuous throughout the length of the casting bed.

All point attachments to concrete piles are attached during casting of the piles and not in field.

5.10.5.2. Splices

Splicing is called for when pile lengths required are too long for trucking or driving in one piece with the available equipment, or if there is a headroom restriction. A pile may have one or more splices. The splice should be capable of resisting stresses induced by driving, and service loads and conditions. Tensile stresses are highest when resistance to driving is low and tensile waves are reflected back up the pile. Moment capacity is particularly important in high seismic zones, in piles possibly subject to impact, and in difficult driving conditions. Not all splices will develop tension and/or moment capacity.

Splices can be generally categorized as follows:

1. Welded: steel plates with attached rebar cast into the pile ends are field welded together. This method is rarely used in the US and Canada as it idles expensive field labor and equipment while the weld is performed.
2. Dowel: holes are cast or field drilled into the top of the bottom pile section. The top pile segment with rebars dowels protruding 1-2 meters from the end is guided into place and the grout or epoxy in the holes cures around the dowels. Although material cost is low, this method is infrequently used in the US and Canada as it requires that the pile top segment be held in place until the epoxy or grout cures. It is useful for "build-ups" -- ex-tending a driven pile to required cut-off elevation with a short precast or cast-in-place section.
3. Sleeve: a "can" of steel 2-3 meters long with a stop in the middle. It slides over the driven section and the top pile section is then lowered into it and driving resumes. The splice is relatively economical and easy to use, but has very little bending and no tensile capacity.
4. Connector ring: a short length of pipe is cast onto each end of the usually square pile. A pipe pile splice is set on the driven section and the top pile segment is set into the sleeve. Intermittent fillet welds can be used to increase the limited tensile strength of the joint. This splice has been used extensively on Bruns piles in the New Orleans area; these splices in the mould are shown in Figure 5-63 and in the field in Figure 5-64.
5. Mechanical: steel plates or castings with attached rebar anchors approximately 2 meters long are precast into the pile ends. Steel bars or "wedges" are inserted into the mating surfaces when the top pile segment is aligned on the driven section. This method is fast and extensively used in the US and Canada for prestressed piles since the 1970's. Proprietary systems are predominant that utilize high-strength steel bars to lock machined plates together. The Sure-Lock mechanical splice is made for all sizes of piling (including some cylindrical piles) and can equal the pile in bending and tension capacities. This splice is shown in Figure 5-65.
6. Post-tensioned: used in large diameter

cylindrical piles. Multiple strand holes are cast longitudinally through the center of the concrete wall. After curing, the segments are aligned and strands are post-tensioned to hold the sections in one long pile section.

7. Wedge: This method is generally used with precast piles. The methods usually utilize wedges driven on plates attached to the pile ends. This puts the connection at the corners to develop resistance at the extreme fibers. This is shown in Figure 5-66.
8. Pinned: These use a pinned connection to secure the splice. These are generally not used in North America.

Most of these types are illustrated in Figure 5-62. Variations in construction with actual splices may be encountered.

Figure 5-62 Commonly Used Prestressed Concrete Pile Splices

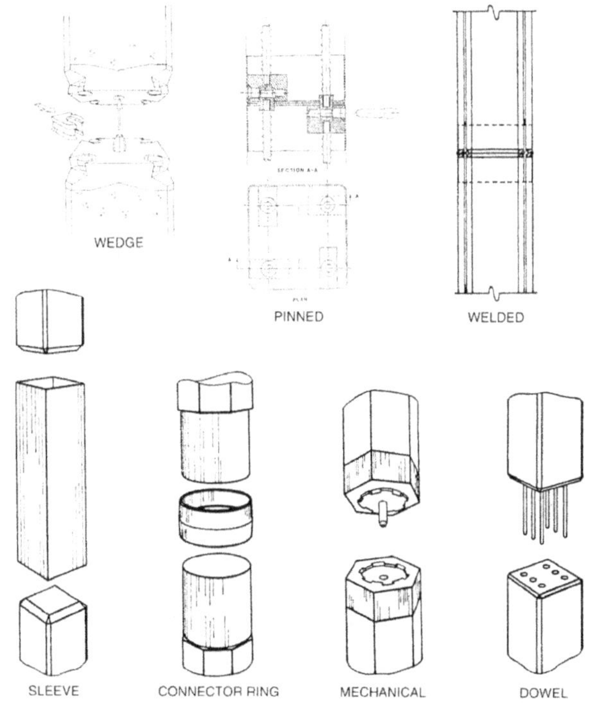

Figure 5-63 Bruns Piles in the Casting Process

Figure 5-64 Finished Bruns Piles

Figure 5-65 Sure-Lock Splice

24" oct. Sure-Lock Pile Splice – inserting the top segment Male into bottom segment Female splice half.

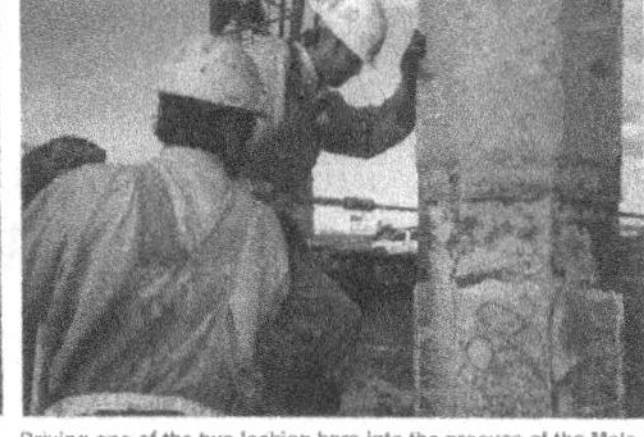

Driving one of the two locking bars into the grooves of the Male & Female plates with an air hammer to form a circular shear key.

Figure 5-66 Wedge Type Splice

Figure 5-67 Sheet Pile Points

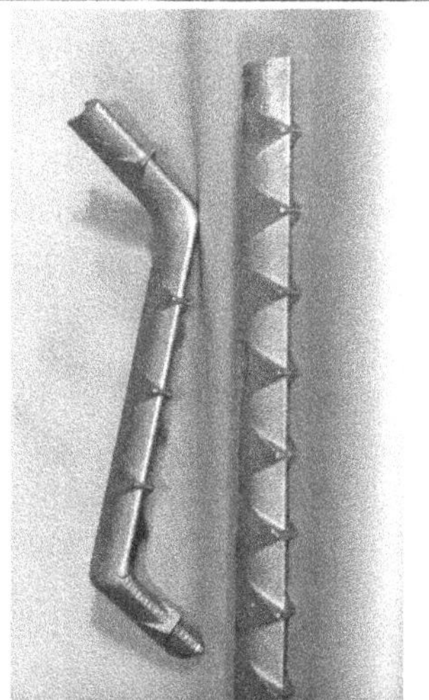

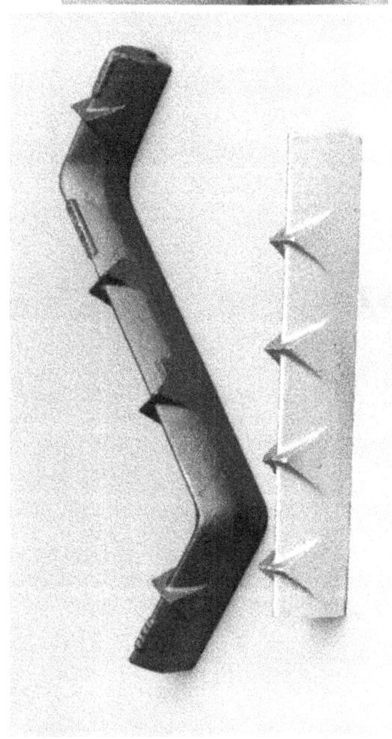

The ability of a splice to develop the strength of the pile, or a reasonable percentage of that strength, depends on close tolerance and proper procedures in making the splice. Careless workmanship or improper field procedures can result in significant deviations from strength and behavior levels desired. Failure of some could occur directly in the joint, while in other cases, failure could occur at the dowels anchoring the splice to the piles. In some splice systems, failure would occur completely outside of the spliced region.

5.10.6. Sheet Piling Points

Cast steel tip protection is increasingly used for steel sheet piles. It is available in straight bars for flat and shallow-web (Larsen and Shallow Arch sections) sheets and for cutting as needed for any "Z" configuration. For more commonly specified Z sections, (e.g. PZ-22 and PZ-27) points are readily available in one piece for quick and easy attachment. They require only intermittent welds that will hold the point firmly in place. Examples of these are shown in Figure 5-67.

Although it is not recommended, steel sheet piles have been known to be extended by welding. As of this date, a satisfactory splicer accessory for sheet piling is not known to be available.

5.10.7. Composite Pile Splices

Although not widely used, the most common type of composite pile is a timber/corrugated shell pile. Timber is used for the most of the pile length, and then topped with a corrugated shell, which is later filled with concrete. A proprietary transition splice is available for use in this application. Its upper end welds to the shell and lower "barbed end" is driven into the top of the timber pile, as shown in Figure 5-68.

Figure 5-68 Timber/Corrugated Shell Splice

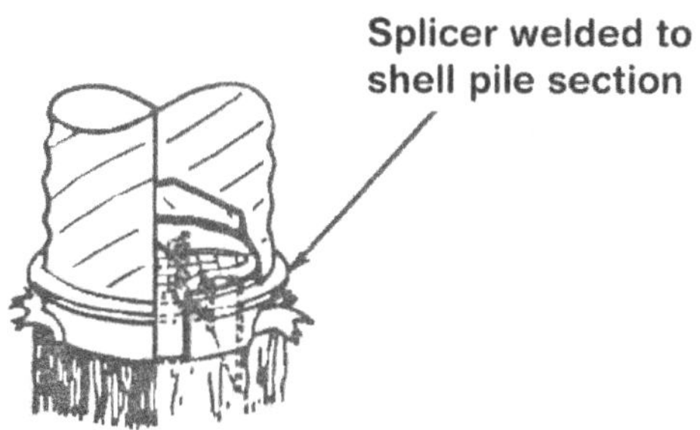

Similar to the timber/corrugated shell "combination" pile, a steel pipe/corrugated shell pile transition accessory for splicing pipe piles to shell piles is also available. The splicer is welded to the shell pile section and then driven fit or welded to the pipe pile section. A steel Monotube can be spliced to a steel pipe pile by welding a transition coupling to both sections. This is shown in Figure 5-69.

Figure 5-69 Steel Pile/Corrugated Shell Splice

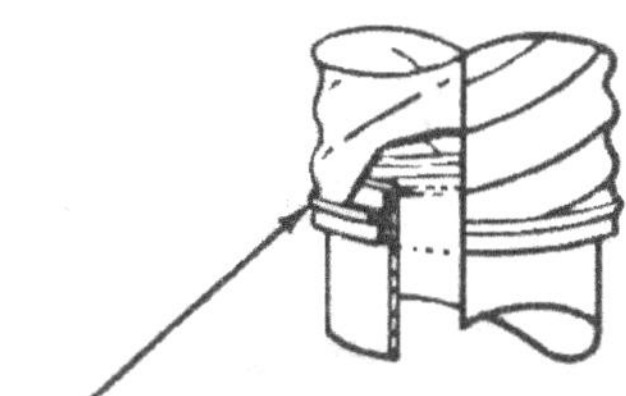

Simple splicers can provide for connection from low-cost untreated timber piles to concrete-filled shell for use above the water table. The timber pile is driven to ground level. A shaped casting with "wedges" for driving into the wood is welded to uniform diameter shell. This is pulled up on a pipe as a mandrel and set on the timber pile. It is then driven as a unit to the desired depth. An expanding mandrel is not required. The pipe mandrel stays attached to the hammer and is removed. The shell is then filled with concrete when convenient.

Corrugated shell piles have a depth limit around 100 feet (the maximum length that can be installed with a practical height of leads.) Pipe pile sections can be driven first and spliced for penetration to any depth. A pipe-to-shell transition splicer provides for changing from pipe to lower cost shell. The splicer is gas welded to the shell and can have a drive-fit connection to the pipe for a quick, no-weld extension.

5.10.8. Conclusion

Points and splicers are necessary accessories to pile installation. Readily available point reinforcement will assist in pile penetration and a fully dependable pile. These accessories should be considered for every driven pile and specified for use in soils where possible obstructions may be encountered. When piles go through soft materials then contact hard rock it is essential that tip reinforcement be required. Remember to specify exactly what is acceptable to allow the contractor to provide the best bid.

The method of splicing piles should always be specified. For almost every installation fabricated splicers, installed as recommended by the supplier, should be completely satisfactory. Detail what should be acceptable and the welding required - if any - so the contractor can accurately estimate his cost.

5.11. Permafrost Areas

Frozen soil often has high strength similar to rock. Installation of piles with impact hammers or even vibratory drivers may be impractical and alternative installation methods may be necessary.

The responsibilities of the persons installing the piles are:

(1) To obtain the necessary penetration.

(2) To use methods that will permit the soil around the pile to refreeze, quickly. The pile capacity (both resistance to downward loads and resistance to frost heave) depends

on support from the surrounding soil that, in turn, requires that the soil adhere (adfreeze) to the pile. Until this freeze develops, only a fraction of the pile capacity is developed.

(3) To minimize disturbance to the permafrost, the ground cover, and the site in general. Any disturbance will upset the heat balance in the ground possibly causing (unless remedial measure are taken) thaw and refreezing with consequent settlement, followed by heave, which will affect buildings, pavements, and utilities. It may take years for the heat balance to be restored if significant site disturbance occurs.

5.11.1. Site Preparation

The site should be prepared to promote equipment mobility and later access in case additional work is necessary.

Disturbance of the ground surface during the construction season frequently increases the depth of thaw. This causes the surface to loose trafficability. The ground surface should be covered with a blanket of gravel or broken stone and the work performed from on top of this blanket. Access to all of the equipment should be limited to the immediate vicinity of the area where the installation is to occur.

5.11.2. Method of Installation

Piles in permafrost areas are not driven to resistance, but rather to some specified toe elevation. Refreeze of the soil around the pile must occur as quickly as practical to develop the required pile capacity. This capacity comes from adherence of the soil to the pile and must support the downward loads and have sufficient resistance to frost heave. Methods of installing piles that can accomplish adequate pile capacity is by drilling or by steam jetting.

5.11.2.1. Drilling

This method consists of dropping pile sections into a bored hole that is made with an auger or with a rotary drill and slurry.

Truck mounted augers are usually applicable in silts, clays, and some sands. A rotary drill or local prethawing may be required in coarse sands and in soils containing cobbles.

After placing the pile sections, the excavation is backfilled with soil-water slurry that bonds to the pile on freezing. The slurry should be placed in the hole at near freezing temperature to minimize refreezing time.

5.11.2.2. Steam Jetting

This method consists of thawing a vertical shaft of soil by a steam jet, pressing or driving the pile into the shaft of thawed soil, and allowing the thawed soil to refreeze.

A 0.75" to 1.25" (19 mm to 32 mm) pipe, open or slightly crimped at the point to provide better jetting action is used with steam pressures of 100 psi (6.9 bar) to 200 psi (13.8 bar). A chisel bit may facilitate penetration in gravel. Water should be added to the jetted shaft in dry soil to promote thawing. If the thawed shaft of material is of small diameter, then the steam jets should be worked alongside the pile as the pile is advanced down the thawed shaft. The thawed volume of soil should be minimized to reduce refreezing time.

5.11.2.3. Seating the Pile

The pile toe should be seated firmly on the bottom of the excavation or thawed shaft of soil to obtain partial capacity and to be able to support the weight of the construction while the refreeze is developing between the soil and the pile.

5.11.2.4. Cautions

(1) In steam jetting method, carefully limit the volume of thawed zone in order to minimize time for refreezing.

(2) In steam jetting or dry auger methods, slurry should be placed into hole at as near freezing temperature as feasible in order to minimize time for refreezing.

(3) In steam jetting or dry auger methods, set the pile firmly on bottom of hole to obtain partial capacity to support partial weight of construction while

waiting for adfreeze to develop.

5.12. Sheet Pile Installation

5.12.1. Overview

Sheet piles are low displacement piles. The basic mechanics of installation and extraction are the same as those of H-piles. There are some important differences, however:

- The effect of the interlocks, which adds to the driving resistance; and
- The fact that, sheet piles are generally not used as bearing piles; thus, the determination of actual bearing capacity is not necessary and the monitoring of the driving equipment's performance is not so rigorous.

Without bearing capacity to consider, the basic objective of the contractor driving sheet piles is twofold: to drive the sheet piles in the geometrical arrangement called for, and to drive them properly to the desired depth. The implementation of these simple objectives is the complex and multi-step process described below.

To insure successful sheet pile installation, the first thing to do is to conduct a complete site soils investigation. This should indicate the location of underground obstacles, such as boulders, rock outcroppings, and rock slopes, and of course the basic depth of hard strata, whether they are rock or dense soil. All of these can stop driving prematurely, damage sheet piling, or skew the wall from the plumb orientation that is normal for sheet pile walls.

5.12.2. Before Driving

5.12.2.1. Storage

Sheet piling may arrive at the project site by rail, ship, barge, or most often by truck. Hopefully, the shippers have packed the sheets into lifts of convenient size for transfer to ground storage at the project site. Sheets can be damaged if roughly handled or carelessly stored. The yard crew should observe the general condition of the shipment before unloading since the transportation company may bear some responsibility for damage in transit. Regardless of whether the material is to be stored near the project or loaded directly onto a barge deck, a plan for blocking lifts and accessing the correct sheets should be made in advance.

In general, piling should be stored on level ground or blocked up in a manner to bring the pile level. Blocking should spaced at short intervals to prevent excessive sag in the lifts, and the sheeting should be 3'-5' (1-1.5m) off the ground. Lighter sheets, aluminum sheets, or those with low section modulus (such as flat sheets) will require closer blocking. Overhang at the ends should be less than 2' (0.6m), except where vibratory hammers are used to pick the sheets up, in which case the overhang should be 3'-4' (1-1.2m).

Lifts stored on site should not exceed the weight as shipped. When lifting, slings should be positioned so there is no concentration of weight at any point. When lifting single pieces of sheet piling from one end only, care should be taken not to bend the sheet and cause a permanent set. One guide for blocking is shown in Figure 5-70.

Figure 5-70 Stacking of Sheet Piling

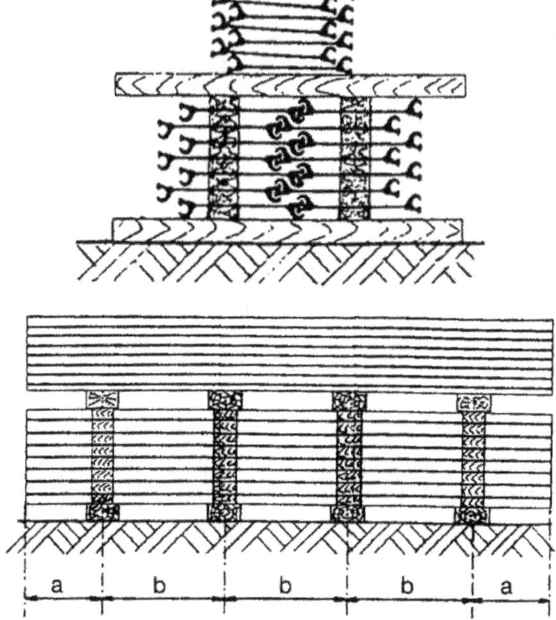

5.12.2.2. Special Features

Sheet piles are normally provided with standard handling holes at the point of manufacture (see Figure 5-71). Z-types sheets require a hole in each end, while U-shaped, arch web, flat profile, and most lightweight sheets need only a single hole. These are standard and customers must advise if they do not want the holes. The holes are designed to accommodate a shackle (clevis) pin or any number of proprietary "quick-release" shackles.

Some suppliers will furnish sheets in pairs, which sometimes facilitate handling compared with single piles, and in any case, sheets are usually driven that way. Some pairs may be crimped or welded together at the mill or by the supplier when specified by the design engineer.

Figure 5-71 Sheet Pile Point Installed with Handling Hole for Sheet

5.12.2.3. Field Inspection of Materials

Space should be left between stacks to permit inspection of the piling as part of the delivery process. Some agencies require this inspection before the piling is used and it is especially important on cellular cofferdams where interlock dimensions are critical.

The interlocks of Z-type sheets should be visually scrutinized or gauged for kinks and incorrectly sized parts that might affect the threading operation later. Interlocks of flat sheets should be checked for open slot and undersize thumb. While "within-tolerance" dimensions are generally not furnished to the purchaser, any signs of non-uniformity in the locks of any sheets should be reported to the supplier immediately.

Piles should be examined for excessive camber and/or sweep prior to use even though the supplier has previously made this inspection. Camber describes a deviation from straightness along the strong direction of a beam, and is often artificially applied to the beams. Sweep is a deviation in the weak direction, and sometimes occurs accidentally in sheet piling during production. A small amount of sweep will not affect threading two sheets except when the sheets are comparatively short.

There are no general specifications (such as ASTM) covering the straightness or interlock dimensions of steel sheet piling. Sheet piling is sold under an "implied" service specification, which is met by each manufacturer. Generally, the sheets are warranted to slide to grade of their own weight when interlocked. This means both interlock dimensions and straightness must meet the mill's internal tolerance measurements.

Used and rental sheet piling should be examined closely so that any question regarding the shipment can be recorded with the supplier immediately. Notes should be made of lengths received, condition of the interlocks, any alloy markings and all other important conditions.

When using flat sheets previously employed in cellular construction, one might assume that the interlocks have been tested in usage. This is partially correct; however, rigid inspection of the interlocks is required by most governmental agencies and this is also good practice on private work.

On new shipments, the quantity received should be compared with the order. Mill test reports should compare with identification numbers (heat numbers) found on the piling and the test reports verified against the grade specified. Sheets should be identified according to length if this information is not effectively contained on the steel or is illegible.

5.12.3. Templates

Correct geometrical arrangement is very important in sheet pile installation, especially in structures such as cofferdams where the sheet walls (meet themselves" as a closed structure, or with tied bulkheads where the sheets must line up with the tie rod location. Failure to assure this will result in time and money consuming solutions.

Accurate alignment can be assured by use of a template, where an external structure can guide the placement of the piles. "All sheet piles shall be driven true to line and suitable temporary wales, templates, guide frames and bracing shall be used" is a typical requirement of many sheet pile bid specifications. Sheets are generally set in place before driving, which allows the contractor to make any necessary adjustments before the piles are fixed into the soil.

5.12.3.1. Templates for Bulkheads

Templates for straight walls can be simple or complex. The ideal template would consist of horizontal beams providing both inside and outside support near the ground level and again at a point below the elevation of the planned initial drive. This template should be long enough to permit installation of a representative "panel" of sheets, say ten to twenty piles. Some small clearance is provided between template wales and the faces of the sheets and blocking utilized to secure the sheets in a plumb position. The horizontal wales are supported on temporary piles or fabricated steel frames. Once the initial panel has been set and pinned, the template is advanced. The rear of the wales may now be attached temporarily to the previously pinned lead piles. An example of a template is shown in Figure 5-72.

Figure 5-72 Template for Sheet Piles

Many contractors use guide wales on one side of the wale rather than the inside-outside described previously. Templates utilizing only one wale are sometimes used for low walls. Sometimes a shallow trench is excavated to position the base of the wall. Templates are made up of structural steel or timber with steel being the more common type.

5.12.3.2. Templates for Box Cofferdams

The template for box cofferdams in water is generally the future internal bracing system. A minimum two-ring system with the top ring above water is recommended. The frame or cage is hung from cables on spud piles. In tidal water or where rapid changes in the elevation of water outside the cofferdam are possible, the guide wales should be designed to accommodate this potential difference in head between the inside and outside water levels.

Some contractors have successfully used spud barges as single wale guides for cofferdams in shallower water, floating the barge out after three sides were completed (don't wait until the wall is finished).

5.12.3.3. Templates for Cellular Cofferdams

Good templates result in good cofferdams and this applies especially to cellular construction. Ring templates are generally fabricated from either structural steel or pipe. The rings are dimensioned to a diameter somewhat less than the theoretical diameter, depending upon the section of sheet pile being used. Two-ring templates provide two points of guidance for the sheets. The rings are spaced from ten to twenty feet apart with the wider spread used on deeper cells. Single ring templates have been used successfully for shallower construction. Ring templates and their use are shown in Figure 5-73.

Figure 5-73 Ring Template

Templates are generally supported on spud piles driven through wells in the frame. Tennessee Valley Authority constructors employed a floating template for a large cofferdam and others since have used this method. A timber deck on the top ring provides a working platform and a nailing surface for the blocking. Extra large templates may have to be fabricated in halves or segments for assembly in place. Smaller templates must also be provided for setting the connecting arcs between main cells. These can be relatively simple.

Diaphragm type cells require templates with two curved ends and two straight walls to the approximate theoretical dimensions of the cell. Since these frames also serve to stabilize the cell until fill is placed on both sides of the straight walls, additional templates should be provided which are essentially bracing frames.

5.12.3.4. Installation without Templates

Although a template is the optimal way to drive sheet piles, there are many cases where one is not used. Such cases include:

2) Walls which are not so geometrically critical;
3) Where the permanent wale system was built in advance and can be used as a guide,
4) Where there is not enough space to erect a guide wale.

Some types of sheet piling (especially Larssen) are traditionally picked up and driven in one at a time without a template. In such cases, sheets should be set skillfully with the crane, and driving should be closely monitored to prevent skewing of the sheeting.

5.12.4. Setting and Driving

5.12.4.1. Setting

Some suppliers will furnish sheets in pairs that sometimes facilitate handling compared to single piles. Some of these pairs may be crimped or welded together at the mill (upon order) when specified by the designing engineer.

For handling sheet piles in pairs tack welding, or a bolt through the interlocks, will hold them together for safe and convenient driving and pulling. Cutting one side of an interlock at the top corner of a sheet back 4 to 6 in. will make the start of threading easier.

Piles should be picked off individual lifts with some skill since long, flat sheets particularly can be permanently damaged during pick-up. (Z-type piles should be stored so that the crane line can reach the pick-up ends of opposing sheets in the wall.) With either U- or Z-type piles, the piling stacks can jam together during transit. Even under the best of conditions there is little room between sheets in a stack. Specialized lifting shoes can be obtained to facilitate pile picking. Crane booms should be of sufficient length to handle what amounts to double length sheet piling. This is not as much a problem in water as it might be on land.

Generally, all sheet piles in a section should be set, if practical, before any are driven. Then all should be driven a few feet at a time, say from 5 ft. to one-third of their length, before any are driven to their full depth. If some sheet piles strike an obstruction, move to the next piles that can be driven; come back later to the piles that resist driving. With interlock guides on both sides, and perhaps a heavier hammer, it may be possible to drive the sheets to the desired depth. If sheets adjacent to those being driven tend to move down below required depth, they should be stopped by welding or bolting to the guide wales. It generally should be better to fill in with a short length at the top rather than trying to pull the sheet back up to plan location.

Some tricks of the trade-immersing the sheet in water (where depth permits) to lubricate the locks; installing a bolt in the hole of the interlock to help keep it open.

The "traditional" way of establishing a lift point on the head of the sheet was to use a conventional shackle; however, a more convenient method is to use a ground release shackle as shown in Figure 5-74.

Figure 5-74 Ground Release Shackle and Its Use

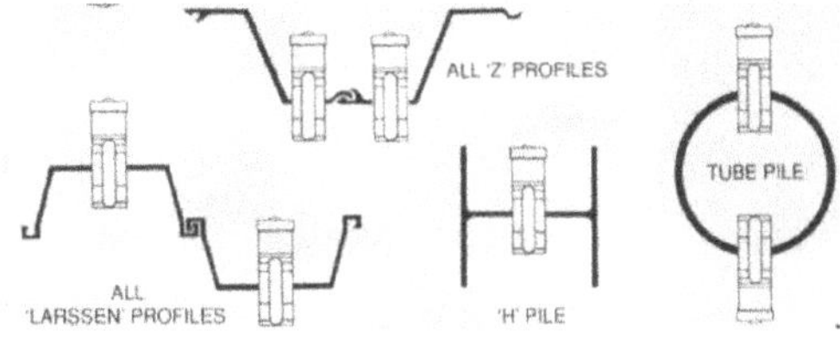

As is the case with conventional shackles, these use the lifting hole at the head of the sheet. Ground release shackles that attempt to hold the sheet with friction should be avoided.

Once lifted, they should be set (setting is alternatively called "threading" or "pitching") in loose pairs so that adjustments for threading can be made. Threading the piles into the interlocks before final setting can be a very tricky operation, especially in a high wind. The first sheet or pair is lifted and temporarily secured to the frame. The second sheet is lifted above the first and threaded therein, either by a member of the pile crew straddling the first pile and guiding the interlocks together (see Figure 5-75), or with the use of a pile threader, which eliminates the person, as shown in Figure 5-76. The danger of the first method is obvious: if the crewmember on the sheeting falls off, serious injury can result. Some tricks of the trade include immersing the sheet in water (where depth permits) to lubricate the interlocks, removing several inches of interlock finger (not recommended with rented piling), or installing a bolt in the hole of the interlock to help keep it open when it is necessary to drive the female interlock first.

Figure 5-75 Manual Threading of Sheet Piling

Figure 5-76 Pile Threader

Figure 5-77 Operation of Pile Threader

Following Figure 5-77, the method of operation is as follows:

1. The threader is pre-loaded by pulling the wires and vice block against the spring pressure and cocking.
2. The pile to be pitched is lifted until the bottom end is about one meter (39") above ground level. The threader is then clamped onto the pile with the vice. The relative position of the pile is governed by a stop plate to suit the pile sections being used.
3. The pile with the threader attached is then lifted by the crane to the last pile in the panel. By means of rollers, the lower part of the threader is then clamped to the last pile of the panel so that the interlocks are adjacent to each other. The clamping operation is designed to permit free movement of the pile and threader in a vertical direction but prevent movement in any other direction. This attachment is normally done at the top frame level.
4. The cocking device inside the vice slide tubes are then released.
5. The crane lifts the pile to be pitched until its bottom edge is above the previous pile in the panel. The spring pressure on the vice block forces the pile across into the locking position. As the pile is lowered it interlocks. The threader is then unclamped at the top frame level.

Wall dimensions found on project drawings are based on theoretical dimensions determined by multiplying the number of sheets by the catalogue dimensions of each sheet. While mill dimensions are carefully controlled, in a long run of wall it may be possible to lose or gain several feet due to positioning of the piles, crowding or stretching of the interlocks or other factors. One method to deal with this problem is to first mark the template for each pair of piles. This will indicate whether the line of piling being set is gaining or losing wall length. Another method is to insert timbers between the piles and the wales to realign the sheets and thus shorten or length their effective length. This is shown in Figure 5-78.

Figure 5-78 Correction for Wall Length

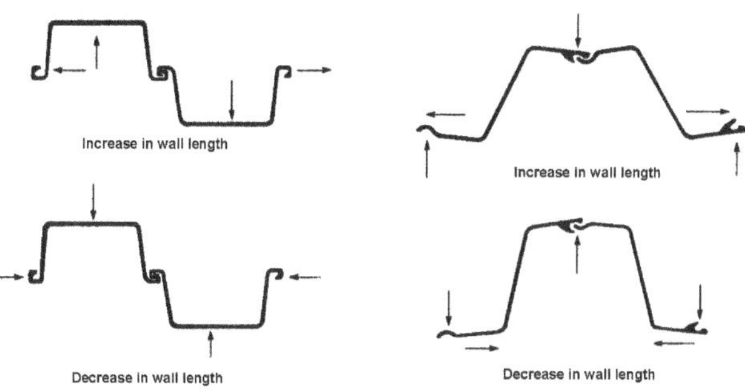

5.12.4.2. Driving

Sheet piling tends to lean in two directions unless properly guided during installation. Guide templates described previously help control plumbness of the wall; proper installation procedures help maintain plumbness of individual sheets along the line of the wall.

Sheet piles may "gain" over their nominal width as they are installed as there may be a tendency to lean in the direction of driving; or there may be other deviation. It is best to mark on the supporting guides the planned edge location of each pile so any change is noted and can be corrected while small. Making on two levels of guides, or frequent use of a plumb device, is essential.

When one thinks of driving sheet piles, one generally thinks of driving them down a row, pair by pair (at least - sheet piles today are seldom driven one at a time). This means that there are two possible directions which one can choose to drive them. Generally, sheet piles should be driven with the male interlock, ball, or thumb pointing in the direction of the driving. This minimizes the filling of the interlocks with soil, which will increase the driving resistance. Fine sand will clog interlocks and could produce symptomatic hard driving in which the entering interlock must extrude the sand before progressing. Continuing this for any length of time could damage the interlock and should be discontinued. Water jetting will generally assist in penetrating these materials. A vibratory driver/extractor is generally the best choice for installation.

Funnel-like guide devices may also be used. It is generally recommended that Z-type piles be driven with the ball end leading so that the socket slides over the ball rather than the ball slide into the socket. It is common practice now to supply sheets pre-threaded as doubles at the mill. In this case, the first set-up would involve a complete pair of piles.

When the lodging or packing of material into the interlocks is a concern during driving, the following suggestion may help. (1) placing a bolt in the bottom end of the forward thumb and finger or (2) on Z sheets, drive with the ball ahead.

There are no firm rules for installation; however, experience indicates that once the sheets are set and are ready to be driven, they should be "gang" or "step" driven, that is to say the sheets should be driven a short length, and then another short length, and so on in sequence until the desired depth is obtain. The steps can be set up for the whole wall or the wall can be divided in sections and each step is a "step behind" the one behind it. This minimizes the risk of an undriven sheet cocking with a fully driven neighbor, with the potential of skewing the entire wall. It is important here not to make the steps too long driving relative to its neighbor.

There are situations where it is advantageous to break the pattern of "down the row" driving. In these cases, one can first drive every other pair, or perhaps every fourth or fifth pair. This can be done when a set wall needs initial stabilization. Another way of obtaining this goal is to "panel" drive the sheets. In this method, the first pair is plumbed in both directions and partially driven. Succeeding sheets in the first "panel" are threaded until five or ten pairs are in place. The last pair in this group is plumbed and partially driven. The dimensional location of this last pair should be verified against the theoretical. Following partial driving of the end pair, the hammer works on successive pairs proceeding back to the beginning of the wall. The last pair set is not driven to grade since it will now be the first pair in the next panel. The "panel" method affords a relatively simple method of correcting problems before they become serious (see Figure 5-79). An actual example of panel driving in progress is shown in Figure 5-80.

In difficult soil conditions, the panel driving method should be modified by the staggered driving of the sheets. This technique is shown in Figure 5-81. In some cases, the sheets driven first need toe reinforcement; these sheets always lead the driving.

Figure 5-79 Panel Method of Driving

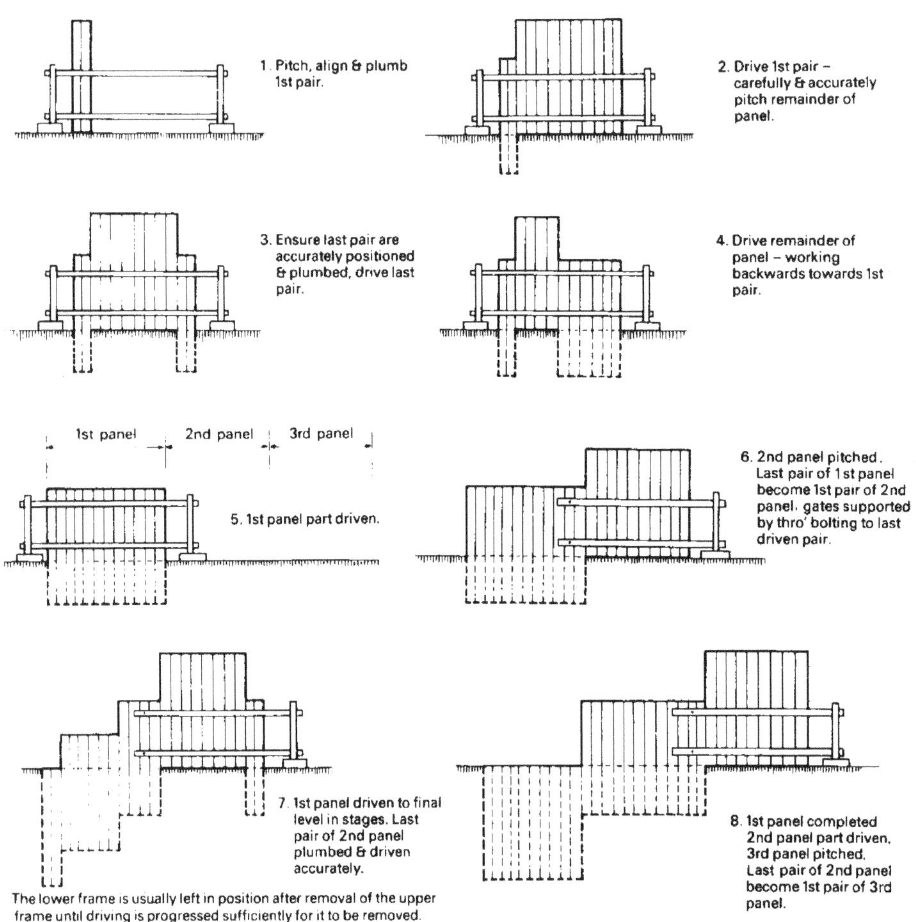

Figure 5-80 Panel Driving in Progress

Figure 5-81 Staggered Method of Driving

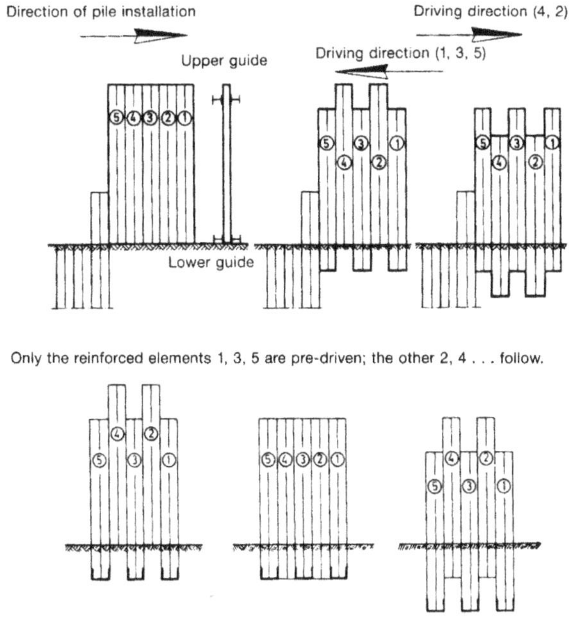

Figure 5-82 Obstructions in Sheet Pile Driving

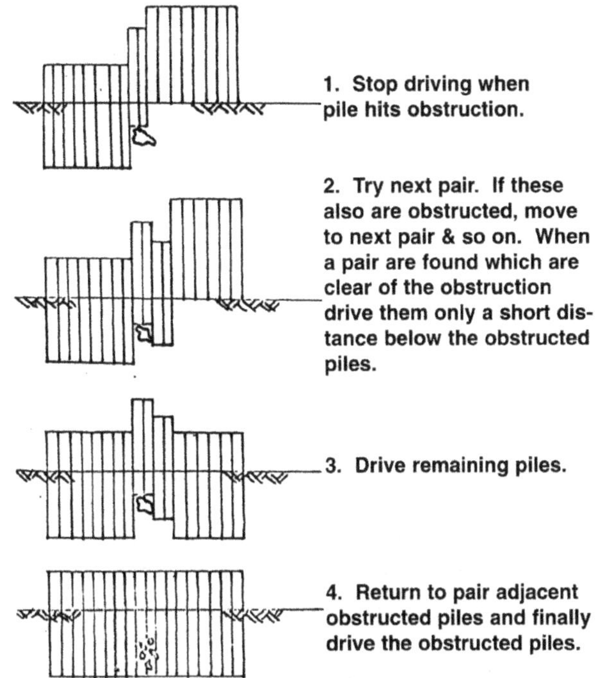

Underground obstacles should be removed before driving begins; however, should these be encountered, cease driving. If this is not possible, or it is difficult to monitor the exact time of such an encounter, then one can bevel the toes of the sheets, weld various guide or reinforcement plates, or use pile points. Some special techniques one can use in driving are shown in Figure 5-82.

Water jetting is an option to assist in driving; it is discussed extensively elsewhere.

5.12.4.3. Skewing

Skewing is a serious concern during pile driving, as it can affect the arrangement of the sheet piling as well as its vertical inclination. Z-type piles are especially prone to skewing. Keeping sheet piles plumb is particularly important since correcting it is not simple. In addition to the methods already described, other measures can be taken to prevent or minimize pile skewing. This first is an accurate and rigid template, especially one that is closely formed to the piles. Other methods involve beveling the pile toe or adding angled guide plates, both of which use wedge action to act as a rudder, or to use tapered piles. These should be set to have the wedge action push the pile in the opposite direction of the skewing.

Another method is to use a wire rope to pull the sheets back into plumb, as shown in Figure 5-83. The hammer can also be placed off center in conjunction with this method.

Figure 5-83 Correction of Longitudinal Lean

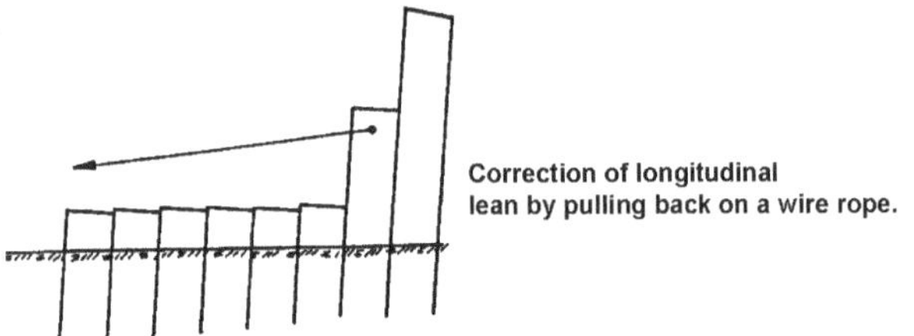

5.12.4.4. Box Cofferdams

Box cofferdams on land or water invariably include four corner sections, and it is therefore logical to begin at one of the corner. On land, the panel method used with bulkhead walls is also appropriate for temporary land cofferdams. In water, the template is generally also the bracing cage and had been prefabricated and supported on spud piles at the site. In flowing streams, the upstream wall will act as a diversion for construction of the other walls, if constructed first. The corner pile and at least one adjacent pile are driven (as with land walls) to provide a solid point. Subsequent piles in the run of wall are customarily not driven until the other corner has been reached to insure closure of the four walls. Driving should proceed in stages again working back from the leading piles.

5.12.4.5. Cellular Cofferdams and Flat Sheets

Cellular cofferdams are generally built for deep-water applications and thus require longer sheets. The flat sheets generally utilized for cellular work offer some special handling and installation problems not associated with Z- or arched sections. Pickup requires a special technique to prevent permanent bending. Extra long booms are often required, particularly if the cells are being built on land or in shallow water. Wind can be a factor at all stages.

Cellular cofferdams can be built from shore with the equipment moving from cell to cell, or from barges. In all cases, one or more templates should be fabricated to the approximate dimensions of the cell as designed.

Some contractors paint a line on the sheets a constant distance from the toe but above the water line, particularly where individual piles of inconsistent length are involved. In this way, a profile of the cell's base is obtained.

Fabricated tees or wyes are set first and become control points for setting the remaining sheets. These should be carefully plumbed and tacked into the bottom. Sheets are then "spun" in both directions from the fabricated sections toward the center of the arc. Alternate sheets for blocked against the template. Guy lines help keep sheets vertical and in proper position.

Closure of the arc is best made with two sheets at the center of the arc. These have been interlocked and are then handled with two lines so that when pulled up, one remains partially interlocked with its neighbor. The closure sheet is lifted together and locked with the last sheet in the other half of the arc. None of the sheets other than the fabricated pieces has been driven at this stage. It may now be necessary to pull some of all of the sheets in multiples and "shake them out" so that they run freely. This operation of threading sheets between the four fabricated piles should be repeated three more times. Before driving and fabricated pieces any further and certainly before filling the cell, one or two sheets in the connecting arc should be threaded with each tee to the stem of each tee or wye pile. If this is done later, there is a good chance these will not be able to follow the contour of the tee and will drive out of the interlock.

Some difficulty has been experienced in setting long sheets in deep overburden. This can be dealt with by providing more holding points in the arcs. Wide-flange beams are added to some of the intermediate sheets to stiffen them; these are pinned similarly to the tees and closure made between these points also.

Splicing of flat sheets is generally permissible provided the splices are located above the water line and are staggered at least five feet. The splices on the outer cofferdam wall should be bead welded to seal them but consideration should be made about eventual extraction of the sheets and a full strength splice may be preferable. Splices on the inside wall needs not be sealed except for subsequent extraction. Any handling or erection holes on the exterior walls should be patched while they are accessible.

Cofferdams using wye type connectors offer some design advantages but are somewhat more difficult to construct because of the small radius of the connecting arc. Tee connected arcs offer a flatter arc and require a slightly smaller swing angle between sheets. Normally the main cells provide no problems in closure other than described previously.

When the theoretical closure angle exceeds 7°, the supplier should be contacted for recommendations or guarantees of "swing". The longer the sheets, the more difficult it is to obtain this swing. In some cases, it may be feasible to plan on ordering bent piles to accomplish these closures; otherwise, connecting arcs are built using the same procedures as used for the main cells. Difficult closures are facilitated by utilizing a bridle to pick up a number of connected sheets and shaking them until a closure is made.

As for actual driving, the establishment of plumbed strong points at the connectors and at intermediate piles within the arcs will help keep all piles vertical. Driving should be in stages without excessive lead from one pair of piles to its neighbors. If driving in the first stage proceeds in one direction from connector to connector, the next stage in that panel should be from the opposite direction. Some contractors drive a pair of sheets advance a pair sheets, then skip the next pair in an alternating pattern of stage driving. The practice of continuous driving in one direction is to be avoided since it will undoubtedly contribute to the leaning of all piles in one direction and possible interlocking problems with the interlocks, which will show up when the cofferdam is dewatered.

Driving through beds of boulders is virtually impossible even though thicker sheets may be specified for the purpose. Even if penetration is obtained, extracting the sheets later should be a costly process. Large masses of boulders should be removed in advance.

Fine sand will clog interlocks and could produce symptomatic hard driving in which the entering interlock must extrude the sand before progressing. Continuing this for any extended length could result in interlock damage and should be discontinued. Water jetting will generally assist in penetrating these materials.

5.12.4.6. Limited Headroom

If sheeting is driven under a bridge or other structure, normal pitching of the pile can be difficult if not impossible. Although the sheets could be driven in sections and added on to, this can be expensive. A more sensible alternative is shown in Figure 5-84; it involves assembling the sheets into panels with the sheet joining the two panels made shorter than the others. The panels can then be threaded with the short sheet already completely into the ground; the initial threading takes place at or near ground level, giving maximum use of the headroom. The short sheet is then made to complete length using a welded or fish plated splice.

Figure 5-84 Pitching in Panels with Limited Headroom

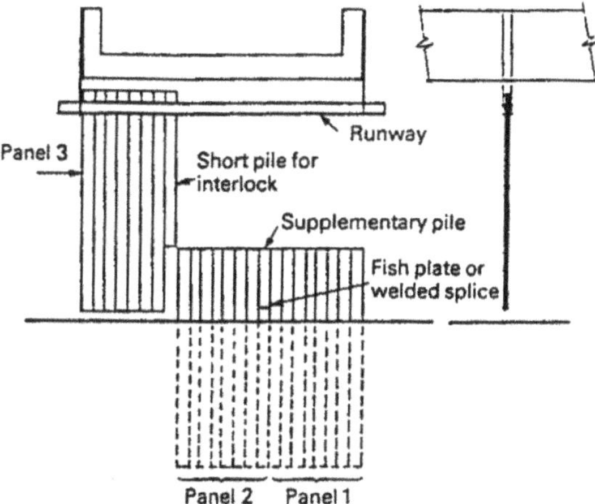

It should be noted that, with vibratory driven sheet piles, specially configured hammers are available to assist in low headroom situations.

5.12.5. Alignment of Sheet Piles

As has been described in detail, proper alignment of sheet piles is important to their successful installation. The following are guidelines for typical sheeting installations and may vary with specific situations:[20]

- Deviation normal to the wall line at the pile heads should be no more that 50 mm.
- Deviation at the finished level from nominal level should be no more than 20 mm at the pile heads and 120 mm at the pile toes.
- Deviation of verticality normal to the line of piles should be no greater than 1% of the driven depth.
- Deviation of verticality along the line of piles should be not greater than 1% for piles driven by a "pitch and drive" method or 0.5% for panel driven piles.

5.12.6. Driving and Extracting Equipment

Presently there are two types of equipment which are used for driving and extracting sheet piling, impact and vibratory. Each has advantages and disadvantages depending upon the application. Although there are no hard and fast rules for the application and use of each type, some guidelines can be followed, tempered of course by experience. Most of the information on equipment in this section concerns driving of sheet piling. Extraction is discussed in 5.8.12.

One problem that many designers do not consider is the ability of a given sheeting design to be installed. Table 5-16 shows a general "rule of thumb" guide for drivability of sheet piling configurations. The table is based on sheeting no wider than 500 mm (19.7"); adjustments should be made for wider sheeting.

Table 5-16 Sheeting Configuration for Drivability

SPT minimum Dominant N value	Wall modulus (cm^3/ m)	
	Low-yield steel	High yield steel
0- 10	500	
11- 20		500
21- 25	1000	
26- 30		1000
31- 35	1300	
36- 40		1300
41- 45	2300	
46- 50		2300
51- 60	3000	
61- 70		3000
71- 80	4000	
81-140		4000

Where N represents the Standard Penetration Test value.
"Dominant" means the average of the high values for the soils to be penetrated.

Where piles are to be driven only to a toe-hold in rock, the N value shall be divided by a factor of 4 for that stratum only.

5.12.6.1. Vibratory Drivers

The reputation of vibratory drivers for fast, efficient, and relatively quiet operation was largely established on sheet piles; indeed, the first project vibratory drivers were used on was to drive sheet piles for the Gorky hydroelectric project in the USSR in 1949. Equipment, usage techniques, and productivity, have continuously increased since then, making vibratory hammers, generally, the driver of choice for the installation of sheet piles.

A vibratory hammer driving sheet piling is shown in Figure 5-85. The clamp, attached to the pile, connects the pile to the hammer. Generally, these hydraulic grippers make a firm, frictional connection between the hammer and the pile. Most modern clamps have the capability of clamping onto at least two sheet piles at a time, using a jaw with two clamping faces and a space between them for the interlock.

Figure 5-85 Vibratory Hammer Driving Sheet Pile

5.12.6.2. Drivability of Vibratory Hammers

Although vibratory hammers are the most effective tools available to drive sheet piling, their ability to drive piling is extremely variable with the soil. Generally, vibratory hammers perform best in cohesionless soils and progressively worse as the cohesive nature of the soil becomes more pronounced. There are exceptions to this rule, and of course, the basic soil resistance is a very important factor in determining whether a vibratory hammer should to be used and what size of hammer is needed. Experience with the local conditions becomes very important in this determination. However, no matter what the soil resistance is, as a rule of thumb the dynamic force of the hammer should be at least nine (9) times the vibratory weight of the pile, hammer and clamp combined. Methods to estimate vibratory drivability are discussed in Chapter 4.

One factor that is almost unique to sheet piling is the resistance of the interlocks to driving. This is caused primarily by soil which has seeped into the interlocks, and secondarily by steel-on-steel friction. This can vary from 1-100 kN/m (69-6900 lb/ft,) the force per unit length of interlock resistance during driving. Coarse soils tend to cause higher interlock resistances than fine ones do.

It is equally important to know where a vibratory hammer is not suited for a particular project. Generally, if the pile is penetrating the soil at a rate less than 1 ft/min (5 mm/sec), the contractor should switch to an impact hammer.

5.12.6.3. Special Procedures for Vibratory Hammers

Vibratory hammer procedures should be varied somewhat depending upon whether the sheeting is set and then driven (as with a template) or whether the sheeting is driven and set one at a time. Figure 5-86 shows a hydraulic vibratory hammer installing sheet piling; Figure 5-87 shows one driving flat sheets in a cellular cofferdam. The sheeting wall is set using a template and then driven to the desired depth. This practice requires that the vibratory hammer be no wider at the throat than about 14" (355mm), as the hammer must clear the adjacent piles. When the hammer is lowered to the pile, make sure the exciter is positioned parallel to the pile and that the full length of the jaw will make contact with the pile when clamped. When clamped, the operator should check that the pile is firmly gripped by the jaws.

Figure 5-86 Hydraulic Vibratory Hammer Driving Sheet Piling

Figure 5-87 Vibratory Hammer Driving Flat Sheeting

An alternate method of driving sheeting with a vibratory hammer is to set the sheets and then drive them and, as a rule, drive them one at a time. The vibratory hammer is used as a pile gripper, the hammer being lowered first and attached to the pile. Some piles, such as aluminum piling and some light steel sections, cannot be lifted in this way, as they will buckle under the load.

In either case, to permit lifting of the hammer, a wire rope should be secured from the crane line to the lifting hole or pin on the suspension. In choosing the wire rope for any unit, a generous safety factor should be used. Several turns of a smaller diameter cable will usually last longer than one turn of a large diameter cable. Make sure that the wire rope assembly you use has at least double the capacity of the suspension.

Once the hammer is clamped to the pile and adequate clamping pressure has been reached, the hammer is ready to vibrate. Engage the "start" control. The exciter case and pile should begin vibrating. In driving, the combined weight of the hammer and pile will force the pile into the ground. As the driving resistance increases, the drive pressure or amperage will increase until it reaches the maximum power output of the power pack. The crane hook should be lowered so that the crane does not inhibit the penetration of the pile; on the other hand, there should not be so much free cable that the sheet being driven might buckle under the weight of the vibratory. The cable can be completely slackened, however, during the final stage of driving.

Should skewing be encountered, the vibrator can be pulled in the opposite direction of the skewing to compensate for this. This is dependent upon the nature of the suspension and the direction in which it is laterally loaded. If simple lateral loads are not enough, the pile can be extracted and redriven, frequently without completely disengaging the interlocks from the neighboring pile.

5.12.6.4. Impact Hammers

Impact driving is the oldest way of driving sheet piles, and in some cases, it is the only way possible to do the project. This is especially true with piles facing high soil resistances. As with other types of piling, there are three types of impact pile drivers; air/steam (Figure 5-88), diesel (Figure 5-89) and hydraulic (Figure 5-90). These are discussed in detail in Chapter 3.

Figure 5-88 Air/Steam Hammer Driving Sheet Piling

Figure 5-89 Diesel Hammer Driving Sheet Piling

Figure 5-90 Hydraulic Impact Hammer

Since most sheet piling does not act as bearing piling, experience is the first and best teacher for choosing the size of an impact hammer to drive sheet piling. The next criterion used is the ratio of the ram weight to the pile weight; as a general rule, the pile should weigh no more than 25% of the ram for air/steam hammers or 50-100% for diesel hammers.

Wave equation type analyses have not been generally employed in the prediction either the drivability or bearing capacity of sheet piling; however, as sheet piling find more and more use as bearing pile, these methods will find more use with sheet piles. Wave equation analyses of sheet piles, which do not take into account interlock friction, should be regarded as incomplete. Wave equation analysis to solely assess driveability can be a very effective tool for hammer sizing.

No matter what method is used for determining drivability and/or capacity, there are blow count limits that should be observed. The hammer should not be allowed to exceed the manufacturer's rated capacity, which in most cases is 120 blows per foot.

The driving rigs used for impact driving of sheet piles are generally similar to those for other piles; however, the continuous nature of sheets poses some special problems in the use of this equipment.

Leaders for these hammers generally fall into one of three categories - U-type (American), tubular leaders (European), and H-pile spuds. Although they are the most stable, U-type leaders pose some special problems with sheet piles because they wrap around the hammer and thus can get in the way of the sheeting wall. This can be circumvented by using a very short set of leaders and lowering them with the hammer (as shown in Figure 5-88). In some cases, the geometrical layout of the sheets and the configuration of the driving cap allow the use of these leaders in the conventional manner.

Tubular leaders and H-pile spuds guide the hammer from the rear, and thus are more suited for driving sheet piles; however, for safety reasons they cannot be used with open air/steam hammers such as the Vulcan, Conmaco or Raymond hammers because of the weight of the hammers and the difficulty of affixing these hammers to this type of leaders.

An alternative in some cases to leaders are hammer "pants," which are guides, fixed to the hammer in a rigid way and slid over the sheets, which allows the pile to act as the hammer guide. This method, which obviates the need for leaders, is especially popular with smaller air driven sheeting hammers such as the Vulcan DGH-900 and the MKT 9B3, and of course hydraulic hammers such as the Dawson hammers. The sheets should be securely set before this method is used. Driving sheet piles wild (i.e., with neither leaders nor pants) is unsafe.

As with vibratory driving, sheet piles are best driven two (or more) at a time. It is especially important for the sheet pile heads to be flat and square to afford the maximum contact with the drive head and thus the most efficient transfer of impact energy. It is also important to place the impact hammer's center axis in the center of the sheets being driven and to have the hammer driving the piles straight, although in some cases the hammer can be angled to counteract pile skewing.

5.12.7. Summary and Tips for Successful Sheet Pile Installation

Table 5-17 shows a summary of helpful tips for successful sheet pile installation

Table 5-17 Sheet Piling Driving Tips

- First, review your Soil Borings/SPT's to anticipate "possible" driving conditions.
- Inspect your sheets. Work only with undamaged sheets.
- Piles should be straight or driving should be difficult. All sheets should be inspected for excessive "camber" or "sweep".
- Interlocks must always be free of any dirt, sand, mud or other debris.
- Tight interlocks can be just as difficult to drive and keep plumb as loose interlocks. Check with the manufacturer as to "within tolerance" specifications.
- Always set-up a template system. In addition

to driving a straight wall, a template or guide system will also aid in keeping sheet pile plumb when excessive driving conditions exist or when an obstruction is encountered. In short, a good template system leads to good sheet pile construction.

- A "two level" template or wale system is generally suggested. In tough driving and/or when driving "deeper walls", a front as well as back wale or bracing system is suggested.
- Always lift sheets properly. Utilize the lifting "handling" holes or proper "pick-up" points.
- Always protect the interlocks. Sheets with damaged interlocks are extremely difficult to drive and in some cases may affect the integrity of the wall.
- The crane boom length must always be long enough to thread additional sheets. This is normally at least twice the length of the sheets being driven, and then some.
- Whenever possible, attempt to drive sheets with the male interlock, ball or thumb etc. leading. This will aid in eliminating the possibility of the sheet developing a "soil plug". (The interlock filling with soil, sand or mud etc.) If a sheet should be driven with the female or socket end leading, a bolt or other object may be inserted at the bottom of the sheet to keep the interlock free from filling up with soil or other debris during driving.
- When utilizing a vibratory driver/extractor always make sure all "jaw teeth" are on the sheet prior to clamping and driving. The grip should be equally divided (centered) between the two sheets. By the same token, when utilizing an impact hammer, ensure that the hammer being utilized has the proper head (helmet) for the type of sheet being driven as well as being centered on both sheets prior to driving.
- When threading (sometimes referred to as "pitching"), aligning and plumbing the first two sheets, always drive this first pair carefully and accurately, periodically assure the sheets are plumb. This includes proper lateral, longitudinal as well as vertical alignment.
- Whenever possible it is recommended to drive sheets in pairs.
- Some contractors will notch (cut-out) a portion (approx. 3-4 inches) of an interlock "face" as to aid in threading additional sheets.
- Never rush the pile buck! Threading sheets can be a difficult and dangerous task. When "longer sheets" are driven, a contractor may choose to utilize a pile-threading device.
- When threading sheets "letting the sheet freefall" and drop to aid in penetration as a result of its own weight will generally cause the sheet to set "out of plumb". This is not recommended.
- Thread only the remainder or portion of wall "protected" by the template system.
- When threading, aligning and driving the last pair of sheets, once again periodically stop to make sure the sheets are plumb. This method is sometimes referred to as "panel" driving the sheets. In this method the first pair is plumbed and partially driven. Succeeding sheets in the first "panel" are then threaded until five or ten pairs are in place. The last pair in this group is then plumbed and also partially driven. Following partial driving of this end pair, the hammer works on successive pairs proceeding back to the beginning of the wall. The last pair set is not driven to grade since it will now be the first pair in the next panel.
- When there are no rules for installation experience has indicated that once the sheets are set and ready to be driven, they should be "Gang Driven". This is sometimes referred to as "Driving in steps". That is to say the sheets should be driven a short length, and then another short length, and so on in sequence until the desired depth is obtained. The steps can be set up for the whole wall or the wall can be divided in sections and each step is a "step behind" the one behind it. This minimizes the risk of an undriven sheet cocking with a full driven neighbor, with the potential of skewing the entire wall. It is important here not to make the steps too long or to drive a sheet pile too

far ahead of its neighbor.

- Whether it is referred to as panel driving, gang driven, driving in steps or whatever, sheet piling has a tendency to lean in the direction in which the wall is being driven. Start with a good template system. If your wall starts "walking" (creeping) this should be corrected immediately! Even if it requires going back and extracting or partially extracting a few sheets with tugging, pulling or jacking, in the end it should be well worth the effort.
- Never overdrive. Are sheets bending, bouncing or vibrating with no penetration? If so, you're either overdriving, hit an obstruction or need a larger vibratory driver/ extractor or impact hammer.
- Stop driving when a sheet hits an obstruction. Try next pair - if these are also obstructed move on to the next pair or until such time an unobstructed pair is found. When these are located, drive these only a short distance below the obstructed piles then move on to the next pair. Return to the obstructed sheets and attempt to drive later.
- Some contractors recommend not driving a sheet more than 1/3 its length before driving the adjacent pile.
- Cellular cofferdams require all sheets to be set and "closed" prior to any driving.
- And finally...never rush the pile foreman!

5.12.8. Vibration and Noise Control

As environmental concerns become more pronounced, more attention should be given to both the noise sheet pile driving produces and the ground vibrations that emanate from a sheet pile being driven. Both of these can be dealt with but there are limitations with the methods used.

With impact hammers, there is not much that can be done to control the ground vibrations they produce, short of reducing the energy output of the machine. If the piles can still be driven after this, then all is well. As for noise, the two main sources of noise (in general order of amplitude) are the crack of the exhaust and the impact on the pile. Using adequate and fresh cushion, which will also lengthen the life of the hammer, best controls pile impact noise.

With vibratory hammers, the most important thing to reduce noise is to remove all loosely hanging objects (such as pile shackles) from the vibrating hammer and pile, thus eliminating their rattle. Most vibratory hammer exciters are reasonably quiet, and if necessary sound deadening can be fitted to the power pack. As for ground vibrations, if these are a problem the best solution is to use a vibrator with an operating frequency higher than 2000 RPM/33.3 Hz. This generally will take the vibration frequency above that of the soil resonance.

In some cases, vibratory driving will produce an intense interlock noise. This is generally due to lateral vibrations in the sheet pile towards the pile top, which is a function of lack of lateral rigidity accentuated by the lack of alignment fixity of the vibrator itself, especially with vibratories hanging off of a crane (as opposed to those which ride in leaders or on a mast.) These vibrations not only can make noise, they can also significantly degrade the drivability of the hammer-pile system. These vibrations can take place with or without the noise, and can also induce additional vibrations in the ground.

If the vibrations generated by impact or vibratory driving are unacceptable, then pile jacking should be considered.

5.12.9. Pile Toe Modification

Whether driving with impact or vibratory equipment, there are times when the toe resistance of the pile encounters will either damage the pile or impede the necessary penetration of the pile. This is especially important when underground obstacles are encountered, when lightweight sheet piles are used, or in any case where driving can either tear the sheet or skew it. This is even true with vibratory equipment, especially in cohesive soils where the vibratory achieves penetration by turning the pile into an impact ram and beating its way through the soil. In these cases, some modification of the pile toe is necessary. This is discussed in detail in 5.10.6

5.12.10. Pile Rentals and Returns

Steel sheet piling driven for temporary protection usually is pulled and reused. Most companies which rent or buy back sheet piling for temporary uses have similar conditions for the return of their material. In general, the sheets should be returned in the same general condition as received to avoid penalty charges. If singles were received then singles should be returned as well. Interlocks and surfaces should be steam cleaned of soil, holes filled, badly deformed ends trimmed square, new handling holes cut (if the others were removed), and sheets blocked at stacked in lifts of equal length. Badly deformed material should be separated from good material for possible scrapping or salvage, and to avoid a dispute over what is good and what is not. The lessee should understand well the return conditions at the time of rental.

Prevention of damage begins with adequate design and planning. Since cofferdams generally are temporary, there may be a tendency to use a higher stress value for steel than would be permitted for a permanent structure. This may be satisfactory if all factors are considered. Steel sheet piles may have been previously used and have deteriorated in some way. Because project forces expect the installation to be temporary, they may do a less than perfect job of assembly or bracing. Even with the best intentions, sheet piles may strike obstructions that cause damage. Unplanned loads may be put on bracing; cranes may be operated immediately along the outside, causing surcharge loads not considered in design. There may be impact from unexpected sources. Successful foundation contractors take all factors into consideration and require that installation of sheet piles and bracing be as carefully done and inspected as any part of the permanent work.

A contractor pulling piles for reuse may elect to pull and store in doubles, since the use of double sheets is a time saver. Cleaning interlocks and conditioning in advance of the next use will avoid needless problems. If cuts are made on the sheets, the new lengths should be painted on for future identification. Sheets placed in long term storage should be blocked as if they were new material.

If high strength material is involved, the identification of this material should be maintained since it generally sells for a premium. If it cannot be positively identified for the next use, it should be treated as regular grade for design purposes.

5.12.11. Welding and Splicing

Regular grade (ASTM A-328) sheet piling is a specification designed to accommodate stresses from driving and service loads. It has never been considered ideal steel for fabrication, although it was satisfactory for the riveted construction popular for many years. Welding and burning should be conducted on this steel with some caution, particularly in cold weather or around water.

Sheet piling can be welded by regular methods if good shop and field practice is followed. Since other grades of steel are being offered which meet the minimum requirements of ASTM A-328, the supplier should be asked for a recommended welding procedure when in doubt. The purpose of the following is to establish some general welding parameters for sheet piling. The welding techniques are well established and generally standardized.

5.12.11.1. Qualification of Welders

It is recommended that welding operators be certified in accordance with the requirements of Section IX of the ASME Boiler Code or ASTM A-488. Performance Qualification of Welders under these documents is simple, testing the ability of the welder to produce a "sound" weld in small carbon-steel plate.

5.12.11.2. The Welding Process

There are many welding processes available today, from completely automatic "electroslag" processes to semi- automatic "Metal-Inert-Gas" (MIG) process. Only one process should be considered here, the "Shielded-Metal-Arc-Welding" (SMAW) process. This is the oldest, simplest and most versatile process available today.

The current for the SMAW process is usually DC Positive (Reverse Polarity). Any type of welding current generators may be used. Most welding machines have "rectifiers" with a capacity of at least 400 amps and a 100% duty-cycle.

Only low-hydrogen electrodes should be used for this process. In the AWS-ASTM Classification the approved electrodes would be: EXX15, EXX16, and EXX 18. It is important to keep these electrodes moisture-free after removing from the container. They should be stored in an oven at 250°F (120° C) and used within one (1) hour after removal. The absorption of moisture in the electrode coating will result in hydrogen pick-up in the weld deposit and likely produce cracking.

Standard welding techniques should be used. The bead should be of the "stringer" type. Heat input should always be kept as low as is practical. The welding pass sequence should be selected with the idea of keeping welding stresses to a minimum. This can be done by controlling the heating and cooling rate of the process. The welding sequence passes should be separated as much as is practical; the interpass should be kept low by the use of small-diameter electrodes and low welding current.

5.12.11.3. Welding Parameters

The keys to high quality welding can be summarized in three (3) factors: 1) the selection of the proper electrode for the particular base metal to be repaired, 2) the establishment of the correct preheat temperature to (a) maintain a low hardness in the Heat-Affected-Zone (HAZ) and (b) minimize welding stresses, and 3) the selection and use of a suitable interpass temperature to reduce the effects of heating and cooling on the components. This is important since post-welding heat treatment is not always practical and thus excessive hardness and stresses should be minimized by the correct selection of the welding parameters.

The selection of the proper electrode is based on the composition and mechanical properties of the metal to be welded.

The composition of welding electrodes contain the same alloys as the sheeting; i.e. manganese, nickel, chromium, molybdenum and vanadium. However, the carbon and silicon contents are lower than the base metal. Therefore, the alloy content of the rod should be higher to achieve the same mechanical properties.

Ideally, the composition and mechanical properties of the electrode should match the metal being welded. When this is not possible, it is more important to coordinate the mechanical properties than the composition. This is true of carbon/low alloy steels, but not of corrosion and heat-resisting high- alloy steels.

The selection of the preheat temperature should be based on three (3) factors, listed in order of importance:

- Composition and hardenability of the base or parent metal.
- The feasibility of post welding heat treatment.
- The size and configuration of the part to be welded.

All of these factors require a higher preheat temperature. The preheat can range from 100° F (40° C) for unalloyed steel to 400° F (200° C) for steels of high hardenability. The temperature is always listed as "minimum" and may be higher, if desired. If possible, the entire part should be preheated in a temperature-controlled furnace. Otherwise, localized heating with torches may be used. There are several precautions that should be observed:

1) The heating must done slowly (100° F/hr or 38° C/hr),
2) The part should be heated throughout the section,
3) The area preheated should be at least 12" (30 cm) from the edge of the cavity.

The temperature of the part can be checked by use of "Temp- Stiks", a temperature-sensitive marker that melts at the designated temperature. The temperature of both sides of the part should be checked to insure complete heating throughout the section.

The interpass temperature is the temperature of the base metal measured 2" (5 cm) from the weld, between weld passes in a multi-pass welding operation. It is always listed as a maximum, and is usually less than 200° F (95° C) above the preheat temperature.

The primary purpose of controlling the interpass temperature is to minimize the welding induced stresses. A reasonable interpass temperature will also prevent the parent metal from becoming so hot that it "anneals" the weld deposit, causing lower hardness and strength.

5.12.11.4. Welding Procedure Specifics

As there is some variation of the material composition of sheet piling, no welding should be done without first obtaining the recommended welding procedure for the specific piling from the manufacturer or supplier. Once this is obtained, it should be strictly followed. This is also the case with the equipment used on the jobsite.

5.12.12. Burning

Burning is a common practice to remove excess length, to cut handling and erection holes and to shape special areas such as pipe cutouts on bulkheads. If in doubt about good burning practice, the supplier should be asked for his or her recommendations and these be followed to the letter. In cold weather, cuts should be preheated and potential stress concentrations should be avoided. One supplier suggests following AWS practice and preheating to temperatures listed for "low hydrogen" practice in welding.

5.12.13. Special Considerations for Timber Bulkheads

The proper sequence for bulkhead construction is:

- Drive all round timber piles, vertical and battered; set or drive all posts.
- Using bolts, attach the horizontal wales for the sheet piles and the anchorage system.
- Drive sheet piling.
- Complete all bolted connections and install tie rods.
- Place the backfill; where passive resistance anchorage systems are used, be certain to place fill over the anchors before backfilling behind the sheet-pile wall.

Establish accurate survey lines as a control for the construction of waterfront and shore protection structures, and anchorage systems. Exercise care in locating positions of round piles and posts that support horizontal timbers that should be used as driving guides for sheet piles. In all cases, use a driving guide for sheet piles, preferably the permanent horizontal wale that is attached to the vertical round piling or posts.

It is particularly important to drive the first few sheet piles accurately vertical in all directions. The wall should be plumb and the sheet piles must not be inclined within the plane of the wall. One of the common problems facing piling contractors is "creep," the tendency for successive sheet piles to lean more and more in the direction of construction of the wall. The wall can be perfectly plumb, yet piles can lean; this error in alignment tends to accumulate and, if left uncorrected, can create considerable difficulties in driving successive piles. As sheet-pile driving proceeds, place the tongue of each new sheet in the forward position and the groove in tight contact with the tongue of the sheet previously driven. Keep the joints between piles as tight as practicable. Remember that the maximum allowable opening at joints is 1/2 (13 mm) for splined and Wakefield piling, and 1/4 (6.4 mm) for tongue and groove piling. If wider joints appear after the sheet piles have been spiked to the outer wale, cover with treated timber lath to prevent the backfill material from gradually filtering through the cracks and being lost.

Wherever passive-resistance anchorage systems are used, the anchorage should be well covered with a mound of earth before backfill material is deposited to any appreciable depth against the piling. Otherwise, the pressures generated by the backfill may disrupt the sheet piling and the anchorage system.

Use a predominately granular material for backfill adjacent to the sheet piling and over the anchorage system. Shoreward of the anchorage, a poorer quality filling material may be used unless it is objectionable from the standpoint of foundation support for shore structures.

If a hydraulic method is used for backfilling, provide sufficient drainage to permit rapid escape of water at the ends of the construction area, both to prevent formation of pools and to maintain as low a free water level in the backfill as possible. Although the bulkhead is designed to hold earth, it may not be designed to resist water pressures that can be generated during hydraulic filling.

One method of facilitating drainage is to provide openings through the sheet piling above the level of the outside wale. Space these openings at intervals of about 60 (18.3 m) to supplement the escape of drainage water. The final 3 (914 mm) of backfill adjacent to the sheet piles should be put into place by earthmoving equipment to avoid hydraulic pressures at the upper parts of the bulkhead. The hydraulic discharge line should be parallel to the bulkhead alignment, not directed at it, and should be located at least 100 (30.5 m) behind the bulkhead sheet piling.

At a construction site, the natural conditions that exert the most influence on the design of any waterfront structure are water level variation, wave action, and type of soil. Ice conditions are a special consideration for locations subject to the effects of solid ice sheets, floating ice fields, or large icepacks.

Often the natural soil is capable of providing the necessary resistance for the lower ends of the sheet piling in seawalls, bulkheads, and groins. However, if the bottom soil is soft silt, mud, or soft clay, it should be removed and replaced with granular materials.

In most cases, earth fill is required above the existing ground line for some distance shoreward from the face of the sheet piling. The filling material for a sufficient width to encompass the anchor system should be predominantly granular in nature, even though it may be necessary to transport it from a considerable distance.

5.13. Project Completion - Moving Off Site

5.13.1. Planning

Well before completion of driving, the superintendent should start planning project completion details. While it may be too early to finalize plans, the general superintendent will advise on the probable disposition of the rig and reassignment of personnel. In many cases, it may be necessary to plan alternatively for temporary storage on the site and for loading out to another location.

The superintendent should discuss with the client plans for moving off so as to be certain that a suitable area should be free as near as possible to the point where driving should be completed.

At this time all equipment should be inspected and the need for repairs discussed with your general superintendent. He will provide directions on items that should be shipped to yards or shops for major repairs and those that can be repaired at the site. Any parts that should be required for project site repairs should be ordered so that they should be available when needed.

In the absence of some specific instructions to the contrary, it is customary to store the basic rig at the site unless it is to be shipped to another project. This includes the crane, boiler and boiler platform, boom, leaders, hammer extensions, spotter and mats. An exception might be when no storage space is available at the site or a storage yard is located nearby.

Hammers, cores and drills may be stored on the site but it is frequently more convenient to ship them to a yard or shop as they often need repair and may be required independently of the rig. If stored at a remote location, it is usually expensive to mobilize a crew when they have to be shipped. Excess shells are normally sent to the nearest yard, unless your general superintendent instructs otherwise.

Special items that are unlikely to be used with this particular rig again, such as followers, jet pumps, long moonbeams and pile lifting brackets, will normally be loaded out to the nearest storage yard so that they should be more readily available when needed. Tool boxes, office trailers and light equipment may be stored with the rig unless there is a security problem, in which case they may have to be moved into a yard.

If the rig is to be moved to another project, ascertain exactly what equipment items should be required so that unneeded items can be sent to a storage yard. One should avoid leaving miscellaneous items at a project site as the cost of sending a crew in to load them later may exceed by several times the cost of trucking them to a yard.

Before making definite plans for trucking a rig to a new project, be sure you understand what unloading facilities should be available at the new site and any unusual restrictions, such as low clearance or low-load-limit bridges. If you are not familiar with the truck size and weight limitations prevailing in the area, obtain these from your general superintendent and/or local trucking company.

At this time, the superintendent should make arrangements for disposal of cut-off shells and other debris, if this has not already been done. He should also give notice of the anticipated date that he will release any rented equipment, as may be required in various rental agreements.

As soon as plans for disposition of equipment are finalized and approved by your general superintendent, order the necessary trucks for loading out equipment and for the use of a service crane, if this should be required. If railroad cars or a barge is required, it may take 10 days or longer to mobilize them.

5.13.2. Completing Work Assignments

As the driving nears completion, double check, by reviewing the piling plan and by a physical check in the field, that all piles should be driven and that all other work required under the contract will have been completed. Any contract work, which, at this point, is to be omitted, must have been confirmed in writing, so that there is no possibility of a misunderstanding at a future date. While it is always desirable to have concreting follow the driving as closely as possible, it is particularly important that there be no unnecessary lag at the end of the project. Unless you have specific instructions to the contrary from your general superintendent, the rig must not be moved off or dismantled until all piles have been concreted and accepted by the client.

The specifications and the contract documents should be reviewed at this time and the need for any contingent items, such as retapping, be resolved with the resident engineer. If not all piles can be cut off, be sure it is understood who will do it when they are uncovered and who will dispose of the waste shell. Any matters that cannot be resolved satisfactorily should be referred to your general superintendent in ample time so that action can be taken before you leave the site.

Pile cut-offs are very easily recognized and they should not be left lying around the site. While cut-offs can occasionally be sold, it will very often be necessary to pay someone to haul them away.

5.13.3. Storage on the Site

A storage site should be conveniently located near where driving should be completed, should be accessible to trucks for loading out when required and not have to be vacated within the anticipated storage time.

It is usually preferable to walk the equipment from the driving site to the storage site standing up. If the hammer and core are to be shipped out, this should normally be done as soon as driving and concreting are completed.

Unless it is anticipated that the rig should be driving at an early date on the same site, the rig should be prepared for loading out. If not shipped out, the hammer and core should be stored so that they can be loaded without moving other rig components; leaders should be lowered and disassembled into truckable sections. Standing leaders are attractive to children and could fall as result of ground settlement or high wind.

A careful inventory should be made and some photographs taken to show how the equipment is stored. These should be forwarded to your general superintendent with a letter explaining the storage arrangements.

5.13.4. Loading Out

While a local trucker may be of considerable help in planning your move, the superintendent must familiarize be familiar with size and weight restrictions to accomplish the move off as economically as possible. When moving to another project, it is important that items of equipment arrive at the new site in the order in which they should be needed. It may be necessary to arrange for extra trailers so that all items can be loaded without rehandling and still arrive at the new site in the order needed for erection.

Whether a crane is used for loading out will depend, to a great extent, on how far the rig should be disassembled to comply with trucking requirements. If the cats have to be removed from the crane, one usually finds that it pays to have a helper crane. If the crane has any distance to travel, it will very often expedite the move to get the crane on the road as early as possible, and finish loading with a light crane. Trailers with heavy and wide loads move slowly as they are often restricted from moving during certain hours, at night or over weekends.

In planning loads, effort should be made to combine heavy and bulky items. As an example, mats can often be shipped with boom or leader sections in order to utilize the weight capacity of the truck without having an overwidth or overheight load. In some areas, several truckloads can be combined on a single bill of lading to get a more favorable rate. Valuable or breakable items should be removed and stored in toolboxes or otherwise protected.

As a general rule, the labor costs for loading, as well as crane rental, are charged to the project that is loading out. Freight and trucking is charged to the new project.

5.13.5. Repairs

Whenever possible, the superintendent should make the necessary repairs to equipment before it is shipped or stored. Avoid shipping faulty equipment to a new project. However, if time does not permit you to get parts and make repairs before shipping, make sure that the parts are ordered for the new location and advise your general superintendent so that work can be accomplished without delaying the new project. All equipment should be lubricated before it is stored or shipped.

5.13.6. Final Records and Reports

The superintendent should assure the final pile reports and work order slips are prepared and signed promptly. Be certain that all pay items such as test piles, etc., have been included. Check that load test reports and comparative driving records have been completed and submitted:

1. Numbered Piling Plan
2. Clearance form, signed by client
3. Project Ending Report
4. Final billing data sheet
5. Shell Report
6. Transfers covering all equipment sent to storage or shipped.

Before leaving the site, the superintendent should be sure that he has submitted all delivery tickets covering material received and that any invoices for local purchases have been approved and submitted for payment. Instructions should be left at the local post office to forward mail to the district or regional office. Finally, advise your general superintendent the status of any matters not completely resolved before you leave the site.

5.13.7. Transportation

The efficient and economical movement of the rigs, components and parts from one location to another is essential to the successful outcome of a project. The superintendent must plan all major moves with the help of his or her general superintendent who can advise him on the best local methods. The following comments are intended to acquaint the superintendent with some of the problems he may encounter.

Figure 5-91 Loading Out

5.13.7.1. Local Trucking

Trucking is the most common and, usually, most economical method of moving both complete rigs and major components due to minimum of rehandling and the close control that can be exercised over movements.

Advice of the general superintendent should be sought in selecting a local trucker, as adequate equipment and reliable service have more effect on final cost than do trucking rates.

On local moves, trucks are usually rented by the hour or day, rates depending on size of truck, and accessories provided. Moves of major items, like cranes, are often quoted on a lump sum basis; the trucker may also furnish all the required labor.

Limitations on size and weight of loads on short moves are often governed by practical factors such as overhead clearances, weight limits on bridges and capacity and type of vehicles. The local trucker can best advise on what is feasible.

5.13.7.2. Long Distance Trucking

Interstate "heavy hauler" charges are based on weight, mileage, type of cargo, and the vehicle used, with an allowance of two hours free time at each end for loading and unloading.

The home office should normally arrange long moves and services as rate vary widely between different truckers.

Figure 5-92 Loaded Truck

State laws that are often strictly enforced govern limitation on size and weight of loads. While laws vary from state to state, as a general rule, any item of equipment, up to 8' (2.44 m) wide, 9' (2.74 m) high, 40' (12.2 m) long and weighing under 40,000 pounds (18,100 kg) can be moved anywhere in the United States, without special permits.

Overwidth permits up to 12' (3.7 m) can usually be obtained without difficulty and even wider loads can be moved by permit in some areas. Overlength permits up to 60' (18.29 m) are usually available, for telescopic trailers.

While the legal height limit is 13'-6" (4.1 m) in most states, there are many low bridges on the highways which will not provide this clearance. While permits can usually be obtained for overheight loads, this requires a survey of the route or detailed knowledge of local conditions. Not too many problems are encountered up to 14' (4.3 m) but this is about the practical limit, for a move of any distance.

The above heights are measured from the roadway, so allowance should be made for the clearance under the load. This is usually about 6" (152.4 mm) for a crane on a beam trailer, 24" (609.6 mm) to 42" (1066.8 mm) for a lowboy (depends on type and tire size) and 4'-6" (1.38 m) for a platform or telescopic trailers.

Weight limitations, even with permits, often govern long truck moves and the limit in the more restrictive states is about 60,000 pounds (27,200 kg) of cargo.

Non-permit loads can move at normal highway speeds, at any time, and will cover 300 to 500 miles per day. However, with permit-loads, various restrictions are imposed and frequently these special loads can travel only during daylight hours from Monday to Friday. In congested areas, permit-loads cannot move during morning and evening rush hours and sometimes are even more severely restricted. Permit-loads may require escort vehicles that add to overall cost. Heavy loads, on lowboy trailers with small tires, may not average more than 150 to 200 miles per day.

In any event, a heavy-hauler is always preferred for loads over 20,000 pounds (9,068 kg) and, if time is a factor, on much smaller loads as well.

5.13.7.3. Motor Freight

"Motor Freight" is distinguished from heavy hauling in that it is a service provided by common carriers that operate between fixed terminals on regularly scheduled routes. While open type trailers are sometimes available, most freight is carried in van-type trailers.

While legal limits on motor freight carriers are the same as on the heavy haulers, the practical limits are what can fit through the rear doors and into a van trailer. This is usually about 7'-6" (2.29 m) wide by 7'-6" high and not over 38' (11.6 m) long.

Motor freight is built around handling by forklift truck and palletized loads usually move through terminals most quickly. Motor freight shipments will travel, on the average, 50 to 200 miles per day.

Where time is not any factor, motor freight is an economical method of moving equipment components and parts weighing from 100 pounds (45 kg) up to 15,000 pounds (6800 kg).

5.13.7.4. Ocean Freight

Most international (and occasionally domestic) movements of equipment are by steamship. The most significant fact about ocean freight charges is that they are often based on volume or "cube" rather than weight. Except for a few items, like air/steam hammers and cores, volume rather than weight usually fixes shipment charges. A loading out for an air steam hammer being shipped by ocean freight is shown in Figure 5-93.

Figure 5-93 Loading Out Ocean Freight

In addition, extra charges are made for cargo over 35' (10.7 m) in length and "heavy lift" charges are assessed on heavy single pieces, usually starting at about 9,000 pounds (4081 kg). Accordingly, it is usually economical to disassemble all major items, where possible, before loading. Containerisation is always preferable for ocean freight.

Almost every major world port has heavy lift equipment that can handle any of the pile driver components. However, in less developed areas, careful advance planning is needed to insure that facilities are available to unload and assemble heavy equipment and transport it to the project site.

While most other carriers are fully responsible for the safe delivery of the cargo entrusted to them, water carriers have limited liability. Before any item is shipped by water, you must know the items to be shipped, their estimated value, the name of the vessel or steamship line and when the shipment will depart. In many cases, one must get authorization for the move from United States Government authorities. Documents should be prepared to comply with regulations of both exporting and importing country.

5.13.7.5. Barge Movement

It is frequently possible to move major items of equipment by barge without disassembly. However, no such movement should be planned without a great deal of planning, including insurance considerations. Adequate time to conduct such a survey should be allowed.

5.13.7.6. Air Freight

Domestic airfreight costs are roughly twice the cost of corresponding ground transportation. Considering the substantially increased speed of movement, it should be preferred whenever time of delivery is of any consequence.

While airfreight service is available anywhere there is regular plane service, the usual passenger plane will not take airfreight that weighs over 200-250 pounds (91 - 113 kg) per piece and cannot fit in the regular baggage compartment. However, about fifty U.S. airports, as well as many large overseas cities, are served by air freight planes which will handle almost anything that can go on a regular truck.

International air freight, except on packages up to 25 pounds (11.3 kg), is usually prohibitively expensive, frequently many times more than ocean freight and usually more than the value of the item shipped. Except in urgent situations, it should be used sparingly for out type of freight.

Except on international shipments, requiring documentation, it is preferable to deal directly with the airfreight department of one of the scheduled airlines rather than through an airfreight forwarder.

5.13.7.7. Package Freight

For packages under 70 pounds (32 kg), services such as United Parcel Service or FedEx Ground are available to most places in the United States, rates are reasonable and service is excellent. One or two day service can be expected to nearby points and delivery within one week can be made to most points in the continental United States, depending upon the service requested. For an operation where this kind of freight is common, UPS can set up a computer and label printing set-up for instant weighing, calculation of rates, package labeling and billing.

Parcel Post service is available worldwide and reaches almost any place where mail is delivered. Movement is generally by rail, truck or steamship and service, while cheap, is generally slow. Within the United States, service can be considerably expedited by the use of "Priority Mail." Tracking is available with "Delivery Confirmation."

Most bus lines offer an express service, limited to packages of up to 100 pounds (45 kg). Service is fast and rates are reasonable. Packages should be delivered to the bus terminal and arrangements made to pick up at the bus terminal at the destination.

5.13.8. Loading

The exact procedures to be followed for loading equipment will depend on the type of equipment and where it is being sent.

Learn as much as you can about the transportation equipment available, capacity, height, etc., and about highway restrictions. Discuss the move with the trucker and with the receiving project. You can then plan the loading to minimize loading, transportation and unloading costs.

5.13.8.1. Loading Trucks

1. The heaviest item should be loaded on the centerline of the vehicle. The center of gravity of several heavy items should be as near as practicable to the centerline.
2. A truck should normally loaded so that the center of gravity of the load is in the middle of the body, halfway between the rear of the cab, and the tailgate. This will place 70-80% of the load on the rear axle. Either too much or too little load on the front axle will cause steering problems.
3. Trailers are usually loaded so that weight is equally distributed between the fifth wheel and the rear wheels, unless highway restrictions require a different load distribution.
4. When loads are to move some distance at highway speeds, they should be secured to prevent shifting. It is particularly important that heavy items be adequately chained. blocked, or cabled so they cannot neither move forward in the event of a panic stop nor swing.
5. The shipper shares responsibility for proper securing of the cargo. Do not leave the driver to do this alone. The carrier will refuse to move loads that cannot be carried safely.
6. Diesel fuel tanks, hydraulic reservoirs, cooling systems and lube oil are normally left full.
7. Understand size, weight restrictions applying to the move, and be sure the load complies. This is particularly critical on high loads like boilers.
8. Use of extendible trailers will often permit over length items to be moved in one piece.
9. Time can often be saved and trucking costs reduced if the trucker will piggyback extra trailers to the loading site. You can have the extra trailers available for loading while the other trailers are being moved.

Figure 5-94 Loading Step Taper Core (note tag line to control swing)

Figure 5-95 Loading Hammer

Figure 5-96 Hammer Tied Down

5.13.8.2. Cranes

For local moves, cranes can sometimes be shipped fully assembled, depending on the size of the machine and local restrictions. If weight alone is the problem, remove the counterweight; if this is not enough you can remove the two crawlers plus the boom butt, back hitch and gantry. Where the combined weight of the upper and lower works is too high, the cab may have to be separated from the carbody. This is a time consuming operation and requires a large helper crane. As a last resort, one sometimes has to remove drums, engines or other components to keep the weight under legal limits. If overwidth alone is the problem, you will have to remove the crawlers and turn the carbody around so that the longest dimension is parallel with the body of the vehicle. When available, a beam trailer is preferred for hauling a crane, since the overall height of the load can be held to about 6" (152.4 mm) higher than the crane itself.

The crane should be located on the trailer or car so that the center of gravity is within 1" (25.4 mm) of the centerline and halfway between the trucks of a rail car. On a highway trailer, the load should be positioned so that the load on the fifth wheel and on the rear axles will comply with highway restrictions.

When loading a trailer with removable gooseneck, the front of the trailer can be lowered to the ground for loading. For any other type trailer, or for a rail car, you will need a suitable ramp. If you are going to leave the crawlers attached, it is better to load over the end. If the crawlers are to be removed, you can usually load over the side or end, whichever is most convenient. This will depend sometimes on the construction of the crane. In any case, you will have to block up under the end or both edges of the car or trailer; leave the blocking in place until the crane is in its final position. Be sure you have the car or trailer chocked to prevent movement in either direction. If the crane is to be loaded with counterweight, you may have to leave the boom on until you have climbed the ramp; otherwise, the crane may be too tail-heavy. The counterweight may be left on if the capacity of the car or trailer and weight restrictions permit; in such case the railroad will require that it be blocked up from the car floor.

When jacking up to remove crawlers or to swing the carbody, be sure the jack has adequate bearing to prevent breaking through the deck. Blocking should be located so that loads are not concentrated at a single point.

In loading any crane, the project should be easier if the machine is clean. The day before loading (perhaps while crew is unbolting leaders), have a person clean mud from the undercarriage, particularly around the bolts that attach the crawlers. Wire brush the heads and threads and apply some oil to the threads. Long handled garden spades make good tools to remove mud from a rig. Four 3/4" (19.1 mm) x 20' (6.1 m) slings should be used to remove cats, with the eyes of the slings slipped over the ends of pads. Be sure you have proper size box wrenches on hand to fit the bolts. You may need jacks or come-alongs to pull the cats free of the carbody. If there are many bolts to be removed, it may pay to have an impact wrench available. If you must separate the upper and lower works, consult your crane instruction manual or local representative; be sure you understand how to do it properly.

Figure 5-97 Crane Loading Out

5.13.8.3. Other Equipment

The following should be considered when loading any equipment:

- If proper load distribution can be maintained, load the heaviest items of a mixed load over the axles.
- Combine heavy and bulky items to minimize the number of loads. Mats, hammers and core should normally be mixed with boom and leader sections.
- Valuable or easily damaged items should be removed and boxed, preferably blocked by heavier items.
- Exhaust pipes and other openings should be closed to keep out rain. Protect other items from the weather as required. Equipment being sent overseas may be exposed to salt water spray for many weeks.
- Do not place cans or gasoline or flammable liquids or bottles of propane or other gases in toolboxes. Empty cans should be flushed with water before shipping.
- When shipping by common carrier, (whether truck, rail, air, or water) all fluids should be drained. This is not only because of fire hazard but also because of risk of leakage and damage to other cargo.
- Always tag equipment being shipped to indicate if tanks or cooling systems are drained, what maintenance has been performed and notation of condition of the item or any repairs needed.
- Remove projecting bolts and rods and pad sharp corners that are likely to damage adjacent cargo; otherwise, carrier may bend them or cut them off.
- Provide lifting eyes or clearly mark lifting points for easily damaged items such as boom and leader sections that can be damaged by lifting by the diagonals.
- Counterweights shipped on edge, or similar items with a high center of gravity, should be properly blocked to prevent tipping.

5.14. Storing Equipment

Equipment generally deteriorates more rapidly in storage than when in active use. Accordingly, all plans for storage of company equipment must have the approval of your general superintendent in advance. An example of stored equipment is shown in Figure 5-98.

Figure 5-98 Stored Pile Hammers

How long a piece of equipment can remain in dead storage and still be usable will depend on the type of equipment, the storage site, the climate, and care taken in preparing for storage. Under average conditions, equipment will probably not be usable without major overhaul if stored for longer than the following:

Hydraulic Drills, Power Packs, Power Shell Wrenches, etc.	1 year
Diesel Engines and Jet Pumps	2 years
Steam Engines, Air/Steam Hammers, Boilers, Cranes and Hoists	4 years
Leaders, Booms, Followers, Cores, Spotters	5 or more years

The above assumes that the minimum protective measures outlined in this section are taken.

Standard items of construction equipment, such as hose, rope, hand tools, compressors, welders, small pumps, trucks, etc., should not be stored, except for short periods. They should either be sold locally or shipped to the nearest yard for disposal; consult your general superintendent.

5.14.1. Selecting a Site

To select a local storage site, the assistance of your general superintendent and home office should be sought. A good site should have the following attributes:

1. Access for heavy trucks on all-weather roads.
2. Protection from natural hazards, such as flooding, high tides and salt spray and a safe distance back from slopes and areas to be excavated or subject to erosion.
3. Controlled access to prevent theft and vandalism and keep children away.
4. Reasonable assurance that equipment will not have to be moved prior to end of anticipated storage period.
5. Reasonably level, well-drained area, preferably graveled.
6. Sufficient area so that equipment is not so crowded that it should be piled one piece or another.
7. Covered storage for hose, tool, boxes, hydraulic and electrical equipment or other items subject to damage by the weather.

5.14.2. General Considerations for Long Term Storage

1. Within the limits of the site available, effort should be made to afford the maximum protection from deterioration or loss.
2. When possible, all hand tools, power tools, light equipment, hose, rope and cable and tires should be stored indoors or under cover. All gasoline and diesel engines, if not stored indoors, should be covered with tarpaulins securely wired down.
3. While the rusting of exposed parts is easily seen, the corrosion of finely finished interior surfaces can be even more serious and may render the equipment useless in a short time, if not protected against.
4. Equipment should be stored off the ground, on blocking, to keep it dry, to make it easier to load out and for painting and maintenance, should this be necessary.
5. All items should be so located on the site that it is possible to load out individual pieces, without moving major components. This is particularly true of vehicles, pumps, welders, hammers, cores and followers.
6. When landing individual units, they should be positioned to that number plates are visible for inventory purposes and controls are accessible for starting and testing.
7. All equipment should be thoroughly cleaned before storing. Dirt frequently contains chemical salts that are very corrosive. Non-working surfaces should be sandblasted and painted, or at least given a coat of steam cylinder oil. Exposed threads should be greased.
8. All equipment should have oil changed and be thoroughly lubricated in accordance with appropriate lubrication instructions. Particular care should be exercised at this time to insure that all old grease, moisture and dirt are forced out.
9. Minor repairs should be completed and loose covers and guards securely bolted or wired in place.
10. Unless otherwise indicated, water and gasoline should be drained from equipment; diesel fuel, lubricant and hydraulic oil tanks should be left full. Air tanks and lines should be bled and drains left open.
11. In seacoast areas, in very humid climates or when equipment is to be stored for an extended period the use of special rust proofing compounds and coatings is recommended. Most of the major oil companies market such products and will recommend types suitable for the climate.
12. All equipment should be unloaded, blocked or dogged so that it cannot be moved accidentally.
13. Brake and clutch bands should be in the released position (blocked, if necessary) and protected from the elements.
14. Unguarded equipment should have stairways and ladders removed or boarded up to minimize attraction to children.
15. Door and windows should be locked; if there

is danger of vandalism, all windows should be boarded over.

16. Review operator's manuals for any specific storage recommendations.

5.14.3. Engines

Engines should be prepared for storage as recommended in the maintenance instructions relating to that particular engine. In the absence of such instructions, the following may be used as a guide:

1. Crankcase oil should be drained while engine is still hot from running, new oil filter installed, strainer cleaned, crankcase refilled with fresh oil and engine run for 5 to 10 minutes to circulate oil. The rust inhibitors in new oil will help protect the engine and new oil should be free of acid or other harmful contaminants.
2. Drain and flush radiator and refill with permanent type anti-freeze. For the United States, 50% solution should be considered standard. In tropical areas, radiators are best left drained and tagged if there is any chance the engine may be shipped out. Radiator and hoses should be examined for leaks.
3. Blow out radiator cores; clean screens and the fins on air-cooled engines, compressors, etc.
4. Drain gasoline tanks, fuel pump strainers, carburetors and all fuel lines on gas engines. The fuel tanks of diesel engines should be filled.
5. Remove covers and spray rocker arms and push rods with engine oil.
6. Remove spark plugs and spray or squirt engine oil into each cylinder. Turn engine over by hand so cylinder walls should be coated. Dip plugs in oil and replace.
7. Remove air cleaner covers; clean; clean or replace filters; refill oil cup with fresh oil and reassemble. Clean air compressor intake and crankcase breathers.
8. Release tension on all belts
9. If engine does not have an exhaust pipe cover, the exhaust should be covered with a small can securely taped or wired in place.
10. Transmissions and torque converters should be drained, refilled with fresh oil and operated for 5 to 10 minutes; filters should be cleaned or element replaced. Clutches should be disengaged; block clutch in disengaged position, if necessary. Coast exposed machined surfaces with grease.
11. All openings to the engine and accessories should be plugged or sealed with plastic securely taped in place.
12. Storage batteries should be removed, cleaned, recharged as necessary, water level maintained, protected from freezing or excessively high temperatures and recharged periodically.
13. Replace and secure side panels of engine enclosures.
14. Two tags should be placed on the equipment (one at the operator's station) indicating the maintenance performed, condition of the engine, whether radiator is drained, etc. In non-English speaking areas, the tag should be in both English and local language.
15. Arrange to have engine run periodically while in storage, preferably once a month. If this is not feasible, they should at least be turned over by hand while additional oil is squirted in spark plug openings.

5.14.4. Cranes

1. Machine should be on dry, firm, level ground and free to walk and swing for loading out other items in storage.
2. Do not leave a crane in an excavation that could be flooded.
3. If the crane cannot be placed on a firm well-drained surface, put it on mats or timbers to keep it out of the mud. This is particularly important when freezing weather is expected.
4. Thoroughly clean the machine, removing all mud from the undercarriage. Steam cleaning is desirable but the crane at least should be hosed down with fresh water. It is difficult to assemble components or do a good lubrication

project, on a dirty machine. When draining boiler, the hot water from the blow-off can be used for cleaning equipment.

5. Lower loads, engage travel and swing locks and drum pawls; block any movement that does not have its own lock. Service brakes and clutches should then be released.
6. When space permits or if boom is on a live drum, it should be lowered to ground. Buckets should be stored upright in open position or tilted so that rainwater does not collect in them.
7. Carefully inspect the crane, including undercarriage, gantry, topping lift and boom for breaks, loose bolts, cracks, missing parts, etc. Clearly, mark or tag anything that cannot be corrected immediately.
8. Thoroughly lubricate the machine in accordance with manual furnished with the crane. Keep lubricants off brake and clutch bands.
9. Process engine for storage as detailed above. Protect hydraulic system as required. Bleed air system; drain all tanks and lines and leave drains open.
10. Counterweights or other removed accessories should be stored safely. Replace pins and bolts in their holes and secure with cotter pins or nuts.
11. Wire rope should be lubricated as it is wound on drums by brushing each layer with an open gear lubricant.
12. Replace any broken windows with safety glass; close and lock all doors. Exposed hoisting drums should be covered with a tarpaulin, and wired down. When likely to be damaged, windows should be boarded or welded up.
13. Crane should be started up once a month while in storage; walk machine back and forth a short distance to circulate oil in gear cases and free seal surfaces. One should welcome the opportunity to have the crane used occasionally by a competent operator, as this will definitely prolong its life.

5.14.5. Pile Driving Equipment

1. Clean mud and dirt from all surfaces and hose down with fresh water. This is particularly important if equipment has been operating in seacoast area or heavy smog areas. Steam cleaning is even better.
2. Leaders and booms should be lowered and stored on timbers off the ground and in a position so that water will drain.
3. Pockets where water can collect are usually provided with weep holes; check that these are open; drill or cut them, if necessary.
4. Scrape or wire brush rust spots and touch up with paint.
5. Machined surfaces should be cleaned, coated with grease and wrapped or covered with plastic.
6. Lubricate all sheaves, pins and sliding surfaces.
7. Replace all pins in their holes and secure.
8. Lubricate wire ropes; dirt should be wire brushed and a penetrating type of oil applied with a brush. Oil should be thin enough to penetrate between the wires. Use an open gear lubricant, or any good lube oil; do not use old crankcase oil.

5.14.5.1. Pile Hammers

1. Store in cradles blocked up off ground, allowing sufficient space for inspection and minor repair; keep accessible for loading out without moving other equipment. Such a cradle or skid is shown in Figure 5-99[21].

Figure 5-99 Pile Hammer in Skid

2. Coat sliding machined surfaces with grease and wrap with plastic or tape, if exposed.

This includes slide bars, hammer columns, hammer piston rods and exposed portion of lower cylinders on differential hammers.

3. Lubricate sheaves and operating linkages, being certain to force out old grease, water and dirt.
4. If practicable, the last operation of the hammer should be with dry steam with an excess of steam cylinder oil.
5. Pour one or more gallons of steam cylinder oil into hammer after it is stored on the ground, distributed between the intake, exhaust port and auxiliary exhaust (on single acting hammers). Operate valve back and forth, so that oil will pass into cylinders. This oil will coat areas where water is likely to collect during storage.
6. Plug or cover all openings securely to prevent rain and dirt from entering inside the hammer.
7. Enclosed hammers should be drained and plugs replaced.
8. Hydraulic hammers should be filled with hydraulic oil and all connections tightly capped.
9. All hoses should be removed and stored under cover.

5.14.5.1. Pumps

1. Store off the ground and accessible for inventory and maintenance.
2. Remove drain plugs, open clean-outs and thoroughly flush pump with fresh water; this is particularly important if pumping salt water.
3. Remove packing, clean around shafts and fill packing space with grease.
4. Completely drain casing; tilt pump if necessary to get all of the water out. Replace drain plugs; liberally spray interior of casing with lube oil; close all openings including inlet and outlet, to retain oil and keep out rain.
5. Clean exterior of pump, remove all mud from casing, shafts, couplings and frame; use steam, water jet or wire brushes if necessary.
6. Prepare engine for storage as detailed previously.
7. Lubricate pump; coat threads, shafts, piston rods and other machined surfaces with grease and wrap with plastic; paint other exterior surfaces or coat with steam cylinder oil.
8. Remove heads of piston pumps, clean, coat walls of cylinders with oil and replace heads.
9. Make minor repairs and report any parts needed or major repairs required.

5.14.6. Hydraulic Equipment

Whenever possible, hydraulic components should be stored in inside heated storage. If stored outdoors:

1. Store units on blocking up off the ground away from dampness.
2. Store power packs upright so that oil will flood pumps and valves. Other units should be kept filled with oil in whatever way is practical.
3. Securely cap all openings to keep out moisture, dust and dirt.
4. Protect threads and other projecting parts from physical damage.
5. Unless the hydraulic oil has been very recently changed, drain the system, install new filter elements, fill with fresh oil and operate for 10 minutes to circulate the oil. Fresh oil should better protect the interior surfaces.
6. If practical, shroud hydraulic components with plastic taped up to keep out moisture. Units should at least be covered by tarpaulin.
7. Do not leave any hydraulic components disassembled and open to the atmosphere.

5.14.7. Electrical Equipment

1. Generators, motors, welders, transformers and similar electrical equipment are very easily damaged if moisture enters the windings. The only proper storage for idle electrical equipment is indoors in a heated

building.

2. If not stored out of doors, place in driest available location, cover, close all openings and do whatever is possible to keep out moisture. Keep dirt away from commutators and other moving parts. A light coating of Vaseline on exposed contact surfaces will retard corrosion.
3. When outside for more than a few weeks, moisture is almost certain to enter the windings due to condensation of the moisture in the air. This lowers the resistance of the insulation and increases the danger of short circuit when the unit is returned to service.
4. The chances of breakdown can be reduced by thoroughly drying out the windings before the unit is returned to service by one of the following:
    a. Send unit to a motor repair shop for baking out.
    b. Move the unit into a dry heated area, open up and allow drying out slowly for several weeks. This is cheapest, if time permits.
    c. Place the unit in an enclosure, heated to 150-160°F, and bake out for 10-12 hours. Be sure to leave a vent for moisture to escape and take care not to overheat or set afire.

5.14.8. Motor Vehicles

Experience has shown that most motor vehicles stored more than a month or so usually wind up as junk. If motor vehicles should be stored, the following can extend their life:

1. Wash the vehicle, particularly the under body to remove dirt and road salt.
2. Lubricate and change oil and filters in accordance with the manufacturer's recommendations.
3. Check cooling system; be certain radiator is protected for the lowest possible temperature for the area, even if it is summer.
4. Check wheels to prevent accidental movement.
5. If vehicle is blocked up (preferable), tire pressure should be maintained at 50% of normal. If tires are loaded, they should be maintained at normal inflation.
6. Inspect tires and repair any cuts where cord is exposed. Clean off any oil or grease. Tires should be protected from sunlight so it is best to remove tires and store indoors. The next best is to shroud tires with opaque plastic film, taped in place.
7. Protect engines as detailed previously. Drain gas tanks and lines; leave diesel systems full; depress clutch and block; release brakes.
8. Be sure trucks are parked so that water does not puddle in the body. Dump bodies should be blocked up slightly so water will run out.
9. Repair broken windows and doors. Close and lock all doors and windows.
10. Sandpaper any rust spots on the body and touch up with paint. Coat bright work with water resistant grease.
11. Be sure that winches or other equipment mounted on the vehicle are properly protected; cover with tarps or plastic, if exposed.
12. Clean tool compartments and be sure drain holes are open. Clean and lubricate jack, chains, cables and tools; lock up or remove to safe storage.
13. Remove battery to some place where it can be properly maintained and kept charged.

If the vehicle is not worth the above steps, dispose of it while it still runs. If you store for it, you will probably have to pay someone to haul it away.

5.14.9. Marine Equipment

1. Storage sites for marine equipment present special problems; discuss situation with your construction manager well in advance.
2. Determine if insurance or safety inspections are required to arrange to have these made.
3. If equipment is not to be dry docked, it should at least be thoroughly hosed down with fresh water. Mud containing salt will cause extremely rapid destruction of metal parts.

4. Thoroughly lubricate all equipment on board. Give particular attention to linkages, sheaves and pins that have been exposed to salt water.
5. Check cleats, bollards, chocks and fenders and make repairs, if necessary.
6. Pump out all compartments; remove debris; flush salt-water tanks with fresh water. Make emergency repair from inside, if tank is holed through. Spray inside of tank with rust retarding oil.
7. Check all manholes, using new gaskets, if needed to make them watertight.
8. Rafts and small craft should be removed from water and stored on deck of large units, securely lashed down.
9. Life preservers should be inspected, cleaned, dried and stowed out of the weather. Life rings should be left in place on deck unless it is necessary to remove them to prevent theft.
10. When practicable, leaders and booms should be lowered to decks and made secure.
11. Small gear should be stored below deck or inside deckhouses. Anything remaining on deck should be secured.
12. Anchor cables should be washed with fresh water as brought aboard, lubricated, wound on drums and covered.
13. Manila lines, except as needed during storage, should be cleaned, dried and stored below deck.
14. Deck engines and similar equipment should be covered with tarpaulins, securely wired down; be certain that frictions and brakes are protected from weather. Crossheads, piston and valve rods, etc. should be coated with grease and wound with plastic strips.
15. Engines, compressors, welders, generators, jet pumps, etc., should be prepared for storage as previously noted, covers installed and securely wired in place. Heat exchangers should be drained and drain cocks left open. Air tanks should be blown down and all drains left open.
16. Compasses, radios and navigation equipment should be brought ashore for storage.
17. Deckhouses should be boarded up or otherwise made secure.
18. A portable pump should be available for pumping out.
19. Except as they should be removed for security purposes, walkways and ladders required for access to equipment and between equipment should be made safe and left in place.
20. An experienced person with knowledge of local tides, currents and weather should check the provisions for mooring or anchoring. Anchor lines should be made fast to bits or cleats and anchor winch drums slacked off.
21. Arrange for security guard service. Perhaps for a larger storage site they will require a vehicle such as is shown in Figure 5-100[22].

Figure 5-100 Security Service Vehicle

5.14.10. Reporting

When storage is completed, send to the local office, to your general superintendent, or to a regional office:

1. Complete and detailed inventory of everything at the site with an indication of its condition.
2. Rough sketch of site showing means of access.
3. Photographs that will indicate general storage conditions, accessibility of particular items, etc.

4. Details of arrangements for the use of the site including name, address and telephone number of owner or his or her local representative, length of time one may use the property, rental, if any; any obligations undertaken or other arrangements for use of the property.
5. Name, address and telephone numbers of local truckers, crane owners, riggers, etc., who might be needed to assist in loading out the equipment or individual pieces.
6. Instructions for reaching the site from nearest major city or major highway; location of limited-load bridges, low overhead trucks.
7. List of repairs and replacement parts needed.
8. Arrangements for periodic starting of engines and other required maintenance.

5.14.11. Summary

1. Clean thoroughly.
2. Lubricate and change oil.
3. Remove unnecessary loads.
4. Protect from weather.
5. Coat exposed machined surfaces with water resistant oil or grease.
6. Paint non-working surfaces.
7. Inspect and report.

Photo Credits

- Dawson Construction Plant (Mr. Mark Lee): Figure 5-74, Figure 5-76, Figure 5-77 and Figure 5-90.
- GRL and Associates/Pile Dynamics, Inc. (Mr. Mohamad Hussein): Figure 5-48, Figure 5-58, Figure 5-60, Figure 5-66.
- International Construction Equipment (Mr. Dick Morris): Front page photo.
- Sure-Lock (Mr. Glenn Lockie): Figure 5-65.
- Vulcan Foundation Equipment: Figure 5-1, Figure 5-2, Figure 5-28, Figure 5-33 and Figure 5-92.
- Don C. Warrington, P.E.: Figure 5-3, Figure 5-34, Figure 5-36, Figure 5-38, Figure 5-39, Figure 5-49, Figure 5-72, Figure 5-85, Figure 5-86, Figure 5-87, Figure 5-88, Figure 5-93, Figure 5-96, Figure 5-99 and Figure 5-100.

Footnotes

[1] He will appreciate the experienced help - if he already knows too much to consider suggestions he probably should be assigned to other activities.

[2] Wiss, J.F. (1981). *Construction Vibrations: State-of-the-Art*. American Society of Civil Engineers, ASCE, Journal of the Geotechnical Engineering Division, February, 167-181.

[3] Lacy H.S. and Gould, J.P. (1985). "Settlement from Pile Driving in Sands." *Vibration Problems in Geotechnical Engineering*. American Society of Civil Engineers, ASCE Special Technical Publication, New York, 152-173.

[4] Lukas, R.G. and Gill, S.A. (1992). "Ground Movement from Piling Vibrations." *Piling-European Practice and Worldwide Trends*, The Institute of Civil Engineers, Thomas Telford House, London, 163-169.

[5] Woods, R.D. (1997). *Dynamic Effects of Pile Installations on Adjacent Structures*. NCHRP Synthesis 253, National Cooperative Highway Research Program, Transportation Research Board, Washington, D.C.

[6] There may be references to ASTM or other standards, which should be reviewed. There may be reference to other parts of the specifications, such as "Concrete"; be sure these are reviewed.

[7] Supervisor must sign this paperwork directly below the last employee listed to comply with Federal regulations.

[8] Improved plow steel, fiber core.

[9] Much of this section is taken from Occupational Safety and Health Administration, *OSHA Technical Manual*. TED 1-0.15A. Washington, DC: Occupational Safety and Health Administration, 1999.

[10] For a maximum excavation depth of 12 ft.

[11] These slopes and other recommendations in this section assume the use of fixed leaders. With swinging or underhung leaders, the sway of the leaders presents additional complications.

[12] Calcium chloride admixture should be carefully controlled to prevent permanent damage to the concrete. At no time should the amount used be greater than 2% of the cement content, by weight.

[13] Source: Florida Department of Transporation.

[14] It is suggested that this weld should be a minimum of 5/16 in. and increased when attaching to material of more than 0.55 inch metal thickness.

[15] Piles can be dragged either way so point reinforcement

may be attached at alternate ends in the stack.

[16] For metal thicker than 9/16 the weld should be 3/8 in. and greater for above 3/4 in. steel. Clips should be temporarily welded to one length at the splice for quick positioning of the next length. This will also aid in preventing the pile from moving should a sudden wind gust or lurch of the crane occur. For this direct welding it is desirable to have an additional small crane or cherry picker handling the extension and holding the pile while it is firmly attached. This additional piece of equipment will permit greater efficiency for the pile driver and crew, while maintaining a safe attachment procedure.

[17] The American Wood preservers Institute (AWPI) recommends banding timber piles with heavy metal strapping at the driving end prior to driving to prevent the start of splitting.

[18] The PCI (Prestressed Concrete Institute) suggests an H pile "stinger" for tip protection and extension of prestressed concrete piles. With cast steel point protection on the H Pile, firm embedment can be obtained as well as a foothold on rock at an angle to the axis of the pile.

[19] In this case, these are usually referred to as "stingers"; these wre discussed in Chapter 2.

[20] Installation of Sheet Sheet Piles, Technical European Sheet Piling Association, 1993.

[21] It should be noted that, although the hammer shown has its ram blocked in the center between the cylinder and the base, it is very common - and mandatory for some manufacturers - to block the ram touching the base, so that the ram cannot fall when the hammer is stood up. Diesel hammers usually have a plug to prevent ram movement and this should be used during shipment.

[22] Perhaps not.

Chapter 6. Test Methods for Driven Piles

6.1. Overview

The ultimate bearing capacity of a pile is limited by either the structural strength of the pile shaft or the capacity of the supporting soil or rock. The structural capacity of the pile is determined by the allowable loads on the pile based on structural considerations alone, and is generally governed by codes. The capacity of the pile-soil system may be evaluated from static methods based on soil mechanics and standard geotechnical test methods; however, the uncertainties associated with these methods may make the foundations designed solely with them uneconomical. To address this problem, the purpose of in situ pile load testing is:

- To develop criteria to be used for the design and installation of the pile foundation; or
- To prove the adequacy of the pile-soil system for the proposed pile design load.

Many different procedures have been proposed for conducting load tests. The main differences are in the selection of loading systems, instrumentation requirements, magnitude and duration of load increments, and interpretation of results. Some innovative test procedures, which are potentially applicable to piles and drilled shafts, include the Osterberg load cell and high strain dynamic testing. The Osterberg procedure involves installation of a non-retrievable hydraulic jack at the pile or shaft base. The jack reacts against the larger of the base or skin resistance to cause a failure condition of the weaker resistance. High strain tests involve the use of heavy drop weights or explosive devices (the Statnamic procedure) to create strain and acceleration data that are used to predict capacity.

The three most important methods of in situ verification are as follows:

- Static load testing. This consists of applying loads of known magnitude to the pile head and measuring its corresponding displacements. Static load testing can be axial (along the centerline of the pile) or lateral (perpendicular to the pile's centerline.)
- In-situ Dynamic Testing. With this type of testing, a rapid load (usually with an impact pile hammer) is applied to the pile. Dynamic measurements and analyses of force and motion of the pile head are measured and analyzed.
- Statnamic™ testing. This involves the use of a force that is not as rapid as the one used in dynamic testing. By doing this, the designer seeks to induce a load rate that is closer to the static load of piles in use.

The principles, applications, and interpretation of these methods are discussed in this chapter as well as the reliability and accuracy of their results. It should be emphasized that none of the methods discussed here can be expected to yield quantitative information on the magnitude or time rate of long term settlement of the entire structure. It is primarily for these reasons that careful subsurface investigations must precede the pile design. Test results and geotechnical analyses must mutually complement each other. Since most pile foundations consist of a number of piles acting as a unit, consideration must also be given to the relationship of the capacity of a single pile to that of the whole group.

6.2. Static Load Tests

A static load test is conducted to measure the response of a deep foundation under applied load. Conventional static load test types include axial compressive, axial tensile and lateral load testing. The cost and engineering time associated with a load-testing program should be justified by a thorough engineering analysis and foundation investigation. Load tests are possible on either single elements or groups but due to cost considerations, only single element tests are performed on production projects.

Static load tests provide the most accurate means of determining deep foundation capacity and if properly designed, implemented and evaluated, should pay for themselves on most projects. Depending on availability of time and on cost considerations, the load-testing program may be included either in the design or in the construction phase. Dynamic load tests, performed in conjunction with static load tests, greatly increase the cost-effectiveness of a pile load test program and should be specified whenever piles installed by impact driving are load tested.

6.2.1. Prerequisites for Static Load Testing

Static load testing is not a substitute for an adequate foundation investigation program. In the planning stage of any load test program, the following will be required:

- Adequate subsurface exploration that results in a well-defined subsurface profile.
- Adequate soil/rock testing to determine engineering properties.
- Static analysis results to rationally select foundation type and length, as well as the load test site(s).

6.2.2. Advantages of Static Load Testing

Static load testing offers several advantages:

- Allows a more "rational" design. The load transfer can be determined much more reliably by applying a test load to a foundation element than from the results of laboratory tests or based on assumptions.
- Allows use of lower factor of safety. Many foundations are designed using a factor of safety of 3. Testing allows the engineer to use a lower factor of safety that translates into cost savings.
- Improved knowledge regarding load transfer has the potential of permitting an increase in the design load and a reduction in the foundation number or length (for friction elements) with a corresponding savings in foundation costs.
- Verifies that the design load can be attained at selected tip elevation.

The reasons often cited for not load testing include:

- Costs involved.
- Delays to contractor if done as part of construction contract.
- Delay of project if done in the design phase.

The cost of performing a load test should always be weighed against the benefits to be obtained. A load test costing $100,000 could be considered inexpensive if cost savings in the millions resulted. Delay of a project during the design or construction phase is most likely to occur in those instances where the decision to perform load tests is made at the last minute. The need for design phase load tests should be addressed in the early stages of the design phase, and construction phase load tests should be clearly specified in the contract documents. In this way, the load tests are incorporated into the schedules and unforeseen delays are minimized.

6.2.3. When to Perform a Static Load Test

The decision whether or not to initiate a load test program on a particular project will be influenced by several factors. The following criteria can be used to assess when load testing can be effectively utilized:

- When the potential for substantial cost savings is readily apparent. This is often the case on large projects, either to determine whether friction pile lengths can be reduced, or whether allowable pile stresses can be increased for end-bearing foundations.
- When safe load carrying capacity is in doubt, due to limitations of an engineer's experience or unusual site or project conditions.
- When soil or rock conditions vary considerably from one portion of a project or another.
- When the design load is significantly higher than typical design loads.
- When time related pile-soil capacity changes are anticipated (i.e., setup or relaxation)
- When using concrete friction piles - so that piles can be cast long enough to avoid costly and time consuming splicing during construction.
- When new, unproven pile types and/or pile installation methods are utilized.
- When existing foundations will be utilized to support a new structure carrying heavier loads.
- When a reliable assessment of uplift resistance for lateral behaviour is important.
- When, during construction, the load carrying

capacity of a pile by hammer formula or dynamic analysis differs from the estimated ultimate load at the anticipated tip elevation (for example, H-piles that "run" when driven into loose to medium dense sands and gravels).

6.2.4. Effective Use of Static Load Tests

6.2.4.1. During Design

On major projects, the benefit to construction of conducting a load test program in the design phase should be considered. The subsurface profile must be adequately defined to determine the optimal number and locations of load tests as well as the area over which each test can be considered representative for driving of production elements. A design phase static load test program will require highway agencies to prepare and let a construction contract. The unit cost per test will be significantly higher than for tests performed during construction (particularly if over water testing is involved), due to the mobilization of men, materials and equipment to install a small number of piles. For maximum benefit, the design load test program should be completed at least a year before project advertisement to permit foundation and structural engineers to optimise final design.

Design phase load tests offer several advantages:

- Allow load testing of alternate foundation types and selection of most economical foundation.
- Installation information can be made available to bidders - this should reduce their bid "contingency,"
- Greatly reduce potential for claims arising from pile driving or shaft installation problems, especially for piles that are difficult to splice.
- Maximize cost savings for foundations (e.g., permit lower factor of safety, permit changes in design load and number of elements, reduce number of orders-on-contract).

6.2.4.2. During Construction

Typically, the primary purpose of load tests performed during construction is to verify that the design load does not exceed allowable capacity (proof testing), particularly if set-up or relaxation is anticipated. For drilled shaft and piles installed other than by driving with an impact hammer (e.g., vibrated or auger cast), load tests during construction can be used to confirm that both the soil and the structural foundation element can safely sustain the design load.

Construction phase load tests are also commonly used to determine final tip elevation of production piles after test drive (indicator) piles are evaluated at estimated length.

6.2.4.3. Limitation of Load Tests

A load test performed on a single pile does not:

- Account for long-term settlement
- Take into account downdrag from settling soils
- Take into account the effect of group action
- Eliminate the need for an adequate foundation investigation.

The above must be considered when using load test results to design or analyse deep foundations.

Many load tests performed today are "proof" tests, which are designed to prove that the pile can safely hold the design load or to determine the design load. Proof tests do not determine the ultimate capacity so that the pile is often designed to support a higher load than necessary and can cause foundation costs to be greater than necessary. Proof tests are not adequate when the soil strength may deteriorate with time such as from frequent cyclic loads in some soils. Coral sands, for example, can cause cementation that can degrade from cyclic loads.

6.2.4.4. Categories of load tests

Types of load tests performed are proof tests, tests conducted to failure without internal instrumentation, and tests conducted to failure with instrumentation.

- Proof tests are not conducted to a bearing capacity failure of the pile but usually to twice the design load.
- Tests conducted to failure without instrumentation determine the ultimate pile capacity but do not indicate the separate components of capacity of end bearing and shaft resistance.

- Tests with internal instrumentation, such as strain gauges mounted on reinforcement bars of drilled shafts or mounted inside of pipe piles, will determine the distribution of load carried by shaft friction as a function of depth and will also determine the end-bearing capacity when conducted to failure.

6.2.4.5. Selecting and timing load tests

Load tests are always technically desirable, but not always economically feasible because they are expensive. These tests are most frequently performed to assist in the design of major structures with large numbers of piles where changes in length, size, and type of pile and installation method can provide significant cost savings. The costs of load tests should be compared with potential savings when using reduced safety factors permitted with the tests. Factors to be considered before considering load test are:

- Significance of structure: The type and significance of a structure could offset the added cost of load tests for a complex foundation when the consequences of failure would be catastrophic.
- Subsurface conditions: Some subsurface investigations may indicate unusual or highly variable soils that are difficult to define.
- Availability of test site: Testing should not interfere with construction. Load tests should be conducted early after the site is prepared and made accessible. The contractor must wait for results before methods and equipment can be determined and materials can be ordered. Advantages of completing the testing program prior to construction include discovery of potential and resolution of problems, determination of the optimum installation procedure, determination of the appropriate type, length and size of the piles. Disadvantages include increased design time to allow for load tests and testing conditions and data extracted from a test site used in the design may not simulate actual construction conditions such as excavation, groundwater, and fill. Problems may also occur if different contractors and/or equipment are used during construction.
- Location: Test piles should be located near soil test borings and installed piezometers.
- Timing: Load tests of driven piles should be performed after one or more days have elapsed to allow dissipation of pore water pressures and consideration of freeze or relaxation.

6.2.5. Application of Test Load

Three methods are commonly used to apply load to test piles: jacking against reaction piles, jacking against a dead load and dead load applied directly to the pile.

- Jacking against reaction piles (Figure 6-1) is usually the most economical method of loading. This is particularly true where other pay-piles can be used as reaction piles. It is the most practical method of testing batter piles or testing piles for uplift. Load in reaction piles is generally limited to 25% to 50% of the load in the test pile but this will depend on soil conditions. Reaction piles should be located at least five feet away from the pile to be tested and this distance may need to be increased under certain conditions; check your plans and specifications. Provisions should be made to detect any upward movement of the reaction piles. For moderate loads, consider wood reaction piles cabled together with the test beam running under the cables. When concrete piles are used, they should be properly reinforced to take the tension; at least one-half of the reinforcing should extend to the pile toe.

Figure 6-1 Jacking against Reaction Pile

- Jacking against a dead load platform (Figure

6-2) is a very desirable method if an adequate amount of dead load is readily available. This is frequently an economical method where a number of tests are to be conducted and the dead load can be moved easily from one pile location to another. Always allow at least 15% additional dead load, over the required load, to avoid tipping and to allow for eccentric loading.

Figure 6-2 Dead Load Platform

- A dead load directly applied to the pile at one time was a commonly used method. It is still adequate where the total load is moderate, the load can be applied in known increments, and the load should be applied only once and then removed. It is prohibitively expensive when recycle loading is used. Since the load should be balanced on the pile, increments should be added or removed with care to avoid tipping the platform. It is also necessary to place safety cribs at the corners with loose wedges so that the platform cannot possibly overturn, should it tip. Wedges should be monitored to be sure they stay loose as load is added.

For either of the last two methods, the total available dead load (including the platform) should be adequate for the test. Concrete blocks or piles (Figure 6-3), steel billets (Figure 6-4), water or other ballast (Figure 6-5), the driver, sand or dirt may be used. Unless otherwise specified, test piles should be loaded to a minimum of twice the design load and to soil-pile failure whenever possible. Pile tests conducted to failure optimize the test results. The failed piles in most situations may still be used as a production pile.

Figure 6-3 Dead Load of Concrete Piles

Figure 6-4 Steel Billet Dead Load

Figure 6-5 Dead Load of Ballasted Barges

Figure 6-6 Hydraulic Load Testing Equipment

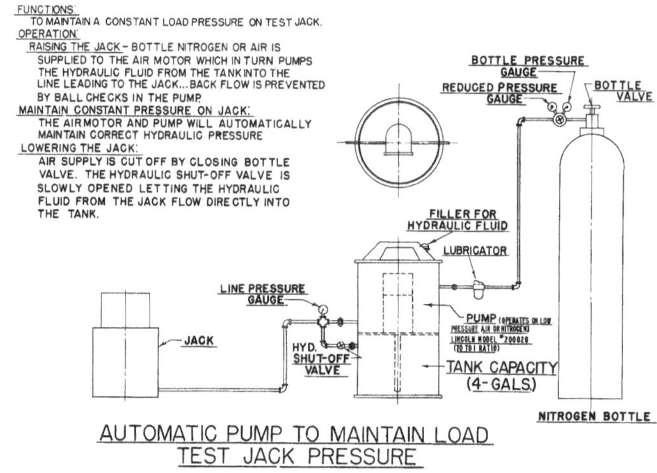

6.2.6. Jacks and Load Cells

Most load tests are conducted with hydraulic jacks to apply the load, and load cells to measure the load. A typical hydraulic jack set-up is shown in Figure 6-6. The hydraulic jack ram or the load cell should have a spherical head to minimize eccentricity between the jack and the loading frame. Both the jack (with pressure gage) and the load cell should be calibrated by a qualified lab to include calibration curves. During the load test both the load cell and the jack pressure gage should be read and compared. In the event it is later discovered or determined that the load cell has malfunctioned, the pressure gage readings will then be available. It is also important during the test to continually monitor the load cell to ensure that the load increment is being maintained at a constant value. The load has a tendency to decrease due to pile penetration into the ground, deflection of the test beams, and loss of hydraulic fluid from leaking valves, etc.

6.2.7. Pile Head Displacement Measurements

Measurement of pile head displacement can be made by use of a surveyor's level to read a rod or scale attached to the pile. Measurements are made before and after each load increment to the nearest 0.001-foot. Two reference bench marks should be selected that will not be disturbed during the tests. It is recommended that this method be used at least as a check against other settlement measurements.

Dial gauges, with .001-inch graduations, are typically employed in pairs, one on each side of the pile. They should be firmly supported independently of test pile or test load with the stems bearing on the test plate or on clips welded to the pile.

Gauges with 2" (50.8 mm) travel are typically required. Gauge stem extensions and gauge blocks may also be used to extend the range of the gauges. Readings are taken before, immediately after each load application, and at intervals thereafter. Typical gauge set-ups are shown in Figure 6-7 and Figure 6-8.

Figure 6-7 Dial Gauge Set-up: Jacking Against Dead Load

Figure 6-8 Use of Dial Gauges with Steel Scale, Wire and Mirror

Figure 6-9 Combined Vertical and Lateral Loading

Because of their extreme sensitivity, dial gauges should be used with caution. Specifications sometimes require that no additional load be applied "until all settlement has stopped." In order not to delay tests, it is common to apply some tolerance to such a requirement. It is often considered that if the rate of movement is less than 0.01" (0.254 mm) per hour, settlement has ceased for practical purposes.

Dial gauges are also extremely sensitive to temperature changes. The test set-up should be shaded from both direct sunlight and heat and insulated in severely cold weather or when wide temperature variations are expected.

A third method of measuring settlement of the test pile is to attach to the pile a mirror on which is mounted a steel machinist scale. Across the face of the scale, a wire is stretched between two stakes independent of the test pile or load. Tension is maintained in the wire by running one end through a pulley and attaching a weight to the end. By lining up the wire with its image in the mirror, consistent readings can be made. Movements of 0.01" (0.254 mm) can be measured by this method.

Because of the possibility of damage or loss of one measurement system, two independent systems should always be used which will permit resetting the primary system while still providing for continuous observation. Any deflection measurement system should be adequately supported and protected against accidental disturbance.

6.2.8. Types of Axial Load Tests

Axial compressive load tests should be conducted and recorded according to ASTM D 1143. The quick load test described as an option in ASTM D 1143 is recommended for most applications, but this test may not provide enough time for some soils or clays to consolidate and may underestimate settlement for these soils. The standard load test takes much longer and up to several days to complete than the quick load test and will measure more of the consolidation settlement of compressible soils than the quick load test procedure. However, neither the standard test nor the quick test will measure all of the consolidation settlement. The cyclic load test will indicate the potential for deterioration in strength with time from repeated loads. Procedures for load tests are presented:

6.2.8.1. Quick load test

The "Quick Load Test Method" is the recommended method for the static load testing of piles on highway projects. Originally developed by the Texas Highway Department, it is allowed as an optional load test procedure by ASTM D-1143. This method requires that a test load of 300% of the design load be applied in increments of 10 to 15 percent of the design load with load, gross settlement, and other pertinent data recorded immediately before and after the addition of each increment of load. After an increment of load is added, the load is maintained constant for a time interval of 2-1/2 minutes before the next increment is added. The test normally last 3-5 hours.

The Quick Load Test Method offers the following advantages:

- The load test can be performed in 1-2 hours, versus over 100 hours in the "standard" method, with resultant savings in time and money.
- Construction delay to the project caused by load testing is greatly reduced.
- Full-scale load testing on smaller projects is feasible because of reduced time and costs.
- Simplicity of the testing procedure ensures standardization of the test and easy interpretation and utilization of the results.

Similar advantages can be achieved by using the constant rate of penetration (CRP) load test procedure, which is also described in ASTM D-1143. In a CRP test, the load is applied to cause the pile head to settle at a predetermined constant rate, usually approximately 0.01" (0.254 mm) per minute to 0.1" (2.54 mm) per minute, depending on whether the subsurface conditions are cohesive or granular, respectively. The duration of the test is usually 1 to 4 hours, depending on the variation used. The particular advantage of the CRP test is that it can be conducted in less than one working day. A disadvantage is that ordinary pumps with pressure holding devices like those used for "slow" tests are difficult to use for the CRP test. A more suitable pump is one that can provide a constant, non-pulsing flow of oil. Appropriate references should be consulted relative to the CRP test, if it is utilized.

6.2.8.2. Standard load test

Load is applied in increments of 25% of the design load and held until the rate of settlement is not more than 0.01 inch/hour but not longer than 2 hours. Additional load increments are applied until twice the design load is reached. The load is then removed in decrements of 50, 100 and 200% of the design load for rebound measurements. This is a proof test if no further testing is performed. A preferred option of the standard load test is to reload the pile in increments of 50% of the design load until the maximum load is reached. Loads may then be added at 10% of the design load until plunging failure or the capacity of the equipment is reached. This option is recommended to evaluate the ultimate pile capacity. The duration of this procedure can exceed 70 hours or longer, especially when cyclic loading is included.

6.2.8.3 Repeated load test

The standard load test is initially performed up to 150% of the design load, allowing 20 minutes between load increments. Loads are removed in decrements equal to the load increments after 1 hour at the maximum applied load. Load is reapplied in increments of 50% of the design load allowing 20 minutes between increments until the previous maximum load is reached. Additional load is then applied and removed as described in ASTM D 1143. This test is useful to determine deterioration in pile capacity and displacements from cyclic loads.

6.2.8.4. Tension test

Axial tension tests may be conducted according to ASTM D 3689 to provide information on piles that must function in tension or tension and compression.

Some advantages to this are:

- A direct comparison of tension and compression on the same subsurface profile;
- Cost savings in nor having to drive an additional pile; and
- Information on piles that must function in both tension and compression under operating conditions.

Some disadvantages are:

- Residual stresses may significantly affect the results;
- Remolding of the soil may take place during the first test; and
- A minimum waiting period of 7 days is therefore required following installation before conducting this test, except for tests in cohesive soil where the waiting period should not be less than 14 days. A waiting period is also generally required between the compression and tension test.

Appropriate references should be consulted relative to residual stresses and the necessary waiting period.

6.2.9. Conducting Load Tests

6.2.9.1. Technical Specifications

Plans and specifications for the pile test should be developed generally in accordance with the referenced ASTM D1143 Standard and should be specifically modified as needed to satisfy the particular project requirements and subsurface conditions. Technical specifications should include the following as a minimum:

a) Type, size, length, and location of pile(s) to be tested

b) Size and capacity of pile driving equipment

c) Driving criteria and any special installation methods required

d) Types of tests to be conducted and maximum testing capacity necessary

e) Required testing equipment and instrumentation, including calibration, to be furnished

f) Testing procedures to be followed

g) Data to be recorded and reported

h) Report format to be followed

i) Provisions for additional tests

j) Payment schedule and schedule of bid items

Provisions should be made during the pile tests to determine the net settlement of the pile (i.e. the total settlement less the elastic compression of the pile and soil). This is required to develop a net settlement (i.e. the pile toe movement) versus load curve to determine pile capacity. Loading and unloading the pile in cycles by employing a telltale located at the pile toe may determine net settlement.

6.2.9.2. Planning

1. Discuss the proposed tests with your general superintendent who should be familiar with local procedures and who can assist you in obtaining necessary test equipment.
2. Secure and review the contract, specifications and applicable local codes so that you thoroughly understand the test requirements.
3. Review the contract and specifications to learn how many and what type piles are to be tested, when they are to be tested and, if several piles are involved, how they are distributed as to location, etc.
4. Plan tests for minimum overall cost consistent with contract requirements. Tests should involve minimum interference or delay to other project operations. Where more than one test should be conducted, consideration should be given to performing two or more tests simultaneously. The added cost of additional test equipment may be more than

offset by the saving in project time.

5. Plan how you will drive the test piles; the equipment needed, driving criteria, observations to be made and recorded, etc. The blow count observed for test and reaction piles is always logged for the entire length, including obtaining the final driving resistance in blows per inch.
6. What is maximum test load to be applied, how is it to be applied (increments, cycling, etc.) and what method should be used?
7. Determine how settlement readings are to be taken and procedures specified for observing and recording.
8. Try to schedule work so that the need for night, weekend and holiday work is kept to a minimum consistent with lowest total cost.
9. Make certain you know how one should be paid for the test work so that proper records are kept.
10. Find out if special measurements, tests or reports should be made and who is going to make them.
11. After the test load set-up and procedures are determined, review them with your general superintendent and with the engineer or other authority that must approve the tests. Any differences should be adjusted before testing begins.

6.2.9.3. Driving Test Piles

1. Determine, before driving, what is the criteria for stopping the test pile. Determine who is going to stop driving. Do not over-drive the test pile.
2. Do not drive any shell test pile until you have made all arrangements to concrete the pile within a maximum of 2 hours after driving.
3. Prepare a driving log of each test pile or reaction pile and record all data. This information is essential for later interpretation of the load test results. This driving log should be completed for several typical piles, even when load tests are not made. An example of a driving log is shown in Figure 6-10.
4. Prior to lifting the pile, mark the test pile or pile shell every foot with paint stick or other easy-to-read method. Number the marks at increments of one foot[1], starting at the toe.
5. Record the penetration under the weight of the hammer and core or pile (if any) and the number of blows per foot for each foot. Record total number of blows, final penetration in blows per inch, depth of pre-excavation or jetting, interruptions to driving, re-tapping, etc.
6. The type of pile, hammer, hammer cushion, follower and methods used for installing the test pile should be identical to those specified for the structure. Adequate steam, air or hydraulic pressure should be maintained during driving. The Engineer will normally not allow any modification of driving equipment or driving criteria from that used on the test pile. Any intended modifications should have the Engineer's written approval prior to driving. In any event, driving a heavier or lighter test pile, or to drive it longer or shorter, or easier or harder, will make it difficult to relate the test results to the actual project. While this may be done deliberately to test several types, it should not be done just because the proper pile or equipment is not immediately available. It should not be done without written approval.
7. For concrete piles, the concrete mix selected should be such that the pile will have adequate strength by the time loading begins, considering pile size, soil conditions, curing time, etc. Keep in mind that the test pile will usually be subjected to twice its normal load within a few days after driving. Additional concrete strength can be obtained by increasing cement content by 1 to 1-1/2 bags per cubic yard, by the addition of 1% calcium chloride, or both. Prepare at least six test cylinders at time of pouring; cylinder breaks should show 80% of 28-day strength before test is started. Frequently, making extra cylinders so that two can be broken every day until 80% strength is secured can save time. When the test pile is precast, no attempt should be made to handle or drive it until the concrete has developed adequate strength.

Figure 6-10 Pile Driving Log

PILE DRIVING LOG

STATE PROJECT NO.: Bridge #1 DATE: 5-29-98

JOB LOCATION: Bogalusa

PILE TYPE: 457 mm PCC LENGTH: 15 m BENT/PIER NO.: 1 PILE NO.: 1

HAMMER: D-30-32 ENERGY/BLOW: 99.9 kJ OPERATING RATE: 36-52 BPM HELMET WEIGHT: 14.5 kN

REF. ELEV.: 109.5 m PILE TOE ELEV.: PILE CUTOFF ELEV.: 108.3 m

PILE CUSHION THICKNESS AND MATERIAL: 190 mm of plywood

WEATHER: sunny TEMP.: 80° START TIME: 8:23 am STOP TIME: 8:58 am

METERS	BLOWS	STROKE / PRESSURE	REMARKS	METERS	BLOWS	STROKE / PRESSURE	REMARKS
0 - 0.25	W.O.P			8.00 - 8.25	25		
0.25 - 0.50	W.O.P			8.25 - 8.50	21	51 BPM	
0 50 - 0.75	W.O.P			8.50 - 8.75	23		
0.75 - 1.00	W.O.P			8.75 - 9.00	26		
1.00 - 1.25	W.O.P			9.00 - 9.25	22	51 BPM	
1.25 - 1.50	W.O.P			9.25 - 9.50	21		
1.50 - 1.75	W.O.H			9.50 - 9.75	23		
1.75 - 2.00	W.O.H			9.75 - 10.00	24	51 BPM	
2.00 - 2.25	W.O.H			10.00 - 10.25	22		
2.25 - 2.50	5		Fuel #2	10.25 - 10.50	26		
2.50 - 2.75	6	52 BPM		10.50 - 10.75	30	44 BPM	
2.75 - 3.00	8			10.75 - 11.00	34		
3.00 - 3.25	10			11.00 - 11.25	40		
3.25 - 3.50	12			11.25 - 11.50	51	43 BPM	
3.50 - 3.75	17	50 BPM		11.50 - 11.75	38	42 BPM	Fuel #4
3.75 - 4.00	22			11.75 - 12.00	41		
4.00 - 4.25	30	49 BPM		12.00 - 12.25	42	42 BPM	
4.25 - 4.50	21	47 BPM	Fuel #3	12.25 - 12.50	53		
4.50 - 4.75	24			12.50 - 12.75	58	41 BPM	
4.75 - 5.00	27			12.75 - 13.00	65		
5.00 - 5.25	29			13.00 - 13.25	77	40 BPM	
5.25 - 5.50	31	45 BPM		13.25 - 13.50	80	40 BPM	
5.50 - 5.75	32			13.50 - 13.75			
5.75 - 6.00	32			13.75 - 14.00			
6.00 - 6.25	35	45 BPM		14.00 - 14.25			
6.25 - 6.50	31			14.25 - 14.50			
6.50 - 6.75	25			14.50 - 14.75			
6.75 - 7.00	21	47 BPM		14.75 - 15.00			
7.00 - 7.25	18			15.00 - 15.25			
7.25 - 7.50	20			15.25 - 15.50			
7.50 - 7.75	19	51 BPM		15.50 - 15.75			
7.75 - 8.00	22			15.75 - 16.00			

6.2.9.4. Test Procedure

1. Cut off the head of the pile at the proper elevation. Cap the pile with a steel plate, adequate to transfer the load; a 2" (50.8 mm) thick plate is normally adequate for loads up to 200 kips (890 kN) and a 3" (76.2 mm) plate for loads up to 400 kips (1779 kN).
2. Level the plate carefully and grout it (use pure cement) to the head of a concrete pile or weld it to the head of a steel pile.
3. If the pile is located in a small excavation,

an extension column may be needed to facilitate loading and to provide a practical working area, or the pile may be temporarily extended to ground level.

4. Where the direct load method is used, make provision for determining or measuring the load increments. Where a hydraulic jack or load cell is used, a recent calibration chart should be available for relating pressures to actual forces. Inspect hydraulic lines for leaks before starting test.
5. Determine which trade union, if any, normally operates the jack and monitors the test set-up.
6. With the jacking method, make provision to maintain hydraulic pressure overnight or over the weekend when indicated. When more than one jack is used, they should be connected to a common manifold to equalize pressure.
7. Usually a period of 3 to 5 days is allowed between driving and testing, to permit the soil around the pile to gain strength that was temporarily lost during driving.
8. Load and unload the pile using the sequence and timing specified in the test requirements.
9. Record data relating to the test clearly and completely. A sample report form is shown in Figure 6-11. It is important that all spaces on the report be completed, even though the answer may seem obvious. Particular care should be exercised to assure all units of measurement be indicated (inches, feet, millimeter, short ton, metric ton, etc.). It is typical to refer to reports many years later, when alterations to a structure are contemplated.

6.2.10. Interpretation of Axial Load Tests

Interpretation of load test results begins with an evaluation of the load-movement curve to determine the allowable capacity of the pile. The term allowable capacity, when used in this manual, is defined quite simply as the ultimate capacity divided by a specified factor of safety.

The problem arises as to what constitutes failure or ultimate capacity of a test pile. Ideally, failure would be defined as that point along the load-movement curve where the test pile experiences continuous movement at no increase in load (plunging in the case of an axial compressive test). In practice, particularly in cohesionless soils, plunging is not often observed, and therefore a means of defining failure load is required. For a failure definition to be useful it must be based on some mathematical rule and generate a repeatable value that is independent of scale relations and the opinion of the individual interpreter. Numerous rules for determining failure loads based on the results of load tests, particularly for axial compressive tests, can be found in the literature.

Proper definition is imperative to communicate the results of load tests to others. The following are some general rules to follow when interpreting load test results:

- The term “failure load” should be avoided and “interpreted failure load” used if failure (i.e. plunging of pile during axial compression test) has not occurred.
- The method of defining the interpreted failure load should be stated (e.g. Davisson Method, etc.)
- Any attempt to extrapolate an interpreted failure load that is greater than the maximum test load should be avoided. The failure load should be established by measurements and not from hypothetical models.
- When a pile has been loaded to a percentage of the design load or to the capacity of the loading system and ''interpreted failure” has not occurred, the term “maximum test load” should be used.
- In selecting the allowable capacity for a pile, the method used should be clearly stated (e.g. “allowable capacity is equal to one-half the maximum test load”).

6.2.10.1. Elastic Shortening

In order to aid in the interpretation of an axial compressive load test, the theoretical elastic shortening of the pile should be determined. This can be accomplished with the following expression:

Figure 6-11 Load Test Report

STATE DEPARTMENT OF HIGHWAYS
AND PUBLIC TRANSPORTATION
FORM 1302 REVISED
8/76

RECORD OF FOUNDATION TEST LOAD

Loading No. ____

TEXAS QUICK TEST LOAD METHOD

County ____________ Control ____________ Structure ____________
Highway No. ____________ Project ____________ Structure No. ____________

Bent No. ________ Foundation No. ________ Sta. ________ Rt. ______ Lt. ______
Foundation Size & Type ____________ Total Length ________ Design Load ________
Foundation Tip Elevation ____________ Effective Penetration ________ Ground Elevation ________
Hammer Type & Size ____________ Dynamic Resistance ____________
Time Test Began ________ Date ________ Resident Engineer ____________

Time	Time Inter-val	Load Added	Total Load	Extensometer Readings		Total Gross Settlement - Inches		
Min.	Min.	Tons	Tons	Dial 1	Dial 2	Dial 1	Dial 2	Average

Remarks: ____________________________

District ____________
Date ____________
By ____________

Equation 6-1:

$$e_s = \frac{Q}{A_p E_p} L$$

(end bearing pile)

Where[2]

- e_s = elastic shortening, inches or millimeters
- Q = applied load, kips or kN
- L = pile length, inches or meters
- A_p = cross-sectional area of pile, in^2 or m^2
- E_p = elastic modulus of pile material, ksi or MPa
  - E_s = elastic modulus of steel
  - E_c = elastic modulus of concrete

For piles composed of two materials, such as a concrete filled pipe pile or a heavily reinforced drilled shaft, the area of the steel is converted to an equivalent concrete area:

Equation 6-2:

$$A_{trans} = A_c + A_s \frac{E_s}{E_c}$$

The value of e_s computed is the amount of elastic shortening a pile undergoes for the condition when the full load reaches the pile tip (i.e. zero skin friction). In the case of a pile that derives some or all of its support from skin friction, the elastic deformation of the pile will be less than that of a pile whose support is strictly in end bearing. The elastic deformation of a pile, taking into account, in a simplified manner, the distribution of applied load to the surrounding soil vice skin friction, can be determined by the following expression;

Equation 6-3:

$$e_s = \left(Q_p + \alpha_s Q_s\right) \frac{L}{A_p E_p}$$

Where

- Q_p = point load transmitted to the pile tip, kips or kN
- Q_s = shaft friction load, kips or kN
- α_s as follows:
  - 0 for no shaft friction (end bearing pile)
  - 0.5 for uniform distribution of shaft friction
  - 0.67 for triangular distribution of shaft friction starting from zero friction at pile head to a maximum value at pile point.
  - 0.33 for triangular distribution of shaft friction starting from a maximum at pile head to zero at the pile point

Utilizing Equation 6-1 for the case where the pile is a friction pile (no end bearing) and where the skin friction is uniformly distributed throughout the length of the pile one can see that the elastic deformation will be exactly one-half of that obtained for an end bearing pile.

For piles less than 35 feet (11 m) long, the exact amount of elastic shortening that occurs is relatively unimportant but does become an important consideration in evaluating apparent axial pile movement under test load for longer piles.

The cross sectional area of a pile (A_p) and the modulus of a pile (E_p) are important in evaluating the results of load tests. These values must be accurately determined to reliably determine pile load distribution with depth, the elastic shortening of a pile and the load transfer characteristics, based on telltale or strain gage data.

The area and elastic modulus of specified steel piles (H-piles, unfilled pipe) for all practical purposes could be considered constant and are well known. For concrete filled pipe piles, the value of the concrete modulus is uncertain because stress levels vary from location to location. Another part of the difficulty in properly interpreting the results of this kind of data is determining the elastic modulus of concrete (steel piles' modulus is known.) Concrete is non-linear in its stress-strain characteristics and this does effect the results from strain gauges.[3] Further uncertainties in the pile properties arise when instrumentation represents a major portion of the pile cross-section. Therefore the cross-sectional area of the pile should be taken into consideration when selecting the amount of instrumentation.

6.2.10.2. Load Transfer

When an axial load is applied to a vertical pile, a certain percentage of the applied load is transferred into the soil surrounding the pile shaft with the remainder being transmitted through the pile and into the tip. This transmission of applied load to the foundation soil through skin friction and end bearing resistances is commonly referred to as load transfer. The distribution, particularly in skin resistance, is dependent on the following:

- Initial state of stress and condition of soil
- The strength and compressibility of the soils and the changes that occur with pile installation and with time
- The material, size, shape and length of the pile
- The method of pile installation (e.g. driving, vibrating, boring, jacking, etc.)
- The interactions between the pile and the soil that are time dependent (negative skin friction, disturbed soil adjacent to pile, relaxation, creep, groundwater fluctuations)

If the information needed from a load test is to be used for design of production piling, an evaluation of the load transfer characteristics of the pile-soil system is required. Load transfer characteristics help the designer identify properly the significant characteristics of the supporting soils.

The determination of load transfer with depth along the pile shaft is achieved by use of internal instrumentation. Instrumentation can consist of telltales (strain rods), electrical strain gages or load cells, or some combination thereof located at various depths along the pile shaft, as shown in Figure 6-12.

Figure 6-12 Instrumentation for Determining Load Transfer and Skin Friction

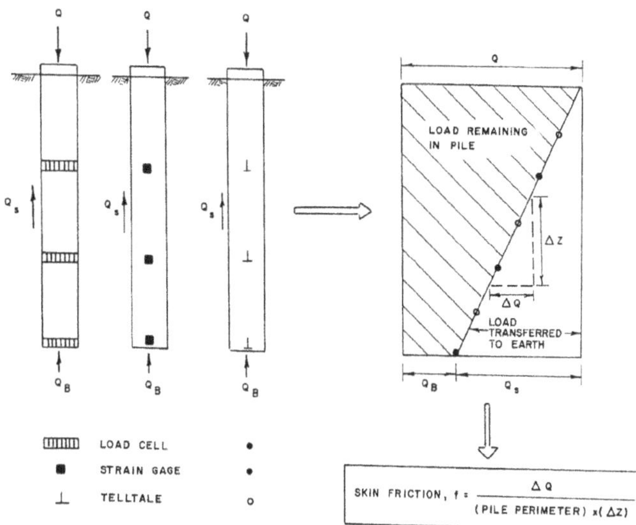

6.2.10.3. Methods of Interpretation

As noted earlier, there are many empirical and arbitrary methods available to determine the axial capacity of a pile from load test data. The most commonly used one is Davisson's Offset Limit Method. This method has been suggested for application to the results of the Quick Load (QL) test. The Davisson Method is implemented as follows:

1. Calculate elastic compression of pile (e_s) when considered as a free column by Equation 6-1. With a variable load, this is obviously a linear function of Q.
2. Plot this function on a graph (see Figure 6-13.)
3. Plot the failure criterion line. This line is parallel to the elastic compression line, offset downward by a given distance on the displacement axis. This offset depends upon the pile diameter or width, and is

Equation 6-4:

$$S_{off} = 0.15 + \frac{D}{120} \quad (D < 24")$$

$$S_{off} = \frac{D}{30} \quad (D > 24")$$

(English Units, inches)

Equation 6-5

$$S_{off} = 4 + \frac{D}{120} \quad (D < 610 \text{ mm})$$

$$S_{off} = \frac{D}{30} \quad (D > 610 \text{ mm})$$

(SI Units, millimeters)

Where S_{off} = offset from elastic compression line, inches or millimeters. Pile diameter or width D must be in inches. The equation for the failure criterion line is thus

Equation 6-6:

$$S_f = \frac{QL}{A_p E_p} + 0.15 + \frac{D}{120} \quad (D < 24")$$

$$S_f = \frac{QL}{A_p E_p} + \frac{D}{30} \quad (D > 24")$$

(English Units, inches)

Equation 6-7:

$$S_f = \frac{QL}{A_p E_p} + 4 + \frac{D}{120} \quad (D < 610 \text{ mm})$$

$$S_f = \frac{QL}{A_p E_p} + \frac{D}{30} \quad (D > 610 \text{ mm})$$

(SI Units, millimeters)

4. Plot pile head total movement vs. applied load.
5. Where observed load-movement curve does not intersect failure criterion, the factor of safety will be applied to the maximum test load to determine the allowable capacity.
6. Apply factor of safety to interpreted failure load" to determine the allowable capacity. Allowable capacity is normally determined by dividing the "interpreted" failure load by some suitable factor of safety. The factor of safety most commonly used is 2.0, but larger factors of safety may be required:
   - For friction piles in clay, where group settlement may control
   - Where total settlement that can be tolerated by the structure is exceeded
   - For piles installed by methods other than impact driving
   - Where soil conditions are extremely variable
   - Where limited load testing will be performed

Figure 6-13 Typical Test Plot And Interpretation Using Davisson's Method

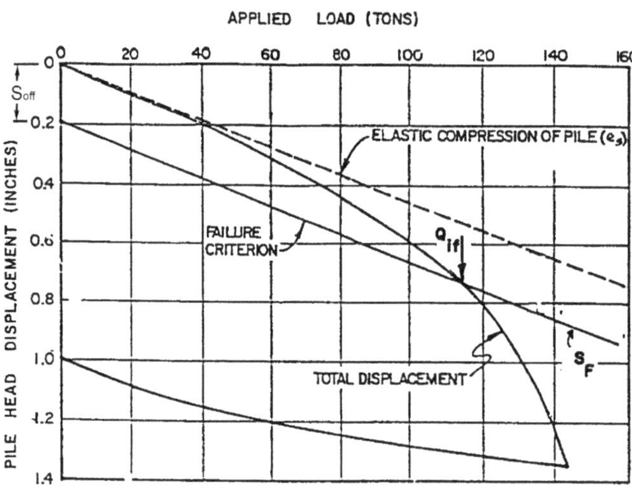

Layered soils may cause the test piles to have a different capacity than the service piles if the test piles have tips in a different stratum. Consolidation of a cohesive layer supporting the toe load may also cause the load to be supported by another layer. The support of a pile could change from friction to end bearing or the reverse depending on the strata.

6.2.11. Tensile Load Tests

Tensile load tests are performed to determine axial tensile (uplift) load capacities of piles. The uplift capacity of piles is important for pile groups subjected to large overturning moments. Hence, the importance of determining pile uplift capacity has greatly increased in recent years, particularly with regard to seismic design issues. The basic mechanics of the test are similar to compression load testing, except the pile is loaded in tension.

6.2.11.1. Tension Test Equipment

ASTM D-3689 describes The Standard Method of Testing Individual Piles Under Static Axial Tensile Load by the American Society of Testing Materials. Several alternative systems for (1) applying tensile load to the pile, and (2) measuring movements are provided in this standard. Most often, tensile loads are applied by centering a hydraulic jack on top of a test beam(s) and jacking against a reaction frame connected to the pile to be tested. The test beam in turn is supported by piles or cribbing. When a high degree of accuracy is required, the primary means of measuring the load applied to the pile should be from a calibrated load cell with the jack load recorded from a calibrated pressure gage as backup. A spherical bearing plate should be included in the load application arrangement.

Axial pile head movements are usually measured by dial gages or LVDT's that measure movement between the pile head and an independently supported reference beam. For tensile load testing, ASTM requires a longer travel length and higher precision for movement measuring devices than in a compression load test. For tensile testing, ASTM requires that the dial gages or LVDT's have a minimum of 75 mm of travel and a precision of at least 0.025 mm. A minimum of two dial gages or LVDT's mounted equidistant from the center of the pile and diametrically opposite should be used. Two backup systems consisting of a scale, mirror, and wire system should also be provided with a scale precision of 0.25 mm. The backup systems should be mounted on diametrically opposite pile faces and be independently supported systems. Additional details on load application, and pile head load and movement measurements may be found in ASTM D-3689. A photograph of a typical tension load test arrangement is presented in Figure 6-14.

Figure 6-14 Tension Test Arrangement on Batter Pile

6.2.11.2. Tension Test Loading Methods

Several loading procedures are detailed in ASTM D-3689. The quick loading procedure is recommended. This procedure requires that load be applied in increments of 10 to 15% of the pile design load with a constant time interval of 2 1/2 minutes, or as otherwise specified between load increments. Readings of time, load, and gross movement are to be recorded immediately before and after the addition of each load increment. This procedure is to continue until continuous jacking is required to maintain the test load, or the capacity of the loading apparatus is reached, whichever occurs first. Upon reaching and holding the maximum load for 5 minutes, the pile is unloaded in four equal load decrements, which are each held for 5 minutes. Readings of time, load, and gross movement are once again recorded immediately after, 2 1/2 minutes after, and 5 minutes after each load reduction including the zero load. Additional optional loading procedures are detailed in ASTM D-3689.

It is generally desirable to test a pile in tensile loading to failure, particularly during a design stage test program. If construction stage tensile tests are performed on production piles, the piles should be redriven to the original pile toe elevation and the previous driving resistance upon completion of the testing.

6.2.11.3. Presentation and Interpretation of Tension Test Results

The results of tensile load tests should be presented in a report conforming to the requirements of ASTM D-3689. A load-movement curve similar to the one shown in Figure 6-15 should be plotted for interpretation of tensile load test results.

Figure 6-15 Typical Tension Load Test Load-Movement Curve

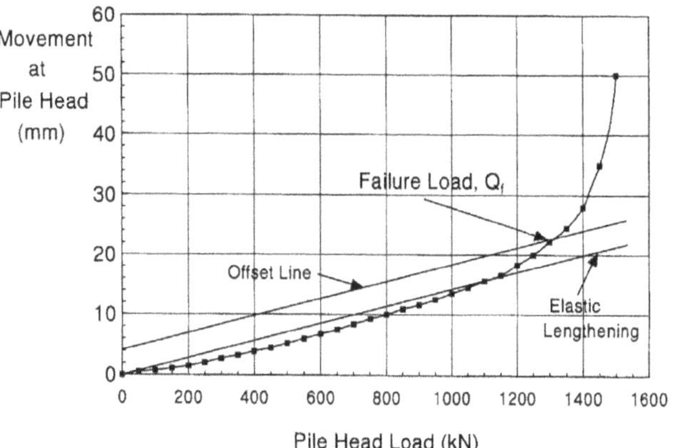

A widely accepted method for determining the ultimate pile capacity in uplift loading has not been published. Some acceptance criteria for uplift tests have included a limit on the gross or net upward movement of the pile head, the slope of the load movement curve, or an offset limit method that accounts for the elastic lengthening of the pile plus an offset.

Due to the increased importance of tensile load testing, it is recommended that the elastic lengthening of the pile plus an offset limit be used for interpretation of test results. For tensile loading, the suggested offset is 4.0 mm. The load at which the load movement curve intersects the elastic lengthening plus 4.0 mm is then defined as the tensile failure load. The uplift design load may be chosen between 1/2 to 2/3 of this failure load.

6.2.12. Lateral Load Tests

Lateral load tests are performed on projects where piles are subjected to significant lateral loads. The importance of determining pile response to lateral loading has greatly increased in recent years, particularly with regard to special design events such as seismic and vessel impact. This need has also increased due to the greater use of noise walls and large overhead signs. The primary purpose of lateral load testing is to determine the p-y curves to be used in the design or to verify the appropriateness of the p-y curves on which the design is based.

6.2.12.1. Lateral Load Test Equipment

ASTM D-3966 describes The Standard Method of Testing Piles Under Lateral Load by the American Society of Testing Materials. Several alternative systems for (1) applying the lateral load to the pile, and (2) measuring movements are provided in this standard. Most often, lateral loads are applied by a hydraulic jack acting against a reaction system (piles, deadman, or weighted platform), or by a hydraulic jack acting between two piles as shown in Figure 6-16. The primary means of measuring the load applied to the pile(s) should be from a calibrated load cell with the jack load recorded from a calibrated pressure gage as backup. ASTM requires a spherical bearing plate(s) be included in the load application arrangement unless the load is applied by pulling so that the pile head is free to rotate.

Figure 6-16 Lateral Load Test Setup

The loads are applied as near the ground surface as convenient. Note the support that is required for the hydraulic jack and strut arrangement.

Lateral pile head movements are usually measured by dial gages or LVDT's that measure movement between the pile head and an independently supported reference beam mounted perpendicular to the direction of movement. For lateral load testing, ASTM requires the dial gages or LVDT's have a minimum of 75 mm of travel and a precision of at least 0.25 mm. For tests on a single pile, one dial gage or LVDT is mounted on the side of the test pile opposite the point of load application. A backup system consisting of a scale, mirror, and wire system should be provided with a scale precision of 0.25 mm. The backup system is mounted on the top center of the test pile or on a bracket mounted along the line of load application.

It is strongly recommended that lateral deflection measurements versus depth also be obtained during a lateral load test. This can be accomplished by installing an inclinometer casing on or in the test pile to a depth of 10 to 20 pile diameters and recording inclinometer readings immediately after application or removal of a load increment held for a duration of 30 minutes or longer. Lateral load tests in which only the lateral deflection of the pile head is measured are seldom justifiable.

6.2.12.2. Lateral Test Loading Methods

Several loading procedures are detailed in ASTM D-3966. The standard loading procedure requires that the total test load be 200% of the proposed lateral design load. Variable load increments are applied with the magnitude of load increment decreasing with applied load. The load duration is also variable, increasing from 10 minutes early in the test to 60 minutes at the maximum load. Upon completing the maximum test load, the pile is unloaded in four load decrements equal to 25% of the maximum load with 1 hour between load decrements.

A modified lateral loading schedule makes the recommended loading increment be 12.5% of the total test load with each load increment held for 30 minutes. Upon reaching and holding the maximum load for 60 minutes, the pile is unloaded and held for 30 minutes at 75, 50, 25 and 5% of the test load.

Readings of time, load, and gross movement are recorded immediately after each change in load. Additional readings are taken at 1, 2, 4, 8, 15 and 30 minutes. This procedure is followed during both the loading and unloading cycle.

6.2.12.3. Presentation and Interpretation of Lateral Test Results

The results of lateral load tests should be presented in a report conforming to the requirements of ASTM D-3966. The interpretation and analysis of lateral load test results is much more complicated than those for compression and tensile load testing. Figure 6-17 presents a typical lateral load test pile head load-movement curve. A lateral deflection versus depth curve similar to the one shown in Figure 6-18 should also be plotted for interpretation of lateral load test results that include lateral deflection measurements versus depth. The measured lateral load test results should then be plotted and compared with the calculated result as indicated in Figure 6-18.

Figure 6-17 Typical Lateral Load Test Pile Head Load-Deflection Curve

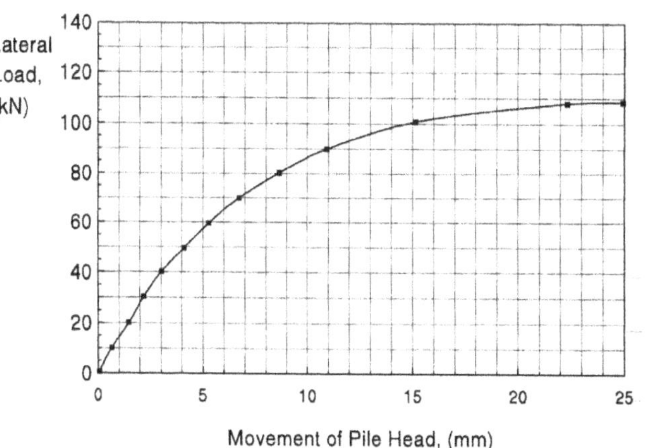

Figure 6-18 Comparison of Measured and COM624P Predicted Load-Deflection Behavior versus Depth

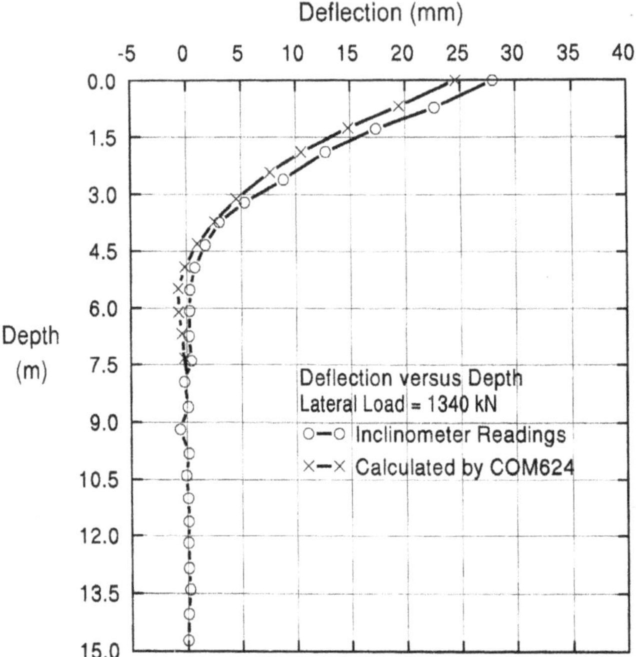

Based upon the comparison of measured and predicted results, the p-y curves to be used for design (design stage tests), or the validity of the p-y curves on which the design was based (construction stage tests) can be determined.[4]

6.3. Statnamic™ Testing

Statnamic™ testing, developed as a joint venture of Berminghammer Foundation Equipment of Ontario, Canada, and TNO of the Netherlands, is an alternative method of load testing piles. It seeks to avoid the cumbersome aspects of static load testing (both in terms of the physical set-up and the time it takes) and the shortcomings of dynamic testing (mostly related to high loading rate and effects of stress wave propagation.) The method is applicable both to driven piles and to cast-in-place piles of all kinds as well. With the latter, the stresses are lower than in high-strain dynamic testing, which reduces the possibility of pile damage.

The basic concept of Statnamic™ testing is to generate a controlled fuel detonation at the pile head, either axially or laterally depending on the load being tested. This fuel detonation produces a pile head force that, although certainly much shorter than static load testing, is many times longer than the stress-wave travel time in the pile. This enables the pile to be tested with the pile mass acting as a single unit, as would be the case in a static load test. Only the velocity induced resistance of the soil and pile inertia should be removed to produce a meaningful test.

6.3.1. Overview of the Testing Apparatus

Figure 6-19 Statnamic™ Testing Device

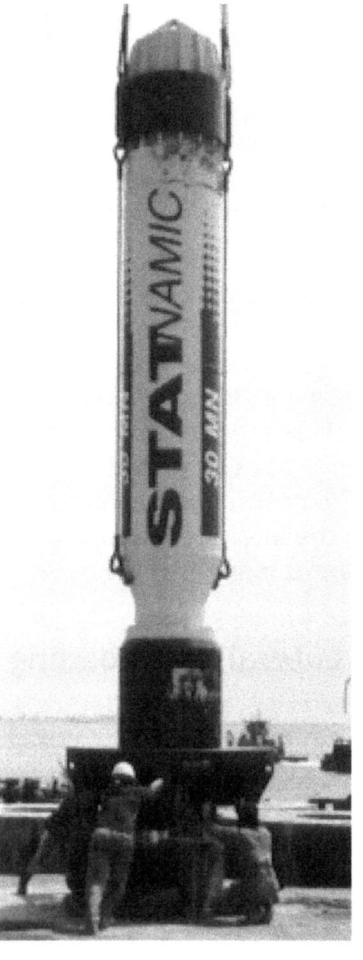

A typical Statnamic™ testing device is shown in Figure 6-19. The device consists of a pressure chamber and a reaction mass. To produce the load in the pile head, fuel is burned within the pressure chamber. Since the pressure of the fuel produces forces both upward and downward, it is necessary to have a reaction mass (generally about 5% of the desired ultimate load) to insure that the desired downward force is induced in the pile. This downward force has a typical duration of 100 ms. The pile velocity induced by this force is generally below 1 m/s (3.28 ft/sec) and the maximum pile acceleration is around

1-3 g.

Figure 6-20 Cutaway View of the Statnamic™ Device

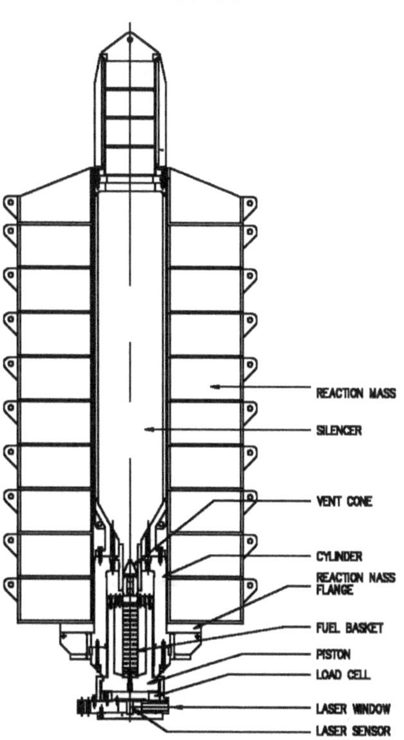

The details of the device are shown in Figure 6-20. The pressure chamber consists of a piston and cylinder. A fuel cavity is within the piston to burn the fuel. The cylinder is placed on top of the piston; the cylinder performs a number of functions, including containing the explosion, muffling the exhaust with a baffle system, and including a platform for the reaction masses. The reaction masses are placed on the cylinder after its installation. This modular assembly enables a relatively small crane to set the assembly in place.

The fuel is burned in the pressure vessel, both throwing the reaction mass(es) upwards around 2.5 m (8') and inducing a pile head force on the pile. The maximum pressure generated by the fuel burning is around 680 bar (9860 psi.) The exhaust gasses are vented upwards through the baffle system.

The force induced by the apparatus is dependent upon its the physical size. The devices are made in a number of sizes and rated according to their maximum loading capability. The devices vary in rated peak force from 0.6 MN to 30 MN (135 to 6750 kips.) The fuel itself is classified as U.N. Number 0431, Class 1.4G.

Figure 6-21 Statnamic™ Device with Canister for Gravel Catching and Reaction Weights

The reaction mass is hurled upwards; return of this mass would result in an impact similar to a pile driver, which would negate the benefits of the relatively long force of the Statnamic™ fuel combustion and complicate the instrumentation. To catch the reaction mass before it does this, two methods are employed. The first and earliest to be used was the gravel catch. A Statnamic™ device with a canister for a gravel catch is shown in Figure 6-21 along with the reaction weights. Once the device and weights are in place, the canister is closed the area around the bottom of the cylinder is filled with gravel. As the reaction masses travel upwards, the gravel moves in and fills the area it vacated. When the reaction mass returns it impacts the gravel, which dissipates the impact. This is generally used for devices with a testing force greater than or equal to 16 MN (3600 kips.)

Figure 6-22 Hydraulic Catching Mechanism

The second method employs the use of a mechanical or hydraulic catch, which simply takes hold of the reaction masses at the top of their stroke and prevents their return. The catching mechanism is mounted in a frame, which also assists in guiding the reaction masses. Such a catching system is shown in Figure 6-22. A mechanical catching system also exists for devices up to 16 MN (3600 kips.)

Typical sizes and specifications for Statnamic™ devices are shown in Table 6-1.

6.3.2. Instrumentation

Instrumenting a Statnamic™ test involves measuring three variables. These variables and the technique used to measure them are shown in Table 6-2.

Table 6-2 Measurement Parameters for Statnamic™ Testing

Quantity to be measured	Method of measurement
Pile head loading induced by combustion	Load cell in the lower part of piston.
Displacement of pile during loading	Photovoltaic sensor in piston base. A stationary laser source 10-20 m (30-60') from pile provides stationary datum.
Acceleration of pile during loading	Accelerometer mounted in piston base or pile head.

All of these variables are taken as a function of time; the data is digitized at a rate of 4-5 kHz. It can be stored in a file for later processing; the theory behind the processing is described in the next section. The data can be reduced and analyzed using the Microsoft Excel Spreadsheet SAW R3, or by software commercially available from Berminghammer or TNO.

Table 6-1 Typical Sizes for Statnamic™ Units

Load	3 MN Size	4 MN Size	8 MN Size	16 MN Size	30 MN Size
Maximum Load	3.0 MN	4.0 MN	8.0 MN	16.0 MN	30.0 MN
Minimum Load	0.7 MN	1.0 MN	2.0 MN	4.0 MN	7.5 MN
Fuel Charge for Max. Load	1.2 kg	1.5 kg	3.5 kg	11.5 kg	25.0 kg
Fuel Charge for Min. Load	0.3 kg	0.8 kg	1.2 kg	4.0 kg	8.0 kg
Piston Assembly					
Diameter	215 mm	253 mm	368 mm	508 mm	711 mm
Height	1115 mm	996 mm	1130 mm	1294 mm	1777 mm
Base Dimension	578 mm	645 mm	892 mm	1170 mm	1402 mm
Mass	355 kg	480 kg	800 kg	2020 kg	5600 kg
Cylinder/Silencer Assembly					
Diameter	406 mm	406 mm	559 mm	762 mm	1016 mm
Height	4235 mm	5502 mm	5093 mm	6803 mm	5757 mm
Reaction Mass Flange Diameter	798 mm	1219 mm	1321 mm	1524 mm	2032 mm
Mass	1100 kg	2300 kg	3270 kg	8160 kg	17,500 kg
Pressure					
Operating Pressure for Max. Load	93.1 MPa	93.1 MPa	93.1 MPa	93.1 MPa	93.1 MPa
Reaction Mass					
Total Mass for Max. Load	15,300 kg	20,400 kg	40,800 kg	81,500 kg	153,000 kg

6.3.3. Theory of Operation

Although the Statnamic™ impulse is considerably longer than that of high strain dynamic analysis, it still is short enough to induce accelerations and velocities in the pile that affect the analysis of the data. Some kind of dynamic analysis is necessary for proper interpretation of the results.

Figure 6-23 Diagram of the Physical System of a Statnamic™ Test

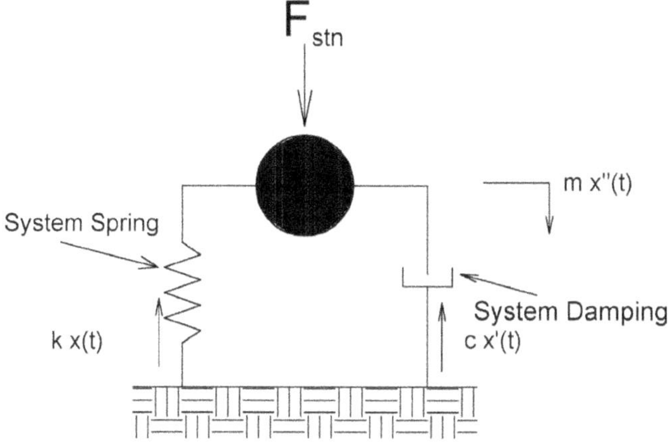

Consider the model shown in Figure 6-23. Summing the forces on the pile, we have

Equation 6-8:

$$mx''(t) = F_{stn}(t) - kx(t) - cx'(t)$$

Where

- m = mass of pile, kg
- x(t) = displacement of pile as a function of time, m
- $F_{stn}(t)$ = Statnamic™ force, N
- k = system spring constant of pile, N/m
- c = system damping force, N-sec/m

The soil is assumed to be acting as one unit, thus both k and c describe properties for the entire soil/pile system. The values of k and c are also not necessarily constant, and in fact, k cannot be constant when pile failure is induced.

To better understand this equation, consider typical the force-time and displacement-time plots, shown in Figure 6-24.

Figure 6-24 Typical Force-Time and Displacement-Time Curves for Statnamic™ Force

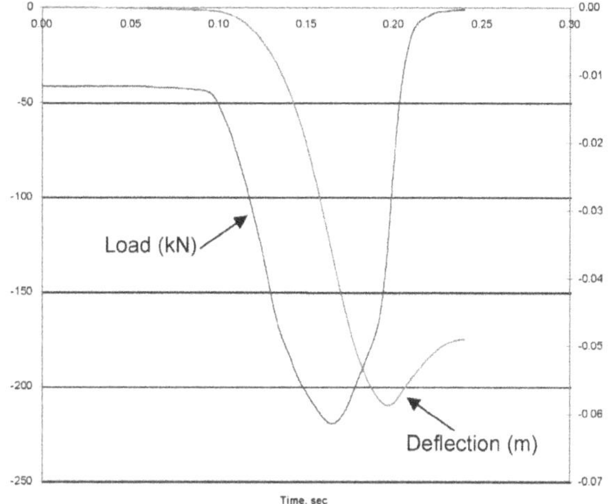

The curve with the downward peak to the left is the force-time curve for the Statnamic™ load; its axis is on the left. The curve with the downward peak to the right is the displacement-time curve and its axis is on the right.

It is first noteworthy that there is a time lag of about 40 ms between the peak force and the peak displacement. If there were no dynamic effects - either soil damping or inertial effects - the two peaks would take place at the same time, as the only resistance would be the elastic (and at failure, plastic) static resistance of the soil.

If we numerically integrate and apply Equation 6-8, we can plot the various forces against the displacement and obtain the information shown in Figure 6-25.

Figure 6-25 Force-Displacement Curves

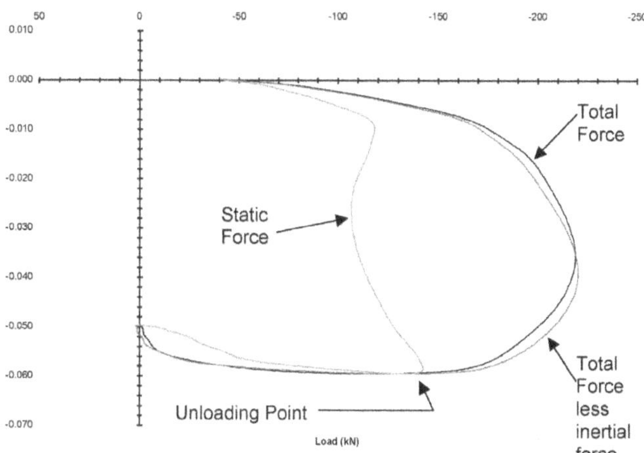

The inertial force and total force are very similar; however, when we take the effects of the dampening force out, the static force is considerably smaller than the other forces for much of the loading history. This is due to the dynamic nature of the test. The static force and the other forces come together at the bottom of the curve. At this point, the displacement is at a maximum and the velocity is zero; the effects of soil resistance are likewise zero. This is referred to as the "Unloading Point." This point, in conjunction with the static force-deflection curve, is used to determine the static capacity of the pile.

We have noted that the pile is considered a solid mass for purposes of analysis. This holds only if

Equation 6-9:

$$t_{stn} \geq \frac{12L}{c}$$

Where

- t_{stn} = time of Statnamic™ combustion force, sec.
- L = pile length, m
- c = acoustic speed of pile material, m/sec

For a typical Statnamic™ impulse period of 100 msec, this translates into a pile length of approximately 32 m (104') for concrete piles and 42 m (140') for steel piles. For longer piles Statnamic™ testing can be used if either accelerometers are imbedded in the pile toe or strain gauges are imbedded along the pile shaft.

6.3.4. Lateral Load Testing

While most of the attention above has been given to axial load testing of the pile, Statnamic™ testing is also used to laterally test foundations as well. Such a test is shown in Figure 6-26. The test apparatus is identical because the force generation is not dependent upon gravity, and the problem of reaction mass catching is simpler as well. The use of the data in lateral load testing is different than in axial testing and is being developed more fully.

Figure 6-26 Lateral Load Test Using Statnamic™ Device

6.3.5. Other Considerations

While Statnamic™ represents an important development in pile load testing, end users should be aware of other considerations.

Since velocity induced soil characteristics are significant factors in Statnamic™ data, soils with lower damping - principally cohesionless soils - are better suited for Statnamic™ analysis. In addition to having higher damping, cohesive soils are more prone to either increases in pore water pressure or thixotropic relaxation with pile movement, which means that some consideration for set-up should be made for these soils.

High-strain dynamic testing can provide the user an indication of the distribution of soil resistance along the pile shaft. This type of data is not available in Statnamic™ analysis, as it is in traditional static load testing. This information may or may not be necessary depending upon the requirements of the foundation; however, this information can be obtained from Statnamic™ testing if strain gauges are employed.

6.4. Dynamic Measurements and Analyses

6.4.1. Background

Dynamic pile load testing is the estimation of static axial compressive pile capacity from dynamic measurements of pile strain and acceleration. A method to accomplish this, called the Case Method, was developed at Case Institute of Technology in a research project funded by the Ohio Department of Transportation and the Federal Highway Administration.

Dynamic pile load testing is generally used in conjunction with the computer programs WEAP (Wave Equation Analysis of Piles) and CAPWAP (Case Pile Wave Analysis Program). WEAP predicts, based on assumed soil resistance distribution and soil damping parameters, the driving resistance at which the required pile capacity will be attained, as well as the driving stresses in the pile and provides information that can be used to evaluate the adequacy of the proposed pile driving system. CAPWAP is used to analyze the data from a dynamic pile load test in order to determine (1) a better estimate of the static pile capacity and (2) the soil damping parameters and soil resistance distribution acting on the pile during the test.

Advantages and disadvantages of dynamic testing are presented in Table 6-3.

6.4.2. Dynamic Testing Equipment

A typical dynamic testing system consists of a minimum of two strain transducers and two accelerometers bolted to diametrically opposite sides of the pile to monitor strain and acceleration and account for nonuniform hammer impacts and pile bending. The reusable strain transducers and accelerometers are generally attached two to three diameters below the pile head. Almost any driven pile type (concrete, steel pipe, H, Monotube, timber, etc.) can be tested with the pile preparation for

Table 6-3 Advantages and Disadvantages of Dynamic Pile Testing

Advantages	Disadvantages
Dynamic tests provide information on the complete pile installation process. Test results can be used to estimate pile capacity, to check hammer and drive system performance, to monitor driving stresses, and to assess pile structural integrity.	Dynamic testing to determine the ultimate static pile capacity requires that the driving system mobilize all the soil resistance acting on the pile. Shaft resistance can generally be mobilized at a fraction of the movement required to mobilize the toe resistance. However, when driving resistances approach 100 blows per quarter meter, the full soil resistance is difficult to mobilize at and near the pile toe. In these circumstances, dynamic test capacities tend to produce lower bound capacity estimates unless a larger hammer or higher stroke can be used to increase the pile net penetration per blow.
Many piles can be tested during initial driving or during restrike in one day. This makes dynamic testing an economical and quick testing method. Results are generally available immediately after each hammer blow.	Dynamic testing estimates of static pile capacity indicate the capacity at the time of testing. Since increases and decreases in the pile capacity with time typically occur due to soil setup/relaxation, restrike tests after an appropriate waiting period are usually required for a better indication of long term pile capacity. This may require an additional move of the pile driving rig for restrike testing.
On large projects, dynamic testing can be used to supplement static pile load tests or reduce the overall number of static tests to be performed. Since dynamic tests are more economical than static tests, additional coverage can also be obtained across a project at reduced costs. On small projects where static load tests may be difficult to justify economically, dynamic tests offer a viable construction control method.	Larger diameter open ended pipe piles or H-piles which do not bear on rock may behave differently under dynamic and static loading conditions. This is particularly true if a soil plug does not form during driving. In these cases, limited toe bearing resistance develops during the dynamic test. However, under slower static loading conditions, these open section piles may develop a soil plug and therefore a higher pile capacity under static loading conditions. Interpretation of test results by experienced personnel is important in these situations.
Dynamic tests can provide information on pile capacity versus depth, capacity variations between locations, and capacity variations with time after installation through restrike tests. This information can be helpful in augmenting the foundation design, when available from design stage test pile programs, or in optimizing pile lengths when used early in construction test programs.	
When used as a construction monitoring and quality control tool, dynamic testing can assist in early detection of pile installation problems such as poor hammer performance or high driving stresses. Test results can then facilitate the evaluation and solution of these installation problems.	
On projects where dynamic testing was not specified and unexpected or erratic driving behavior or pile damage problems develop, dynamic testing offers a quick and economical method of troubleshooting.	
Results from dynamic testing and analysis can be used for driving criteria development including wave equation input parameter selection and refinement of wave equation results as described in Chapter 4.	

each pile type slightly varying.

A close up view of a strain transducer and an accelerometer bolted to a Monotube pipe pile is shown in Figure 6-27. The individual cables from each gage are combined into a single main cable that in turn relays the signals from each hammer blow to the data acquisition system on the ground. The data acquisition system, such as the Pile Driving Analyzer shown in Figure 6-28, conditions and converts the strain and acceleration signals to force and velocity records versus time.

Figure 6-27 Strain Transducer (left) and Accelerometer (right) Installed on a Monotube Pile

Figure 6-28 Pile Driving Analyzer

Figure 6-29 shows both pile and analyzer in the field. The concrete pile to be tested during driving has been positioned in the leaders for driving. It has been prepared by having concrete anchors drilled and installed before lifting into position. After it has been set in the leaders, a member of the pile crew climbs the leads and then bolts the gages to the pile at this time. Piles to be tested during restrike can be instrumented at any convenient location and the climbing of the leads is usually not necessary. Pile preparation and gage attachment typically requires 10 to 20 minutes per pile tested. After the gages are attached, the driving or restrike process continues following usual procedures. Most restrike tests are only 20 blows or less.

Figure 6-29 PDA Field Test Set-up

Older dynamic testing systems required multiple components for processing, recording, and display of dynamic test signals. In newer dynamic testing systems, these components have been combined into one PC computer based system. During driving, the Pile Driving Analyzer performs integrations and all other required computations to analyze the dynamic records for transferred energy, driving stresses, structural integrity, and pile capacity. Numerical results for each blow for up to nine dynamic quantities are electronically stored in a file that can be later used to produce graphical and numeric summary outputs. In this system, force and velocity records are also viewed on a graphic LCD computer screen during pile driving to evaluate data quality, soil resistance distribution, and pile integrity. Complete force and velocity versus time records from each gage are also digitally stored for later reprocessing and data analysis by CAPWAP.

Data quality is automatically evaluated by the Pile Driving Analyzer and if any problem is detected, then a warning is given to the test engineer. Other precautionary advice is also displayed to assist the engineer in collecting data. The capabilities discussed in the remainder of this chapter are those included in these newer systems.[5]

The latest generation of dynamic test equipment is the PAL-R, as shown in Figure 6-30. This unit can be operated remotely (off site) by the test engineer via a cell phone connection. In a remote test, basic pile information is entered into the unit by on site personnel and then the unit is connected to the test engineer's location. For site where cell phone service is not available, the data can be stored and later sent over a telephone line. Remote testing can be particularly cost effective on remote construction sites or when large numbers of production piles are to be tested.

Figure 6-30 Pile Driving Analyzer—Model PAL-R

6.4.3. The Case Method

The Pile Driving Analyzer records strain-time and velocity-time histories for the hammer blow being analyzed. Pile head force (F) is computed from the strain and the velocity (v) is obtained by integrating the acceleration; all of these are functions of time. Using wave propagation theory and assuming a uniform elastic pile and ideal plastic soil behavior, the total soil resistance R active during the pile driving can be calculated by the expression:

Equation 6-10:

$$R = \frac{F(t_1) + F(t_2) + Z\left(v(t_1) + v(t_2)\right)}{2}$$

Where

- $t_2 = t_1 + 2L/c$
- t_1 = a selected time (often the first velocity maximum) during the hammer blow
- v_1 = pile head velocity at time t_1
- v_2 = pile head velocity at time t_2
- Z = pile impedance = Mc/L
    - L = the pile length
    - M = the pile mass
    - c = the wave transmission speed in the pile material.

This total resistance R is the sum of static S (displacement dependent) and dynamic D (velocity dependent) components. To estimate the static resistance of the pile, the dynamic resistance component is eliminated and the static resistance is

Equation 6-11:

$$S = R - D = R - J_c\left(v(t_1) + F(t_1) - R\right)$$

Where J_c = dimensionless Case-damping factor (Table 6-4.)

Table 6-4 Case Damping Constants for Various Soil Types[6]

Soil Type	Case Damping Constant J_c
Clean Sands	0.10 to 0.15
Silty Sands	0.15 to 0.25
Silts	0.25 to 0.40
Silty Clays	0.40 to 0.70
Clays	0.70 to 1.00

Figure 6-31 presents typical plots of pile head force and velocity histories during a hammer blow along with Case Method capacity interpretations. The damping factor, J_c, can be solved directly from the above equations if a static load test is substituted for S.

Figure 6-31 Typical Plots of Pile Head Force and Velocity with Case Method Calculations

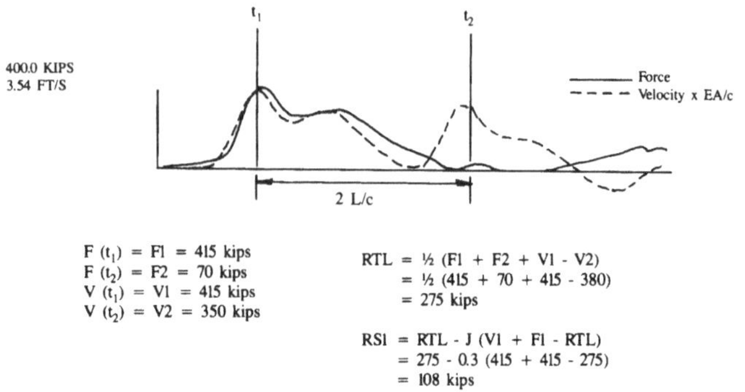

As is the situation with any dynamic method, soil set-up must be taken into consideration when evaluating the results. Also, the deflection of both pile shaft and toe must be sufficient to exceed the elastic limit (quake) of both in order to properly evaluate the ultimate static capacity of the pile.

On site, a pile may be tested either during initial installation, or during restrike. Over 30 dynamic measured or computed quantities are available for each hammer blow in real time. The quantities of most interest for routine applications are:

1. Case Method axial pile capacity
2. Maximum energy delivered to the pile, ram impact velocity and hammer or pile cushion stiffness for hammer performance evaluations.
3. Maximum compressive and tensile forces during driving to check for potentially damaging stresses.
4. Indication of location and extent of damage for pile integrity evaluation.

Figure 6-32 shows the correlation of ultimate static pile bearing capacity determined by static load testing and dynamic load testing during the original Case research project. It should be kept in mind that significantly different pile capacities may often be determined from the results of a single static pile load test, depending on the method of interpretation used. The correlation would not be as good for open-ended pipe piles, H-piles and precast cylinder piles. In general, dynamic load tests on these types of piles underestimate the static pile capacity.

Figure 6-32 Correlation of Ultimate Static Pile Capacity by the Case Method and Static Load Tests

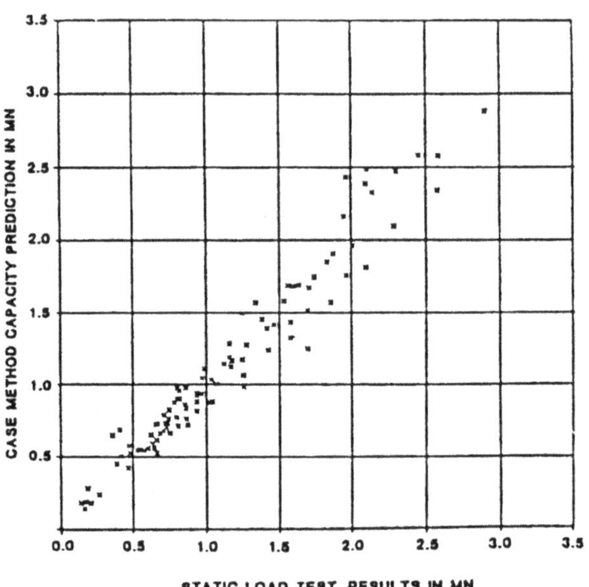

6.4.4. The CAPWAP Method (CAse Pile Wave Analysis Program)

CAPWAP is a computer program for a more rigorous evaluation of static pile capacity, the relative soil resistance distribution, and soil quake and damping characteristics. A CAPWAP analysis is performed on an individual hammer blow that is usually selected from the end of driving or beginning of restrike. As such, a CAPWAP analysis refines the Case Method dynamic test results at a particular penetration depth or time. CAPWAP uses wave equation type pile and soil models; the Pile Driving Analyzer measured force and velocity records are used as the head boundary condition, replacing the hammer model.

In the CAPWAP method depicted in Figure 6-33, the pile is modeled by a series of continuous pile segments and the soil resistance modeled by elasto-plastic springs (static resistance) and dashpots (dynamic resistance). The force and acceleration data from the Pile Driving Analyzer are used to quantify pile force and pile motion, which are two of the three unknowns. The remaining unknown is the boundary conditions, which are defined by the soil model. First, reasonable estimates of the soil resistance, distribution and quake and damping parameters are made. Then, the measured acceleration is used to set the pile model in motion. The program then computes the equilibrium pile head force, which can be compared to the Pile Driving Analyzer determined force. Initially, the computed and measured pile head forces will not agree with each other. Adjustments are made to the soil model assumptions and the calculation process repeated.

Figure 6-33 Overview of the CAPWAP Procedure

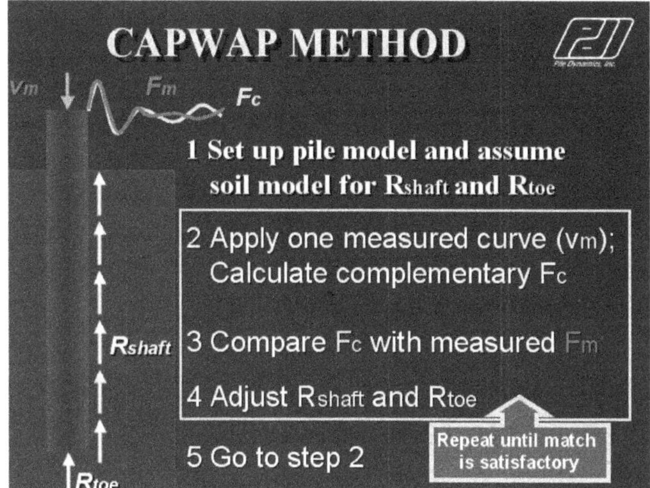

With each analysis, the program evaluates the match quality by summing the absolute values of the relative differences between the measured and computed waves. The program computes a match quality number for each analysis that is the sum of the individual match quality numbers for each of these four zones.

Through this trial and error iteration adjustment process to the soil model, the soil model is refined until no further agreement can be obtained between the measured and computed pile head forces. The resulting soil model is then considered the best estimate of the static pile capacity, the soil resistance distribution, and the soil quake and damping characteristics. An example of the final CAPWAP result summary is presented in Figure 6-34. A summary of the stress distribution throughout the pile is also obtained as illustrated in Figure 6-35. Lastly, CAPWAP includes a simulated static load-set graph based on the CAPWAP calculated static resistance parameters and the elastic compression characteristics of the pile.

Figure 6-34 CAPWAP Final Results Table

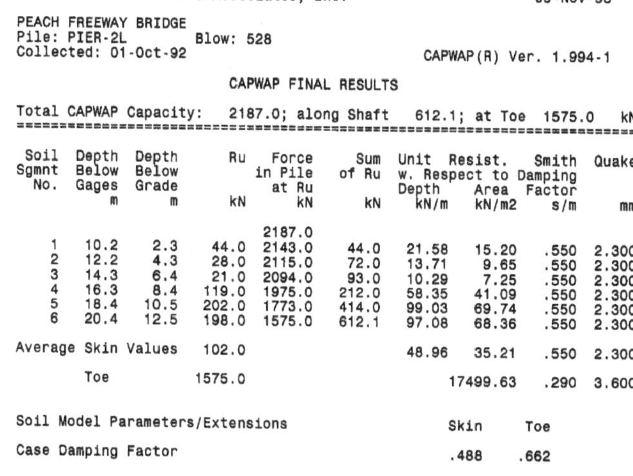

Goble Rausche Likins & Associates, Inc. 09-Nov-95

PEACH FREEWAY BRIDGE
Pile: PIER-2L Blow: 528
Collected: 01-Oct-92 CAPWAP(R) Ver. 1.994-1

CAPWAP FINAL RESULTS

Total CAPWAP Capacity: 2187.0; along Shaft 612.1; at Toe 1575.0 kN

Soil Sgmnt No.	Depth Below Gages m	Depth Below Grade m	Ru kN	Force in Pile at Ru kN	Sum of Ru kN	Unit Resist. w. Respect to Depth kN/m	Unit Resist. w. Respect to Area kN/m2	Smith Damping Factor s/m	Quake mm
				2187.0					
1	10.2	2.3	44.0	2143.0	44.0	21.58	15.20	.550	2.300
2	12.2	4.3	28.0	2115.0	72.0	13.71	9.65	.550	2.300
3	14.3	6.4	21.0	2094.0	93.0	10.29	7.25	.550	2.300
4	16.3	8.4	119.0	1975.0	212.0	58.35	41.09	.550	2.300
5	18.4	10.5	202.0	1773.0	414.0	99.03	69.74	.550	2.300
6	20.4	12.5	198.0	1575.0	612.1	97.08	68.36	.550	2.300
Average Skin Values			102.0			48.96	35.21	.550	2.300
Toe			1575.0				17499.63	.290	3.600

Soil Model Parameters/Extensions	Skin	Toe
Case Damping Factor	.488	.662

Figure 6-35 CAPWAP Stress Distribution Table

Goble Rausche Likins & Associates, Inc. 09-Nov-95

PEACH FREEWAY BRIDGE
Pile: PIER-2L Blow: 528
Collected: 01-Oct-92 CAPWAP(R) Ver. 1.994-1

EXTREMA TABLE

Pile Sgmnt No.	Depth Below Gages m	max. Force kN	min. Force kN	max. Comp. Stress kN/cm2	max. Tension Stress kN/cm2	max. Trnsfd. Energy kN- m	max. Veloc. m/s	max. Displ. cm
1	1.0	3174.0	-72.5	18.848	-.430	32.31	4.4	1.737
2	2.0	3195.0	.0	18.973	.000	31.13	4.3	1.680
4	4.1	3214.2	.0	19.086	.000	29.83	4.3	1.560
6	6.1	3235.5	.0	19.213	.000	28.49	4.2	1.440
8	8.2	3273.7	.0	19.440	.000	27.20	4.2	1.320
10	10.2	3342.5	-60.9	19.848	-.362	25.92	4.1	1.200
12	12.2	3257.1	-83.6	19.342	-.497	23.56	4.0	1.080
14	14.3	3252.0	-140.1	19.311	-.832	21.75	3.9	.960
16	16.3	3379.0	-126.5	20.065	-.751	20.15	3.6	.850
18	18.4	3182.0	-49.7	18.895	-.295	17.11	3.4	.730
19	19.4	2718.8	.0	16.145	.000	14.02	3.3	.680
20	20.4	3005.1	.0	17.845	.000	11.97	3.0	.640
Absolute	16.3			20.065		(T=	24.7 ms)	
	14.3				-.832	(T=	41.2 ms)	

CAPWAP is a proprietary computer program of Goble, Rausche, Likins and Associates, Inc. and the program software is available from the developer. Alternatively, analysis of dynamic test data can be obtained from the developer or other consulting engineers who have acquired program licenses.

6.4.5. Application of Dynamic Testing Methods

Dynamic testing is specified in many ways, depending upon the information desired or purpose of the testing. For example, a number of test piles driven at preselected locations may be specified. In this application, the test piles are usually driven in advance of, or at the start of, production driving so that the information obtained can be used to establish driving criteria and/or pile order lengths for each substructure unit. Alternatively, or in addition to a test pile program, testing of production piles on a regular interval may be specified. Production pile testing is usually performed for quality assurance checks on hammer performance, driving stress compliance, pile integrity, and ultimate capacity. Lastly, dynamic testing can be used on projects where it was not specified to troubleshoot problems that arise during construction.

The number of piles that should be dynamically tested on the project depends upon the project size, variability of the subsurface conditions, the availability of static load test information, and the reasons for performing the dynamic tests. A higher percentage of piles should be tested, for example, where there are difficult subsurface conditions with an increased risk of pile damage, or where time dependent soil strength changes are being relied upon for a significant portion of the ultimate pile capacity.

On small projects, a minimum of two dynamic tests is recommended. On larger projects and small projects with anticipated installation difficulties or significant time dependent capacity issues, a greater number of piles should be tested. Dynamically testing one or two piles per substructure location is not unusual in these situations. Regardless of the project size, specifications should allow the engineer to adjust the number and locations of dynamically tested piles based on design or construction issues that arise.

Restrike dynamic tests should be performed whenever pile capacity is being evaluated by dynamic test methods. Restrikes are commonly specified 24 hours after initial driving. However, in fine-grained soils, longer time periods are generally required for the full time dependent capacity changes to occur. Therefore, longer restrike times should be specified in these soil conditions whenever possible. On small projects, long restrike durations can present significant construction sequencing problems. Even so, at least one longer-term restrike should be performed in these cases. The longer-term restrike should be specified 2 to 6 days after the initial 24-hour restrike, depending upon the soil type. A warmed up hammer (from driving or restriking a non-test pile) should be used whenever restrike tests are performed.

When a consultant performs dynamic testing, the requirements for CAPWAP analyses should be specifically addressed in the dynamic testing specification. On larger projects, CAPWAP analyses are typically performed on 20 to 40% of the dynamic test data obtained from both initial driving and restrike dynamic tests. This percentage typically increases on smaller projects with only a few test piles, or on projects with highly variable subsurface conditions.

It is often contractually convenient to specify that the general contractor retain the services of the dynamic testing firm. However, this can create potential problems since the contractor is then responsible for the agency's quality assurance program. Some agencies have contracted directly with the dynamic testing firm to avoid this potential conflict and many large public owners have purchased the equipment and perform the tests with their own staff.

6.4.6. Presentation And Interpretation Of Dynamic Testing Results

The results of dynamic pile tests should be summarized in a formal report that is sent to both the construction engineer and foundation designer. The construction engineer should understand the information available from the dynamic testing and its role in the project construction. As discussed in Chapter 9, numerous factors are considered in a pile foundation design. Therefore, the foundation designer should interpret the dynamic test results since many other factors; (downdrag, scour, uplift, lateral loading, settlement, etc.) may be involved in the overall design and construction requirements.

Construction personnel are often presented with dynamic testing results with minimal guidance on how to interpret or use the information. Therefore, it may be helpful to both construction personnel and foundation designers to familiarize themselves with the typical screen display and information available during a dynamic test. Figure 6-36 presents a typical Pile Driving Analyzer display[7] for a 356 mm square prestressed concrete pile driven with a diesel hammer having a maximum rated energy of 89.6 kJ.

Figure 6-36 Typical Dynamic Test System Screen Display

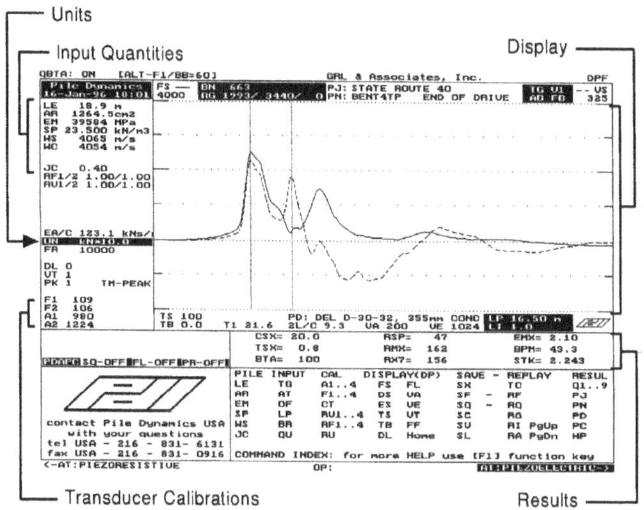

The main Pile Driving Analyzer input quantities are displayed in the upper left corner of the screen and include the following:

- The pile length below gages, LE;
- The pile cross sectional area at the gages, AR;
- The pile elastic modulus, EM;
- The unit weight of the pile material, SP;
- The pile wave speed, WS; and
- The Case damping factor, JC.

The lower left corner includes input quantities for display scales and transducer calibrations and is generally of little interest except to the test engineer. Construction personnel reviewing field results should, however, note the units indicator, UN, in this area of the screen. The force units are noted to be in ''kN * 10" or kilonewtons times 10. This means any forces (but not stresses), capacity, or energy results displayed in the numerical results area must be multiplied by 10.

The screen is dominated by the graphical display of force (solid line) and velocity (dashed line) records versus time. This display will change for each hammer blow. The first vertical line represents time t, in the Case Method calculations and corresponds to the time of impact as the waves pass the gage location near the pile head. The second vertical line represents time t_2 in the Case Method calculations and corresponds to the time when the input waves have traveled to the pile toe and returned to the gage location or time 2L/C.

An experienced test engineer can visually interpret these signals for data quality, soil resistance distribution and pile integrity. As discussed earlier, soil resistance forces cause a relative increase in the force wave and a corresponding relative decrease in the velocity wave. Therefore on a pile with a uniform cross section, the separation between the force and velocity records between times t, and t_2 indicates the shaft resistance. The magnitude of separation is also indicative of the magnitude of the soil resistance at that depth. Toe resistance is indicated by the separation between these records near and after time t_2.

The Pile Driving Analyzer searches for convergence between the force and velocity records beginning at the time of the sharp rise in the records prior to time t, and continuing for a time interval of 2L/C thereafter. If convergence between the force and velocity records occurs prior to the rise in the velocity record preceding time t_2, a cross sectional reduction or pile damage is indicated. The degree of convergence between the force and velocity records is expressed by the BTA integrity value as a percentage of the approximate reduced cross sectional area.

Three letter codes displayed below the graphical records identify numerical results from Case Method computations. In the example given in Figure 6-36, the first column of results provides information on the driving stresses and pile integrity. The compression stress at the pile head, CSX, is 20.0 MPa and the calculated tension stress, TSX, is 0.8 MPa. These calculated stress levels are below the recommended driving stress limits for a prestressed concrete pile given in Chapter 11. Pile integrity, BTA, is calculated as 100%, indicating that no damage is present.

The middle column of results includes computations for the standard Case Method capacity, RSP, and maximum Case Method capacity, RMX, both calculated with the input Case damping factor, JC, of 0.4. These results are 470 and 1620 kN respectively, when adjusted by the units multiplier. As noted earlier, a damping factor at least 0.2 higher is usually used with the maximum Case Method as compared to the standard Case Method. Therefore, the capacity using the RMX equation with a damping factor of 0.7 labeled RX7 was calculated and indicated a capacity of 1560 kN. From the force and velocity records in the example, the experienced test engineer would note that the resistance is delayed in time, based upon the separation between the force and velocity records occurring after time t_2. Therefore, the maximum Case Method equation should be used for capacity evaluation, and from the capacity results noted above, a Case Method capacity of 1560 kN would be chosen.

The final column of numerical results includes the transferred hammer energy, EMX, which is 21 kJ; the hammer operating speed in blows per minute, BPM, which is 43.3; and the calculated hammer stroke for the single acting diesel hammer of 2.24 meters.

Depending upon the hammer-pile combination, average transferred energies as a percentage of the rated energy range from about 25% for a diesel hammer on a concrete pile to 50% for an air hammer on a steel pile. Hence, the transferred energy of 21 kJ is 23% of the rated energy and is therefore slightly below average. The performance of a hammer and driving system can be evaluated from a driving system's rated transfer efficiency, which is defined as the energy transferred to the pile head divided by the manufacturer's rated hammer energy. Figure 6-37 presents transfer efficiencies for selected hammer and pile type combinations expressed as a percentile. In this graph, the average transfer efficiency for a given hammer-pile combination can be found by noting where that graph intersects the 50 percentile. Histograms of the transfer efficiencies for each of these hammer and pile types are also presented in Figure 6-38. The histograms may be useful in assessing drive system performance as they provided the distribution and standard deviation of drive system performance for a given hammer-pile combination at the end of drive condition.

Figure 6-37 Transfer Efficiencies for Select Hammer and Pile Combinations

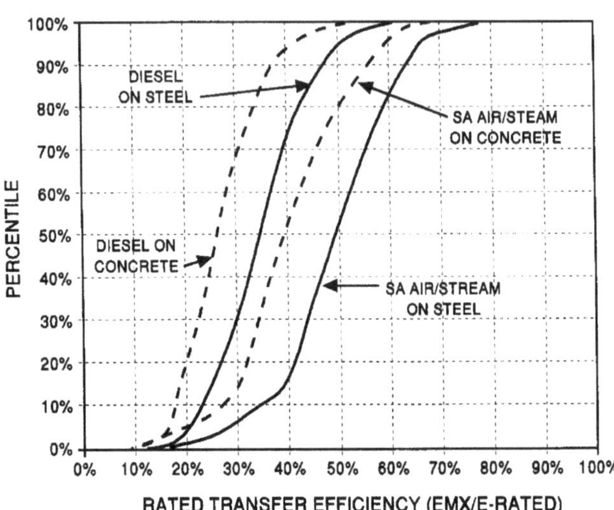

In the field, construction personnel should check that the calculated driving stresses CSX and TSX, are maintained within specification limits. Drive system performance indicated by the transferred energy, EMX, should be within a reasonable range of that predicted by wave equation analysis or recorded on previous tests at the site. If significant variations in energy are noted, the reasons for the discrepancy should be evaluated. The recorded hammer speed should be compared to the manufacturer's specifications. Capacity estimates should be compared with the required ultimate pile capacity. In soils with time dependent changes in capacity, this comparison should be based on restrike tests and not end-of-initial driving results.

A force and velocity record for a 406 mm x 13 mm wall closed end pipe pile is presented in Figure 6-39. As can be seen from the input properties, the pipe pile is 29.1 meters long below gages. A visual interpretation of the signal would indicate the pile has developed moderate shaft resistance over the lower portion of the pile with the majority of the pile capacity due to toe resistance. Note that an intermediate vertical line labeled D has also appeared between the two vertical lines corresponding to the pile head, t_1, and pile toe, t_2. Convergence between the force and velocity records before time 2L/C, as noted by the D line, indicates a pile impedance reduction or damage. A warning box has also appeared on the screen asking the test engineer if damage is occurring. For the

Figure 6-38 Histograms of Transfer Efficiency for Diesel and Single Acting Air/Steam Hammers

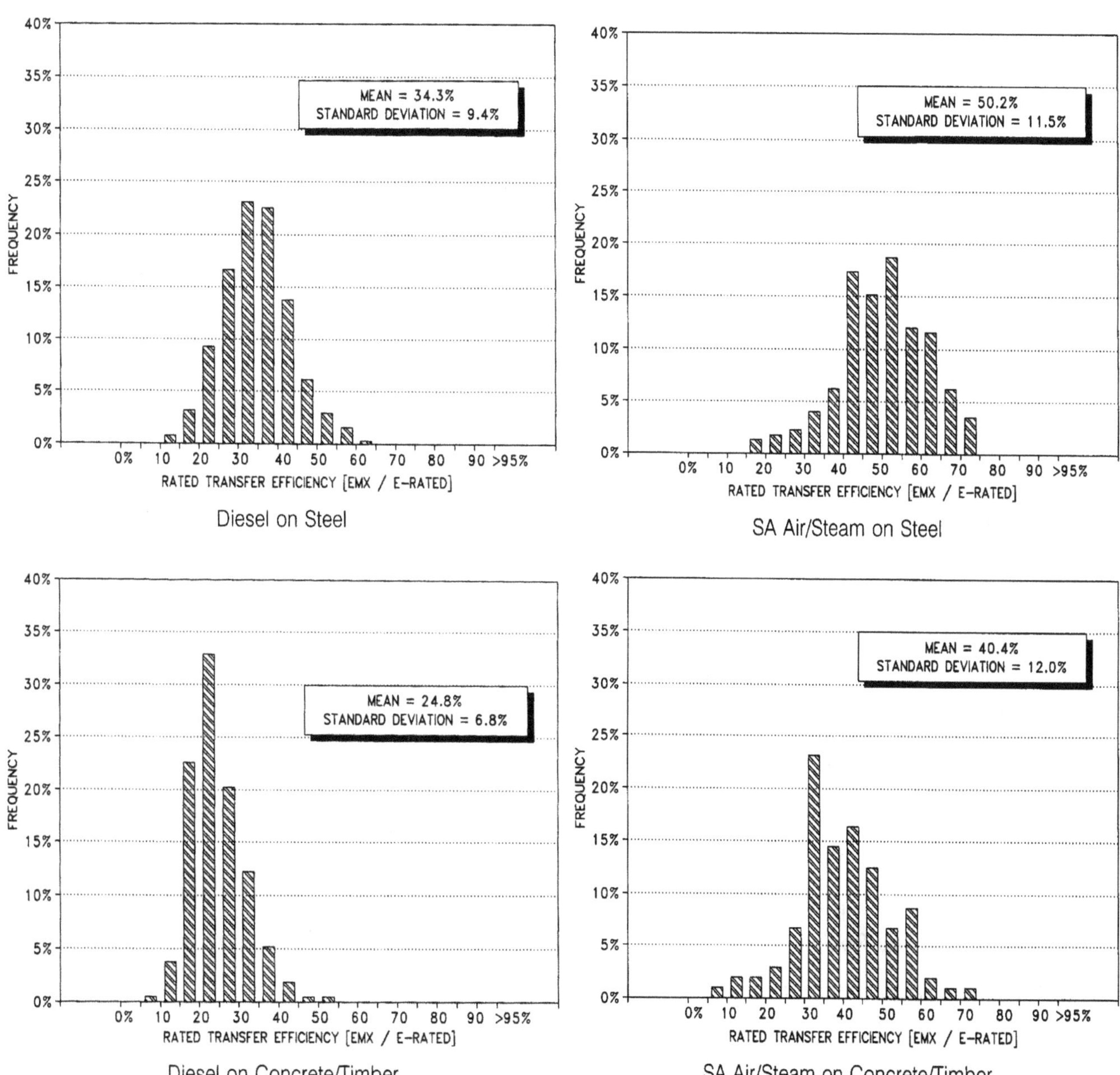

example shown, damage was occurring at a depth of 14.9 meters below gages due to a welding problem at the pile splice.

Figure 6-39 Force and Velocity Record for Damaged Pile

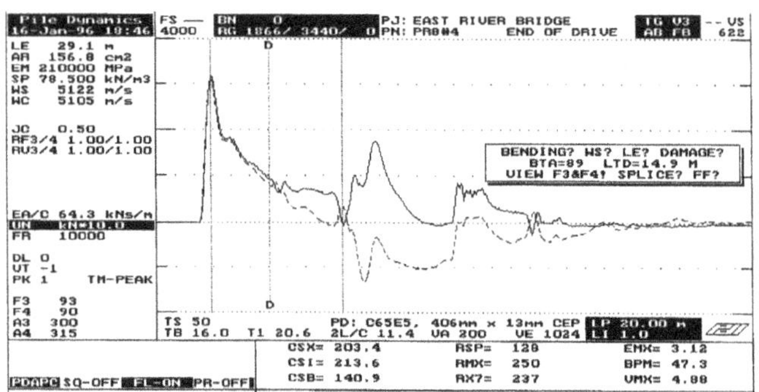

In Figure 6-40, a force and velocity record for a HP 360 x 132 H-pile is presented. This record is typical of a pile driven to rock. Note the strong separation in the force and velocity records at time 2L/C (second vertical line). The compression stress at the gage location, CSX, is 211 MPa. This is within the recommended driving stress limit of 223 MPa for A-36 steel given in Chapter 11. The Pile Driving Analyzer can also compute an estimate of the compression stress at the pile toe, CSB. This quantity may be helpful in driving stress control for piles to rock. For the record shown, CSB is calculated to be 232 MPa, which is above the recommended driving stress limit. Therefore, a slight reduction in hammer stroke at final driving may be necessary. The CSB quantity is an approximate value. A better assessment of the compression stresses at the pile toe could be gained from CAPWAP or wave equation analyses.

Figure 6-40 Force and Velocity Record for H-pile to Rock

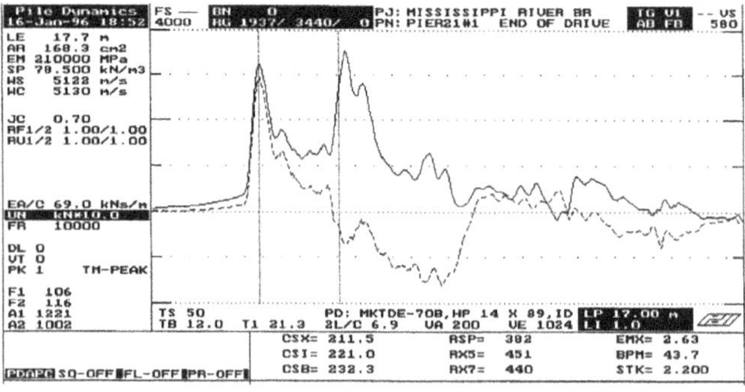

Comparing the dynamic test numerical results versus pile penetration depth and corresponding driving resistance can obtain additional insight into the pile and soil behavior during driving. Dynamic testing systems typically assign a sequential blow number to each hammer blow. By comparing the pile driving records with these blow numbers, numerical and graphical summaries of the dynamic testing results versus pile penetration depth and driving resistance can be prepared. An example of a numerical summary of the dynamic testing results versus depth for a 610 mm octagonal concrete pile is presented in Table 6-5 with accompanying graphical results presented in Figure 6-41. Construction personnel can then easily compare these results to project requirements.

Table 6-5 Typical Tabular Presentation Of Dynamic Testing Results Versus Depth

Proj: PDAPLOT EXAMPLE — Increment = (depth) — Pg1
Pile: PRESTRESSED CONCRETE — FileName = FGCTP5.MDF
Desc: DIESEL HAMMER — BL# 2 to 1349 31-Oct-94

CSX: Max Measured C-Stress — RMX: Capacity - RMX
TSX: Max Computed T-Stress — BPM: Blows Per Minute
EMX: Max Transferred Energy — STK: Stroke (O.E.Diesels)

BL# end	bl/m	depth m	TYPE	#Bls	CSX MPa	TSX MPa	EMX kN-m	RMX kN	BPM bl/min	STK m
11	18	8.00	AVG	6	8.48	2.05	18.85	285	56.1	1.439
34	23	9.00	AVG	9	12.04	2.92	27.03	584	49.6	1.696
64	30	10.00	AVG	23	14.67	3.63	27.40	1107	47.6	1.845
92	28	11.00	AVG	12	15.25	3.43	27.98	1210	47.5	1.846
123	31	12.00	AVG	15	15.62	4.08	28.03	1059	47.6	1.843
166	43	13.00	AVG	14	15.93	5.33	30.79	577	48.1	1.801
199	33	14.00	AVG	11	17.29	3.92	30.55	1143	47.1	1.882
225	26	15.00	AVG	8	17.62	4.96	33.44	875	47.5	1.850
245	20	16.00	AVG	7	18.29	6.23	33.91	586	47.4	1.857
266	21	17.00	AVG	6	18.03	6.47	33.83	340	48.1	1.802
289	23	18.00	AVG	8	18.09	6.65	35.90	284	48.0	1.808
322	33	19.00	AVG	16	18.25	6.14	36.47	529	47.1	1.880
370	48	20.00	AVG	24	20.80	5.75	39.59	1104	45.9	1.989
409	39	21.00	AVG	20	25.14	7.97	45.27	1137	45.3	2.045
453	44	22.00	AVG	22	26.33	7.91	45.81	1124	45.5	2.026
493	40	23.00	AVG	20	26.31	8.10	46.87	1003	45.6	2.014
568	75	24.00	AVG	30	15.65	1.52	33.30	1040	47.3	1.863
609	41	25.00	AVG	13	16.63	3.56	34.01	931	47.8	1.824
641	32	26.00	AVG	15	17.01	4.48	32.45	893	48.3	1.780
668	27	27.00	AVG	14	18.19	5.50	33.36	857	48.2	1.793
696	28	28.00	AVG	14	19.54	6.30	36.02	864	47.5	1.847
730	34	29.00	AVG	16	19.88	6.67	35.87	816	48.2	1.794
759	29	30.00	AVG	14	20.50	6.51	39.53	870	47.1	1.876
785	26	31.00	AVG	13	23.97	8.49	44.45	868	46.9	1.896
820	35	32.00	AVG	18	25.61	9.52	45.19	761	47.2	1.868
857	37	33.00	AVG	18	27.16	9.16	48.37	861	46.4	1.939
893	36	34.00	AVG	18	28.13	8.13	51.74	1056	45.8	1.994
933	40	35.00	AVG	20	27.48	9.15	49.39	868	46.5	1.933
971	38	36.00	AVG	19	26.64	9.04	45.58	767	47.3	1.867
1012	41	37.00	AVG	41	25.49	7.91	38.45	899	48.9	1.739
1050	38	38.00	AVG	38	24.55	9.00	36.31	796	49.4	1.699
1078	28	39.00	AVG	14	24.27	9.75	36.11	671	49.5	1.692
1106	28	40.00	AVG	10	22.62	8.52	35.04	596	49.6	1.689
1199	93	41.00	AVG	49	15.60	1.67	28.10	948	49.4	1.703
1235	144	41.25	AVG	18	17.01	0.60	31.84	1672	47.6	1.839
1275	160	41.50	AVG	18	16.88	1.16	30.38	2072	47.7	1.835
1329	216	41.75	AVG	26	19.70	1.50	40.17	2434	44.7	2.115
1349	200	41.85	AVG	9	33.30	6.11	67.31	2623	41.5	2.451

Figure 6-41 Typical Graphical Presentation of Dynamic Testing Results versus Depth

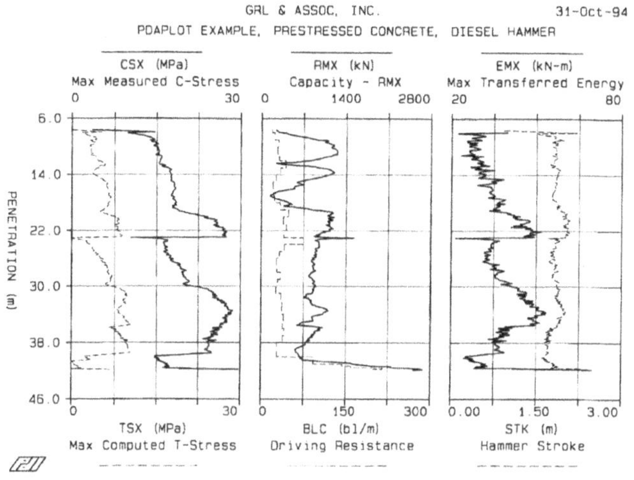

6.5. Low-Strain Integrity Test Methods

Up until now, all of our discussion on dynamic testing methods have concentrated on high strain dynamic testing methods and their applications. This section will discuss low strain integrity testing methods that can be used on driven pile foundations. These low strain methods may be used to evaluate pile length or integrity of piles with a high impedance (EA/C), such as solid concrete piles or concrete filled pipe piles. Additional details on low strain methods including equipment requirements and analysis of measurements may be found in ASTM D-5882 Standard Test Method for Low Strain Integrity Testing of Piles. Low strain integrity methods are not applicable to steel H-piles or unconcreted pipe piles.

6.5.1. Pulse Echo Method

Pulse echo pile testing consists of applying a low strain impact to the head of a pile, and monitoring the resulting pile head response. A small hand-held hammer (0.5 to 4 kg) is employed to deliver a clean impact to the pile head. An accelerometer, temporarily attached to the pile head, records pile head response as the generated low strain stress wave propagates down the pile length. Any changes in pile impedance (determined by the cross sectional area, the elastic modulus of the pile material and the stress wave speed of the pile material) along the pile shaft will generate a partial reflection of the downward travelling stress wave, thus identifying pile damage. At the pile toe a significant change in impedance would also occur, therefore allowing determination of pile length. The accelerometer records the magnitude and arrival time of the reflected waves. For undamaged piles, if a toe reflection is apparent, then it is possible to reasonably estimate an unknown pile length based upon an assumed wave speed.

The returning analog signals are captured and digitized by a portable high accuracy analog to digital data acquisition system. A display panel presents the record of one or more (averaged) blows for review and interpretation. Typically, the acceleration versus time data is integrated to a velocity versus time record to facilitate record evaluation.

This test method can also be used in cases where the pile length is known but the pile integrity is in question. In this application, a clearly indicated toe signal, together with a fairly steady velocity trace between the impact time and toe reflection, are signs of a sound pile. Strong velocity reflections before the expected toe signal are the result of changes in pile cross section and indicate pile damage.

Pulse echo integrity records of velocity versus time are presented in Figure 6-42 and Figure 6-43 for two 305 mm square prestressed concrete piles. These records were obtained after a slope failure occurred during construction and the integrity of the driven piles was questioned. Figure 18.24 shows an amplified record for an undamaged 16.3 m long pile. Note the record drops below the origin at a depth 5 m, which corresponds to soil resistance effects. A clear toe signal is apparent in the record at a depth of 16.3 m.

Figure 6-42 Pulse Echo Velocity versus Time Record for Undamaged Pile

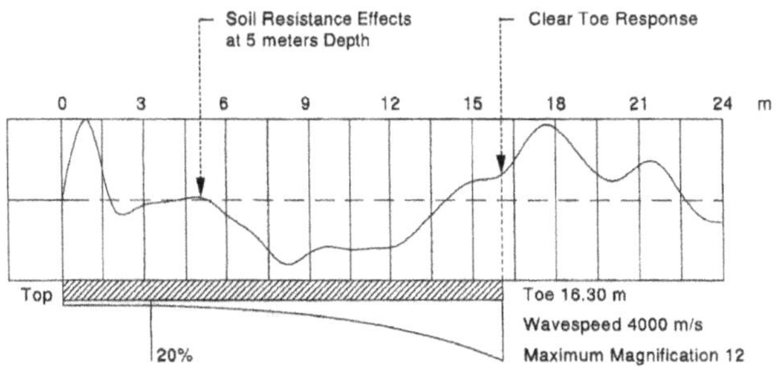

Figure 6-43 Pulse Echo Velocity versus Time Record for Damaged Pile

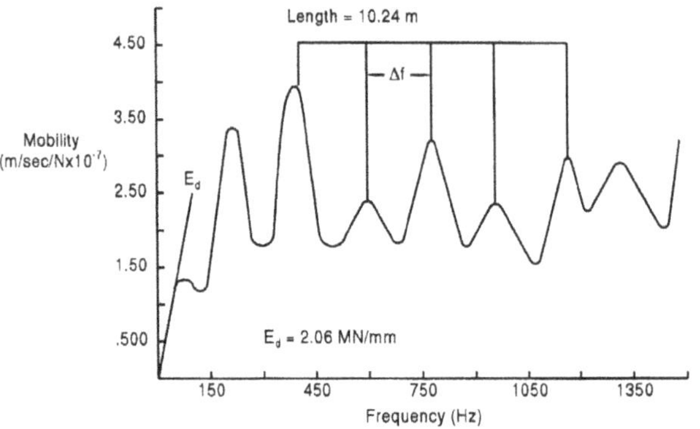

In Figure 6-43, an amplified pulse echo record on a nearby pile is presented. This pile has a clear indication of damage due to the slope movements based on the positive velocity reflection starting at a depth of 4 m.

6.5.2. Transient Response Method (TRM)

In the TRM method, both the pile head response and the impact force are measured. A simple hand held hammer can adequately produce the frequency components necessary to test both well-constructed and defective piles with TRM. The standard TRM plot of the ratio of the frequency velocity spectrum to force spectrum is called "mobility", and is an indication of the pile's velocity response to a particular excitation force at a certain frequency. Figure 6-44 depicts a typical response curve for a TRM test.

Figure 6-44 Typical Response Curve from a TRM Test

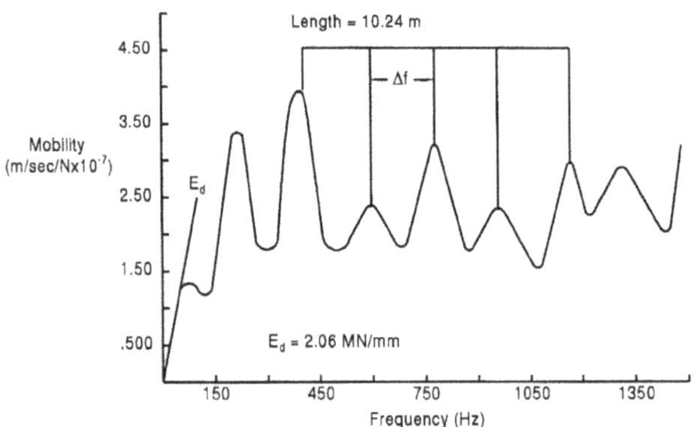

A mobility peak occurs at a frequency indicative of the time when the velocity changes due to a reflection from the pile toe or an intermediate impedance reduction or defect. Mobility peaks occurring at regular intervals are indicative of a dominant frequency Δf. The corresponding length to the pile toe or to a major defect at which the change in frequency occurs is calculated from:

Equation 6-12:

$$L = \frac{C}{2\Delta f}$$

Where:

- C = Wave speed.
- L = Pile length.
- Δf = Change in frequency.

In practice, low frequency (i.e. near static) values are divided by the associated mobility yielding a so-called dynamic stiffness, E_d. This quantity increases with decreasing pile toe response. A low pile toe response is often the result of high soil resistance. A low pile toe resistance may also be caused by highly variable pile properties of internal pile damping, and is therefore only indirectly related to pile capacity. However, E_d is calculated, since it does provide a quantitative result for the evaluation of pile quality. Generally, higher stiffness values (for piles on the same site and of comparable length) indicate piles of higher strength (structural and soil) while lower stiffnesses indicate piles with potential defects or lower soil strength.

6.5.3. Low Strain Applications to Unknown Foundations

Design or construction records on many older bridges are not available. In some cases, the foundation supporting these structures is unknown and therefore the performance of these structures under extreme events such as scour is uncertain. Recent research[8] on the application of non-destructive testing methods to the evaluation of unknown foundations found the pulse echo and transient response methods fair to excellent in their ability to identify the depth of exposed piles and poor to good in their ability to determine the depth of footing or pile cap. These techniques are most applicable when the bridge is supported on a columnar substructure rather than a pier or abutment. Access to the bridge substructure is also generally required for implementation of these techniques.

One method that can be used with piles whose heads are no longer accessible is dispersive wave technology (DWT). This involves striking the pile along the shaft near to the top; the sensors are mounted below where the strike takes place. The results are analyzed and defects can be uncovered with this method as well.[9]

6.5.4. Limitations and Conclusions of Low Strain Methods

The low strain methods can typically be used for integrity or length assessments of pile foundations where the length to diameter ratio does not exceed about 30. For piles with severe cracks or manufactured mechanical joints, the stress wave will generally not be transmitted below the gap. Therefore, the pile integrity or length below this gap cannot be evaluated. Records from piles with multiple or varying (i.e. tapered piles) cross sectional areas can also be difficult to interpret. For piles of low impedance (H-piles and unfilled pipe piles) low strain methods are generally not suitable. When used for pile length determinations, the length information obtained from a toe signal (or a governing frequency) is only as accurate as the wave speed value assumed in the processing of the records. Wave speed variations of approximately 10% are not uncommon. Some defects can also have secondary and tertiary wave reflections. For example, if an impedance reduction occurs in the middle of the pile, then what may appear to be the pile toe response may actually be a secondary reflection of the mid-pile defect.

The additional force measurement obtained during TRM testing provides supplemental information of cross sectional changes near the pile head, i.e. within the distance corresponding to the impact signal. The minor additional expense of the force measurements is therefore worthwhile whenever questions arise as to the integrity of upper (1.5 m) pile portion.

Using low strain methods, many piles can be tested for integrity in a typical day. Therefore, low strain methods are a relatively economical test method and can provide valuable information when used in the proper application. Low strain testing has been used to assist in evaluating integrity questions on high impedance piles due to construction equipment or vessel impact, pulling on out of position piles, and storm damage. Any low-strain method involves the interaction of the soil with the stress waves that are being propagated in the pile. These methods, as is the case with high-strain methods, should be conducted and analyzed by experienced personnel. This is especially true with long piles and certain soil types, where the results can be inconclusive even under the best of circumstances.

Photo Credits

- Berminghammer Corporation (Mr. Mike Justason): Figure 6-19, Figure 6-20, Figure 6-21, Figure 6-22, Figure 6-23 and Figure 6-26.
- GRL and Associates/Pile Dynamics, Inc. (Mr. Mohamad Hussein): Figure 6-1, Figure 6-2, Figure 6-3, Figure 6-5, Figure 6-14, Figure 6-29, Figure 6-27, Figure 6-30 and Figure 6-33.

Footnotes

[1] Production piles can be marked in increments of 5'. Test piles should be marked every foot.

[2] Adhere to units given carefully! If this is not possible, insure that units are consistent.

[3] For more information on this, see Fellenius, Bengt, "From Strain Measurements to Load in an Instrumented Pile." Geotechnical News, Vol. 19, No. 1, March 2001, pp. 35-38.

[4] Refer to ASTM D-3966, FHWA-IP-84-11, *Handbook on Design of Piles and Drilled Shafts Under Lateral Load* by Reese (1984) as well as FHWA-SA-91-042, *Static Testing of Deep Foundations* by Kyfor et. al. (1992) for additional information on lateral load tests and the interpretation of their results.

[5] Additional information on the equipment requirements for dynamic testing are detailed in ASTM D-4945, *Standard Test Method for High Strain Dynamic Testing of Piles* and in AASHTO T-298-33, *Standard Method of Test for High Strain Dynamic Testing of Piles.*

[6] The Case Method as presented here is best suited for pile capacity evaluation of low displacement piles or piles with large shaft resistances. Other variations of the Case methods can be used when these conditions are not present. While these values of the Case damping factor may be suitable for initial capacity estimates, site specific damping correlations should be developed based on static load tests results or CAPWAP analysis.

[7] Screen shown in the DOS version. Windows version has much of the same information but is presented differently.

[8] FHWA Geotechnical Guideline No. 16 (1998), provides a summary of this NCHRP study.

[9] Slaughter, S.H., and Timm, R.J. "Exploring Unknown Foundation Conditions with Dispersive Wave Technology." *Geo-Strata*, pp. 10-13, January 2001.

[10] Unit conversions done in this book are mainly derived from Raznjevic, Kuzman, *Handbook of Thermodynamic Tables and Charts*. New York: McGraw-Hill Book Company, 1976.

Appendices

A. Units

In the U.S., the construction industry is making a gradual transition from U.S. to SI units, and this book reflects that transition.

The general approach in this book is to furnish both sets of units when practical. With the varied nature of the source material, absolute consistency is difficult. Exceptions are as follows:

- When the original source comes from a code or standard that is presently in U.S. units.
- When the original source is a calculation that depends upon the use of U.S. units rather than a consistent set of units.
- When the original source reflects a technology that was developed in U.S. units, such as Raymond Step-Taper® piles, or conversely one that was developed in SI units, such as Statnamic®. Even with these, in some cases conversions are provided to assist the reader.
- Theoretical formulations, such as vibratory hammer performance and wave equation theory (SI Units) or dynamic formulae (U.S. Units). In these cases, a consistent units approach is adopted.[10]

B. Compilation of Definitions and Terms Related to Piling

- Adapter - Structural fabrication or casting placed in a helmet to adjust to a different type size of pile. Also called filler or insert.
- Allowable Pile Load - The load permitted on a pile. It is computed from the pile capacity (ultimate) by dividing with an appropriate safety factor.
- Anvil - The part of a pile hammer seated directly under the ram, which transits the blow of the ram to the pile. (For diesel hammers, also called "impact block". For other hammers, often called "striker plate.")
- Atomization, Impact - A type of fuel atomization used in a diesel hammer by which the raw fuel injected and trapped between the ram and the impact block is atomized by impact of the ram on the impact block. Impact atomization may also be referred to as "liquid fuel injection."
- Atomization, Injection - A type of fuel atomization used in a diesel hammer in which the fuel is atomized by high-pressure injection.
- Assembly - All hammer parts except ram.
- Assembly Drop - Impact between assembly and helmet when the pile rebounds.
- Batter Pile - A pile driven at an angle.
- Bounce Chamber Pressure - The pressure in the air compressed by the upward moving piston of a closed-end diesel hammer. A gauge connected to the upper chamber by a hose indicates the pressure.
- Bouncing - A strong pile rebound after a large temporary pile penetration, accompanied by cylinder uplift and unstable driving conditions.
- Brace - A structural member used to attach the bottom of the leads to the crane base, and used to position or batter leads in or out, left or right. Can be mechanically, hydraulically, or pneumatically operated; it is used with either fixed leads or semifixed leads. Also called "bottom brace," "a-frame," "spotter," "spreader bar," "spider," "kicker," "stinger," "strut," "apron."
- Caisson A large, deep foundation unit other than a driven or bored pile that is sunk down into the ground to carry a structural unit.
- Capacity The maximum or ultimate soil resistance mobilized by a foundation unit.
- Capacity, bearing The maximum or ultimate soil resistance mobilized by a foundation unit subjected to downward loading.
- Capacity, geotechnical See capacity, bearing.
- Capacity, lateral The maximum or ultimate soil resistance mobilized by a foundation unit subjected to horizontal loading.
- Capacity, structural The maximum or ultimate strength of the foundation unit.

- Capacity, tension The maximum or ultimate soil resistance mobilized by a foundation unit subjected to tension (upward) loading.
- Closed End Hammer - See Diesel.
- Coefficient of Restitution - A number used in the dynamic analysis of collisions indicating the amount of strain energy returned when a body (for example cushion) returns to its original shape. COR = 1 - all energy is returned (elastic), COR = 0 - no energy is returned.
- Combustion Delay - The time between impact and combustion in an impact atomization diesel pile hammer.
- Compression Stroke - Distance traveled by a diesel hammer ram while compressing cylinder gas. Measured from bottom of air intake port to point of impact.
- Cushion - Material such as wood, plastic, compressed paper, aluminum, or rope. Main purpose of a cushion block is uniform distribution of impact forces.
- Cushion, hammer The material placed in a pile-driving helmet to cushion the impact. (Formerly called "capblock").
- Cushion, pile The material placed on a pile head to cushion the impact.
- Cylinder Lift - The point at which the cylinder of a double acting air/steam or diesel hammer begins to lift or float. The point at which the force of fluid in the hammer overcomes the weight of the hammer casing or cylinder.
- Design Load - See Allowable Pile Load.
- Diesel Hammer - A pile hammer that carries its own power source (internal combustion hammer). An open end diesel hammer falls under the action of gravity. A closed end diesel (double acting) compresses air on its upward stroke and therefore runs faster than open-end hammers.
- Differential-Acting Hammer - See External Combustion Hammer.
- Double-Acting Diesel Hammer - See Diesel Hammer.
- Double-Acting Hammer - Hammer that falls under the action of both gravity and power.
- Downdrag The downward pull on a deep foundation unit due to negative shaft friction and expressed in terms of dragload or settlement.
- Dragload The load transferred to a deep foundation unit from negative shaft friction.
- Dragload, maximum The load transferred to a deep foundation unit from negative shaft friction acting from the pile head to the neutral plane.
- Driving Criteria - Requirement for resistance of a pile to penetration, stated in blows per increment of depth. Required toe elevation of pile.
- Driving Log - Record of physical pile properties of a pile, including location on site, hammer type, driving resistance in blows per unit, penetration, time of pile installation, and restrike information.
- Drop Hammer - Usually a cable raised ram that is dropped by releasing the hoist.
- Dynamic method of analysis The determination of capacity, impact force, transferred energy, etc. of a driven pile using analysis of measured strain waves induced by the driving of the pile.
- Dynamic monitoring The recording of strain and acceleration induced in a pile during driving and presentation of the data in terms of stress and transferred energy in the pile as well as of estimates of capacity.
- Eccentric Loading - Load does not bear vertically or axially on pile, perhaps due to pile being driven out of plumb or plan location.
- Elastic Shortening - Reduction in length of a pile, or structural member, due to load. An important factor in evaluating apparent pile movement under test load. "Elastic compression" is a variation of this that takes place when an impact hammer strikes the pile head; this is the compression due to impact solely attributable to the distributed elasticity and mass of the pile.
- Factor of safety The ratio of maximum

available resistance or of the capacity to the allowable stress or load.

- Final Blow Count - The number of blows per inch, foot, or other unit length of measure. Recorded immediately prior to the end of driving of the pile.
- Final Set - The net penetration a pile moves under one blow or a specified series of blows at the end of driving. (Reciprocal of final blow count).
- Fluid Power - Steam, air, combustion gases, or hydraulic oil used to move the ram of the hammer and to drive the pile. Also called "motive fluid."
- Follower - An attachment, shaped like a short pipe, used to extend the upper portion of the helmet. A member interposed between a pile hammer and a pile to transmit blows when the pile head is below the reach of the hammer. Also called "dolly."
- Foundation unit, deep A unit that provides support for a structure by transferring load or stress to the soil at depth considerably larger than the width of the unit. A pile is the most common type of deep foundation.
- Foundations A system or arrangement of structural members through which the loads are transferred to supporting soil or rock.
- Gate - See Pile Gate.
- Grade - The elevation of the natural ground level of any point on a project site.
- Guides - The part of a pile hammer that loads the ram. Often referred to as "columns." The part of the pile leads which forms a pathway for the hammer.
- Groundwater table The top surface of a free body in the ground.
- Hammer Cushion - See Cushion.
- Hammer Efficiency - The ratio of kinetic energy of the ram immediately prior to impact divided by the rated energy.
- Hammer Energy - The project of ram weight and equivalent stroke measured in ft-lbs or kips-ft.
- Hammer Speed - The number of hammer blows per minute.
- Hammer Uplift - See Cylinder Lift.
- Hammer Leads - See Leads, Swinging.
- Helmet - Steel block placed over a pile to prevent damage from driving. It contains a recess on top for hammer cushion. The bottom is formed to accept a specific shaped pile, along with its cushion, if used. Also called "cap" or "driving head."
- Impact Atomization - See Atomization, Impact.
- Impact - The collision and ensuing energy transmission of two bodies.
- Impact force The peak force delivered by a pile-driving hammer to the pile head as measured by means of dynamic monitoring before the measurements are influenced by soil resistance reflections.
- Impact Velocity - The velocity of the ram of a gravity or single-acting steam/air or diesel hammer when it strikes the anvil.
- Inlet Port - The inlet fluid path of external combustion hammers.
- Inlet Pressure - Pressure of motive fluid at the inlet of an external combustion hammer.
- Inspector, Pile - The individual who observes pile installation for conformance with contract requirements.
- Kip - A force unit equal to 1,000 lbs.
- Leaders — See Leads
- Leads - A structure on which a pile hammer travels up and down and with which the hammer and pile assembly is aligned in driving position.
- Leads, Fixed - A lead system which is pivot supported at the crane boom top and by a brace at its bottom.
- Leads, Semifixed - A trade name for a type lead that may be raised or lowered independently of the boom by a cable from the crane. Also called "vertical-travel leads."
- Leads, Swinging - Are hung by their top from one load line of a crane. The second support point is at the ground through spikes.

- Leads, Offshore - Also called "free riding leads," they align hammer and pile, but the pile should be supported by a template or other means.
- Line Oiler - Lubricator inserted in pressure air or steam line to lubricate hammer piston or cylinder.
- Load, allowable The maximum load that may be safely applied to a foundation unit under expected loading and soil conditions and determined as the capacity divided by the factor of safety.
- Load, applied or load, service The load actually applied to a foundation unit.
- Monkey - See Pile Monkey
- Motive Fluid - See Fluid Power
- Needle Gauge - A pressure gauge with a needle stem for measuring air or steam pressure in a pressure hose, by puncturing the wall of the hose with the needle stem.
- Neutral plane The location where equilibrium exists between the sum of downward acting permanent load applied to the pile and dragload due to negative shaft friction and the sum of upward acting positive shaft resistance and mobilized toe resistance. The neutral plane is also where the relative movement between the pile and the soil is zero.
- Overdriving - Driving in a way that damages material in a pile, most often by continuing to pound after penetration of the pile stops.
- Penetration Resistance - The resistance to penetration from driving expressed in blows per increment of length to advance of the pile.
- Pile A slender deep foundation unit, made of wood, steel, or concrete, or combinations thereof, which is either premanufactured and placed by driving, jacking, jetting, or screwing, or cast-in-situ in a hole formed by driving, excavating, or boring. A pile can be a non-displacement, a low-displacement, or displacement type. Installed in the ground to resist or transfer vertical, horizontal, or combination loads imposed upon it.
- Pile Cushion - See Cushion.
- Pile Driver - A rig consisting of leads, hoisting apparatus and a pile. A member of the pile crew.
- Pile Driving Cap - See Helmet
- Pile Gate - A device at the base of the pile hammer leads which is closed around the pile to maintain alignment between the pile and the leads.
- Pile Hammer - General term for machine which drives piling by impact or vibration. Power source may be mechanical, air steam, diesel, or hydraulic.
- Pile head The uppermost end of a pile.
- Pile impedance Z EA/c, a material property of a pile cross section determined as the product of the Young's modulus (E) and area (A) of the cross section divided by the wave speed (c).
- Pile Load - See Driving Log.
- Pile Monkey - Trade name for a mechanical device controlled by the crane operator to position the pile underneath the helmet.
- Pile point A special type of pile shoe.
- Pile Refusal - See Refusal.
- Pile Rig - See Pile Driver.
- Pile Set - See Set.
- Pile shaft The portion of the pile between the pile head and the pile toe.
- Pile shoe A separate reinforcement attached to the pile toe of a pile to facilitate driving, to protect the lower end of the pile, and/or to improve the toe resistance of the pile.
- Pile Skin - The pile's outside surface between head and toe.
- Pile toe The lowermost end of a pile. (Use of terms such as pile tip, pile point, or pile end in the same sense as pile toe is discouraged).
- Pipe Pile - A steel or concrete cylindrical shell of specified strength and thickness. It is driven either open or closed ended and is usually filled with concrete.
- Piston - The ram of a diesel hammer. For internal combustion hammer, pistons are

also used to move the ram.

- Plumb Pile - A vertical pile, i.e., a zero batter pile.
- Pore pressure Pressure in the water and gas present in the voids between the soil grains minus the atmospheric pressure.
- Pore pressure, artesian Pore pressure in a confined body of water having a level of hydrostatic pressure higher than the level of the ground surface.
- Pore pressure, hydrostatic Pore pressure varying directly with a freestanding column of water.
- Pore pressure elevation, phreatic The elevation of a groundwater table corresponding to a hydrostatic pore pressure equal to the actual pore pressure.
- Practical Refusal - See Refusal
- Preadmission - An early opening of the inlet valve of an external combustion hammer. It allows a pressure build up before impact.
- Prefiring - See Preignition.
- Preignition - A combustion in a diesel hammer prior to impact.
- Pressure Omnidirectional force per unit area. (Compare stress).
- R - Symbol for pile capacity in kips. Example: 2R is a term meaning twice design capacity.
- Racking Hammer - Severe bouncing of the casing of a pile hammer during driving operations.
- Raker Pile - See Batter Pile.
- Ram - The moving and driving part of any hammer. It delivers an impact blow to an anvil and to the pile.
- Ram Point - The impacting part of the ram.
- Rated Speed - The specified operation speed for a pile hammer in blows per minute.
- Rebound - Amount of upward movement of the head of a pile following a hammer blow.
- Refusal - The condition reached when a pile being driven by a hammer has a penetration per blow that is smaller than a specified value (e.g. 1/10 inch).
- Resistance - The sum total of all the forces that oppose the penetration of a pile.
- Saximeter™ - Trade name for a device used for remote measuring of the stroke of an open-end diesel hammer or the measurement of the hammer speed.
- Settlement The downward movement of a foundation unit or soil layer due to rapidly or slowly occurring compression of the soils located below the foundation unit or soil layer, when the compression is caused by an increase of effective stress.
- Shaft resistance, negative Soil resistance acting downward along the pile shaft because of an applied uplift load.
- Shaft resistance, positive Soil resistance acting upward along the pile shaft because of an applied compressive load.
- Shaft friction, negative Soil resistance acting downward along the pile shaft as a result of downdrag and inducing compression in the pile.
- Shaft friction, positive Soil resistance acting upward along the pile shaft caused by swelling of the soil and inducing tension in the pile.
- Single Acting Diesel Hammer - See Diesel Hammer.
- Single Acting Hammer - A pile hammer in which fluid lifts the ram on the up stroke and gravity alone acts on the down stroke.
- Stabbing Point, Leads - Fabricated steel points or spike fastened to the very bottom of swinging leads.
- Steel H-Pile - A rolled steel section with web and flange the same thickness. Wide flange and other steel sections are also used as piles.
- Step-Taper® Pile - A cast-in-place pile whose diameter is stepped up in increments, usually in 8 to 12' sections of corrugated, thin-metal shell, and internally driven with a mandrel (a heavy steel pipe).
- Stress Unidirectional force per unit area. (Compare pressure).
- Stress, effective The total stress in

a particular direction minus the pore pressure.

- Test Pile - A pile driven to ascertain driving conditions and probable required lengths. A pile on which a loading test also may be performed to determine the carrying capacity of the soil. A pile that will later be used as a guide for efficient design of pile foundations.
- Timber Pile - A tree trunk, usually driven small end down. Probably the earliest form of pile. Frequently, it is treated with creosote as a preservative. A fabricated or pressed steel tip, and frequently, a steel ring around the head to ensure against damage during driving.
- Toe resistance soil resistance acting on the pile toe.
- Transferred energy The energy transferred to the pile head and determined as the integral over time of the product of force, velocity, and pile impedance.
- Trip - A block in the leads of a drop hammer causing the release of the weight at a predetermined height. Mechanical device used in diesel hammers to both pick up the ram and to start the hammer. Also called "trip block" or "tripping device." A diesel trip can also be referred to as a "crab."
- Ultimate Resistance - The maximum load that a single pile will support; a load such that any increase will cause objectionable progressive settlement and constitute failure.
- Vertical Travel Leads - See Leads.
- Wave speed The speed of strain propagation in a pile.
- Wave trace A graphic representation against time of a force or velocity measurement.
- Working Load - See Allowable Pile Load.

C. Credits and Sources

Various sources of information are footnoted in the text. The following represent works that were used to actually compile the text and figures of this publication.

C.1. Pile Buck Publications

- *Pile Buck Steel Sheet Piling Design Manual.* 1986.
- *Pile Buck Annual.* 1988. Includes material from Chapters 6, 7, 9, 10, and 11.
- *Foundations.* 1992. Includes material from the following sources included in this book:
    - *Design of Pile Foundations*, EM 1110-2-2906, U. S. Army Corps of Engineers, 1991.
    - S.N. Vanikar, *Manual on Design and Construction of Driven Pile Foundations*, FHWA DP-66-1, U.S. Department of Transportation, 1986.
- *Materials and Equipment for Marine Construction.* 1992. Includes material from the following sources included in this book:
    - *Recommended Practice for Design, Manufacture and Installation of Prestressed Concrete Piling.* Precast/ Prestressed Concrete Institute, 1977.
    - F. Raushe, G. Likins, G.G. Goble and M. Hussein, *The Performance of Pile Driving Systems.* FHWA/RD-86/160. U.S. Department of Transportation, 1986.
    - B.V. Hansen, *Aluminum Sheet Piling.* 1990.
- Articles from the Pile Buck newspaper:
    - Warrington, D. "Survey of Methods for Computing the Power Transmission of Vibratory Hammers." Second August Issue 1994.
    - Warrington, D., and Erofeev, L.V. "Russian Impact-Vibration Pile Driving Equipment." Second May Issue, 1995.

C.2. Other Publications

- *California Foundation Manual.* State of California, Department of Transportation, Engineering Service Centre, 1996.
- *Design of Deep Foundations.* Engineering Instructions EI 02C097. U.S. Army Corps of Engineers, 1996.

- *Florida State Highway Specification, Section 455* (Structures Foundations). Florida Department of Transportation, 1991
- Hannigan, P.J., Goble, G.G., Thendean, G., Likins, G.E. and Rausche, F. (1998) *Design and Construction of Driven Pile Foundations.* FHWA-HI-97-013. Two (2) Volumes. Washington, DC: Federal Highway Administration.
- Hunt, H.W., *Design and Installation of Pile Foundations.* Associated Pile and Fitting Corp., 1979.
- Occupational Safety and Health Administration. *OSHA Technical Manual.* TED 1-0.15A. Washington, DC: Occupational Safety and Health Administration, 1999.
- *Pile Driving Equipment,* Engineering Instructions EI 02G001. U.S. Army Corps of Engineers, 1997.
- Prestressed Concrete Institute. *Recommended Practice for Design, Manufacture and Installation of Prestressed Concrete Piling.* PCI Journal, March-April 1993.
- *Raymond Superintendent's Manual.* 1972.
- F. Rausche and G.G. Goble, *Wave Equation Analysis of Pile Foundations, WEAP86 Program.* U.S. Department of Transportation, Federal Highway Administration, 1986.
- S. Silvennionen, *Rautaruukki Steel Products Designer's Guide.* Rautaruukki OY, Raahe, Finland. Third Edition, 1996.
- Southern Pile Council, *Marine Construction Manual*, 1997
- *Structures Inspsection, Part Two.* Prepared by Roy Jorgensen and Associates for Florida Department of Transportation, 1992
- TESPA, *Installation of Steel Sheet Piles,* Technical European Sheet Piling Association, 1993.
- U.S. Department of Transportation (1992). *Static Testing of Deep Foundations.* FHWA-SA-91-042. Washington, DC: Federal Highway Administration.
- Warrington, D.C., *Closed Form Solution of the Wave Equation for Piles.* Master's Thesis, University of Tennessee at Chattanooga, 1997

Pile Driving by Pile Buck

Pile Driving by Pile Buck

www.ingramcontent.com/pod-product-compliance
Lightning Source LLC
Chambersburg PA
CBHW080128220326
41598CB00032B/4991

* 9 7 8 0 5 7 8 0 0 8 7 2 1 *